T0180023

IFIP Advances in Information and Communication Technology

479

IFIP – The International Federation for Information Processing

IFIP was founded in 1960 under the auspices of UNESCO, following the first World Computer Congress held in Paris the previous year. A federation for societies working in information processing, IFIP's aim is two-fold: to support information processing in the countries of its members and to encourage technology transfer to developing nations. As its mission statement clearly states:

IFIP is the global non-profit federation of societies of ICT professionals that aims at achieving a worldwide professional and socially responsible development and application of information and communication technologies.

IFIP is a non-profit-making organization, run almost solely by 2500 volunteers. It operates through a number of technical committees and working groups, which organize events and publications. IFIP's events range from large international open conferences to working conferences and local seminars.

The flagship event is the IFIP World Computer Congress, at which both invited and contributed papers are presented. Contributed papers are rigorously refereed and the rejection rate is high.

As with the Congress, participation in the open conferences is open to all and papers may be invited or submitted. Again, submitted papers are stringently refereed.

The working conferences are structured differently. They are usually run by a working group and attendance is generally smaller and occasionally by invitation only. Their purpose is to create an atmosphere conducive to innovation and development. Refereeing is also rigorous and papers are subjected to extensive group discussion.

Publications arising from IFIP events vary. The papers presented at the IFIP World Computer Congress and at open conferences are published as conference proceedings, while the results of the working conferences are often published as collections of selected and edited papers.

IFIP distinguishes three types of institutional membership: Country Representative Members, Members at Large, and Associate Members. The type of organization that can apply for membership is a wide variety and includes national or international societies of individual computer scientists/ICT professionals, associations or federations of such societies, government institutions/government related organizations, national or international research institutes or consortia, universities, academies of sciences, companies, national or international associations or federations of companies.

More information about this series at http://www.springer.com/series/6102

Daoliang Li · Zhenbo Li (Eds.)

Computer and Computing Technologies in Agriculture IX

9th IFIP WG 5.14 International Conference, CCTA 2015
Beijing, China, September 27–30, 2015
Revised Selected Papers, Part II

 Springer

Editors
Daoliang Li
China Agricultural University
Beijing
China

Zhenbo Li
China Agricultural University
Beijing
China

ISSN 1868-4238   ISSN 1868-422X (electronic)
IFIP Advances in Information and Communication Technology
ISBN 978-3-319-83919-6   ISBN 978-3-319-48354-2 (eBook)
DOI 10.1007/978-3-319-48354-2

Printed on acid-free paper

This Springer imprint is published by Springer Nature
The registered company is Springer International Publishing AG
The registered company address is: Gewerbestrasse 11, 6330 Cham, Switzerland

Preface

The 9th International Conference on Computer and Computing Technologies in Agriculture (CCTA 2015) was held in Beijing, China, during September 27–30, 2015.

The conference was hosted by the China Agricultural University (CAU), Agricultural Information Institute of Chinese Academy of Agricultural Sciences (AIICAAS), China National Engineering Research Center for Information Technology in Agriculture (NERCITA), China National Engineering Research Center of Intelligent Equipment for Agriculture (NERCIEA), International Federation for Information Processing (IFIP), Chinese Society of Agricultural Engineering (CSAE), Chinese Society for Agricultural Machinery (CSAM), Chinese Association for Artificial Intelligence (CAAI), Information Technology Association of China Agro-technological Extension Association, Beijing Technology Innovation Strategic Alliance for Intelligence Internet of Things, Industry in Agriculture, Beijing Society for Information Technology in Agriculture, China (BSITA), Group of Agri-Informatics, Ministry of Agriculture, China, Sino-US Agricultural Aviation Cooperative Technology Center, and the Club of Ossiach. It was sponsored by the National Natural Science Foundation of China (NSFC), Ministry of Agriculture, China (MOA), Ministry of Science and Technology, China (MOST), Ministry of Industry and Information Technology, China (MIIT), State Administration of Foreign Experts Affairs, China (SAFEA), Beijing Administration of Foreign Experts Affairs, China, Beijing Municipal Science and Technology Commission (BMSTC), Beijing Natural Science Foundation, China (BNSF), Beijing Association for Science and Technology, China (BAST), Beijing Academy of Agriculture and Forestry Sciences, China (BAAFS), Dabeinong Education Foundation, China, and the Global Forum on Agricultural Research (GFAR).

In order to promote exchange and cooperation among scientists and professionals from different fields and strengthen international academic exchange, the Joint International Conference on Intelligent Agriculture (ICIA) included the 8th International Symposium on Intelligent Information Technology in Agriculture (8th ISIITA), the 9th IFIP International Conference on Computer and Computing Technologies in Agriculture (9th CCTA), and the AgriFuture Days 2015 International Conference (AgriFuture Days 2015). These events provided a platform, for experts and scholars from all over the world to exchange techniques, ideas, and views on intelligent agricultural innovation. Nine International Conferences on Computer and Computing Technologies in Agriculture have been held since 2007.

The topics of CCTA 2015 covered the theory and applications of all kinds of technology in agriculture, including: intelligent sensing, monitoring, and automatic control technology models; the key technology and model of the Internet of Things; agricultural intelligent equipment technology; computer vision; computer graphics and virtual reality; computer simulation, optimization, and modeling; cloud computing and agricultural applications; agricultural big data; decision support systems and expert system; 3s technology and precision agriculture; the quality and safety of agricultural products; detection and tracing technology; and agricultural electronic commerce technology.

We selected the 122 best papers among the 237 papers submitted to CCTA 2015 for these proceedings. All papers underwent two reviews by two Program Committee members, who are from the Special Interest Group on Advanced Information Processing in Agriculture (AIPA), IFIP. In these proceedings, creative thoughts and inspirations can be discovered, discussed, and disseminated. It is always exciting to have experts, professionals, and scholars with creative contributions getting together to share inspiring ideas and accomplish great developments in the field.

I would like to express my sincere thanks to all authors who submitted research papers to the conference. Finally, I would also like to express my gratitude to all speakers, session chairs, and attendees, both national and international, for their active participation and support of this conference.

September 2016   Daoliang Li

Conference Organization

Organizers

China Agricultural University
China National Engineering Research Center for Information Technology
    in Agriculture (NERCITA)
Agricultural Information Institute of Chinese Academy of Agricultural Sciences
    (AIICAAS)
China National Engineering Research Center of Intelligent Equipment for Agriculture
    (NERCIEA)
International Federation for Information Processing (IFIP)
Chinese Society of Agricultural Engineering (CSAE)
Chinese Society for Agricultural Machinery (CSAM)
Chinese Association for Artificial Intelligence (CAAI)
Information Technology Association of China Agro-technological Extension Association
Beijing Technology Innovation Strategic Alliance for Intelligence Internet Of Things
    Industry in Agriculture
Beijing Society for Information Technology in Agriculture, China (BSITA)
Group of Agri-Informatics, Ministry of Agriculture, China
Sino-US Agricultural Aviation Cooperative Technology Center

Sponsors

National Natural Science Foundation of China (NSFC)
Ministry of Agriculture, China (MOA)
Ministry of Science and Technology, China (MOST)
Ministry of Industry and Information Technology, China (MIIT)
State Administration of Foreign Experts Affairs, China (SAFEA)
Beijing Administration of Foreign Experts Affairs, China
Beijing Municipal Science & Technology Commission (BMSTC)
Beijing Natural Science Foundation, China (BNSF)
Beijing Association for Science and Technology, China (BAST)
Beijing Academy of Agriculture and Forestry Sciences, China (BAAFS)
Dabeinong Education Foundation, China
The Global Forum on Agricultural Research (GFAR)

Organizing Committee

Chunjiang Zhao	China National Engineering Research Center for Information Technology in Agriculture
Daoliang Li	College of Information and Electrical Engineering, China Agricultural University
Ajit Maru	Global Forum on Agricultural Research
Walter H. Mayer	PROGIS Software GmbH
Nick Sigrimis	Department of Agricultural Engineering, Agricultural University of Athens, Greece
Yubin Lan	Texas A&M University, USA
Liping Chen	China National Engineering Research Center of Intelligent Equipment for Agriculture
Xinting Yang	China National Engineering Research Center for Information Technology in Agriculture
Xianxue Meng	Agricultural Information Institute of Chinese Academy of Agricultural Sciences (CAAS)

Chairs

Daoliang Li
Chunjiang Zhao

Conference Secretariat

Xia Li
Fangxu Zhu
Jieying Bi

Contents – Part II

Contents – Part I

Effects of Waterlogging and Shading at Jointing and Grain-Filling Stages on Yield Components of Winter Wheat

Yang Liu, Chunlin Shi$^{(\boxtimes)}$, Shouli Xuan, Xiufang Wei, Yongle Shi,
and Zongqiang Luo

Institute of Agricultural Economics and Information/Key Laboratory
of Agricultural Environment in Lower Reaches of the Yangtze River, MOA,
Jiangsu Academy of Agricultural Sciences, Nanjing 210014, China
luisyang@126.com, shicl@jaas.ac.cn,
shirleyxuan2008@hotmail.com, 540883728@qq.com,
1039266182@qq.com, luozongqiang@163.com

Abstract. Waterlogging and shading result from continuous rain are the main meteorological disasters for wheat (*Triticum aestivum* L.) production. In order to evaluate the effects of waterlogging and shading on yield components of winter wheat (both independent and combined), pot experiments were conducted using two representative cultivars in local, Ningmai 13 and Yangmai 13. In total, 4 treatments, including CK (control), WA (waterlogging alone), SA (shading alone) and WS (both waterlogging and shading) were established with three duration (5, 10 and 15 d, respectively) at jointing and grain-filling stages. Results showed that, in the case of non-stressed environment, Yangmai 13 got a better production compared with Ningmai 13 (grain yield per plant was 14.25 g and 15.97 g for Ningmai 13 and Yangmai 13, respectively). However, compared with Yangmai 13, Ningmai 13 got a better yield under stresses at jointing stage, while a similar yield was observed when stresses are at grain-filling stage. By comparing wheat yield and its components, the negative effects of the stresses showed a tendency that WA > WS > SA at jointing stage, whereas WS > WA > SA at grain-filling stage. The result demonstrated that shading had a compensative effect on waterlogging at jointing stage while an addictive effect at grain-filling stage. Reduction of wheat production caused by continuous rain depended on the growth stages. Effect of growth stage on grain yield should be considered when waterlogging and shading packages of wheat growth model were established.

Keywords: Winter wheat · Waterlogging · Shading · Yield components

1 Introduction

The winter wheat production in Jiangsu Province covers an area of 2.13 M ha, accounting for approximate 9 % of the overall winter wheat area of China in 2012 [1]. However, most winter wheat in Jiangsu is planted in paddy fields in a rice–wheat rotation [2], which results in poor drainage conditions. Furthermore, continuous rain is

D. Li and Z. Li (Eds.): CCTA 2015, Part II, IFIP AICT 479, pp. 1–14, 2016.
DOI: 10.1007/978-3-319-48354-2_1

frequent during the growth season of winter wheat due to the subtropical monsoon climate [3, 4]. The total rainfall is 500–800 mm during the wheat growth season, which far exceeds requirements for winter wheat production [5]. In addition, frequency of extreme weather events is increasing globally and regionally [4]. Therefore, soils are easily waterlogged in Jiangsu and waterlogging has become a prime limitation for wheat production locally.

Previous experimental results on plants have shown that waterlogging stress restricts root growth, decreases root hydraulic conductance [6], results in leaf senescence [7], shortens the grain filling time [8] and reduces final wheat yield [9, 10]. The influence of waterlogging on yield components of wheat has been studied by many researchers. Grain yield under waterlogging treatments was 10–15 % lower than that under non-waterlogging treatments [6]. Plot experiment showed that waterlogging from anthesis to maturity would accelerate the senescence of flag leaves, leading to a reduction of grain-filling rate and grain weight [11]. Results of a pot experiment indicated that waterlogging reduced the accumulation of starch in the grains, and the allocation of carbon assimilates to grain yield [7]. However, damage from waterlogging depends on the growth stage, waterlogging duration and wheat cultivar [12, 13]. Thus, different growth stages should be considered when establishing experiment.

Shading always accompanies waterlogging during continuous rain events. Similarly to waterlogging, shading could also reduce dry matter accumulation and grain yield [14, 15] by reducing radiation, impairing net photosynthesis in leaves [16] and reducing the LAI (leaf area index) [17]. However, diffuse light increase under shading can compensate for the reduced radiation [18]. The decrease of LAI is compensated by increases of the bottom leaf area, and the reduction in photosynthetic rate (Pn) of flag leaf is compensated by the increase in Pn of the third leaf [17]. Furthermore, shading increases transport of dry matter from organs to grain [19]. Thus, the shading effect depends on the cultivar and the shading level applied [19].

Although the physiology of wheat under waterlogging and shading stress has been studied independently, research on the combined stress is still scant, especially in consideration of different stress durations and different growth stages. Hence, the objective of the present study was to study the independent effects of waterlogging, shading as well as their combined effect on the yield components of winter wheat at jointing and grain-filling stages. The results will advance our understanding of winter wheat yield under continuous rain events and could be used to improve wheat yield evaluation models by calibrating the combined effect of waterlogging and shading.

2   Experiments and Methods

2.1   Experimental Design

The experiment was established during the winter wheat growing season of 2013–2014 at the Experimental Station of Jiangsu Academy of Agricultural Sciences, Nanjing (32° 2′N, 118°52′28″E), China. Two representative local wheat cultivars, Ningmai 13 and Yangmai 13, were grown in plastic pots with height at 20 cm and diameter at 25 cm. Each pot was filled with 12 kg of air-dried soil, and seven small holes (1 cm in diameter) were drilled to drain excess water. The soil contained 13.7 g/kg organic carbon, 54.95 mg/kg available nitrogen, 24.25 mg/kg Olsen-phosphorus and 105.03 mg/kg available potassium. Soils of each pot was pre-mixed with 0.7 g of N, 0.3 g of P_2O_5 and 0.7 g of K_2O, and another 0.4 g of N was applied at jointing stage to each pot. 12 seeds were sown to each pot on 5 November 2013, and then limited to four plants at the 3-leaf stage.

Four treatments were established: CK (control, non-stressed plants), WA (water-logging alone), SA (shading alone) and WS (combined waterlogging and shading). The WA treatment was achieved by reserving a 1 cm water layer above soil surface (pots were placed in an artificial pool and the depth of water layer adjusted manually). For the SA treatment, a black polyethylene screen was fitted about 180 cm above the ground to block about 80 % of the total radiation. The WS treatment was achieved by combining both waterlogging and shading treatments. All treatments started when jointing and grain-filling stage was reached (5 March and 18 April 2014, respectively), and there were three durations (5, 10 and 15 d respectively) for each treatment. At the end of each treatment, excess water was drained and the black polyethylene screen removed.

Two pots of each treatment and duration for both cultivars were retained until maturity (20 May 2014) for measurement of yield components. Generally, spike number per plant (SN), grain number per spike (GN), 1000-kernel weight (TKW) and grain yield per plant (GY) were recorded.

2.2   Statistics Analysis

Significant differences between the treatments was determined by one-way analysis of variance. $P < 0.05$ was used as the standard for significance by least significant difference (LSD). Statistical analysis was performed using the SPSS 16.0.

3   Results

3.1   Effects of Waterlogging and Shading on Wheat Grain Yield at Jointing Stage

All three treatments had a limited effect on SN at jointing stage. Significant difference was seldom shown between treatments, cultivars and treatment durations (Fig. 1).

GN and TKW showed no significant differences for both cultivars under SA (Figs. 2, 3). There was a significant reduction of GN for both cultivars under WA (Fig. 2).

Ningmai 13 showed good tolerance to WA and WS with no significant decreases of TKW. However, TKW of Yangmai 13 significantly decreased under both WA and WS (Fig. 3).

With increased duration of WA treatment, GY of both cultivars decreased (Fig. 4). After 15 d of waterlogging, GY significantly decreased (14.25 to 9.57 g and 15.97 to 9.40 g for Ningmai 13 and Yangmai 13, respectively; $P < 0.05$). The effects of SA and WS on GY depended on the cultivar. GY of Ningmai 13 under SA and WS decreased

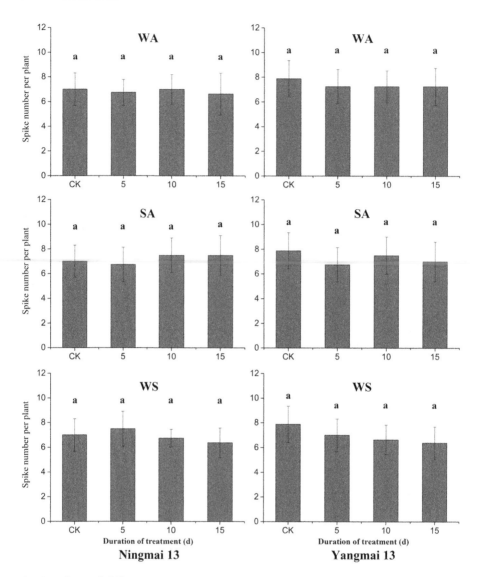

Fig. 1. Effects of different treatments (CK, WA, SA and WS) on spike number per plant at jointing stage. Different letters represent significant differences at $P < 0.05$. Error bars represent the standard deviation.

non-significantly compared with CK. However, there were greater decreases in GY for Yangmai 13 under SA and WS, and decreases were significant when duration reached 10 d. Under all stresses, GY was significantly reduced for Yangmai 13 and was lower than for Ningmai 13 (Fig. 4). Thus, Ningmai 13 had better tolerance to stresses than Yangmai 13 at jointing stage.

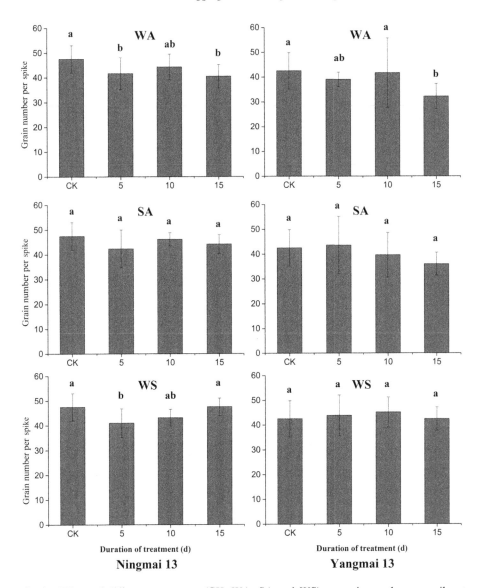

Fig. 2. Effects of different treatments (CK, WA, SA and WS) on grain number per spike at jointing stage. Different letters represent significant differences at $P < 0.05$. Error bars represent the standard deviation.

Wheat yields under different treatments showed an order of CK > SA > WS > WA at jointing stage. Ningmai 13 had a significant yield reduction only under WA, mainly due to the reduction of GN. Yangmai 13 had significant yield reductions under all three treatments but for different reasons: yield reduction was caused by decrease of GN and TKW under WA; by decrease in GN under SA; and by decrease in TKW under WS.

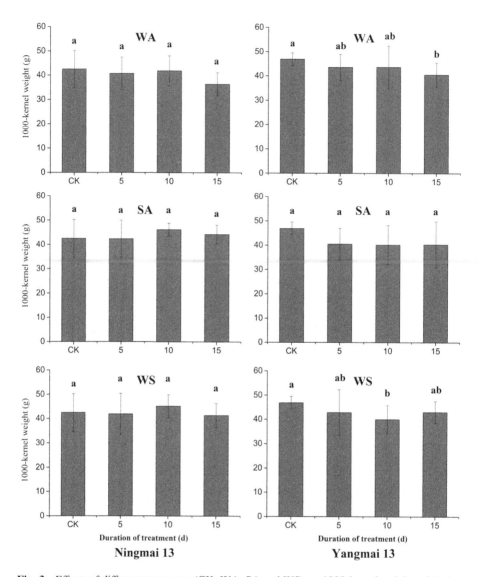

Fig. 3. Effects of different treatments (CK, WA, SA and WS) on 1000-kernel weight at jointing stage. Different letters represent significant differences at $P < 0.05$. Error bars represent the standard deviation.

3.2 Effects of Waterlogging and Shading on Wheat Grain Yield at Grain-Filling Stage

All three treatments at grain-filling stage showed non-significant effects on SN (Fig. 5), which is similar to that at jointing stage (Fig. 1). The reason is mainly due to that SN had been determined before jointing and grain-filling stages. Therefore, waterlogging and shading have a limited effect on SN.

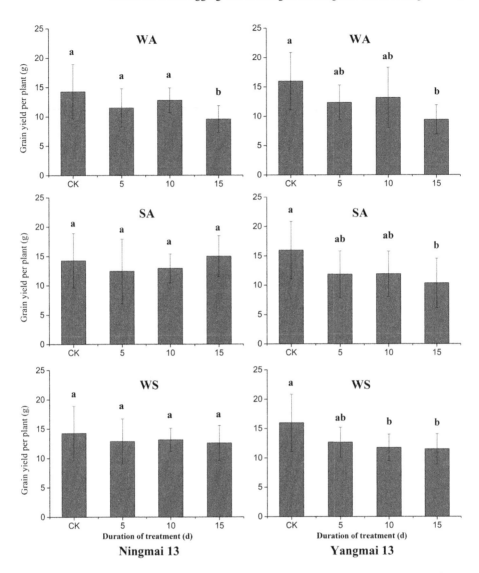

Fig. 4. Effects of different treatments (CK, WA, SA and WS) on grain yield per plant at jointing stage. Different letters represent significant differences at $P < 0.05$. Error bars represent the standard deviation.

Waterlogging and shading showed a weak effect on GN at grain-filling stage. Significant but limited reductions of GN were observed only for Ningmai 13 under WA and SA (Fig. 6).

All three treatments caused reduction on TKW compared with CK at grain-filling stage. And significant reduction was shown after 15 d of stresses (Fig. 7). Moreover, TKW under different treatments showed an order of SA > WA > WS. The results indicated that WA caused a more severe damage to grain-filling compared with SA,

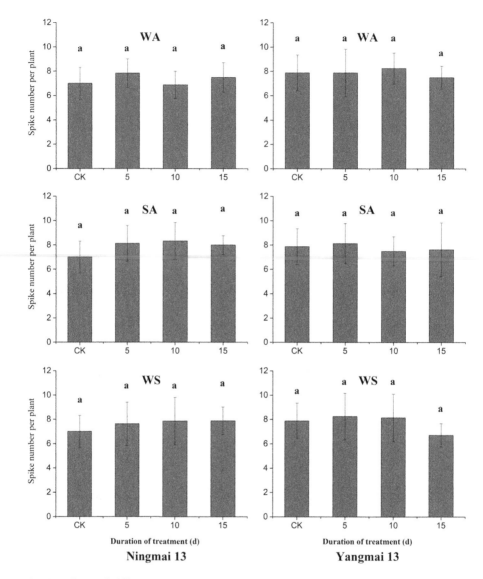

Fig. 5. Effects of different treatments (CK, WA, SA and WS) on spike number per plant at grain-filling stage. Different letters represent significant differences at $P < 0.05$. Error bars represent the standard deviation.

while shading caused an addictive damage when accompanied with waterlogging at grain-filling stage.

GY of both cultivars showed a reduction when suffering waterlogging at grain-filling stage. After 15 d of WA, GY significantly decreased compared with CK (14.25 to 9.2 g and 15.97 to 11.48 g for Ningmai 13 and Yangmai 13, respectively; $P < 0.05$). WS showed a similar reduction of GY compared with WA when the stress

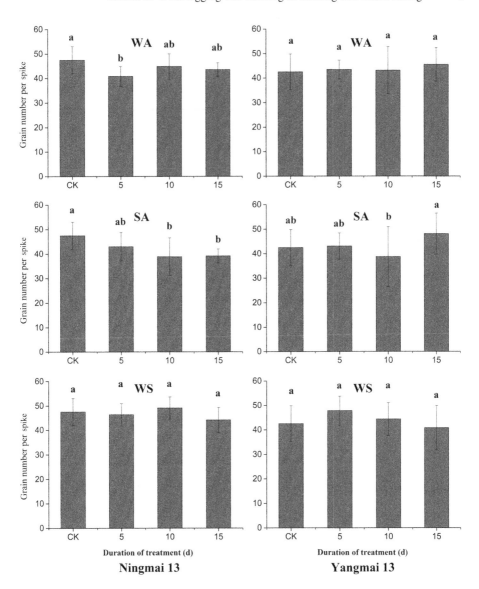

Fig. 6. Effects of different treatments (CK, WA, SA and WS) on grain number per spike at grain-filing stage. Different letters represent significant differences at $P < 0.05$. Error bars represent the standard deviation.

duration is shorter than 10 d, but a larger reduction when duration reached 15 d (Fig. 8).

Wheat yields under different treatments showed an order of CK > SA > WA > WS at grain-filling stage, which was different from the order at jointing stage. Three treatments all caused significant yield reduction for both cultivars when durations were over 10 d. The yield reduction was mainly due to the reduction of TKW.

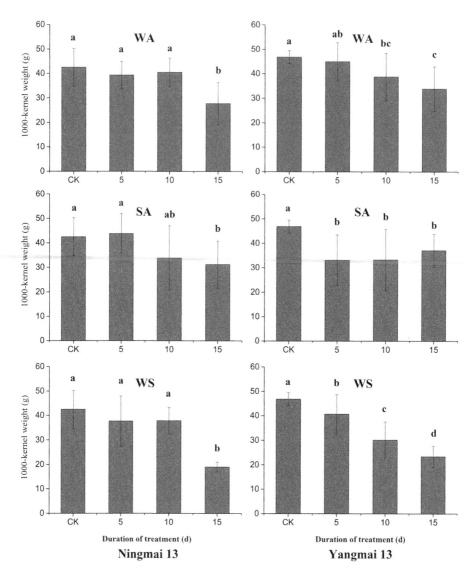

Fig. 7. Effects of different treatments (CK, WA, SA and WS) on 1000-kernel weight at grain-filing stage. Different letters represent significant differences at $P < 0.05$. Error bars represent the standard deviation.

4 Discussion

Waterlogging caused severe yield reduction at both jointing and grain-filling stages and the reduction amplified with increased duration. After 15 d waterlogging treatment, yield of Ningmai 13 and Yangmai 13 was significantly reduced by 28–41 % compared to CK at two growth stages ($P < 0.05$), which is similar to former studies [20].

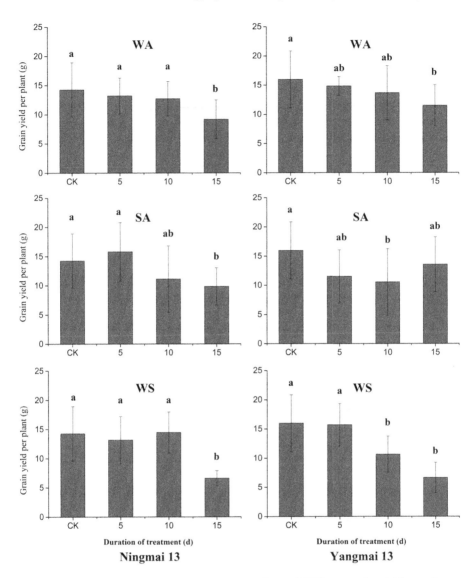

Fig. 8. Effects of different treatments (CK, WA, SA and WS) on grain yield per plant at grain-filing stage. Different letters represent significant differences at $P < 0.05$. Error bars represent the standard deviation.

However, the yield reduction was due to different reasons at both stages. Waterlogging caused yield reduction mainly results from the reduction of GN at jointing stage, indicating that waterlogging at jointing stage would influence the booting stage. At grain-filling stage, SN and GN have been determined. Thus, yield reduction was mainly results from the reduction of TKW. After 15 d WA treatment at grain-filling stage, TKW of Ningmai 13 and Yangmai 13 were 27.76 and 33.88 g respectively, which is lower than CK (42.51 and 46.89 g) and the same treatment at jointing stage (36.39 and 40.57 g). The result indicated that grain-filling process was severely impaired when suffering waterlogging stress at grain-filling stage.

Due to the increases in aerosols and air pollutants, the reduction rate of solar radiation is approximate 2.6 % per 10 years [17]. Shading becomes a constraint of winter wheat production in the Yangtze River Basin [21]. Results of the present study also showed that shading would cause yield reduction of winter wheat. Shading at grain-filling stage induced a greater reduction than at jointing stage. The difference is mainly due to the duration of recovery. After removal of stresses at jointing stage, there was a recovery for more than 60 d until harvest. But at grain-filling stage, the recovery duration was less than 30 d. The shorter recovery duration resulted in the significant reduction of TKW under SA treatment at grain-filling stage, thus the yield reduction was greater than that at jointing stage.

Compared with independent waterlogging and shading stresses, the combined stress (WS) is a more realistic situation in a continuous rain event. Results of combined stress treatment showed that WS at different growth stages induced different effects on wheat yield. WS had a less negative effect compared with WA at jointing stage, whereas a greater negative effect at grain-filling stage. The result indicated a compensative effect of shading on waterlogging at jointing stage, but an additive effect at grain-filling stage. The difference was mainly due to the process of recovery. Recent studies indicate that mechanisms of plant resistance to waterlogging and subsequent recovery are highly different [22]. The compensative effect from shading might be for both morphological and biochemical reasons [17]. Therefore, shading at jointing stage showed a limited effect on wheat yield and had a compensative effect under WS. However, compensative effect of shading was impeded due to natural leaf senescence at grain-filling stage. Thus a different effect from shading was shown at different stages.

5   Conclusions

Independent and combined stresses of waterlogging and shading all caused the yield reduction of winter wheat, and stresses at different stages have different impacts. WS caused a less severe reduction than WA at jointing stage whereas a more severe reduction at grain-filling stage. The results indicated that shading had a compensative effect on waterlogging at jointing stage, but an addictive effect at grain-filling stage. The different effect is mainly due to recovery process of winter wheat after removal of stresses. However, most present wheat growth models are based on the experiment of waterlogging alone, without considering the effect of shading during continuous rain events. Thus, the present study advances the understanding on the physiology of winter wheat under combined stress of waterlogging and shading. Winter wheat growth models could be improved to avoid overestimating or underestimating production losses due to continuous rain.

Acknowledgment. This study was funded by the Special Fund for Agro-scientific Research in the Public Interest (201203032), Jiangsu Province Science and Technology Support Program (BE2012391), and the Fund for Independent Innovation of Agricultural Sciences in Jiangsu Province (CX(12)3055).

References

1. National Bureau of Statistics of China: China Statistical Yearbook 2012. China Statistics Press, Beijing (2013). (in Chinese)
2. Li, C., Jiang, D., Wollenweber, B., et al.: Waterlogging pretreatment during vegetative growth improves tolerance to waterlogging after anthesis in wheat. Plant Sci. **180**(5), 672–678 (2011)
3. Shi, C., Jin, Z.: A WCSODS-Based model for simulating wet damage for winter wheat in the middle and lower reaches of the Yangtze-River. J. Appl. Meteorolog. Sci. **14**(4), 462–468 (2003)
4. Jin, Z., Shi, C.: An early warning system to predict waterlogging injuries for winter wheat in the Yangtze-Huai Plain (WWWS). Acta Agronomica Sinica **32**(10), 1458–1465 (2006)
5. Jiang, D., Fan, X., Dai, T., et al.: Nitrogen fertiliser rate and post-anthesis waterlogging effects on carbohydrate and nitrogen dynamics in wheat. Plant Soil **304**(1–2), 301–314 (2008)
6. Araki, H., Hamada, A., Hossain, M.A., et al.: Waterlogging at jointing and/or after anthesis in wheat induces early leaf senescence and impairs grain filling. Field Crops Res. **137**, 27–36 (2012)
7. Zheng, C., Jiang, D., Dai, T., et al.: Effects of salt and waterlogging stress at post-anthesis stage on wheat grain yield and quality. Chin. J. Appl. Ecol. **20**(10), 2391–2398 (2009)
8. Dickin, E., Wright, D.: The effects of winter waterlogging and summer drought on the growth and yield of winter wheat (*Triticum aestivum* L.). Eur. J. Agron. **28**(3), 234–244 (2008)
9. Malik, A.I., Colmer, T.D., Lambers, H., et al.: Short-term waterlogging has long-term effects on the growth and physiology of wheat. New Phytol. **153**(2), 225–236 (2002)
10. Araki, H., Hossain, M., Takahashi, T.: Waterlogging and hypoxia have permanent effects on wheat root growth and respiration. J. Agron. Crop Sci. **198**(4), 264–275 (2012)
11. Fan, X., Jiang, D., Dai, T., et al.: Effects of nitrogen supply on flag leaf senescence and grain weight in wheat grown under drought or waterlogging from anthesis to maturity. Acta Pedol. Sin. **42**(5), 875–879 (2006)
12. Rasaei, A., Ghobadi, M.-E., Jalali-Honarmand, S., et al.: Impacts of waterlogging on shoot apex development and recovery effects of nitrogen on grain yield of wheat. Eur. J. Exp. Biol. **2**(4), 1000–1007 (2012)
13. Hayashi, T., Yoshida, T., Fujii, K., et al.: Maintained root length density contributes to the waterlogging tolerance in common wheat (*Triticum aestivum* L.). Field Crops Res. **152**, 27–35 (2013)
14. Slafer, G., Calderini, D., Miralles, D., et al.: Preanthesis shading effects on the number of grains of three bread wheat cultivars of different potential number of grains. Field Crops Research **36**(1), 31–39 (1994)
15. Abbate, P.E., Andrade, F.H., Culot, J.P., et al.: Grain yield in wheat: effects of radiation during spike growth period. Field Crops Res. **54**(2), 245–257 (1997)
16. Wang, Z., Yin, Y., He, M., et al.: Allocation of photosynthates and grain growth of two wheat cultivars with different potential grain growth in response to pre-and post-anthesis shading. J. Agron. Crop Sci. **189**(5), 280–285 (2003)
17. Mu, H., Jiang, D., Wollenweber, B., et al.: Long-term low radiation decreases leaf photosynthesis, photochemical efficiency and grain yield in winter wheat. J. Agron. Crop Sci. **196**(1), 38–47 (2010)
18. Greenwald, R., Bergin, M.H., Xu, J., et al.: The influence of aerosols on crop production: A study using the CERES crop model [J]. Agric. Syst. **89**(2–3), 390–413 (2006)

19. Li, H., Jiang, D., Wollenweber, B., et al.: Effects of shading on morphology, physiology and grain yield of winter wheat. Eur. J. Agron. **33**(4), 267–275 (2010)
20. Hossain, M.A., Araki, H., Takahashi, T.: Poor grain filling induced by waterlogging is similar to that in abnormal early ripening in wheat in Western Japan. Field Crops Res. **123** (2), 100–108 (2011)
21. Jin, Z., Shi, C., Ge, D., et al.: Characteristics of climate change during wheat growing season and the orientation to develop wheat in the lower valley of the Yangtze River. Jiangsu J. Agric.Sci. **17**(4), 193–199 (2001)
22. Setter, T.L., Waters, I.: Review of prospects for germplasm improvement for waterlogging tolerance in wheat, barley and oats. Plant Soil **253**(1), 1–34 (2003)

The Measurement of Fish Size by Machine Vision - A Review

Mingming Hao[1], Helong Yu[1], and Daoliang Li[2(✉)]

[1] College of Information Technology, Jilin Agricultural University, Changchun 130118, China
haomingming314@163.com, yuhelong@aliyun.com
[2] College of Information and Electrical Engineering, China Agricultural University,
Beijing 100083, China
dliangl@cau.edu.cn

Abstract. Aquatic products are becoming increasingly popular because of their high nutritional value. Size information is an important parameter that can be used to measure the growth, weight, gender, grading and even species identification of fish. However, size information is a highly tedious and inefficient measure when conducted manually through traditional methods. Machine vision is a non-destructive, economic, rapid and efficient tool; hence, it is suitable to measure fish size. This review introduced methods and results for fish size measurement through machine vision. The paper is organised according to the measurement of body dimensionality: length measurement and area measurement. Simultaneously, the advantages, disadvantages and future trends of the system are discussed. With development in those areas, the size measurement by machine vision technology will become more effective. Machine vision system brings high accuracy and high efficiency and is easier than manual work. The methods reported can help researchers and farmers bring benefits for aquaculture.

Keywords: Machine vision · Fish size · Fish length · Fish area

1 Introduction

Fish and fish products play an important role in meeting human protein demand. To sustainably manage fish resources with high efficiency, fishers have to catch fish appropriately and obtain meaningful information [1]. One important piece of information is fish size [2], which is a key parameter to appraise stock status, provide management advice and enhance economic benefits for an aquaculture enterprise [3]. In fish culture, gathering fish size information to describe and manage the growth of fish is necessary to harvest the stock at the optimum time. Furthermore, length distribution of the fish can be used to predict a range of population-level characteristics [4, 5] and to assess the stock scientifically during rearing [6–8]. Whether for aquaculture or market purposes, fish grading is an important and frequent operation [9]. During rearing, fish need to be graded by size; fishers can separate the fast growers from the slow ones to adjust the distribution of food and management to reduce cost [10–12]. For fish farms to adapt to the market, fish should be sorted by size after harvesting to estimate the value. This step

D. Li and Z. Li (Eds.): CCTA 2015, Part II, IFIP AICT 479, pp. 15–32, 2016.
DOI: 10.1007/978-3-319-48354-2_2

is necessary to set different prices and manage markets conveniently [13–17]. Simultaneously, size information can be used to monitor quality in the food industry, which can help fishers assess the price. When many kinds of fish are reared in one pond or net, identifying different species is necessary; measuring size information is an important means to enable identification [14, 19, 20]. Size information enables identifying different genders [9, 19, 21] and even estimates the age [22] because of the size difference from different individuals of different genders. Size information can also be used in many other measurement issues. Fish weight is also an important and direct parameter to judge whether the fish is satisfactory [23–26]. Fishers or sellers can measure weight directly and use the high correlation between size and weight to estimate the weight [16, 27, 28]. The size information also can help to measure biomass and morphology, such as volume [29] and even estimation of fat content [30]. Therefore, fish size is a meaningful parameter that is necessary to determine the size information.

Given the application of fish size, fishers have realised the importance of size information and started to measure fish size by traditional measurement methods. For length data, fishers sometimes estimate length by eye based on experience. Most use a ruler, which has high accuracy, or a roller grader, which is slightly faster [31]. To measure the area, fishers place the fish on a measuring board, draw the fish shape along its edges, and then calculate the amount of forms to estimate the area [31]. This process can also be conducted through electronic measuring boards. These methods are time consuming [32], labour intensive and inaccurate, with subjective [26] and expensive operation [13]. These methods also cause considerable stress to the fish [4, 33] and raise the risk of related damage [14, 21] or hinder growth by reducing appetite [12, 15] of the fish. Therefore, a simpler, fast and scientific method is urgently needed to collect fish size data [34].

Machine vision systems have been developed and used in various industries [35]. Machine vision is a non-destructive [5, 36], rapid [30], economical [6], consistent [29], objective, repeatable [26], quantitative [37], relatively efficient and robust tool [38, 39]. It can capture images in real time, analyse these pictures and make decisions to deal with certain questions based on the result [26]. Furthermore, machine vision has been used widely in aquaculture, such as health monitoring, disease detection, behaviour detection, species identification and so on. Fish size measurement by machine vision has been developed and has become popular because of its high efficiency [15].

This study reviews methods of machine vision technology in fish size measurement, including fish length and fish area. The aim of this study is to converge all the methods, as well as discuss the advantages, disadvantages and future trends. In addition, the study helps other researchers promote the development of machine vision in aquaculture.

2   Image Acquisition

For a machine vision system that measures fish size, obtaining images by image acquisition equipment is important. The image acquisition process relies on a specific environment. Several environments are discussed.

2.1 Sonar

Sonar is powerful tool used to obtain images in a wide range of areas [8, 40, 41]. It can be equipped on a ship or a fixed facility to measure fish schools or large fish [42]. However, image quality captured by sonar is not very high, and a sonar system is costly [43, 44]. The system is illustrated in Fig. 1.

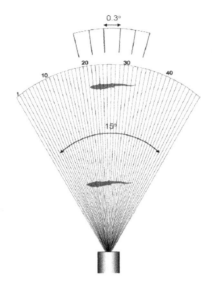

Fig. 1. Image acquisition by sonar [41]

2.2 Single Camera

A single camera can be used to capture images. This camera can obtain a 2D image. This setup can be used when the fish is placed in a net, cage, pond or light box. Alternatively, the fish can go through the light box on a conveyor [32, 45], and the camera can be used under water. A single camera can easily obtain high-quality images. However, it can only capture 2D images. The structure is shown in Fig. 2.

2.3 Stereo Camera

In this system, two cameras are placed side by side, which take and process two images synchronously [46]. The stereo camera can take images on a 3D plane. However, the cost of the system and the demands for system rigor are very high. The structure is shown in Fig. 3.

After the images are obtained by the machine vision system, the colour image is transformed to grey scale or to only its HSV values. The binary image is transformed after grey transformation. The interesting region is segmented from the background in binary images by global threshold [47, 48] or dynamic threshold method [42]. In the stereo-camera system, two image acquisition devices exist; two side-by-side cameras

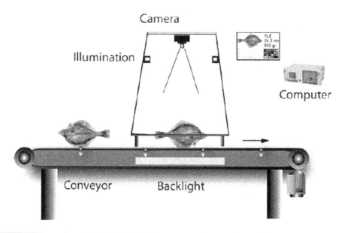

Fig. 2. Image acquisition by single camera [32]

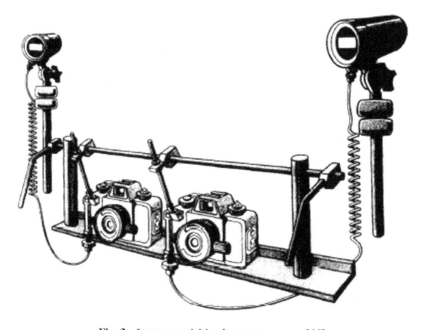

Fig. 3. Image acquisition by stereo-camera [46]

take two images synchronously. Two interesting regions are segmented in two images. Thus, machine vision can take the next step to measure the size information.

3   Length Measurement

Given that fish length is one of the meaningful parameters of size measurement, many researchers have proposed methods of length measurement by machine vision. The methods are described in two fields: 2D and 3D. Methods in 2D mean that the image is 2D and the process is conducted on one plane using one camera. By contrast, 3D uses two or more cameras to capture images. In this method, 3D reduction or processing two or more images is needed. In this study, 3D methods rely on the stereo-camera system. The different methods are described as follows.

3.1   Length Measurement in 2D

The length information can be reviewed in two parts according to the fish body structure. When the fish body structure can be regarded as a straight line and curve, the length can be measured by the linear measurement method and non-linear measurement, respectively.

3.1.1   Linear Measurement

When the fish body can be regarded as a straight line, the length equals the line length. The fish length can be calculated by the relation between the fish body pixel length and image reference scale [37]. The reference scale can be obtained through a square on the colour plate, which captures a photograph of the fish or separately by the same camera [3]. In Fig. 4, A and B are the image length of fish and the colour plate, respectively. A' and B' are the actual length of fish and the colour plate, respectively. Thus, the actual fish length A can be calculated using Eq. (1):

$$A' = A * B'/B. \qquad (1)$$

Thus, the next step is calculating the length of A [37, 48].

Many methods are available to calculate the length of A. For some fish, the body can be regarded as a linear structure; Therefore, the length of A can be described as a line measurement. According that, Hsieh et al. [37] proposed that uses Hough transform to calculate the length of A around the tuna's snout to determine fork length in the image. The longest line measured by Hough transform in the image can indicate the length of the fish. Thus, transforms for every point from image space to Hough space are conducted; The collinear points in the image space are presented in the Hough space. Therefore, the weight of the largest peak is the length of fish in image, called A. The images are taken at different angles from the top and the horizontal direction of fish and corrected by projective transform to reduce the error. Different results are calculated because of the different angles and projective transform. The results were approximately 10.7 ± 5.5 % and 5.6 ± 4.7 % before and after correction by projective transform, respectively. For different angles, the best situation was direction angles between 315–0° and 135–225° and a top-view angle of more than 45° with an error of 2.4 % ± 2.3 %. The average error of testing 600 tuna images was 4.5 % ± 4.4 %.

Fig. 4. Using relation between A and B to calculate fish length [37]

Simultaneously, the best fitting rectangle can also calculate fish length. In the image, the fish is surrounded by an external rectangle, and the area of the rectangle is calculated. Then, we rotate the square constantly and recalculate the area until the minimum area of the rectangle is derived. Finally, the fish length is calculated by the length of the rectangle. Misimi et al. [48] used this method to calculate the length and width of Atlantic salmon and Atlantic cod to estimate the change of size and shape during rigor mortis and ice storage with area, roundness and height information. Balaban et al. [49] described the concrete method to build the rectangle and used this method to calculate the length of four salmon species (pink, red, silver and chum) to weigh the fish stored in a light box through regression analysis with area and width information. The error was less than 0.5 %. Balaban et al. [50] used this method to calculate the ratio of the width/length of Alaska Pollock roe in a light box to identify if roes were single or double so that an appropriate equation could be chosen to estimate the weight. Lee et al. [51] proposed a model of automatic vaccine injection for flatfish by computer vision in a light box. The injection site was different for various sizes; However, the site was related to the length and the width of the flatfish. The best fitting rectangle was used to calculate the length and width for the injection site.

Both the aforementioned methods are used to calculate the length of fish easily and intuitively when fish body is regarded as a linear structure. Otherwise, both these methods must have low accuracy. The first method uses the Hough transform, which can determine a straight line easily and with high accuracy in a fish image. However, the first method has high time complexity. In addition, this method demands high image quality. In the future, the count of points that need transformation will decrease, and the image quality will increase. The second method, which uses the best fitting rectangle, can also calculate the width information and can be used to assist the other method. Furthermore, the first method can ignore the image detail. However, the fins and tail can introduce significant error into the length and width information. In the future, the pre-processing of segment fins and tail will be adapted, such as morphological operations of opening and closing [52, 53].

3.1.2   Non-linear Measurement

For certain types of fish, the body is curved; Therefore, the linear measurement methods shown in Sect. 3.1.1 must have a large error. To solve the non-linear measurement, many researchers purpose some methods. One of the methods is that the length is estimated by the key points, as shown in Fig. 5. The best line is found through several different points, which are the midpoints of the width along the length of the fish body. Strachan [31] used the key points method to measure the length of haddock on a conveyor. The paper determined the fish species as well as its orientation and length. The best line was found to measure the length of the fish image caught in a light box, which was composed of seven different midpoints in the image. Simultaneously, the best fitting rectangle method was used as a supporting tool to measure length. The length was calculated by the sum of distance between each of two adjacent points. A total of 35 haddocks were measured five times, and the error was less than 1 %. Strachan [45] used a similar method to measure fish length to sort round fish and flatfish, which were transformed on a conveyor and with images caught in a light box. The colour and shape information were processed to identify the species. The length information was estimated to sort various species of fish, which was conducted by measuring the line, which was composed of 10 points. The error was less than 4 %. White et al. [32] used this method to identify the species of fish (round fish or flatfish) on a conveyor by the length-width ratio and colour information. Fish length was also calculated by 10 points. The length was measured 100 times in varying positions and rotations, and a result was obtained with less than 0.3 % error. Jeong et al. [54] introduced a vision-based automatic system to measure the body length and the width of flatfish in a light box. The length was estimated by a line composed of five points. The error was less than 0.2 %. Lee et al. [55] used this method to measure the length of sea cucumbers to estimate the weight. The line was composed of 10 points, and the length was estimated by the sum of Euclidean distance of each 2 points. The error of 300 measurements was less than 0.17 %.

For the choice of the key points to compose the line, the best situation is selecting all the midpoints along the body length to connect a line. This method is called image thinning. Yamana et al. [56] used this method to measure the size of Japanese sea cucumbers. Their study found a relationship between length and width information. For length measurement, the error was less than 7 %. Han et al. [57] used a dual-frequency identification sonar (DIDSON) image to count and measure the size of tuna. The best fitting rectangle method was used to segment the tuna from the image, and the image thinning method was used to calculate the length. The error was less than 3 %. Pan et al. [58] used area, perimeter, length and width information to estimate the weight of shelled shrimp in a light box. Image thinning was achieved by the MATLAB function *bwmorph* with parameter *thin*. Simultaneously, the weight was calibrated by an artificial neural

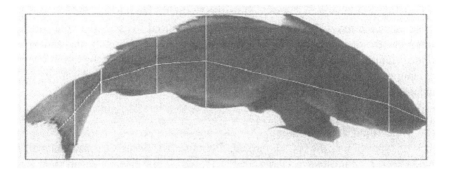

Fig. 5. Key points of fish image. The points are the midpoints of width along the body [31].

network. Fan et al. [59] counted zebra fish fry and calculated seven geometric features, such as area, perimeter and so on. The fish skeleton was described through image thinning, and the best fitting rectangle was used to segment the fish and calculate their width.

Table 1. Detailed information on methods of length measurement in 2D

Method	Object	Environment	Error	Reference
Hough transform	Tuna	–	<5 %	Hsieh et al., 2011
Best fitting rectangle	Salmon, cod	–	–	Misimi et al., 2008(a)
	Salmon	Light box	<0.5 %	Balaban et al., 2010(a)
	Roe	Light box	–	Balaban et al., 2012(a)
	Flatfish	Light box	–	Lee et al., 2013
Key points	Haddock	Light box	<1 %	Strachan, 1993
	Round fish, flatfish	Light box	<4 %	Strachan, 1994
	Round fish, flatfish	Conveyor	<0.3 %	White et al., 2006
	Flatfish	Light box	0.2 %	Jeong et al., 2013
	Sea cucumber	–	<0.17 %	Lee et al., 2014
Image thinning	Sea cucumber	–	<7 %	Yamana et al., 2006
	Tuna	–	<3 %	Han et al., 2009
	Shelled shrimp	Light box	–	Pan et al., 2009
	Zebra fish fry	Tank	–	Fan et al., 2013

Both the aforementioned methods can achieve non-linear measurement of fish length. However, fish cannot avoid curling their bodies, and neither of the methods can solve this situation. For the key points method, the choices for points do not strongly rely on the detail of the image. Furthermore, the choice of special points can ignore the influence of fins. However, determining the position of the point is difficult; Other methods are needed, such as the best fitting rectangle. In the future, a larger number of points will be added to measure linear bent significantly. The image thinning method can describe the body structure clearly no matter the degree of fish body bend. However, the image thinning method has a high demand for image quality. Furthermore, fish fins significantly affect the results. Following the development, image pre-processing will improve, which will enhance image quality. In addition, additional algorithms and theories of eliminating fish fins will be proposed. All detailed information on the aforementioned methods is listed in Table 1.

3.2 Length Measurement in 3D

In recent years, a stereo video camera system was developed for fish length measurement [4]. This system provides the potential and high accuracy in estimating the length of free-swimming fish [5, 21]. Importantly, the stereo-camera system is not stressful and is non-invasive to the fish [4]. The stereo-video camera system is composed of two digital video cameras in underwater housing or two divers with two cameras to take images. The two cameras must be connected and synchronized to reduce error [24]. Fish length information can be calculated by operating a series of processes on two images acquired by the stereo-video camera system.

After pre-processing, fish length can be calculated by the relation between image and real world. Torisawa et al. [60] used the direct linear transformation (DLT) method to calculate the 3D coordinate positions in the real world from the two 2D stereo-images recorded simultaneously by the stereo-camera system. The method transformed the fish head and fish tail coordinates in the image to the real world using Eqs. (2, 3), where u and v were the point coordinates in the image and the coordinates in the real world were described by X, Y and Z, respectively. L_1–L_{11} were the DLT parameters of each camera. Then, we calculated the distance of those two real-world points as the fish length. According to this calculation, two lengths were measured through the stereo-video camera system, and the final length was determined by the average of two lengths. Using this method, the fork length and length frequency distribution for tuna in a net cage were estimated. The error in measuring 107 tuna was less than 5 %.

$$u = \frac{L1X + L2Y + L3Z + L4}{L9X + L10Y + L11Z + 1}. \tag{2}$$

$$v = \frac{L5X + L6Y + L7Z + L8}{L9X + L10Y + L11Z + 1}. \tag{3}$$

Fish length can also be measured by the relation between the fish length in the image, the distance from camera to fish and the distance between a specific point to the camera. Dunbrack [5] proposed a method according to this theory to measure the length of sharks in the ocean, as shown in Fig. 6. LOS was a point coordinates in image on this common visual axis. The specific point S was determined when the Eq. (4) was set up while S was moving from UC to LOS. The fish length was calculated using Eq. (5), where DBC was the distance from camera to fish, L was the fish length in the image and D was the distance from UC to S. The final length was estimated by the average of both values that were calculated by both side camera systems using this method. The error was less than 1.6 % through 13 test measurements.

$$\angle LH'SRH' = \angle LH''LOSRH''. \tag{4}$$

$$FL = DBC\frac{L}{D}. \tag{5}$$

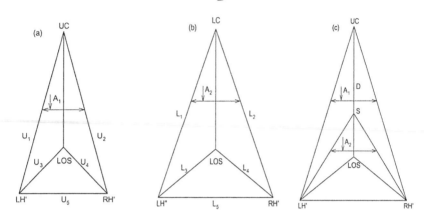

Fig. 6. (a) Upper camera. (b) Lower camera. (c) Angle is used to find S [5]

A specific method can calculate fish length when the distance between the two cameras is known and the optical axes of the cameras are parallel. In addition, the stereo-camera system is a horizontal structure. Through this method, the fish length in each image can be calculated first by the relationship between the focal length, camera distance and the distance from camera to fish, as shown in Fig. 7. The fish length in the image was calculated using Eq. (6), and the final length was calculated by the average of the two image lengths. Finally, the actual length could be calculated by the relation between the image and the real world. Ruff et al. [40] used this method to measure the fish length in a cage. Simultaneously, the image calibration was introduced by a colour plane. The error was less than 1.5 %. Rooij et al. [46] also used this method to measure fish length. Their paper introduced the specific theory and the steps of the method. The error with calibration was less than 3 %. Petrell et al. [61] used this method to measure the length of salmon in a cage and tank; the speed was also measured. Costa et al. [62] used a similar method to measure the length of tuna in a sea cage. The specific theory was described by vector and matrix. Simultaneously, image calibration was introduced. The error corrected by the artificial neural network was approximately 5 %.

$$FL = \frac{zF \times xC}{d(xl') - d(xr')}. \tag{6}$$

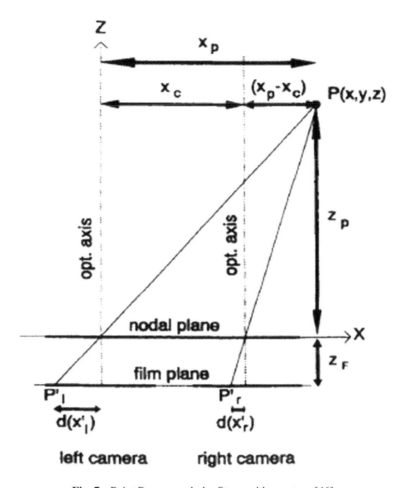

Fig. 7. Point P was caught by Stereo video system [46]

Overall, the three methods can effectively measure length in three dimensions. They have a high accuracy; however, the stereo camera system also has disadvantages. The economic costs have to increase because of the two-camera system. Furthermore, the system must have high synchronicity. The DLT method is easy to achieve because only 11 parameters are needed for calculation. However, finding heads and tails of fish manually or by extra theory is necessary. In the future, an automatic process that does not require manual work to identify fish heads and tails will be proposed. The second method does not rely on the camera constant parameters. Furthermore, this method has a low demand for equipment rigor. However, the angle in both camera systems is highly difficult to calculate and needs a large number of iterations. With development, additional cameras will be equipped to increase the accuracy. The third method only needs two parameters for measurement, which are camera distance and distance from camera to fish. In addition, the method does not need many calculations. However, fish length was indirectly calculated by the fish length in the image; Therefore, it cannot estimate the bend. The third method has high demand for equipment rigor; The two camera systems must be parallel. In the future, a new method and theory on non-linear measurement will be added to the third method.

Detailed information on the aforementioned methods is listed in Table 2.

Table 2. Detailed information on Methods of length measurement in 3D

Method	Object	Environment	Error	Reference
Method 1	Bluefin tuna	Cage	<5 %	Torisawa et al., 2011
Method 2	Shark	Ocean	<1.6 %	Dunbrack, 2006
Method 3	–	Cage	<1.5 %	Ruff et al., 1995
	–	–	<3 %	Rooij et al., 1996
	Salmon	Cage, tank	–	Petrell et al., 1997
	Tuna	Cage	5 %	Costa et al., 2006

Method 1 is the method that uses the direct linear transformation.
Method 2 is the method that shows the relation between the fish length in the image, the distance from camera to fish and the distance between a specific point and the camera.
Method 3 is the method that uses the relation between the focal length, camera distance and distance from camera to fish.

4   Area Measurement

All the 2D and 3D methods are used to measure the length of fish. Fish length is an important parameter; simultaneously, area is a very meaningful parameter. Hence, methods of area measurement should be reviewed.

For a given species of fish, area information has a mathematical relationship with the length of fish. Thus, the area can be calculated by its length information. Newlands et al. [63] used this theory to calculate the area information of tuna schools in the open ocean. The study introduced fish schools as ellipsoids, and the area was calculated by length. The error was approximately 5 %.

Furthermore, the image area can be calculated by the actual pixel number of a fish region after a good segmentation. Then, the real area can be determined according to the relation between the image and real world. Iwamoto et al. [64] used the method to measure the area of a fish egg. The study used real-time flow imaging and the classification system to capture images. Then, the image was processed to identify the species and development stage. Misimi et al. [48] used this method to calculate the fillet area of cod and salmon in a light box. The area was used to estimate roundness, and the error was less than 6 %. Misimi et al. [47] introduced area measurement of salmon in a light box according to this method. The area information was one of the parameters used to grade the fish. Pan et al. [58] calculated the area of shrimp to estimate the weight. The area was calculated using the MATLAB function *bwarea*. Gumus et al. [27] used this method to calculate the area of trout in a light box. The area was used to estimate the weight by establishing a relation model. Balaban et al. [49] measured the area of salmon and used this method to sort them. Simultaneously, the weight and volume dimension were estimated by the area information. Balaban et al. [26] used this method to calculate the area of Pollock in a light box. The area information was used to estimate the weight, and the volume was calculated by the area. Balaban et al. [50] also used this method to measure the area of roe in a light box. In their paper, the weight was estimated by the area, and length information was calculated. For area measurement, the error was less than 2.5 %. Fan et al. [59] used the method to calculate the area to count fry in a light box. The length, width and perimeter were also measured. Back propagation neural network and least squares support vector machine (LS-SVM) were used to count them.

Both the aforementioned methods can calculate the area effectively. For the first method, the area is calculated by the relationship with the length. This method does not need to calculate area directly, but the length must be calculated. Therefore, this method does not rely on image quality. However, this method is achieved through the relationship with length; thus, the demand for accurate length is significantly high. Simultaneously, establishing the relation with length is complex. For the second method, the area is measured by the relation between the actual pixel number of the fish region and the real world. This method is more accurate and more effective than the first. The relation is easy to establish and can be found through the relation between a known object such as a colour plane and real world. However, one pixel can take a significant error; thus, this method strongly relies on image quality. Furthermore, fish fins and tails can influence the result. Therefore, image pre-processing is more important. Additional development will significantly improve the software and the speed of calculation, and new methods of area measurement will be proposed.

All detailed information on the methods is listed in Table 3.

Table 3. Detailed information on methods of area measurement

Method	Object	Environment	Error	Reference
Method 1	Tuna	Ocean	5 %	Newlands et al., 2008
Method 2	Fish egg	–	–	Iwamoto et al., 2001
	Cod and salmon	Light box	<6 %	Misimi et al., 2008(a)
	Salmon	Light box	–	Misimi et al., 2008(b)
	Shelled shrimp	–	–	Pan et al., 2009
	Trout	Light box	–	Gümüş et al., 2010
	Salmon	–	–	Balaban et al., 2010(a)
	Pollock	Light box	–	Balaban et al., 2010(b)
	Roe	Light box	2.5 %	Balaban et al., 2012(a)
	Fry	Light box	–	Fan et al., 2013

Method 1 uses the relationship with the length.
Method 2 uses the relationship between pixel number and real-world area.

5   Discussion and Perspective

This study reviewed the methods of fish size measurement through machine vision. Length and area are important information that can help fishers manage fish scientifically and conveniently. This information could be used to calculate the volume and weight according to their relation; other information could also be calculated. Machine vision is more effective, economical and faster than traditional methods. Simultaneously, machine vision is more accurate; Overall accuracy is 10 % higher than that of traditional methods. The size measurement is based on the two parts, which are fish length and fish area. The methods reviewed in this paper vary according to different platforms for different fields. Length measurement was reviewed according to 2D and 3D platforms; however, the 2D platform was used for area measurement. For length measurement, the methods were reviewed in two parts according to the body structure of fish in the 2D platform. One part considers the shape as a straight line measured as a linear structure, and the other considers the shape as a curve measured as a non-structure. In the 3D platform, the length measurement relies on a stereo-camera system. For area, fish are measured in 2D, and methods are described by the relationship with length and the amount of pixels in the image. All the methods used the image captured by the machine vision system, and the length or area information could be calculated. Given the difference of methods, the best one can be chosen to solve the specific situation and can obtain the best effect.

Machine vision system can help fishers obtain size information of fish, is better than traditional methods and reduces labour. However, the machine vision system also encounters problems such as lack of accuracy. Although the system has low error, it is still unsatisfactory. This is particularly true in special situations such as when the fish body is curved; as a result, the accuracy is reduced. Simultaneously, the image quality captured by the system has a significant effect. A satisfactory image can result inefficient processing. However, the image must sometimes be taken under water; The reflection of light, volatility of water, variations in temperature, water mist and other factors not easy to control could increase the noise in the image and reduce the quality. Therefore, processing the image is difficult. Obtaining the length or area information through machine vision is complex, needs many iterations or vector operations and costs much time to achieve, resulting in poor time efficiency. Although several deficiencies exist, machine vision is also a powerful tool to measure fish size.

Although machine vision has only been developed in recent decades, it has already experienced a qualitative leap. Therefore, the application of machine vision in fish size measurement will also be developed further. The machine system may replace traditional methods to measure fish size. The development of hardware will result in high calculation speed, can solve complex calculation and can enable the adaption of an effective real-time system. The image acquisition device will also improve to capture high-quality images. Thus, processing could be reduced, and efficiency could be increased. The processing that the system uses for the image includes complex key steps, such as image denoising, image segmentation, image enhancement and image recognition. These steps are difficult and will be developed in the next several decades. In the future, these steps must be examined by researchers globally. Such a focus could improve the speed and efficiency of processing. Nowadays, mobile equipment has become widely used. Thus, machine vision systems could be used in mobile devices, which could bring convenience as well as sufficient real-time operations. Fish size information could be obtained directly by an image taken by mobile equipment. Simultaneously, the methods reviewed in this paper could be used in size measurement in other fields, such as leaf, fruit, live stock and even certain industrial equipment. Fish size measurement by machine vision is meaningful and can benefit aquaculture and improve economic value.

6   Conclusion

This study reviewed the methods of fish size measurement based on two parts: length measurement and area measurement. The advantages and disadvantages were discussed and the development was described. This study can be used as a research reference in promoting the aquaculture industry.

Acknowledgment. We would like to express our gratitude to the guidance of professor Daoliang Li. Thanks CrossEdit for polish assistance. Thanks Xi Qiao, Hang Zhang, and Hao Yang for amending and suggesting.

References

1. Torisawa, S., et al.: A technique for calculating bearing and tilt angles of walleye pollock photographed in trawls with digital still-picture loggers. Fish. Res. **77**(1), 4–9 (2006)
2. Lines, J.A., et al.: An automatic image-based system for estimating the mass of free-swimming fish. Comput. Electron. Agric. **31**(2), 151–168 (2001)
3. Chang, S., et al.: How to collect verifiable length data on tuna from photographs: an approach for sample vessels. ICES J. Mar. Sci. **66**(5), 907–915 (2009)
4. Beddow, T.A., Ross, L.G., Marchant, J.A.: Predicting salmon biomass remotely using a digital stereo-imaging technique. Aquaculture **146**(3–4), 189–203 (1996)
5. Dunbrack, R.L.: In situ measurement of fish body length using perspective-based remote stereo-video. Fish. Res. **82**(1–3), 327–331 (2006)
6. Stjohn, J., Russ, G.R., Gladstone, W.: Accuracy and bias of visual estimates of numbers, size structure and biomass of a coral-reef fish. Mar. Ecol. Prog. Ser. **64**(3), 253–262 (1990)
7. Harvey, E., et al.: The accuracy and precision of underwater measurements of length and maximum body depth of southern bluefin tuna (Thunnus maccoyii) with a stereo-video camera system. Fish. Res. **63**(3), 315–326 (2003)
8. Stanton, T.K., et al.: New broadband methods for resonance classification and high-resolution imagery of fish with swimbladders using a modified commercial broadband echosounder. ICES J. Mar. Sci. **67**(2), 365–378 (2010)
9. Costa, C., et al.: Automated sorting for size, sex and skeletal anomalies of cultured seabass using external shape analysis. Aquacult. Eng. **52**, 58–64 (2013)
10. Shieh, A.C.R., Petrell, R.J.: Measurement of fish size in atlantic salmon (Salmo salar l.) cages using stereographic video techniques. Aquacult. Eng. **17**(1), 29–43 (1998)
11. Zion, B., et al.: Real-time underwater sorting of edible fish species. Comput. Electron. Agric. **56**(1), 34–45 (2007)

12. Hufschmied, P., Fankhauser, T., Pugovkin, D.: Automatic stress-free sorting of sturgeons inside culture tanks using image processing. J. Appl. Ichthyol. **27**(2), 622–626 (2011)

13. Zion, B., Shklyar, A., Karplus, I.: Sorting fish by computer vision. Comput. Electron. Agric. **23**(3), 175–187 (1999)

14. Zion, B., Shklyar, A., Karplus, I.: In-vivo fish sorting by computer vision. Aquacult. Eng. **22**(3), 165–179 (2000)

15. Karplus, I., Gottdiener, A., Zion, B.: Guidance of single guppies (Poecilia reticulata) to allow sorting by computer vision. Aquacult. Eng. **27**(3), 177–190 (2003)

16. Arechavala-Lopez, P., et al.: Discriminating farmed gilthead sea bream Sparus aurata and European sea bass Dicentrarchus labrax from wild stocks through scales and otoliths. J. Fish Biol. **80**(6), 2159–2175 (2012)

17. He, H., Wu, D., Sun, D.: Nondestructive spectroscopic and imaging techniques for quality evaluation and assessment of fish and fish products. Crit. Rev. Food Sci. Nutr. **55**(6), 864–886 (2015)

18. Dowlati, M., Mohtasebi, S.S., de la Guardia, M.: Application of machine-vision techniques to fish-quality assessment. Trac-Trends Anal. Chem. **40**, 168–179 (2012)

19. Cadrin, S.X., Friedland, K.D.: The utility of image processing techniques for morphometric analysis and stock identification. Fish. Res. **43**(1–3), 129–139 (1999)

20. Alsmadi, M.K., et al.: Fish recognition based on robust features extraction from size and shape measurements using neural network. J. Comput. Sci. **6**(10), 1088–1094 (2010)

21. Tillett, R., McFarlane, N., Lines, J.: Estimating dimensions of free-swimming fish using 3D point distribution models. Comput. Vis. Image Underst. **79**(1), 123–141 (2000)

22. Bermejo, S.: Fish age classification based on length, weight, sex and otolith morphological features. Fish. Res. **84**(2), 270–274 (2007)

23. Harvey, E., Fletcher, D., Shortis, M.: Estimation of reef fish length by divers and by stereo-video - A first comparison of the accuracy and precision in the field on living fish under operational conditions. Fish. Res. **57**(PII S0165-7836(01)00356-33), 255–265 (2002)

24. Dios, J., Serna, C., Ellero, A.: Computer vision and robotics techniques in fish farms. Robotica **21**(3), 233–243 (2003)

25. Harbitz, A.: Estimation of shrimp (Pandalus borealis) carapace length by image analysis. ICES J. Mar. Sci. **64**(5), 939–944 (2007)

26. Balaban, M.O., et al.: Prediction of the weight of alaskan pollock using image analysis. J. Food Sci. **75**(8), E552–E556 (2010)

27. Gumus, B., Balaban, M.O.: Prediction of the weight of aquacultured rainbow trout (Oncorhynchus mykiss) by image analysis. J. Aquatic Food Prod. Technol. **19**(PII 9286615753), 227–237 (2010)

28. Mathiassen, J.R., et al.: High-speed weight estimation of whole herring (Clupea harengus) using 3D machine vision. J. Food Sci. **76**(6), E458–E464 (2011)

29. Cocito, S., et al.: 3-D reconstruction of biological objects using underwater video technique and image processing. J. Exp. Mar. Biol. Ecol. **297**(1), 57–70 (2003)

30. Stien, L.H., Kiessling, A., Marine, F.: Rapid estimation of fat content in salmon fillets by colour image analysis. J. Food Compos. Anal. **20**(2), 73–79 (2007)

31. Strachan, N.J.C.: Length measurement of fish by computer vision. Comput. Electron. Agric. **8**(2), 93–104 (1993)

32. White, D.J., Svellingen, C., Strachan, N.J.C.: Automated measurement of species and length of fish by computer vision. Fish. Res. **80**(2–3), 203–210 (2006)

33. Booman, A.C., Parin, M.A., Zugarramurdi, A.: Efficiency of size sorting of fish. Int. J. Prod. Econ. **48**(3), 259–265 (1997)

34. Ching-Lu, H., et al.: A simple and effective digital imaging approach for tuna fish length measurement compatible with fishing operations. Comput. Electron. Agric. **75**(1), 44–51 (2011)

35. Zion, B.: The use of computer vision technologies in aquaculture - a review. Comput. Electron. Agric. **88**, 125–132 (2012)

36. Balaban, M.O., et al.: Quality evaluation of alaska pollock (Theragra chalcogramma) roe by image analysis. Part II: color defects and length evaluation. J. Aquat. Food Prod. Technol. **21**(1), 72–85 (2012)

37. Ching-Lu, H., et al.: A simple and effective digital imaging approach for tuna fish length measurement compatible with fishing operations. Comput. Electron. Agric. **75**(1), 44–51 (2011)

38. Harvey, E., Fletcher, D., Shortis, M.: A comparison of the precision and accuracy of estimates of reef-fish lengths determined visually by divers with estimates produced by a stereo-video system. Fish. Bull. **99**(1), 63–71 (2001)

39. Gumus, B., Balaban, M.O., Unlusayin, M.: Machine vision applications to aquatic foods: a review. Turkish J. Fish. Aquatic Sci. **11**(1), 167–176 (2011)

40. Ruff, B.P., Marchant, J.A., Frost, A.R.: Fish sizing and monitoring using a stereo image-analysis system applied to fish farming. Aquacult. Eng. **14**(2), 155–173 (1995)

41. Burwen, D.L., Fleischman, S.J., Miller, J.D.: Accuracy and precision of salmon length estimates taken from DIDSON sonar images. Trans. Am. Fish. Soc. **139**(5), 1306–1314 (2010)

42. Boswell, K.M., Wilson, M.P., Jr, J.H.: Cowan, A semiautomated approach to estimating fish size, abundance, and behavior from dual-frequency identification sonar (DIDSON) data. North Am. J. Fish. Manag. **28**(3), 799–807 (2008)

43. Kang, M.: Semiautomated analysis of data from an imaging sonar for fish counting, sizing, and tracking in a post-processing application. Fish. Aquatic Sci. **14**(3), 218–225 (2011)

44. Hightower, J.E., et al.: Reliability of fish size estimates obtained from multibeam imaging sonar. J. Fish Wildl. Manage. **4**(1), 86–96 (2013)

45. Strachan, N.: Sea trials of a computer vision-based fish species sorting and size grading machine. Mechatronics **4**(8), 773–783 (1994)

46. van Rooij, J.M., Videler, J.J.: A simple field method for stereo-photographic length measurement of free-swimming fish: merits and constraints. J. Exp. Mar. Biol. Ecol. **195**(2), 237–249 (1996)

47. Misimi, E., Erikson, U., Skavhaug, A.: Quality grading of Atlantic salmon (Salmo salar) by computer vision. J. Food Sci. **73**(5), E211–E217 (2008)

48. Misimi, E., et al.: Computer vision-based evaluation of pre- and postrigor changes in size and shape of atlantic cod (Gadus morhua) and atlantic salmon (Salmo salar) fillets during rigor mortis and ice storage: Effects of perimortem handling stress. J. Food Sci. **73**(2), E57–E68 (2008)

49. Balaban, M.O., et al.: Using image analysis to predict the weight of alaskan salmon of different species. J. Food Sci. **75**(3), E157–E162 (2010)

50. Balaban, M.O., et al.: Quality evaluation of alaska pollock (Theragra chalcogramma) roe by image analysis. part I: weight prediction. J. Aquat. Food Prod. Technol. **21**(1), 59–71 (2012)

51. Lee, D., et al.: Development of a vision-based automatic vaccine injection system for flatfish. Aquacult. Eng. **54**, 78–84 (2013)

52. Merkin, G.V., et al.: Digital image analysis as a tool to quantify gaping and morphology in smoked salmon slices. Aquacult. Eng. **54**, 64–71 (2013)

53. Hong, H., et al.: Visual quality detection of aquatic products using machine vision. Aquacult. Eng. **63**, 62–71 (2014)

54. Seong-Jae, J., et al.: Vision-based automatic system for non-contact measurement of morphometric characteristics of flatfish. J. Electrical Eng. Technol. **8**(5), 1194–1201 (2013)

55. Lee, D., et al.: Weight estimation of the sea cucumber (Stichopus japonicas) using vision-based volume measurement. J. Electrical Eng. Technol. **9**(6), 2154–2161 (2014)

56. Yamana, Y., Hamano, T.: New size measurement for the Japanese sea cucumber Apostichopus japonicus (Stichopodidae) estimated from the body length and body breadth. Fish. Sci. **72**(3), 585–589 (2006)

57. Han, J., et al.: Automated acoustic method for counting and sizing farmed fish during transfer using DIDSON. Fish. Sci. **75**(6), 1359–1367 (2009)

58. Pan, P., et al.: Prediction of shelled shrimp wight by machine vision. J. Zhejiang Univ. (Science B) **10**(8), 589–594 (2009)

59. Fan, L., Liu, Y.: Automate fry counting using computer vision and multi-class least squares support vector machine. Aquaculture **380**, 91–98 (2013)

60. Torisawa, S., et al.: A digital stereo-video camera system for three-dimensional monitoring of free-swimming Pacific bluefin tuna, Thunnus orientalis, cultured in a net cage. Aquat. Living Resour. **24**(2), 107–112 (2011)

61. Petrell, R.J., et al.: Determining fish size and swimming speed in cages and tanks using simple video techniques. Aquacult. Eng. **16**(1–2), 63–84 (1997)

62. Costa, C., et al.: Extracting fish size using dual underwater cameras. Aquacult. Eng. **35**(3), 218–227 (2006)

63. Newlands, N.K., Porcelli, T.A.: Measurement of the size, shape and structure of Atlantic bluefin tuna schools in the open ocean. Fish. Res. **91**(1), 42–55 (2008)

64. Iwamoto, S., Checkley, D.M., Trivedi, M.M.: REFLICS: real-time flow imaging and classification system. Mach. Vis. Appl. **13**(1), 1–13 (2001)

Study on Growth Regularity of Bacillus Cereus Based on FTIR

Yang Liu[1], Ruokui Chang[2(✉)], Yong Wei[2], Yuanhong Wang[3], and Zizhu Zhao[2]

[1] College of Agronomy and Resources and Environment, Tianjin Agricultural University, Tianjin 300384, China
782292767@qq.com
[2] College of Engineering and Technology, Tianjin Agricultural University, Tianjin 300384, China
changrk@163.com, 595183963@qq.com, 274636904@qq.com
[3] College of Horticulture and Landscape, Tianjin Agricultural University, Tianjin 300384, China
529007475@qq.com

Abstract. Combing the one-dimensional infrared spectroscopy (FTIR) technology with two order derivative spectrum technology, the growth change rule of bacillus cereus, the common food borne pathogenic bacteria, are analyzed without destruction, It is found that capsule, spore and other structures of bacillus cereus can be identified, based on the two order derivative spectra characteristic absorption peak. Observing shows the symmetric & anti symmetric carboxyl group stretching vibration absorption peaks near 1604.48 cm^{-1} and 1396.21 cm^{-1} gradually weaken from lag phase to the stable phase. With protein amide absorption peak near 1654.63 cm^{-1} tending to be stable in the three stages, the structural changes of cell capsule can be acknowledged. The DPA absorption peak near 1617.98 cm^{-1}, 1384.64 cm^{-1}, and 1560.13 cm^{-1} indicates the presence of bacillus, changing in three stages from lag phase to stable phase. Experiments show that FTIR can distinguish cells' material structure, which lays a theoretical foundation for the related devices of fast detection for bacillus cereus.

Keywords: Bacillus cereus · FTIR · Capsule · Spore

1 Introduction

Widely existing in nature such as in the air and the soil, bacillus cereus is a kind of gram-positive bacilli which can cause food poisoning [1]. Food poisoning is the result of fast growth and reproduction of bacillus cereus owing to improper insulation and long time placement, which gives rise to vomit enterotoxin and diarrhea enterotoxin, causing human gastrointestinal disfunction. Yet, bacillus cereus can produce antimicrobial substances, which inhibit the propagation of harmful microorganisms and degrade nutrient constituent in soil, thereby improving the ecological environment.

The traditional classification and identification of microorganisms is the application of microscopy, biochemical, physiological and the combination. Although these

D. Li and Z. Li (Eds.): CCTA 2015, Part II, IFIP AICT 479, pp. 33–40, 2016.
DOI: 10.1007/978-3-319-48354-2_3

methods are reliable and effective, operation of them is complex, time-consuming, sometimes the result is not as accurate as expected, and furthermore, these methods are difficult to realize the automation and computerization [2]. In the middle of 20th Century, infrared spectroscopy (FTIR) began to be used to distinguish different species of microbe [3]. In 1991, Naum Ann [4] began to apply the Fourier transformation infrared spectroscopy for the discrimation, classification and identification of microorganism. Since then more and more researchers have begun to employ FTIR technology to study the characteristics of various microbe. With high resolution, Fourier transformation infrared spectroscopy (FTIR) can reflect the whole cell component molecular vibration characteristics, such as the characteristics of proteins, nucleic acids and other substances, as well as can quickly select, identify, and classify a large scale of microorganism in the subspecies level. Usually naked eyes can not identify different microorganism in terms of the spectrum which can only be distinguished by extracting special information with statistics, and the combination of different chemical metrology methods [3, 5]. Instead of the traditional dyeing methods, this experiment employs infrared spectroscopy to observe the cell structure of waxy bacillus cereus, and study the growth regularity, which lays a theoretical foundation for the related devices of fast detection for bacillus cereus.

2   Experiments and Methods

2.1   Experimental Material and Instruments

The devices are IR200 Fourier transformation infrared spectrometer, and TCFW-4 type powder pressing machine; what is also needed in the experiments is test tube, Petri dishes, flask, coater, and gun head; the software is infrared spectrum analysis software OMNIC 8.2

The Bacillus cereus is provided by Beijing North Carolina Chuanglian Biotechnology Research Institute; potassium bromide, tryptone, yeast extract, sodium chloride and distilled water are provided by the laboratory of Tianjin University of Agriculture, and is a pure analysis; experimental water is the water distilled twice.

2.2   Experimental Methods

2.2.1   LB Culturing Medium Configuration

After thawed, bacillus cereus in this experiment is cultured in LB medium whose ratio is 1L solid medium with 10 g tryptone, 5 g yeast extract, 10 g sodium chloride, 15 g agar, and 1000 ml distilled water. LB medium is stirred until all solid dissolve; if the PH value is not between 7.0–7.4, the addition of a small amount of sodium hydroxide is allocated to assure of the PH value. Then LB medium is sterilized with high temperature and shaken well; finally the configured LB medium is batched into the triangular flask.

2.2.2 Measurement of the Reproduction Number

As the individual number is required to be calculated to determine the number of microbial breeding, this experiment adopts plate counting method of indirect measurement method, specifically, a definite volume of diluted broth is mixed with appropriate solid medium before the solidification, or is coated on the solidified medium plate. After insulation culture, when colony number on (in) the plate is multiplied by broth dilution degree, the number of bacteria in the original bacteria liquid could be calculated. Usually on a Petri dish of 9 cm diameter, 50~500 colony is appropriate [6].

2.2.3 Growth Regularity of Microbe

Bacterial cells are extremely small and in the whole growth process, complex biochemical and cytological changes occur in different stages. But, the study of this kind of change of a single bacterium is technically difficult. Methods currently available are: (1) ultrathin section of electron microscopy for the observation of bacterial cells; (2) synchronous culture techniques, which tries to make all the cell population be in the same cell growth and cell division cycle, and then analyze various biochemical characteristics of this group to understand the changes of single cells.

Microbial growth curves (growth curve) are groups of regular curves, which appear by inoculating pure microbe with single cell in a small amount of liquid medium with a constant volume in the appropriate temperature and ventilation (no ventilation for anaerobe). According to the growth rate constant of microorganism, the typical growth curves can be divided into lag phase, exponential phase, stationary phase and decline phase. After tracking and detecting for 28 h in this experiment, according to the generating curve, there are three phases: lag phase, exponential phase and stable phase. Specific curves are as shown in Fig. 1.

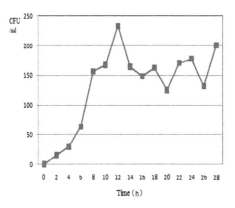

Fig. 1. Bacillus cereus growth curve

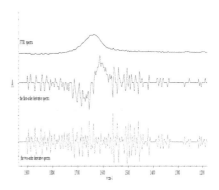

Fig. 2. comparison chart of Infrared spectra, first derivative spectra, and the two order derivative spectra

From Fig. 1, from 0–6 h, the growth curve of bacillus cereus is relatively slow as the lag phase; from 6–12 h, the growth curve is changing rapidly as the exponential phase; from 14–28 h, it tends to be smooth as the stable period.

2.2.4   Detection of FTIR

With doping small amount of cell into potassium bromide tabletting, infrared spectroscope of bacillus cereus is determined by the IR200 Fu Liye transformation infrared spectrometer with deuterated three glycine detector. With Spectral measurement range being 4000 cm^{-1}~400 cm^{-1}, the spectral resolution being 4 cm^{-1}, and signal scan accumulating 48 times, the infrared spectroscope of bacillus cereus is obtained.

3   Results and Discussion

3.1   FTIR Analysis of Bacillus Cereus

Infrared spectroscopy is very valuable in detecting the cell structure of bacillus cereus in the whole growth period. This experiment is to find the changing law of spectral curve for bacillus cereus by collecting one-dimensional infrared spectra and second-order derivation spectra of bacillus cereus. As shown in Fig. 2, with the information extracted from one-dimensional infrared spectra being limited, the second-order derivative spectra are adopted, as it can improve the resolution, increase the amount of information, enhance the information quality, and highlight the spectrum characteristics. Not only can it distinguish peaks overlap, but it can easily distinguish the shoulder peak in the strong ones, and make the implicit information in the infrared spectra outstanding [7].

3.2   The Two Order Derivative Spectrum Analysis of Bacillus Cereus

3.2.1   Composition of Bacillus Cereus

Bacillus cereus is composed of capsule, spore flagellum and other kinds of material structure, among which capsule and spore are the most important.

The capsule is transparent jelly-like material with a certain thickness wrapped with a fixed level over a single cell wall. Its chemical composition includes peptides, proteins, polysaccharides, lipids, lipoproteins, and lipopolysaccharide. The capsule can protect cells from drying and enhance certain pathogen pathogenicity [6].

Bacillus is a round or oval strong resistance dormant structure[6] which is formed in the cells during the later growth of bacillus cereus. In the generation process of bacillus cereus spores, cells producing spores can absorb a large amount of calcium ions and synthesize pyridine carboxylic acid (DPA) two. Therefore, the presence of DPA is the key to determine if there is the spore.

3.2.2   The Second Order Derivative Spectrum Analysis of Bacillus Cereus in Different Periods

In Fig. 2, Bacillus cereus growth curve can be divided into the lag phase, exponential and stationary phases. Concrete analysis is as follows:

(1) Lag phase

Also known as the adaptation period or period of adjustment, the lag phase is the beginning and culturing stage during which the cell number increase slowly after bacillus cereus is inoculated into a new culture liquid. At this stage, it has the following characteristics: (1) the growth rate constant equals to zero; (2) the morphology of the cells increase or become longer, which make many bacteria grow filamentous; (3) Active synthesis of metabolism, accelerate the synthesis of ribosome, enzyme and ATP, which makes enzyme easily inducible; (4) Reaction to the adverse external conditions is sensitive [6].

In the experiment, 0–6 h is the lag phase, second order derivative spectra of 2 h, 4 h and 6 h are as shown in Figs. 3, 4.

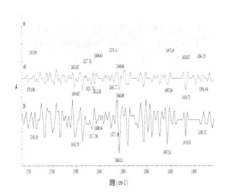 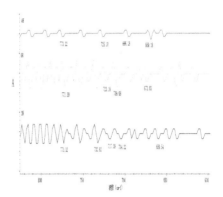

Fig. 3. The second-order derivative infrared spectroscope during in lag phase (1400 cm-1– 1750 cm-1)

Fig. 4. The second-order derivative infrared spectroscope in the lag phase (600 cm^{-1}– 800 cm^{-1})

Figures 3, 4 shows that for Bacillus cereus nearby 1604.48 cm^{-1} and 1394.28 cm^{-1} there appear two strong symmetry and anti stretching vibration absorption peaks of carboxyl, and from the second to the sixth hour, with the passage of time, the absorption peak tends to weaken; while the spectral peak near 1652.70 cm^{-1} and 1560.13 cm^{-1}, is mainly from protein amide belt and protein amide II band, and at this stage the absorption peak is increasing; As the bending vibration of C–H methyl as well as Methylene, the absorption peaks near the 1465.64 cm^{-1} at this time are increasing; Near 771.39 cm^{-1}, 723.18 cm^{-1}, 709.68 cm^{-1}, 673.03 cm^{-1} there exist a series of related vibration absorption peaks which are in stable state; There appear very obvious ester absorption peak in the vicinity of 1735.62 cm^{-1}, the main characteristic absorption peak of a poly beta hydroxybutyrate granules (PHB) [7], and at this stage it is weakening; These characteristics of peak change show the presence of capsule.

The absorption peaks near 1627.63 cm^{-1} and 1394.28 cm^{-1} are stretching vibration absorption peaks of two COO- groups produced by DPA in the spore, and the absorption peak near 1579.41 cm^{-1} is caused by C–N bond in the DPA ring. In the lag phase, three absorption peaks are gradually increasing, and the appearance of the three peaks indicates that this phase has generated spores.

(2) Exponential phase

Exponential phase is also known as logarithmic phase following a lag phase, during which cells divide by geometric rate. It has as the following characteristics: ① with growth rate constant R being maximum, generation time G for cell division each time or the time required for plasma doubling is the shortest; ② cells were balancedly growing and cell elements are well distributed; (3) Enzyme is in activity and metabolism is fast [6]. Figures 5 and 6 show the two derivative spectra collected during the sixth to the fourteenth hour.

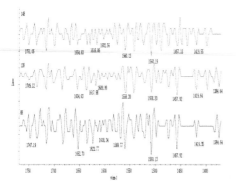 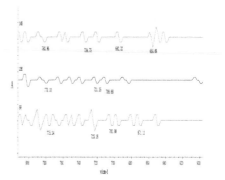

Fig. 5. The second-order derivative infrared spectroscopy in exponential phase (1400 cm^{-1}–1750 cm^{-1})

Fig. 6. The second-order derivative infrared spectroscopy in exponential phase (600 cm^{-1}–800 cm^{-1})

Figures 5 and 6 show in the exponential phase, symmetric and anti symmetric carboxyl stretching vibration absorption peak appear near 1602.56 cm^{-1} and at this stage the absorption peak is weakening; while in the vicinity of 1400.07 cm^{-1}, the absorption peak is gradually increasing; The protein amide I band absorption peak in the vicinity of 1654.63 cm^{-1} tends to increase, in contrast, the amide II band absorption peak around 1560.13 cm^{-1} does not change; Absorption peak at 1457.92 cm^{-1}, generated by the bending vibration of C–H methyl and Methylene, basically hints no change; the absorption peak near 1751.95 cm^{-1} in the ester group has increased; Among a series of stretching vibration absorption peaks, while absorption peak near 734.75 cm^{-1}, 782.96 cm^{-1} is increasing, 655.68 cm^{-1}, absorption peak near 692.32 cm^{-1} is decreasing.

In addition, at this stage, absorption peak of pyridine two carboxylic acid is decreasing near 1560.13 cm^{-1} and 1616.06 cm^{-1}, but shows enhancement tendency in the vicinity of the 1385.56 cm^{-1}, which means at this stage the spores are still rapidly and unstably growing.

(3) Stable phase

Stable phase is also known as the constant regular or higher growth period. Its characteristic is the growth rate constant R is equal to 0, specifically, the number of newly breeding cells and the number of declining cells is the same, or positive growth and negative growth are in the dynamic balance [6].

Figures 7, 8 state that in the stable period symmetric and anti symmetric carboxyl stretching vibration absorption peaks appear in the vicinity of 1396.21 cm^{-1} and 1604.48 cm^{-1}, and decrease in 16–28 h; At this stage, the protein amide I band absorption peak near 1654.63 cm^{-1} shows increasing trend; No change happens for amide II band absorption peak of 1560.13 cm^{-1}; The absorption peak produced by the bending vibration of C–H methyl and Methylene near 1457.92 cm^{-1} has increased; The ester absorption peak near 1749.12 cm^{-1} basically shows no change; In a series of stretching vibration absorption peaks, while absorption peaks near 669.18 cm^{-1}, 723.18 cm^{-1}, and 771.39 cm^{-1} decrease, absorption peak near 707.75 cm^{-1} increases. Figure 7 shows second order derivative spectra of the pyridine carboxylic acid in the stable period, during which absorption peak near the stage of 1617.98 cm^{-1} weakens and there is no significant change for the absorption peak near 1384.64 cm^{-1} and 1560.13 cm^{-1}.

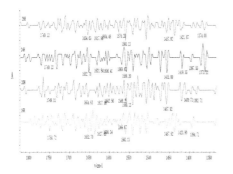 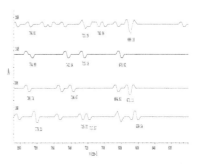

Fig. 7. The second-order derivative infrared spectroscopy in the stable phase (1400 cm^{-1}– 1750 cm^{-1})

Fig. 8. The second-order derivative infrared spectroscopy in the stable phase (600 cm^{-1}– 800 cm^{-1})

4   Conclusion

By calibration experiment and analysis of collected corresponding spectra information, it could be found:

(1) By Fourier transformation infrared spectrometer, the characterization of the main substance of Bacillus cereus spores (characteristic spectrum of capsule and other related substances) could be observed.

(2) By tracking and detecting the growth process of bacillus cereus, collecting the relevant spectral data in different periods, and analyzing growth curve of bacillus cereus, it infers that there is the law of its change in the lag phase, exponential phase and stable phase.

① In the whole growth change interval, for bacillus cereus symmetric and anti stretching vibration absorption peaks of carboxyl group appear nearby 1604.48 cm^{-1} and 1396.21 cm^{-1}, and are gradually weakening; The spectral peaks near 1654.63 cm^{-1} and 1560.13 cm^{-1} are from protein amide I band and protein amide II band, and tend to be stable; In the range of 1457.92 cm^{-1}, bending vibration absorption peaks of the methyl and Methylene also tend to be stable; And an ester absorption peak near 1749.12 cm^{-1} shows a tendency of decrease. According to the analysis of absorption peaks of bacillus cereus, the presence of capsule can be determined in the growth process of bacillus cereus.

② Bacillus cereus has absorption peaks of stretching vibration of the two COO- groups generated by DPA, which appear nearby 1617.98 cm^{-1} and 1384.64 cm^{-1}. The former has increased in the whole process of change, and the latter tends to be stable; Absorption peak near 1560.13 cm^{-1} caused by C–N bond in the DPA ring, tends to be gradually weakening; Absorption peaks caused by DPA appear in the whole growth process of change, and DPA is the evidence determining whether there is the spore for bacillus cereus, it can therefore be concluded that the spore exists and tends to be stable in the stable phase.

In summary, Fourier transformation infrared spectroscopy equipment could quickly identify bacillus cereus spectral changes in different periods, which lays theoretical basis for develop portable devices for rapid detection of bacillus cereus.

Acknowledgements. This research was performed with financial support from National Natural Science Foundation of China under the project No. 31171892, Tianjin Science and technology project 13JCYBJC25700, and Entrepreneurship Training Program of Tianjin Agricultural University under Grant No. 201410061075.

References

1. Morris, J.G.: Bacillus cereus food poisoning. Arch. Intern. Med. **141**(6), 711 (1981)
2. Ci, Y.X., Zang, K.S., Gao, T.Y.: FTIR study of microbes. Chem. J. Chin. Univ. **23**(6), 1047–1049 (2002)
3. Lu, F., Lu, W., Xiao, Z.: A preliminary study of infrared spectrum discriminant method of drug-resistant candida albicans. Chin. J. Anal. Chem. **31**(12), 1532 (2003)
4. Naumann, D., Helm, D., Labischinski, H.: Microbiological characterizations by FT-IR spectroscopy. Nature **351**(6321), 81–82 (1991)
5. Lufeng, C.: The application of Fourier transform infrared spectroscopy combined with chemometrics in the discriminant, classification and identification of microbial. J. Pharm. Pract. **20**(4), 238–240 (2002)
6. Zhou, D.: Microbiology Tutorial, pp. 183–190. Higher education press, Beijing (1992)
7. Cheng, C., Ying, T.: Spectroscopy and spectral analysis **25**(1), 36 (2005)

Soybean Extraction of Brazil Typical Regions Based on Landsat8 Images

Kejian Shen[✉], Xue Han, Haijun Wang, and Weijie Jiao

Innovation Team of Crop Monitoring by Remote Sensing,
Chinese Academy of Agricultural Engineering,
Beijing Chaoyang District Maizidian Street No. 41, Beijing 100125, China
ashenkejian@126.com, hanxue37119@163.com, haijun076481@163.com,
jiaoweijie0502@163.com

Abstract. Considering the spatial distribution and harvest times of Brazil soybean, using Landsat8 data, this paper chose 3 study area, determine the optimum classification images by visually comparison of multi-period images, extract soybean by ISODATA unsupervised classification method and visual correction. The conclusion is that Landsat8 path/row of 225/75 (between Study area3 and Study area2) are soybean transition area of harvest 1 time in a year and harvest 2 times in a year; Classification result can be used for sampling survey of national scale and the full coverage survey of county scale. Soybean classification method can be used to improve the method of low resolution image and to guide other medium resolution image.

Keywords: Soybean transition area · Harvest 2times in a year · Sampling survey

1 Introduction

In recent years, Chinese soybean planting area is gradually reducing, the domestic soybean self-sufficiency ability has been reduced to about 20 %, China mainly imports soybean from United States, Brazil, Argentina [1]. Timely acquisition of the world soybean crop area and its spatial distribution can be used to forecast crop yields, formulate agricultural policies, ensure national food security [2, 3]. Remote sensing has characteristics of wide coverage and low cost, can provided a technical means for fast, accurate and dynamic monitoring of crop area [4].

United States, European Union, China, and other countries have designed crop area survey systems which take advantage of remote sensing and sampling method [5–7], sampling method could solves the contradiction between the accuracy, timeliness and cost. Spot, Rapideye and Landsat images are frequently used as sample image. Considering the free download of landsat8 data, bigger width of Landsat8, this research chooses landsat8 image as data source.

Crop extraction should consider electromagnetic spectrum characteristics, spatial characteristics, time characteristics and auxiliary data [8]. Supervised classification, unsupervised classification, object-oriented method and decision tree classification are

D. Li and Z. Li (Eds.): CCTA 2015, Part II, IFIP AICT 479, pp. 41–47, 2016.
DOI: 10.1007/978-3-319-48354-2_4

frequently used methods. In crop growth cycle, the change of physiological, shape and structure of crop will resulted in seasonal changes of spectral characteristics. Considering to the characteristics of soybean in Brazil, this paper, tried to give transition area of soybean harvest in 1 or 2 times in a year and try to give classify method of image for sampling method.

2   Study Area and Data Source

2.1   Study Area

Most of Brazil belongs to tropical climate, southern parts belongs to subtropical climate. Due to the tropical climate and long growing season, crop cycles are complicated. Below is a month-by-month (Table 1) account of what to expect during growing season. Brazil soybean mainly distribute in central and southern, exists harvest 1 time in a year and 2 times in a year, according to the number of remote sensing images for soybean extraction, three research areas are selected (Fig. 1), the size of each is 40 km × 40 km. Extraction of soybean in study area 1 located in Paraná need 1 time image, which was acquired 2014-01-21; Extraction of soybean in study area 2 located in south Mato Grosso need 2 times images, which were acquired 2013-11-30 and 2014-02-02; Extraction of soybean in study area 3 located in middle Mato Grosso need 3 time images, which were acquired on 2013-11-30/2014-02-02/2014-04-07 (Table 2).

Table 1. Brazil Soybean Month-By-Month Crop Cycle

September	• Early soybean begins in Mato Grosso and central Brazil
October	• Soybean planting in full swing in southern Brazil
November	• Early November is main planting period
December	• Finish planting, early-planted soybeans flowering and setting pods • Begin spraying to control soybean rust
January	• Soybeans flowering and setting pods • Some very early soybeans in central Brazil may be harvested this month • Continue spraying to control soybean rust
February	•Main pod filling month • Early soybeans being harvested • Soybean rust control now focused on later maturing soybeans
March	• Main soybean harvesting month • Critical time for soybean rust to affect late maturing soybeans
April	• Finish soybean harvest
May	• Rains have ended in central Brazil and dry season has started • Scattered rains continue to fall in southern Brazil
June-July-August	• This is the dry season in central Brazil •Occasional rains can occur in southern Brazil

From: http://www.soybeansandcorn.com/Brazil-Crop-Cycles

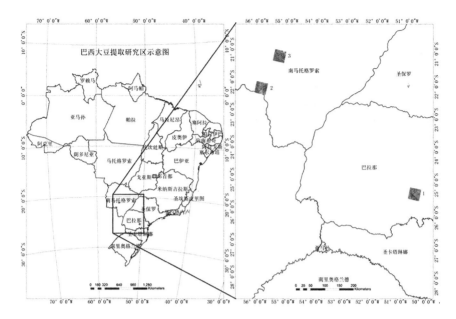

Fig. 1. The sketch map of study area

Table 2. Landsat8 Multi-spectral image

Path/row	Data	Acquired time	Path/row
Study area 1	Landsat8	2014-01-21	221/78
Study area 2	Landsat8	2013-11-30	225/75
		2014-02-02	
Study area 3	Landsat8	2013-11-30	225/75
		2014-02-02	
		2014-04-07	

2.2   Data

landsat8 Multi-spectral image listed as Table 2 are used.

3   Methodology

First classification step is determine the optimum classification images by visually comparison of multi-period images, which need to retrieval same path/row landsat8 images during 2013.09–2014.06. Second classification step is unsupervised classification method and visual correction.

Study area3 harvest 2 times in a year, the extraction of soybean requires three images of different acquired time. The first reason is that same plot harvest 2 times in a year, the second reason is that soybean sowing time of adjacent land vary about 1–2 month. Therefore, classification result of soybean study area3 = time1 + time2 + time3.

Study area2 harvest 1 times in a year, the extraction of soybean requires two images of different acquired time. The reason is that soybean sowing time of adjacent land vary about 1–2 month. Therefore, classification result of soybean study area2 = time1 + time2.

Study area1 harvest 1 times in a year, the extraction of soybean requires one images of acquired time. Classification result of soybean study area1 = time1 (Fig. 2).

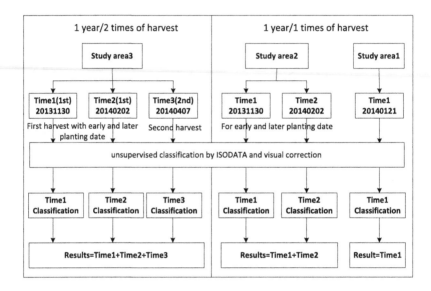

Fig. 2. Flow chart of experiment

4   Results and Discussion

Remote sensing images and classification results of the study area 3 are shown in Fig. 3. Remote sensing images and classification results of the study area 2 are shown in Fig. 4. Remote sensing images and classification results of the study area 1 are shown in Fig. 5. Image features are in Table 3 summarized from Figs. 3, 4 and 5.

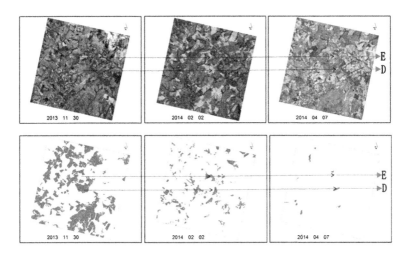

Fig. 3. Study area 3 (3 images)

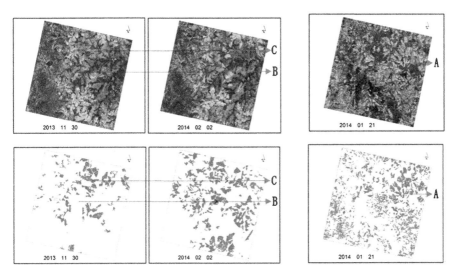

Fig. 4. Study area 2 (2 images)   **Fig. 5.** Study area 1 (1 image)

From Fig. 3 and Table 3, we can see that plot-E which harvest soybean 1 time is bare land on 20131130, soybean on 20140202, bare land on 20140407; plot-D which harvest soybean 2 times is soybean on 20131130, bare land on 20140202, soybean on 20140407. The extraction of soybean is: 639.99 km^2 Result = 483.49 km^2 (Time1) + 145.00 km^2 (Time2) +11.50 km^2 (Time3).

From Fig. 4 and Table 3, we can see that plot-C which harvest soybean 1 time is soybean on 20131130, bare land on 20140202; plot-B which harvest soybean 1 time is bare land on 20131130, soybean on 20140202. The extraction of soybean is: 495.64 km^2 Result = Time1 Union Time2 (in ArcGIS).

Table 3. Image feature

	Plot	Time1	Time 2	Time 3
Study area3	E(1)	Bare land (20131130 blue)	Soybean (20140202 yellow)	Bare land (20140407 blue)
	D(2)	Soybean (20131130 yellow)	Bare land (20140202 blue)	Soybean (20140407 yellow)
Study area2	C(1)	Soybean (20131130 yellow)	Bare land (20140202 blue)	
	B(1)	Bare land (20131130 blue)	Soybean (20140202 yellow)	
Study area1	A(1)	Soybean (20140121 yellow)		

From Fig. 5 and Table 3, we can see that plot-A which harvest soybean 1 time is soybean on 20140121. The extraction of soybean is: 398.04 km^2 Result = Time1.

In this paper, we do not have accuracy assessment, because the lack of high resolution remote sensing data and field data.

5   Conclusions

Considering the spatial distribution and harvest times of Brazil soybean, using Landsat8 data, using Landsat8 data, this paper chose 3 study area, determine the optimum classification images by visually comparison of multi-period images, extract soybean by ISODATA unsupervised classification method and visual correction. The conclusion is that Landsat8 path/row of 225/75 (between Study area3 and Study area2) are soybean transition area of harvest 1 time in a year and harvest 2 times in a year; Classification result can be used for sampling survey of national scale and the full coverage survey of county scale. Soybean classification method can be used to improve the method of low resolution image and to guide other medium resolution image.

Acknowledgment. Funds for this research was provided by National Natural Science Foundation of China (No. 41301506) and Civil Space Project in the 12th Five-Year Plan of China (2011–2015).

References

1. Liu, Z.: Some thoughts concerning development strategy for soybean industry in China. Soybean Sci. **32**(3), 283–285 (2013)
2. Han, L., Pan, Y., Jia, B., et al.: Acquisition of paddy rice coverage based on multi-temporal IRS-P6 satellite AWiFS RS-data. Trans. CSAE **23**(05), 137–143 (2007)

3. Tsiligirides, T.A.: Remote sensing as a tool for agricultural statistics: a case study of area frame sampling methodology in Hellas. Comput. Electron. Agric. **20**(1), 45–77 (1998)
4. Liu, J.: Study on national resources environment survey and dynamic monitoring using remote sensing. J. Remote Sens. **1**(03), 225–230 (1997)
5. Wu, B., Li, Q.: Crop acreage estimation using two individual sampling frameworks with stratification. J. Remote Sens. **8**(6), 551–569 (2004)
6. Shen, K., He, H., Meng, H., et al.: Review on spatial sampling survey on crop area estimation. Chin. J. Agric. Res. Reg. Plann. **33**(04), 11–16 (2012)
7. Gallego, F.J.: Remote sensing and land cover area estimation. Int. J. Remote Sens. **25**(15), 3019–3047 (2004)
8. Jia, K., Li, Q.: Review of features selection in crop classification using remote sensing data. Resour. Sci. **35**(12), 2507–2516 (2013)

Study on Landscape Sensitivity and Diversity Analysis in Yucheng City

Xuexia Yuan[1,2,3], Yujian Yang[4(✉)], and Yong Zhang[5]

[1] Institute of Agricultural Standards and Testing Technology for Agri-Products,
Shandong Academy of Agricultural Sciences, Jinan 250100, China
[2] Shandong Provincial Key Laboratory of Test Technology on Food
Quality and Safety, Jinan 250100, China
[3] State Key Laboratory of Soil and Sustainable Agriculture,
Institute of Soil Science, Chinese Academy of Sciences, Nanjing, China
[4] S & T Information Institute of Shandong Academy of Agricultural Science,
Jinan 250100, China
yyjtshkh@126.com
[5] Shandong Province Land Surveying and Planning Institute,
Jinan 250014, China

Abstract. Landsat ETM image located in Yucheng city in 2002 was interpreted by RS image extraction technology and classification method. Moreover, landscape ecology theories were applied as well as ArcGIS and Fragstats4 to choose the reasonable landscape indices including Contagion Index (CON-TAG), Patch Density (PD), Landscape Shape Index (LSI), Perimeter Area Fractal Dimension (PAFRAC), Shannon's Diversity Index (SHDI), Shannon's Evenness Index (SHEI). The study results showed that correlation between the landscape index was significant at the 0.01 confidence level, the relationship rule was revealed between ecological index based on statistics model. The typical scale effect was selected, including 5 m, 10 m, 15 m, 20 m, 25 m, 30 m, 40 m, 50 m, 60 m, 70 m, 80 m, 90 m, 100 m, 110 m, 120 m, 150 m, 180 m and 210 m. The series results were clarified by PD, LSI, PAFRAC, CONTAG, SHDI and SHEI in response to the different scales, or 18 different scales. The detailed results showed that the decreasing trend was presented from 5 m scale to 210 m scale for each index. Furthermore, we also analyzed the scale effects for different landscape index. Finally, based on image by the change of LSI, PAFRAC, SHAPE-MN and AI on 30 m pixel scale, we emphatically analyzed the LSI, PAFRAC, SHAPE-MN, AI of 12 landscapes. Further, according to the new classification, for the 12 landscapes in Yucheng city, they are Arable-land, Grassland, Traffic and Transmission Land Use, Residential land, Public management and service land, Commercial service land, Garden plot land, Mine and storage land, Woodland, Water and water facility land, Special land and other land we explored and explained the ecological significance of different landscapes in the case city, Especially, landscape sensitivity, fragmentation and complexity of landscape spatial pattern and diversity.

Keywords: Country-level · Landscape sensitivity · Land use · Scale effect

© IFIP International Federation for Information Processing 2016
Published by Springer International Publishing AG 2016. All Rights Reserved
D. Li and Z. Li (Eds.): CCTA 2015, Part II, IFIP AICT 479, pp. 48–59, 2016.
DOI: 10.1007/978-3-319-48354-2_5

1 Introduction

Numerous studies have showed that Landscape is not only the typical scale dependence, but geographical and historical interactions related to ecological system. Obviously, landscape diversity index of Land-use types for grain size and scale changes in response to different sensitivities. At present, land use landscape pattern From Xiamen City, Guangzhou, Shanghai City, Pingyin County in Jinan and Jinghe Watershed landscape had showed landscape diversity and sensitivity of landscape pattern, which had an important impact on landscape pattern of land-use types whether between the landscape pattern index, or between different scales of landscape index [1–4].

From the scale perspective, there is better foundation in Yucheng city, one of the network stations of CAS. With the development of urbanization, information and modernization, land-use landscape pattern and process evolution made rapid changes. Therefore, the case of Yucheng city has great promotion value for the more similar country-level scale in China. Further retrieval of previous literatures have shown that land use landscape diversity and sensitivity of Yucheng City was not retrieved according to the new land use classification standard, the study had few in related landscape analysis, the current research situation is not commensurate with its status for Yucheng city, in the meanwhile, it is not conducive to the landscape of the overall planning and the process of urbanization process in Yucheng City.

2 Materials and Methods

2.1 Study Area Situation

Yellow-Huaihe Rivers Plain is the largest plain in China and an important area of grain, cotton, oil. Yucheng city is the part of Yellow-Huaihe Rivers alluvial Plain and located in the northwest of Shandong province in China, between $116°22'11''–116°45'00''$ E and $36°41'36''–37°12'13''$ N. The total area of the city is 990 km^2, the study area belongs to semi-moisture monsoon climate area and has on average 2639.7 h of sunshine per year. The total radiation of sun is 124.8 K/cm^2. The average temperature per year is 13.1°C, the 200 frost-free days, over 10°C and over 15°C in accumulated temperature are 4441°C and 3898°C in the study area, which provides plenty of thermal conditions, the average rainfall per year is 666 mm. The study area is the part of Yellow-Huaihe Rivers alluvial Plain, the site physiognomy is comparatively complex, there are 7 kinds of landforms in all, including flood land plateau, high land, even land, low-lying land, shallow land, sector crack land and arenaceous river channel. On the basis of topography, landform, parent material and climate, there are two kinds of soil types, Fluvo-Aquic soil and Solonchak. Salt-affected lands are small distributed in the study area, dynamic changes of land use, especially temporal and spatial changes of salt-affected lands, has an important role to improve land quality and promote agricultural sustainable development in the study area [5].

This paper depends on ETM image combined with land-use other data. In order to improve the accuracy of RS image, we referred to the data, 1:50 000 topography map in scale and other spatial maps. The same scale (1:50 000) maps, such as groundwater salinity map, groundwater depth map, soil organic matter content map, soil texture and configuration map. In the study, the author carried out the new land-use classification system, involved in 12 landscapes, they are Arable-land, Grassland, Traffic and Transmission Land Use, Residential land, Public management and service land, Commercial service land, Garden plot land, Mine and storage land, Woodland, Water and water facility land, Special land and other land, respectively.

2.2 Scale Effect System and Landscape Significance

Different landscape index has different ecology significance, the selected landscape index which is applied to analyze the scale effect is illustrated in Table 1 [6].

Table 1. Ecological significance and range of Index

Index	Range	Ecological significance
CONTAG	$0 < CONTAG \leq 100$	Contagion is inversely related to edge density, When edge density is very low, for example, when a single class occupies a very large percentage of the landscape, contagion is high, and vice versa. The index reflects fragmentation and complexity of landscape spatial pattern.
PD	$PD > 0$	Number of patches of a certain landscape element per unit area, the index reflects density degree and difference of landscape spatial pattern.
LSI	$LSI \geq 1$	The index is to measure shape complexity of a certain patch through calculating the deviation of its shape from circle or square of the same area. The more complex and irregular the patch shape is, the higher LSI value is.
PAFRAC	$1 \leq PAFRAC \leq 2$	The index to some extent reflects the degree of human disturbance, and indicates the relationship between shape and area of landscapes consisting of patches, and the index at the landscape level is identical to the class level.
SHDI	$SHDI \geq 0$	The index reflected the diversity of land-use landscape, and it is in response to heterogeneity, and especially sensitive to the non-balanced distribution of all patches, Shannon's index is somewhat more sensitive to rare patch types than Simpson's diversity index.
SHEI	$0 \leq SHEI \leq 1$	The index reflected the diversity of land-use landscape. Shannon's evenness index is expressed such that an even distribution of area among patch types results in maximum evenness. SHDI = 0 when the landscape contains only 1 patch. SHDI = 1 when distribution of area among patch types is perfectly even.

3 Results and Analysis

3.1 Correlation Analysis of Landscape Index

The correlation degree analysis results of six selected index in Yucheng city were showed in Table 2, which disclosed features and changes tendency of land-use landscape spatial patterns.

Table 2. Ecological significance and range of Index

	PD	LSI	PAFRAC	CONTAG	SHDI	SHEI
PD	1					
LSI	0.956	1				
PAFRAC	-0.776**	-0.911**	1			
CONTAG	0.845**	0.961**	-0.976**	1		
SHDI	0.267	0.236	-0.076	0.144	1	
SHEI	0.259	0.227	-0.064	0.135	0.999	1

*: Correlation is significant at the 0.05 level (2-tailed).

**: Correlation is significant at the 0.01 level (2-tailed).

Table 2 summarized the correlation results for the key variables. 6 ecology index (PD, LSI, PAFRAC, CONTAG, SHDI and SHEI) was significant at the 0.01 level, the correlation coefficient was 0.956, −0.776 and 0.845 between PD and LSI, PD and PAFRAC, PAFRAC and CONTAG, respectively. There is significant Correlation between LSI and PD, LSI and PAFRAC, LSI and CONTAG, PAFRAC and CONTAG, the correlation coefficient was 0.956, −0.911, 0.961 and −0.976 at the 0.01 level, especially, the significant coefficient was reflected between SHEI and SHDI, the more significant correlation coefficient was 0.999.

3.2 Sensitivity Analysis and Spatial Pattern of 12 Landscapes

The typical scale effect was selected and explored in the study, including 5 m, 10 m, 15 m, 20 m, 25 m, 30 m, 40 m, 50 m, 60 m, 70 m, 80 m, 90 m, 100 m, 110 m, 120 m, 150 m, 180 m and 210 m. The series results were clarified by PD, LSI, PAFRAC, CONTAG, SHDI and SHEI responding to the different scales, or 18 different scales, the results were clarified in Table 3.

According to Table 3 results, the decreasing trend was presented from 5 m scale to 210 m scale for the index, LSI and CONTAG. The increasing change of PD index was illustrated in the case area from 5 m scale to 210 m scale, furthermore, there was an important point, reflected on 25 m scale, or the decreasing trend of PD index was presented from 25 m scale to 210 m scale. The increasing change of PAFRAC was taken place from 5 m scale to 180 m scale, but on the key 50 m scale, the abruptly

Table 3. Different scale statistics of landscape index

	PD	LSI	PAFRAC	CONTAG	SHDI	SHEI
m	20.214	80.748	1.2236	68.7214	1.4008	0.5637
10m	21.089	79.648	1.2436	66.3961	1.4009	0.5638
15m	22.302	78.447	1.2539	64.4111	1.4007	0.5637
20m	23.438	77.129	1.2645	62.6548	1.401	0.5638
25m	24.333	75.517	1.2734	61.109	1.4012	0.5639
30m	24.232	73.683	1.2836	59.7725	1.4007	0.5637
40m	22.935	70.169	1.2975	57.4559	1.4015	0.564
50m	21.249	66.605	1.3097	55.6154	1.4017	0.5641
60m	19.315	63.080	1.3123	54.1374	1.4019	0.5642
70m	17.267	59.763	1.3179	52.9574	1.4005	0.5636
80m	15.441	56.732	1.3205	51.9548	1.4002	0.5635
90m	13.775	53.954	1.3236	51.0507	1.4022	0.5643
100m	12.327	51.266	1.3272	50.4467	1.3987	0.5629
110m	11.429	49.507	1.3254	49.6355	1.401	0.5638
120m	10.231	47.157	1.3334	49.1382	1.401	0.5638
150m	7.8704	42.043	1.3446	47.6594	1.4044	0.5652
180m	6.1825	38.049	1.3665	46.8699	1.3991	0.5631
210m	4.9978	34.393	1.3276	46.2924	1.3975	0.5624

decreasing change was represented from 180 m scale to 210 m scale. For SHDI and SHEI, obviously, the change rule was consistent, the stable status from 5 m scale to 50 m scale, but the fluctuation was formed from 50 m scale to 210 m scale. Undoubtedly, diversity, fragmentation and hierarchy of landscape located in Yucheng city were interpreted by the change of LSI, PAFRAC, SHAPE-MN and AI landscape index on 30 m scale based on RS image pixel. The detailed tendency of landscape index was shown from Figs. 1, 2, 3, 4, 5 and 6. Obviously, it is crucial for land-use landscape pattern and spatial process evolution in the case study [7–9].

At present, it is widely used that Landscape TM/ETM images were interpreted on country level, whose resolution is 30 m. So 30 m pixel scale was considered and used in the land-use types study in the case region, mainly including Arable-land, Grassland, Traffic and Transmission Land Use, Residential land, Public management and service land, Commercial service land, Garden plot land, Mine and storage land, Woodland, Water and water facility land, Special land and other land. Moreover, the change characteristics of LSI, PAFRAC, SHAPE-MN and AI was calculated by FRAGSTAT4 software for 12 landscapes, the results were referred from Table 4, Figs. 7, 8, 9 and 10.

LSI was applied to analyze shape characteristics of landscape types, The author analyzed the Landscape Shape Index of 12 landscapes, Arable-land, Grassland, Traffic and Transmission Land Use, Residential land, Public management and service land,

PD

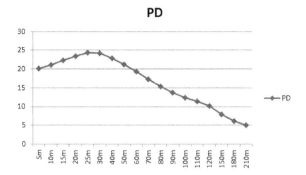

Fig. 1. Sensitivity of PD on different scale

LSI

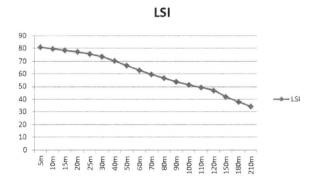

Fig. 2. Sensitivity of LSI on different scale

CONTAG

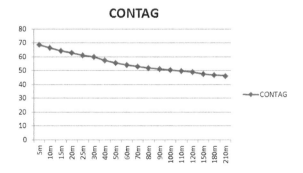

Fig. 3. Sensitivity of CONTAG on different scale

Commercial service land, Garden plot land, Mine and storage land, Woodland, Water and water facility land, Special land and other land. In terms of LSI, the drastically changed from the high value (70.814) to the small value (7.376). Water and water facility land had the high value is 71.814, which meant that Water and water facility land had the most complicate shape and was most influenced by various interventions,

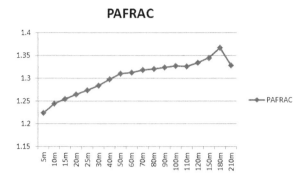

Fig. 4. Sensitivity of PAFRAC on different scale

Fig. 5. Sensitivity of SHEI on different scale

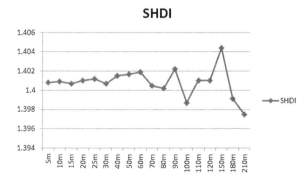

Fig. 6. Sensitivity of SHDI on different scale

which showed the most complex diversity according to LSI ecological significance. Obviously, there are all kinds of Water and water facility land in the case region, including Rivers, water, lake water, water reservoir, ponds, coastal beach, inland beach, ditches, glaciers and permanent snow landform, which explained the complicated

Table 4. Index values of 12 landscapes

	AI	SHAPE_MN	LSI	PAFRAC
Arable-land	93.0069	1.6684	58.8861	1.3356
Grassland	73.6023	1.2772	58.7977	1.2842
Traffic and Transmission Land Use	70.3847	3.552	13.2989	1.5453
Residential land	82.5141	1.4287	56.4615	1.2425
Public management and service land	93.9111	1.6752	7.3756	1.3004
Commercial service land	75.7625	1.2616	62.1716	1.2248
Garden plot land	54.1405	1.1434	25.3056	1.2437
Mine and storage land	56.0564	1.1846	67.5817	1.2749
Woodland	69.2063	1.2298	48.6987	1.2637
Water and water facility land	74.8341	1.3313	71.8142	1.4143
Special land	50.6402	1.1802	17.831	1.2847
0ther land	72.7473	1.2921	63.5423	1.2915

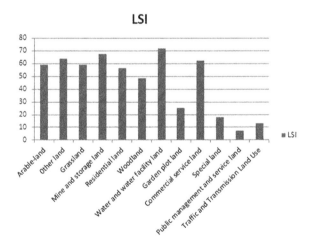

Fig. 7. LSI change for 12 landscapes

landscape types. From the shape matrix perspective, LSI value of Mine and storage land is less than 10, and indicate the landscape types is simple, Mine and storage land had the second most complicate shape, while Public management and service land LSI has the lowest value, which indicated that it had the simplest shapes and were influenced by human interventions [10].

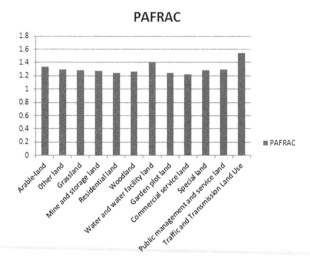

Fig. 8. PAFRAC change for 12 landscapes

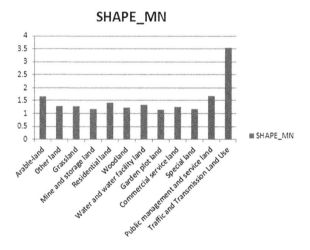

Fig. 9. SHAPE-MN for 12 landscapes

Perimeter-area fractal dimension (PAFRAC) to some extent reflects the degree of human disturbance, and indicates the relationship between shape and area of landscapes consisting of patches, and the index at the landscape level is equal to the class level. For PAFRAC, Traffic and Transmission Land Use had the high value is 1.545, the value decreased step by step, PAFRAC is slight fluctuation centered on 1.2 of 11 landscapes, except 1.6 of Traffic and Transmission Land Use [10].

SHAPE-MN MN (Mean) equals the sum of the corresponding patch metric values, which is divided by the total number of patches. In terms of SHAPE_MN, Traffic and Transmission Land Use had the high value is 3.552, drastically change presented, the

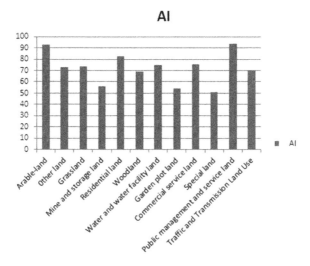

Fig. 10. AI for 12 landscapes

second high value is 1.675 of Public management and service land, the value decreased step by step for other landscapes, Arable-land, Residential land, Water and water facility land, other land, Grassland, Commercial service land, Woodland, Mine and storage land, Special land, Garden plot land.

Area index (AI) does not have a lot of interpretive value about evaluating landscape pattern, but it is important which defines the extent of the landscape. Moreover, many class and landscape metrics computations use total landscape area for. AI of Public management and service land, had the high value is 93.911, and AI of 12 landscapes to some extent presents the variability and gradient characteristics. Basically the kind of change can be divided into three levels in accordance with the area index value, the first gradient change is mainly involved in the three landscapes, Arable-land, Residential land, Public management and service land. The second, Grassland, Woodland, Water and water facility land, Commercial service land, Traffic and Transmission Land Use, other land. The third gradient change included the three landscapes, Garden plot land, Mine and storage land, Special land.

As the precious analysis, the landscape index (LSI, PAFRAC, SHAPE-MN, AI) results showed that they has a certain change and variability for each landscape, especially the landscape is in response to the scale, which indicated the scale effects. From the value of the index perspective, is not steep, or the ecological significance of landscape index indicated has the smooth change, AI has a certain hierarchical characteristics, PAFRAC has a small change, LSI and SHAPE-MN reflected the real landscape, in accordance with own characteristics, and there is no obvious stratification change.

4 Conclusions

In a word, the main contents and conclusions are as follows:

We also definitely understand and investigate the question how changing scale, such as grain size affects pattern analysis. The ecological significance of CONTAG, PD, LSI, PAFRAC, SHDI, SHEI indicated that had a certain scale effects of 18 scales in Yucheng city, though the different degree and different curve. CONTAG changed slightly, which indicated the stable landscape equilibrium. PD has increased from 5 m scale to 30 m scale, but decreased from 30 m scale to 210 m scale, so fragmentation of regional landscapes had presented fluctuation. In terms of LSI and PAFRAC increased from small scale to big scale. SHDI and SHEI changed slightly and showed the balanced landscape diversity, landscape types were evenly distributed, landscape fragmentation and heterogeneity changed slightly [11–13].

We may detect or identify characteristics scales and hierarchical levels to understand and predict ecological phenomena. Based on image by the change of LSI, PAFRAC, SHAPE-MN and AI on 30 m pixel scale, we also emphatically analyzed the LSI, PAFRAC, SHAPE-MN, AI of 12 landscapes. Further, we explored and explained the ecological significance of different landscapes in the case city.

Applying the principles of the landscape ecology, the paper analyzes the landscape diversity and sensitivity of 12 landscapes, points out that landscape sensitivity is the landscape systems response to disturbance at different spatial scales, and reveals that landscape spatial pattern and ecological processes of interaction couple of natural factor and Human disturbance.

Acknowledgements. The work was supported by State Key Laboratory of Soil and Sustainable Agriculture (Institute of Soil Science, Chinese Academy of Sciences) (grant no. 0812201221).

References

1. Geng, Y., Min, Q., Cheng, S., Chen, C.: Temporal and spatial distribution of cropland-population-grain system and pressure index on cropland in Jinghe watershed. Trans. Chin. Soc. Agric. Eng. **24**(10), 68–73 (2008)
2. Xu, L.-H., Yue, W., Cao, Y.: Spatial scale effect of urban land use landscape pattern in Shanghai. Chin. J. Appl. Ecol. **12**, 2827–2834 (2007). (in Chinese)
3. Su, Y., Yang, Y., Liang, Y., Zhao, H.: 3S-based analysis on Guangzhou's landscape pattern. For. Resour. Manage. **6**, 85–89 (2010). (in Chinese)
4. Tang, K.-J., Zheng, X.-Q.: Analysis on the landscape pattern of land use in the middle rolling area of Pingyin county. Res. Soil Water Conserv. **6**, 309–311, 314 (2007). (in Chinese)
5. Homeland planning and land resource investigation report in Yucheng city, Yucheng country government (1991). (in Chinese)
6. Yu, Z.: Landscape Ecology, 3rd edn, pp. 44–64. Chemical Industry Press, Beijing (2008). (in Chinese)

7. Bechtel, A., Puttmann, W., Carlson, T.N., Ripley, D.A.: On the relation between NDVI, fractional vegetation cover, and leaf area index. Remote Sens. Environ. **62**(3), 241–252 (1997)
8. Johnson, G.D, Patil, G.P.: Environmental and ecological statistics series. Vol. 1, pp. 13–21. Landscape Pattern Analysis for Assessing Ecosystem Condition. Springer (2006)
9. Lu, X., Huang, X., Zhong, T., Zhao, X., Chen, Y., Guo, S.: Comparative analysis of influence factors on arable land use intensity at farm household level: a case study comparing Suyu district of Suqian city and Taixing city, Jiangsu province, China. Chin. Geogr. Sci. **22**(5), 556–567 (2012)
10. Fu, B., Chen, L., Ma, K., Wang, Y.: Principle and Application of Landscape Ecology, 2nd edn. Science Press, Beijing (2011). (in Chinese)
11. Zhang, F., Tashpolat, T., Kung, H., Ding, J.: The change of land use/cover and characteristics of landscape pattern in arid areas oasis: an application in Jinghe, Xinjiang. Geo-spatial Inf. Sci. **13**(3), 174–185 (2010)
12. Fang, X., Tang, G., Li, B., Han, R.: Spatial and temporal variations of ecosystem service values in relation to land use pattern in the Loess Plateau of China at town scale. PLoS ONE **9**(10), e110745 (2014)
13. Zeller, K.A., McGarigal, K., Beier, P., Cushman, S.A., Vickers, T.W., Boyce, W.M.: Sensitivity of landscape resistance estimates based on point selection functions to scale and behavioral state: pumas as a case study. Landscape Ecol. **29**, 541–557 (2014). ca sinica **24**(7), 953–956 (2004)

Application and Implementation of Private Cloud in Agriculture Sensory Data Platform

Shuwen Jiang[1,2,3,4], Tian'en Chen[1,2,3,4(✉)], and Jing Dong[1,2,3,4]

[1] Beijing Research Center for Information Technology in Agriculture, Beijing 100097, China
[2] National Engineering Research Center for Information Technology in Agriculture,
Beijing 100097, China
{jiangsw,chente,dongj}@nercita.org.cn
[3] Key Laboratory of Agri-Infomatics, Ministry of Argiculture, Beijing 100097, China
[4] Beijing Engineering Research Center of Argicultural Internet of Things, Beijing 100097, China

Abstract. With the explosive development of the Internet of things technology in recent years, the Internet of things technology is also used more and more widely in modern agricultural production. For mass sensor data was produced by the Internet of things in agricultural production, While big data bring many benefits and unprecedented challenges to users. The Internet of things in agriculture production produces some complexity problem which are mass sensor data's Scale, sensor data's heterogeneity and mass sensor data's operation, distribution of sensor, high concurrency of data is written etc. In the presence of these problems, this paper put forward a kind solution of agricultural private cloud sensor data Platform, which is named "Sensor PrivateClouds Platform" (SPCP). The Private cloud platform including following modules, All of these are distributed sensor data caching module based on cluster of memercached and Nginx load (SensorCache); heterogeneous data adapter of sensor module (SensorAdpter), distributed computing storage module based on hadoop' HDFS (SensorStorage), efficient query module of sensor data warehouse based on the Hive (SensorStore), management module of sensor metadata (SensorManager), parallel sensor data analysis module (SensorNum) based on the map-reduce of the hadoop, cloud service of sensor data module (SensorPublish) based on webservice. The experimental results show that SPCP have had the abilities which are mass sensor data storage, cleaning of heterogeneous sensor data, real-time query and processing of mass sensor data. These abilities provides a feasible solution for the heterogeneous data storage and mass sensor data's query in the Internet of things of agriculture production.

Keywords: Internet of things · Private cloud · Big data · Hadoop · HDFS

1 Introduction

For improving the efficiency of agricultural production, the Internet of things technology also has been widely used in modern agriculture, its main purpose is to collect plenty of sensor data in agricultural production, through the analysis of the sensor data for the

© IFIP International Federation for Information Processing 2016
Published by Springer International Publishing AG 2016. All Rights Reserved
D. Li and Z. Li (Eds.): CCTA 2015, Part II, IFIP AICT 479, pp. 60–67, 2016.
DOI: 10.1007/978-3-319-48354-2_6

agricultural production and utilization to further improve the efficiency of agricultural production. However, in the face of characteristic that are wide area of the production, poor environmental conditions, weak signal of communications in agricultural production, collection and utilization of the sensor data is faced with many challenges. For IOT to compared with Internet, facing the core issues is the storage and query of mass heterogeneous sensor data, processing a large number of sensor's intelligent analysis and work together, complex events such as automatic detection and effective coping [1]. These technology research of problems are relatively limited. Through the application of sensor data in agricultural production, agricultural Internet of things can be found the following problems in the sensory data is:

The big sensor data. Internet of things of agriculture usually contain vast amounts of sensors nodes. Most of these sensors such as GPS sensors, temperature sensors, humidity sensors, these sensors are usually deployed in many different parks or more agricultural production environment in the form of group sensor network [2]. And each sensor can timing acquire the latest sensor data, the sensor data gathering to the storage of sensor data network node. The storage nodes is more than store the recent sensor data and in most cases also need to store a historical data such as 1 year all the historical sensor data value for meeting the needs of the complicated data's processing and analysis. As you can imagine the above data is huge, ordinary server storage for huge amounts of sensory data storage, transmission, query and analysis will be an unprecedented challenge. Heterogeneity of sensory data. The sensor data collection network nodes may include different kinds of sensors such as the environment parameter sensors, geographic information sensor, geological, meteorological sensors, video sensor and so on in a large agricultural production environment. While each kind of sensor also includes many specific sensors such as environment parameters of sensors can be subdivided into soil temperature sensor, a soil moisture sensor, co^2 sensor, light intensity sensor, air temperature sensor, humidity sensor and oxygen sensors. These sensors are not only the structure and function is different, and the format of the sensor data according to the design of the sensor are also different. This will cause the heterogeneity of sensory data [3]. This heterogeneity has greatly enhanced the difficulty of software development and data processing. According to the above problem, this paper proposes a agricultural seneor data Platform based on private cloud which named "Sensor PrivateClouds Platform" (SPCP).

2   Overall Design of Private Cloud in Agricultural Sensor Data Platform

SPCP platform mainly includes the design of Sensor-Cache, which is deploy on a cluster server of MemerCached and Nginx, while is designed as a distributed cache component; Sensor-Adpter which is a adapte component of multi-source heterogeneous sensor data; Sensor-Storage is designed as a distributed storage server based on Hadoop Distributed File System, Sensor-Store is a kind of warehouse which contains sensor meta data, Sensor-Store can efficiently query and compute sensor data via the Hive; Sensor-Manager realize a management system of sensor basic information; Sensor-Num is a component of distributed computing based on the map-reduce of the hadoop; Sensor-Publish is designed as a WebService which realize sensor data's service.

Overall architecture of the SPCP is four levels, First Level is interface level, It contain SensorCache and SensorAdpter. First level mainly complete sensor data's receiving and transforming. Second level is Applications level. It contains SensorPublish and SensorNum. Second level mainly provide private cloud service of analysis and publish. Third Level is manage level which contains SensorManager, it mainly complete management of sensor data and cloud module. Fourth level is storage level which contains SensorStorage and SensorStore. It mainly complete storage of sensor data in private cloud platform. SensorStorage and SensorStore mainly implements the data storage, operation, and query of huge mass data [4]. The platform used the hadoop distributed file system, in order to HBase as non-relational data storage, HIVE for processing and query of mass sensor data. Building sensor data management system on top of hadoop platform structures, the user can manage data access of sensor data node, real-time monitoring, history query and other functions through the web mode. Overall architecture of sensor data network platform is shown below (Fig. 1).

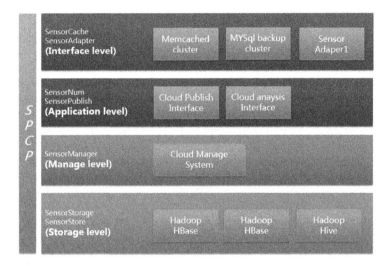

Fig. 1. Overall architecture of sensor data web platform

3 The Design of SensorCache Based on Private Cloud in Hadoop Sensor Data Platform

Sensor data platform use the hadoop distributed file system for data storage and query, now the hadoop platforms use HBase column storage methods for storage, use Hive data warehouse for data query operation, but the experiment proved that real-time latency situation is obvious when the currently used Hive query mass data. It is the problems the entire sensor data platform is facing at present. According to the above problem [5], this paper studies a kind of the cache cluster architecture on top of sensor data platform, cache cluster cache sensor data for a period of time and use the Mysql relational database for sensor data's disaster backup in a period of time. This cache cluster can achieve real-time effect in mass data query. The structure of the cache cluster diagram is shown below (Fig. 2).

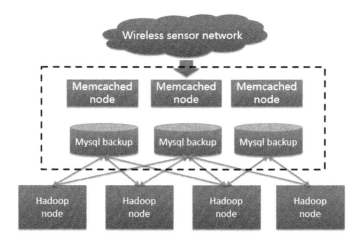

Fig. 2. The overall architecture of the cache cluster diagram

3.1 The Design of the Memcached and Mysql Cluster

The design of Memcached and Mysql cluster is a kind of method based on querying optimization and sensor data backup of Memcached and Mysql cluster. The cluster is used Memcached cluster as sensor data cache service that does not have direct access sensor data from Mysql and liberating the real database. The Mysql so only is used for the history the sensor data's backup, so as to reduce load when the hadoop query mass data. Cluster's cache is make full use of multiple servers network memory, CPU and server memory, its basic process of data access as shown in Fig. 1, When the cache cluster is started, the cache cluster first query out the all sensor data for one day from hadoop data platform by using Hive, and stored the sensor data in the corresponding Memcached server through the distributed algorithm, so that in the next trip, user can obtained the sensor data from the corresponding Memcached server. When number of sensor data reach the specified value, the cache server automatically delete useless cache based on the LRU algorithm, the frequency of accessing database will have obvious drop even zero, it is good for mass data query service. Even database mainly perform sensor data disaster backup, the low configuration database also can easily complete [6].

Sensor data's high concurrent writing is divided into three steps, the first step happens in the cache strategy layer, sensor data will be sent by wireless sensor networks into Memcached nodes via the first layer of the load polling, Memcached node will cache the sensor data as gateway node for the key, sensor data for the value. The second is concurrent sensor data archived Mysql is used to backup history data. Finally, in the time of few users, the cache cluster enable the cache data write to hadoop's sensor data service platform in a distributed file system. The cache cluster meet mass sensor data storage, also has reached the efficient querying requirement of real-time mass sensor data.

4   The Design of SensorManager Based on Private Cloud in Hadoop Sensor Data Platform

Sensor data management system is deployed on a sensor data cluster as a management system. It was mainly used in data access of the wireless sensor network, management of nodes, real-time monitoring of sensor data, query of historical sensor data and simple mass sensor data processing, etc.

Sensor data management system is build with J2EE enterprise architecture. The server used the Spring MVC framework, front end used jquery + backbone. Js + HTML5. Sensor data management system is set up in the hadoop data platform between hadoop cluster and cache strategy cluster, cache cluster mainly complete the sensor data query service as query of real-time and query of short-term history sensor data. Query and analysis of mass sensor data are processed in the hadoop distributed file system [7].

Sensor data management system provides the user for a kind of cloud services of sensor data as entrance door of private cloud. The graph is sensor data management system's operating interface (Fig. 3).

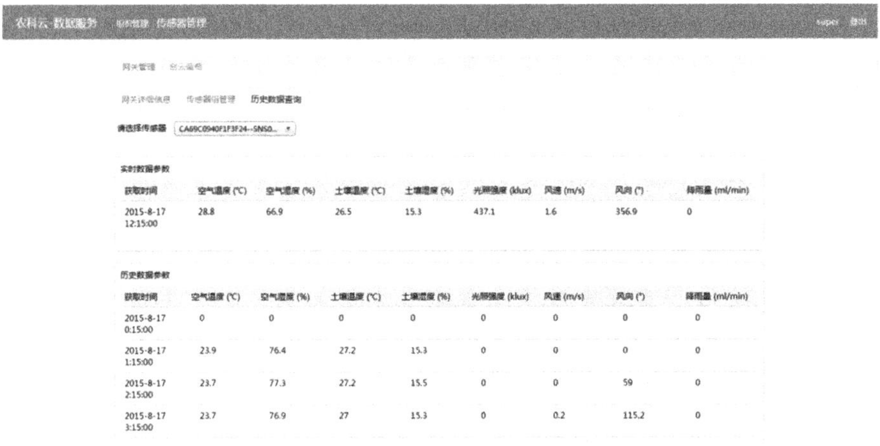

Fig. 3.   Data management system of hadoop sensor data platform.

5   The Design of SensorStorage and SensorStor in Sensor Data Platform

Agricultural sensor data's structure is single and mass. Using relational database for persistence in the early stages that does not take into the characteristics of sensor data's large scale and distributed. With the development of IOT technology, relational database processed data and have a bottleneck. The general solution is that copy tables or distribute storages on different server's partition, but the cost of installation and operation is very high. Through distributed storage system(such as HDFS) technology, the server can dynamically change storage nodes by elasticity features, while the existing storage way of sensor data will not be changed sensor data is distributed on server cluster [8].

Hadoop sensor data storage platform is designed by hadoop open source framework which is used by the hadoop's HDFS system. The platform used of natural database HBase in hadoop, and make a distributed storage of mass sensor data. Its advantage lies in the dynamic increase and decrease of distributed cluster, backuping redundancy and efficient distributed computing.

The sensor data platform adopts 6 server cluster, including one namenode, four data nodes, one manager zookeeper. Sensor data platform obtained history sensor data from the cache cluster and transfer to HBase. Using simple analysis service interface of mass data via the HIVE. Figure 4 for architecture diagram of hadoop sensor data platform.

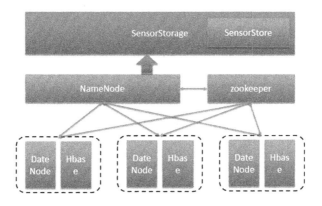

Fig. 4. Architecture diagram of hadoop sensor data platform.

5.1   The Design of NoSql's Sensor Data

Sensor data's storage contains two NoSql relationship tables, the tables respectively contain the table of sensor's meta information and table of sensors original data. The table of sensor's meta information is mainly designed to save sensor's meta basic information. Among the table, SensorId is designed for a column family as a row with the field named sensorInfo. SensorInfo store sensor's meta information by key-value pairs such as sensor-name, sensor-coordinates, sensor-info for a column. Which contains (sensorInfo: id, sensorInfo: net-IP, sensorInfo: net-port, sensorInfo: region servers, sensorInfo: avaCapacity, sensorInfo: location). The table of sensors' original data mainly store original sensors' data of sensor. The table has a raw which contains sensorId and reverse timestamp, field "data" is designed for a family of column which contains(data: temperature, data: carbon-dioxide, data: soil-temperature, data: humidity, data: soil-humidity, data: light) [10].

How big data of sensors can be efficiently read is the key technology in DSM platform. A primary key of "sensorid" is used for a row's identification which describe the table of sensor information. While row key used a id which combines sensor's identification and timestamp for the table of sensor original data. So that we can query realtime data from sensor original data via timestamp. As shown in the Table 1.

Table 1. Design of sensor data table

RowKey	timestamp	timeColumnFamily : sensordata			
	time1	sensordata:tem	sensordata:hum	sensordata:sun	sensordata:co2
sensorGroupid	time2	temValue1	humValue1	sunValue1	co2Value1
	time3	temValue2	humValue2	sunValue2	co2Value2

6 Conclusions

This paper analyzes the problems of big sensor data's acquisition, mass storage and big data analysis in the background of development and application of the IOT and cloud computing technology, with the IOT and cloud computing were wildly applied in the modern agricultural production, these problems brought about the bottleneck is worthy of our attention and solve. To solve the problems existing in the storage of big data, this paper designed a architecture of SPCP, build a wireless sensor network clustering, cache policy cluster and sensor data storage based on hadoop platform. Wireless sensor network realize sensor data acquisition in high concurrency and load balancing, the cache cluster realize efficient real-time query and data backup, data storage platform based on hadoop has realized the huge mass data distributed storage and parallel computing. Experiments show that using SPCP is good at accessing concurrent data, high efficient query and calculation, mass data storage, solved a series of problems brought by the big data.

Acknowledgements. The research work was supported in part by Beijing Natural Science Foundation of China (key project) - Research of adaptive the key technology and application model in agricultural cloud service (4151001) and The Beijing science and technology plan project (D141100004914003). Also with the help of the information engineering department to complete this paper.

References

1. Zhao, Z.-F., Wei, W.-F., Qiang, M.A.: A real-time processing system for massive sensing data. Microelectron. Comput. **29**(9), 10–14 (2012)

2. Ghemawat, S., Gobioff, H., Leung, S.T.: The Google file system. In: Proceedings of the SOSP, pp. 20–43 (2003)
3. Dean, J., Ghemawat, S.: MapReduce: simplified data processing on large cluster. In: Proceedings of the OSDI 2004, pp. 137–150 (2004)
4. Apache Hive. http://hadoop.apache.org/hive/
5. Pike, R., Dorward, S., Griesemer, R., Quinlan, S.: Interpreting the data: parallel analysis with sawzall. Sci. Program. J. **13**(4), 227–298 (2005)
6. Thusoo, A., Sarma, J.S., Jain, N., Shao, Z., et al.: Hive-a petabyte scale data warehouse using Hadoop data engineering. In: Proceedings of the ICDE, pp. 996–1005 (2010)
7. Olston, C., Reed, B., Sirvastava, U., Kumar, R., Tomkins, A.: Pig Latin: a not-so-foreign language for data processing. In: Proceedings of the SIGMOD, pp. 1099–1110 (2008)
8. Chen, T., Xiao, N., Liu, F., Fu, C.S.: Clustering-based and consistent hashing-aware data placement algorithm. J. Softw. **21**(12), 3175–3185 (2010)
9. Patten, S.: The S3 Sookbook: Get Cooking with Amazon's Simple Storage Service. Sopobo (2009)
10. Murty, J.: Programming Amazon Web Service: S3, EC2, SQS, FPS, and SimpleDB. O'Reilly, Sebastopol (2008)

Analysis of Differences in Wheat Infected with Powdery Mildew Based on Fluorescence Imaging System

Shizhou Du[1,2], Qinhong Liao[2,3(✉)], Chengfu Cao[1], Yuqiang Qiao[1], Wei Li[1],
Xiangqian Zhang[1], Huan Chen[1], and Zhu Zhao[1]

[1] Anhui Academy of Agricultural Sciences, Hefei 230031, China
dsz315@sina.com, caocfu@126.com, yuqiangqiao@163.com,
jtlw2007@163.com, xiangqian111@163.com, chenhuanyeah@163.com,
zhaozhu0114@aliyun.com

[2] National Engineering Research Center for Information Technology in Agriculture,
Beijing 100097, China
67076566@qq.com

[3] College of Life Science and Forestry, Chongqing University of Art and Science,
Chongqing 402160, China

Abstract. This study aimed to investigate the variation characteristics of rapid light-response curves of wheat leaves infected with powdery mildew. According to the heterogeneity between two selection patterns of area of interest (AOI), determination of fluorescence induction parameters and fitting of rapid light-response curves were conducted based on fluorescence imaging system in wheat powdery mildew experimental plots. The results showed that relative electron transport rate rETR was reduced with the increase of disease severity level; rETR of the rectangle selection pattern was relatively low. Specifically, the reduction in rETR is mainly influenced by the decrease of absorption coefficient Abs. Among fitting parameters of rapid light-response curves, the potential and the maximum relative electron transport rate, initial slope, light suppression parameter and semi-saturation intensity were reduced with the increase of disease severity level; the heterogeneity of fitting parameters between two selection patterns reflected the "critical state" of leaf fluorescence characteristics. Infected leaves at severe level (80 %) had relatively low light-harvesting capacity and tolerance to strong light, which easily caused light inhibition. According to the lateral heterogeneity analysis of photosynthesis of wheat leaves infected with powdery mildew, there was relatively high heterogeneity between fluorescence parameters of wheat infected leaves, especially in leaves with lesions on the surface.

Keywords: Fluorescence imaging system · Powdery mildew · Rapid light-response curve · Difference

1 Introduction

Wheat powdery mildew (*Blumeria graminis* f. sp. tritici) is a major disease in worldwide wheat production, which is caused by infection of *Erysiphe qraminis* D.C. f. sp. tritici marchal. After infection of *Erysiphe qraminis* D.C. f. sp. tritici marchal., greatly

© IFIP International Federation for Information Processing 2016
Published by Springer International Publishing AG 2016. All Rights Reserved
D. Li and Z. Li (Eds.): CCTA 2015, Part II, IFIP AICT 479, pp. 68–75, 2016.
DOI: 10.1007/978-3-319-48354-2_7

propagated pathogenic bacteria disrupt the moisture transport in wheat leaves and cause the destruction of chlorophyll. With the disease progression, wheat leaves show yellow spots or patchy lesions under severe situations, powder mildew layer is attached to the surface, leading to leaf chlorosis and yellowing or even death [1, 2].

Modulated pulse fluorescence analyzer (Imaging-Pam Mini-version: 24×32 mm) has a relatively large detection window to acquire fluorescence parameters within a range of 24×32 mm [3, 4]. In addition, fluorescence imaging system can clearly distinguish leaf lesion area and non-lesion area due to its visual advantage. Powdery mildew lesion areas on wheat leaves are covered by conidiospores of pathogenic bacteria. Due to the differences in the thickness and status of covering layer, chlorophyll fluorescence characteristics of the lesion areas are relatively complex [5]. Currently, few studies have been reported on the advantages of fluorescence imaging system combining with disease stress. Therefore, in this study, two selection methods of area of interest (AOI) were adopted as data sources which respectively represent the conventional optical fiber probe mode and visualization window mode [6], characteristics of the fluorescence differences in rapid light-response curve between non-lesion area and mixing area of wheat leaves infected with powdery mildew were investigated, to reflect the "critical state" of fluorescence characteristics of infected leaves and explore the dynamic changes in light response of wheat leaves under stress of different levels of powdery mildew, which provided theoretical basis for the in-depth research of fluorescence characteristics of wheat diseases.

2   Experiments and Methods

2.1   Design of Field Trials

During March–June 2013, the experiment was carried out in National Precision Agriculture Experimental Base of Xiaotangshan Town, Changping District, Beijing City, China. Wheat variety Jingshuang 16 was selected as experimental material, which is highly susceptible to powdery mildew. A total of three treatments were designed. After pathogen inoculation on April 5, 2013, wheat powdery mildew was controlled using Triazolone with effective amounts of 240, 120 and 30 g/hm² at booting stage (April 30, 2013) to develop different incidence levels; infected wheat leaves in control group were treated with water. The planting density in experimental plots was 3 000 000 seedlings/hm², with a row spacing of 0.2 m. The rest of the managements were in accordance with the field practices.

2.2   Leaf Selection

Wheat heading stage (May 10, 2013) is the outbreak period of wheat powdery mildew in experimental plots. Wheat leaves infected with different incidence levels of powdery mildew (top second leaf) in various experimental plots were collected as experimental samples. Based on the percentage of the coverage area of lesion hyphal layers on infected leaves accounting for the total area of leaves, disease severity levels (SL) of wheat leaves were classified into four levels, including mild level (10 %), moderate level (50 %), severe level (80 %) and control level (0 %, with no symptom in the whole plant), respectively. Ten leaves at each level were collected.

2.3 Fluorescence Data Acquisition

Based on the pulse amplitude modulation (PAM) techniques, a modulated pulse fluorescence analyzer (Imaging-Pam Mini-version: 24 × 32 mm, Heinz Walz GmbH, Effeltrich, Germany) was used to detect the wheat leaves with disease severity levels according to the chlorophyll fluorescence induction kinetic curve and rapid light response curve [7].

2.4 Rapid Light-Response Curve Fitting

PS II relative electron transport rate was calculated in accordance with the formula: rETR = (Fm'−F) /Fm'·PAR·0.5 Abs. Specifically, PAR indicated the photosynthetic active radiation; 0.5 was the approximate PS II distribution coefficient of light energy; Abs indicated the absorption coefficient. Rapid light-response curves (RLCs) are variation curves of relative electronic transport rate (rETR) with photosynthetic active radiation (PAR). Rapid light-response curve was fit with the least square method using SPSS19.0 software and drawn using Origin 8.0 software.

2.5 Fluorescence Parameter Selection

Modulated pulse fluorescence analyzer has the advantage of image visualization, which provides a variety of data selection patterns of area of interest (AOI). Therefore, lesion areas on wheat leaves infected with powdery mildew could be clearly distinguished. In this study, four sample points were selected in the non-lesion area based on a circle pattern, with a radius of 1 mm, detected values of the four sample points were averaged; four points were selected in the mixing area based on a rectangle pattern, with the length and width of 8 × 28 mm.

3 Results and Discussion

3.1 Changes in Rapid Light-Response Curves of Wheat Leaves Infected with Powdery Mildew Based on Different Selection Patterns

The relative electron transport rate rETR based on different selection patterns rapidly increased with the increase of photosynthetic active radiation PAR, while rETR increased slowly and tended to decline when PAR reached $600 - 1\ 000\ \mu mol/m^2s$. As shown in Fig. 1, the control (0 %) remained relatively high relative electron transport rate rETR; with the increase of severity level, rETR was gradually reduced. Differences in rETR between two selection patterns were gradually enhanced with the increase of severity level, indicating the negative response of the expansion of leaf lesion area to rETR value. At the same severity level, the relative electron transport rate based on a rectangle pattern was lower; for instance, at severe level (80 %), rETR based on a rectangle pattern was only a half of control (0 %).

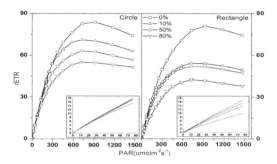

Fig. 1. Rapid light-response curves of wheat leaves infected with powdery mildew based on different selection patterns (left: selected based on a circle pattern; right: selected based on a rectangle pattern)

3.2 Changes in Characteristic Parameters of Rapid Light-Response Curves

As shown in Fig. 2, fitting parameters Pm, α, β and Ik were constantly reduced with the increase of severity level, which were all higher with a circle sampling pattern in non-lesion area; changes in fitting parameters with the disease severity level slightly varied between two selection patterns.

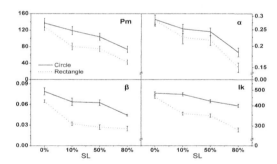

Fig. 2. Changes in characteristic parameters of rapid light-response curves

Variation coefficients of the maximum potential relative electron transport rate Pm in non-lesion area and mixing area were 24.7 % and 43.2 %, respectively. Pm in mixing area at severe level (80 %) was only 33.1 % of normal leaves. Initial slope α reflected the level of light-harvesting capacity of leaves, and the variation coefficients of two selection patterns were 17.6 % and 24.9 %, respectively. Light suppression parameter β reflected the slope of the decline curve, and the variation coefficients of two selection patterns were 22.6 % and 43.8 %, respectively. After powdery mildew infection, β value in mixing area declined greatly and varied slightly between infected leaves, suggesting that the dynamic regulation ability of PS II reaction center under strong light was reduced

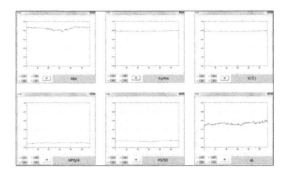

Fig. 3. Heterogeneity of imaging fluorescence parameters of healthy wheat leaves (linear)

after infection, resulting in insufficient capacity of heat dissipation of excess light energy. Semi-saturation light intensity Ik reflected the capacity of leaves to bear strong light, and the variation coefficients of two selection patterns were 8.4 % and 19.4 %, respectively. Ik value in mixing area varied slightly between mild level (10 %) and moderate level (50 %); Ik value was greatly reduced at severe level (80 %), indicating low light-harvesting capacity of leaves and relatively low tolerance to strong light, which could easily cause light inhibition or even light damage.

3.3   Heterogeneity of Chlorophyll Fluorescence Imaging at Different Severity Levels

Conventional fluorescence modulation equipments mostly use optical fiber as signal conductor, such as PAM-2100/2500 and Mini-PAM, which can only be adopted to detect partial photosynthetic activity of leaves. Different parts of leaves have different tissue structures and chlorophyll contents. Therefore, fluorescence characteristics in different parts of the same leaf have lateral heterogeneity, but using fiber optic probes to obtain fluorescence characteristics of a point is difficult to reflect photosynthetic characteristics of an entire leaf. Based on plant chlorophyll fluorescence imaging techniques, the lateral heterogeneity of photosynthesis of wheat leaves infected with powdery mildew was analyzed, which was represented by variation coefficient (CV). Fluorescence imaging monitoring system can even applied to detect the invisible damages in early period, thus revealing the stress state and demonstrating the damage mechanisms.

In healthy wheat leaves (Fig. 3), Abs of leaf veins was slightly reduced, while other fluorescence parameters showed relatively high consistency, with variation coefficients lower than 2 %, suggesting the homogeneity of healthy wheat leaves.

Fluorescence parameters of wheat leaves infected with powdery mildew (SL = 80 %) showed significant heterogeneity. To be specific, fluorescence parameters Fv/Fm (CV = 5.3 %) and PS/50 (CV = 7.8 %) had relatively low heterogeneity; the variation coefficients of other fluorescence parameters Abs, Y(II), NPQ/4 and qL were 9.32 %, 15.12 %, 21.37 % and 25.43 %, respectively. As shown in Fig. 4, fluorescence parameters Fv/Fm, Y(II), PS/50, qL and Abs has consistent variation trends, while NPQ/4 had a contrary variation trend with other fluorescence parameters, which indicated that there was relatively high heterogeneity between fluorescence parameters of wheat leaves infected with powdery mildew, especially in leaves with lesions on the surface. Relative electron transport rate rETR (expressed as PS) is an expression way of photosynthetic rate. A large number of studies have shown that rETR has a linear relationship with photosynthetic oxygen evolution rate or CO_2 assimilation rate before light saturation. Under the same light conditions, rETR is mainly affected by leaf absorbance Abs and PS II actual quantum yield $Y(II)$. $Y(II)$ of wheat leaves infected with powdery mildew had relatively great variation coefficients, suggesting that rETR changes were greatly influenced by $Y(II)$.

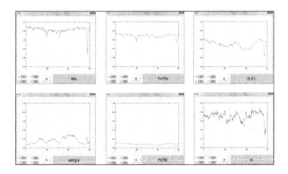

Fig. 4. Heterogeneity of imaging fluorescence parameters of wheat leaves infected with powdery mildew (linear)

Heterogeneity of imaging fluorescence parameters of plant leaves infected with wheat powdery mildew pathogen was mainly produced by the combined action of powdery mildew cleistothecium on leaf surface and damaged photosynthetic systems. Fluorescence techniques provide lossless probes for revealing plant physiology, and the fluorescence imaging system can accurately characterize the physiological characteristics of plant leaves infected with wheat powdery mildew pathogen.

4   Conclusions

In this study, based on the fluorescence imaging system, two selection patterns were adopted to characterize the heterogeneity of rapid light-response curves of wheat leaves infected with powdery mildew, thereby approximately reflecting the "critical state" of fluorescence characteristics of leaves. Rapid light-response curves of non-lesion area of infected leaves vary due to the different severity levels, suggesting that there are differences in PS II system between different severity levels, which may be related with the blocked PS II electron transport caused by the destruction of moisture and chloroplast in leaves, thus leading to the reductions in actual photochemical quantum yield $Y(II)$ and relative electron transport rate rETR [10]. The rectangle selection pattern of wheat leaves infected with powdery mildew covers lesion area and non-lesion area, and the physical damage and inhibition degree become serious, leading to differences in fitting parameters of two selection patterns, which is consistent with the characteristics of other crops under disease stress [11]. Fluorescence parameters of all the pixels in the line (with width of 3) of infected leaves at moderate level (50 %) were shown in Fig. 5. The results showed that: under the same light intensity, the actual photochemical quantum yield Y(II) of leaves is reduced with the increase of photosynthetic active radiation PAR, especially in the lesion area.

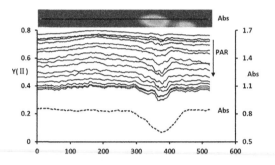

Fig. 5. Heterogeneity analysis of infected leaves based on fluorescence parameters Y (II) and Abs (SL = 50 %)

Relative electron transport rate rETR is an expression way of photosynthetic rate. Results confirm that rETR has an ideal linear relationship with photosynthetic oxygen evolution rate or CO_2 assimilation rate before light saturation. During the detection process, the light adaptation time under different PAR gradients is very short (10 s), resulting in low interference effects on the natural photosynthetic state of samples [12]. Under the same light intensity, rETR is mainly affect by actual quantum yield Y(II) and absorption coefficient Abs. The heterogeneity of Y(II) (CV = 1.45 %-6.37 %) and Abs (CV = 9.17 %) under different light conditions is shown in Fig. 5, which suggests that absorption coefficient Abs of wheat leaves infected with powdery mildew has relatively great effects on rETR, reflecting the heterogeneity of rETR between two selection patterns, which is mainly due to the differences in red band absorption ability of powdery mildew spores on leaf surface and mesophyll cells, and this conclusion has been verified by leaf spectrum detection [13].

Acknowledgements. Funds for this research was provided by Outstanding Youth Fund of President of Anhui Academy of Agricultural Sciences (13B0203); National Science and Technology Support Program of China (2011BAD16B06, 2012BAD04B09).

References

1. Morgounov, A., Tufan, H.A., Sharma, R., et al.: Global incidence of wheat rusts and powdery mildew during 1969–2010 and durability of resistance of winter wheat variety Bezostaya 1. Eur. J. Plant Pathol. **132**(3), 323–340 (2012)
2. Liu, L.Y., Song, X.Y., Li, C.J., et al.: Monitoring and evaluation of the diseases of and yield winter wheat from multi-temporal remotely-sensed data. Trans. Chin. Soc. Agric. Eng. **25**(1), 137–143 (2009)

3. Pedro, S., María, J.Q.: Assessment of photosynthesis tolerance to herbicides, heat and high illumination by fluorescence imaging. Open Plant Sci. J. **3**, 7–13 (2009)
4. Pedro, S., María, J.Q.: Assessing photosynthesis by fluorescence imaging. J. Biol. Educ. **45**(4), 251–254 (2011)
5. Elisa, G., Angeles, C.: Applications of chlorophyll fluorescence imaging technique in horticultural research: a review. Sci. Hortic. **138**(1), 24–35 (2012)
6. Lucia, G., Sauro, M., Elena, D.: Effects of ozone exposure or fungal pathogen on white lupin leaves as determined by imaging of chlorophyll a fluorescence. Plant Physiol. Biochem. **45**(10–11), 851–857 (2007)
7. Olaf, K., Jan, F.H.S.: The use of chlorophyll fluorescence nomenclature in plant stress physiology. Photosynth. Res. **25**(3), 147–150 (1990)
8. Schreiber, U., Gademann, R., Ralph, P.J., et al.: Assessment of photosynthetic performance of prochloron in lissoclinum patella in hospite by chlorophyll fluorescence measurements. Plant Cell Physiol. **38**(8), 945–951 (1997)
9. Platt, T., Gallegos, C.L., Harrison, W.G.: Photoinhibition of photosynthesis in natural assemblages of marine phytoplankton. J. Mar. Res. **38**(4), 687–701 (1980)
10. Kuckenberg, J., Tartachnyk, I., Noga, G.: Detection and differentiation of nitrogen-deficiency, powdery mildew and leaf rust at wheat leaf and canopy level by laser-induced chlorophyll fluorescence. Biosyst. Eng. **103**(2), 121–128 (2009)
11. Stephen, A.R., Julie, D.S.: Chlorophyll fluorescence imaging of plant-pathogen interactions. Protoplasma **247**(3–4), 163–175 (2010)
12. Deng, P.Y., Liu, W., Han, B.P., et al.: A comparative study in photosynthesis heterogeneity of Viola baoshanensis and V. yedoensis. Acta Ecol. Sin. **27**(7), 2983–2989 (2007)
13. Huang, W.J., David, W.L., Niu, Z., et al.: Identification of yellow rust in wheat using in-situ spectral reflectance measurements and airborne hyperspectral imaging. Precis. Agric. **8**(4–5), 187–197 (2007)

Research on Video Image Recognition Technology of Maize Disease Based on the Fusion of Genetic Algorithm and Simulink Platform

Liying Cao, Ying Meng, Jian Lu, and Guifen Chen[✉]

School of Information Technology, Jilin Agricultural University,
Changchun 130118, China
{caoliying99,guifchen}@163.com

Abstract. In order to improve the segmentation accuracy of maize disease leaves with genetic algorithms and reduce segmentation time, this paper proposed a video image recognition technology of maize disease based on the fusion of genetic algorithm and Simulink simulation platform. The technology firstly uses Simulink simulation platform to process the real-time video data captured, including sharpening, segmenting and smoothing, to improve image clarity and quality; Secondly, it uses genetic algorithm to generate optimization model to determine the optimal image of maize diseases; Finally, it fuses genetic algorithms and Simulink platform to analyze and recognize these optimal images. The study results of maize big-spot disease images show that image grey scale values changes after the process of the fused optimal algorithm so that the characteristics of maize diseases are high lightened and the recognition rate of maize disease video image is improved remarkable. The algorithm provides a valid basis for the identification and the diagnosis and treatment of maize disease.

Keywords: Maize big-spot disease · Video image · Genetic algorithm · Simulink platform

1 Introduction

Computer image processing technology is an important component in the field of artificial intelligence, and between man and computer basic theory and application technology provides a specific interface. But the application of image processing technology in agricultural engineering research starts late in our country, mainly in the crop disease diagnosis [1, 2], agricultural product quality detection [3, 4], crop growth status monitoring [5, 6], agricultural crops intelligent classification [7], etc. The present study show that using image processing technology not only can detect, stem diameter, leaf area, leaf circumference petiole Angle of crops such as external growth parameters, can also with the fruit surface color and fruit size to judging the fruit maturity, and crop water lack of fertilizer, and so on and so forth [8]. Computer image processing technology in crop production and research of information collection and has a large amount of information, high speed and high precision of significant characteristics and advantages, and solve some manual measurement is difficult to solve the problem.

© IFIP International Federation for Information Processing 2016
Published by Springer International Publishing AG 2016. All Rights Reserved
D. Li and Z. Li (Eds.): CCTA 2015, Part II, IFIP AICT 479, pp. 76–91, 2016.
DOI: 10.1007/978-3-319-48354-2_8

Domestic started relatively late in the image recognition and processing, but the study of foreign theory made certain optimization, also has obtained certain research results. More mature by identifying plants in the static image texture and color features combined with neural network to achieve for the identification of crop nutrient deficiency. Domestic only video image processing technology was applied to road traffic and dynamic video processing of events, but applied to crop disease monitoring information has not been reported to see.

2   The Research Method

2.1   Simulink

Simulink is a visualization simulation tool of MATLAB, it is a kind of block diagram design based on MATLAB environment, is to realize the dynamic system modeling, simulation and analysis of a software package that is widely used in linear systems, nonlinear systems, digital control and digital signal processing (DSP) in the modeling and simulation. Simulink continuous sampling time and discrete sampling time can be used or two mixed sampling time, it also supports multi-rate system, also is the system of the different parts have different sampling rate. In order to create the dynamic system model and Simulink provides a model block diagram of the graphical user interface (GUI), the creation process can complete just click and drag the mouse operation, it provides a more rapid and straightforward way, and users can immediately see the results of simulation of the system.

2.2   Genetic Algorithms

Genetic algorithm is also in the field of computer science and artificial intelligence to solve the optimization of a heuristic search algorithm, is a kind of evolutionary algorithms. This heuristic is often used to generate useful solutions to optimization and search problems. Evolutionary algorithm was originally borrowed some phenomenon in evolutionary biology and developed, these phenomena, including heredity, mutation, natural selection and hybridization, etc. Genetic algorithm (ga) in the case of wrong selecting fitness function is likely to converge to local optimum [1], and cannot achieve the global optimal.

In genetic algorithm, the above several characteristics together in a special way: based on the parallel search of chromosome group, with a nature of speculation selection operation, switching operation and mutation operation. This particular combination differentiate between genetic algorithm with other searching algorithm.

2.3   Video Analyses

Video analysis, IVS (Intelligent Video System), Video analysis technology is the use of computer image analysis technology vision, through the background and target separation and analysis in the scene and track targets within the camera scene. Video content analysis technology based on visual surveillance camera video image analysis, and has the capability of the background to the variety of the filter, by establishing the model of human activity, with the help of the computer's high speed computing power using a variety of filters, ruled out monitoring scene African human interference factors, accurate judgment in video surveillance images in various human activities.

2.4 Image Classification

Color is a visual features on the surface of the object, the color of each object has its own particular characteristics, to take some similar color to the same object characteristics, so we can according to the characteristics to distinguish object. The color with color in a feature in image classification can be traced back to the Swain and Ballard color histogram method is put forward. Due to the size of the color histogram is simple and with image, rotation changes not sensitive, got the attention of researchers is widely, now almost all image database system based on the content classification to classify color classification method as an important means, and put forward many improved methods, summarized the main can be divided into two categories: the global color feature index and local color feature index.

3 The Genetic Algorithm with Simulink Platform Integration of Video Image Recognition Technology of Corn

3.1 Data Mining Based on Genetic Algorithm

Using the matlab software platform to maize disease image for processing object, by using the genetic algorithm coding, selection, crossover and mutation in the four basic programming operation, find out the objective function, according to the objective function to find the fitness, according to the fitness retains the image of the optimal selection as the candidate solution, give up the other candidate solutions. After iterative cycle, choose the biggest fitness individual images as the optimal solution output. The specific process as shown in Fig. 1.

3.2 Data Collection and Data Processing

Using GPS to collect boundary figure, spatial data such as sample point; Using high-definition camera fetching looks like the corn field, of video image. Establish a spatial database, relational database, video image library and model library.

3.3 SUMILINK Platform for Data Processing

Simulink is a visualization simulation tool of MATLAB, it is a kind of block diagram design based on MATLAB environment; realize the dynamic system modeling,

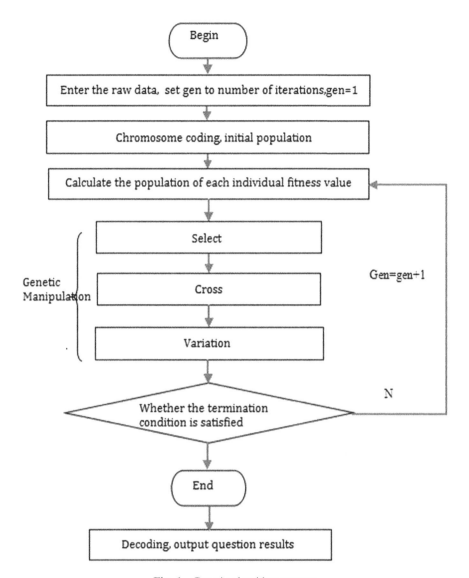

Fig. 1. Genetic algorithm process

simulation and analysis of a software package. In the MATLAB/SUMILINK Video and Image Processing Block set module library can also be used for the Image Processing.

First of all, establish related model, so that the image is processed with the facts, as shown in Fig. 2.

(1) the image transformation and geometric transformation

Before more complex image processing, need to convert the image first, according to the requirements of each image is converted to the required type; Images of

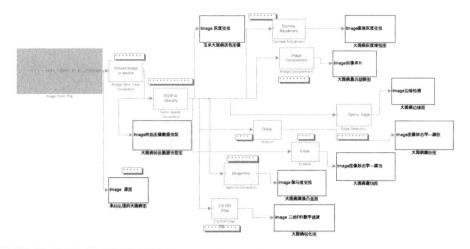

Fig. 2. The image processing model

Fig. 3. Corn big spot pathogen figure

geometric transformation is the image in the size, location and shape of transformation, the collected in the image rotation, translation, mirror to determine the optimal image, in order to deal with. Figure 3 for corn big spot without image processing:

Like respectively to rotation and translation: in the face of corn big spot diagram

(a) The translation of the image

Image of translation is the most simple geometry transform one of the most common transformation, it is a picture of all the points on the image according to the given offset in a horizontal direction along the x axis, in the vertical direction along the y axis moving, the image the same as the original image size after translation. Set (x_0, y_0) to a point on the original image Δx, image level Δy, amount of translation (x_0, y_0) for vertical translation (x_1, y_1), the points after the translation Coordinates will be, the mathematical relationship between them as shown below, processing of the image as shown in Fig. 4:

$$x_1 = x_0 + \Delta x$$

$$y_1 = y_0 + \Delta y$$

Fig. 4. Corn big spot pan figure

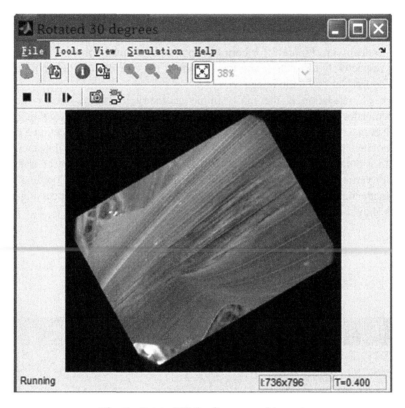

Fig. 5. Rotate 30° the figure corn big spot

(b) The rotation of the image

Image rotation transform belongs to the location of the image transform, usually in the center of the image as the origin, all the pixels in the image rotation Angle of the same angel. Rotation, tend to change the size of the image. Angel values greater than zero, according to the counterclockwise; Angel value less than zero, according to the clockwise, Fig. 5 images for angel too values:

(2) The signal processing module library

Analysis and enhanced module library (Analysis Enhancement) of Block Matching), it can be collected by the image sequence or video frames used for motion estimation, also used to remove redundant information, between the video frames in video compression; Edge Detection, Edge Detection), you can use the Sober operator, Roberts operator, Prewitt operator or Canny operator to find objects in the image Edge, reduce the amount of data, and eliminated can think irrelevant information, retained the image important structural properties; Conversion module library (Conversions) automatic threshold (Auto threshold) can be obtained in the gray image into binary image. The use of the rest of the sons of several module library module of image processing.

(a) Roberts operator edge detection

For discrete image $f(x, y)$, edge detection operator is used the image of vertical and horizontal difference gradient approximation operator, namely:

$$\nabla f = (f(x,y) - f(x-1), f(x,y) - f(x,y-1))$$

In edge detection, the calculation ∇f of each pixel in the image, then the absolute value, the final threshold operation can be achieved. Robert operator calculation formula is:

$$R(i,j) = \sqrt{[f(i,j) - f(i+1,j+1)]^2 + [f(i,j+1) - f(i+1,j)]^2}$$

Roberts operator is composed of the following two templates:

$$\begin{bmatrix} 1 & 0 \\ 0 & -1 \end{bmatrix} \begin{bmatrix} 0 & 1 \\ -1 & 0 \end{bmatrix}$$

The Roberts operator for processing image is shown in Fig. 6:

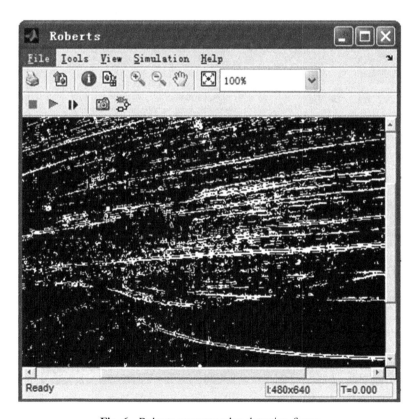

Fig. 6. Roberts operator edge detection figure

(b) Prewitt operator edge detection

For complex images, Roberts operator cannot get better image edges, and the need to adopt more sophisticated 3 * 3 operator, can use Prewitt operator.

Is a kind of first order differential operator edge detection, using pixel point up and down, left and right neighboring points of gray level difference, at the edge

The extreme edge detection, remove part of the false edge, has a smooth function to noise. Its principle is in the image space by using two direction template and image neighborhood convolution to complete, the two direction template a horizontal edges detection, a vertical edge detection. For digital images $f(x,y)$, Prewitt operator are defined as follows:

$$G(i) = |[f(i-1,j-1) + f(i-1,j) + f(i-1,j+1)] - [f(i+1,j-1) + f(i+1,j) + f(i+1,j+1)]|$$
$$G(j) = |[f(i-1,j+1) + f(i,j+1) + f(i+1,j+1)] - [f(i-1,j-1) + f(i,j-1) + f(i+1,j-1)]|$$

Than $P(i,j) = \max[G(i), G(j)]$ or $P(i,j) = G(i) + G(j)$

Prewitt operator is composed of the following two templates:

$$\begin{bmatrix} -1 & -1 & -1 \\ 0 & 0 & 0 \\ 1 & 1 & 1 \end{bmatrix} \begin{bmatrix} -1 & 0 & 1 \\ -1 & 0 & 1 \\ -1 & 0 & 1 \end{bmatrix}$$

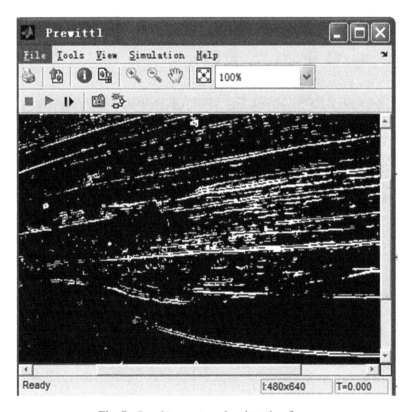

Fig. 7. Prewitt operator edge detection figure

The Prewitt operator for processing the image shown in Fig. 7:

(c) Sobel edge detection operator

The size of the Sobel operator and Prewitt operator is the same size, is 3 * 3 matrix. Horizontal and vertical, respectively, with image plane convolution, can be calculated the brightness of the horizontal and vertical difference approximation. If S on behalf of the original image, G_x and G_y representing the longitudinal and transverse edge detection of image, its formula is as follows:

$$G_X = \begin{bmatrix} -1 & 0 & 1 \\ -2 & 0 & 2 \\ -1 & 0 & 1 \end{bmatrix} G_y = \begin{bmatrix} -1 & -2 & -1 \\ 0 & 0 & 0 \\ 1 & 2 & 1 \end{bmatrix}$$

Each pixel of the image of the transverse and longitudinal gradient approximation can be used in combination with the following formula, to calculate the size of the gradient.

$$G = \sqrt{G_x^2 + G_y^2}$$

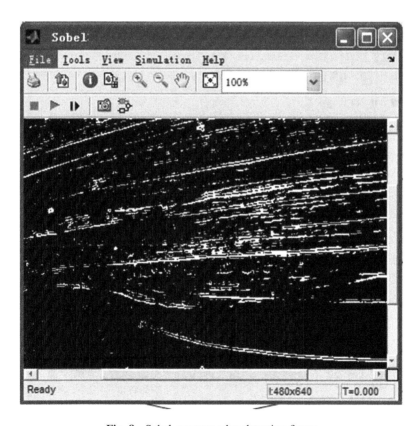

Fig. 8. Sobel operator edge detection figure

And then can use the following formula to calculate gradient direction.

$$\theta = \arctan\left(\frac{G_x}{G_y}\right)$$

By the Sobel operator for processing the image shown in Fig. 8:
(d) The Canny operator edge detection
Canny use of hysteresis threshold.

Hysteresis threshold need two threshold - high threshold and low threshold. Assuming that the image is the important edge in the continuous curve, so that we can track the given the fuzzy part of the curve, and there will be no composition curve to avoid the noise pixels as edges. So we start with a large threshold, which will identify real edge we're pretty sure, using the export in the direction of the information, we started from the real edge at the edge of the image tracking the whole. At the time of tracking, we use a smaller threshold, so that you can trace curve of the fuzzy part until we go back to the beginning.

Once this process is complete, we will get a binary image, each point of said is an edge point.

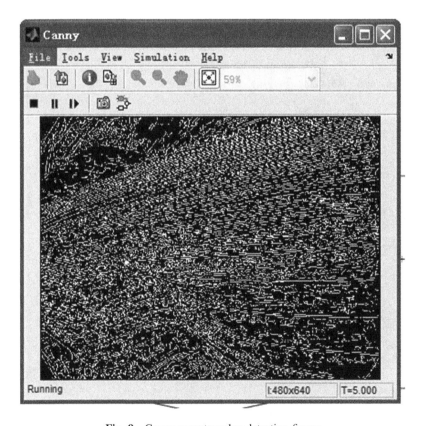

Fig. 9. Canny operator edge detection figure

A achieve sub-pixel accuracy improvement implementation is in the direction of the gradient on the edge of the detection of the second order directional derivative to zero.

$$L_x^2 L_{xx} + 2L_x L_y L_{xy} + L_y^2 L_{yy} = 0$$

Is it in the direction of the gradient direction of the third order derivative conditions meet the symbol

$$L_x^3 L_{xxx} + 3L_x^2 L_y L_{xxy} + 3L_x L_y^2 L_{xyy} + L_y^3 L_{yyy} < 0$$

The Canny operator for processing images as shown in Fig. 9

(3) image enhancement of Simulink implementation

Hd camera and remote monitoring system for video quality may not be used, so in video image enhancement is conducive to further analysis of the image. Image smoothing to highlight the image of wide area, the low frequency part, trunk or suppress image noise and interference of high frequency component processing, make

Fig. 10. Corn big spot binarization figure

the image brightness flat gradient, gradient decrease mutations, improve image quality. Image sharpening is contrary to the image smooth, used for compensating the contour image, enhancing image edge and gray level jump part, makes the image becomes clear. To combine two methods of processing, can improve the quality of the image better for analysis.

(a) The image binarization processing is to the point of the image gray level set to 0 or 255, or the entire image showing a clear black and white effect. About 256 brightness levels of gray image through appropriate threshold selection and can still reflect the overall image and local features of binary images.

The binary processing of image is shown in Fig. 10:

(b) The original image grey value, make the black white and white black. To obtain the main visual effect is a negative effect of the image. The complementary processing of image is shown in Fig. 11:

Fig. 11. Corn big spot complement graphs

3.4   Maize Disease Image Recognition Experiment Result Analysis

For maize disease image recognition research, in view of the common disease types, this paper select maize large spot type test research, which based on Simulink platform

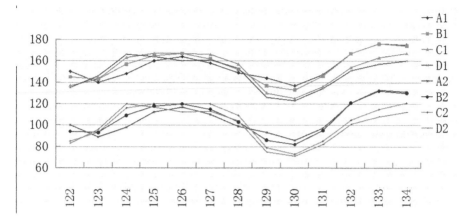

Fig. 12. Gray level and direct gray-scale transformation after the original image

building model, after for corn big spot image processing, using the multiple line chart comparison can obviously see the image gray level transformation, differences before and after. To gray-scale transformation technology of the processed image is direct gray-scale transformation methods, corrosion technique the processed image and corn big spot gray image to do comparative analysis, selected pixel values of 343 ~ 340 line corresponds to 122 ~ 134 columns of gray value contrast, one of the A1 is the original line 340, A2 line 340 for after processing, and so on.

Using the above methods of corn after the big spot gray level image processing, test results as shown in Fig. 12, shown in Fig. 13

From the multiple line chart 3.1, using the gray scale transformation technology of direct gray-scale transformation method after processing the image grey value is decreased obviously, and reduce useless blurs the image grey value, instead, to enhance

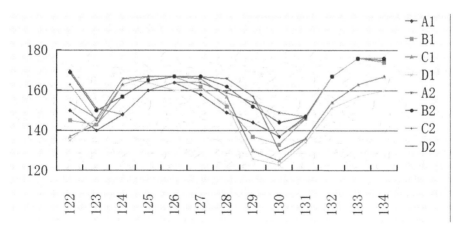

Fig. 13. Gray original image and the image after the corrosion

the image resolution highlighted the disease characteristics, is conducive to the identification of diseases.

From the point of double line chart 13, using corrosion technology processing after image grey value increase. With direct gray-scale transformation processing figure, to the contrary, but corrosion technology can accurately extract the characteristics of large spot disease of maize, to facilitate the identification of diseases.

4 Results and Discussion

(1) Using the Simulink platform, the algorithm established the reorganization model of maize disease video images, improved image resolution, high lightened disease characteristics and increased the recognition rate of maize diseases.

(2) This paper fused genetic algorithms and Simulink platform effectively and built a model with remote video image analysis and classification reorganization of maize diseases. It gained the texture, color and geometry characteristics of maize diseases through field video image analysis and then established analytical model for all kinds of maize diseases. It monitored effectively the real-time video images of field diseases, processed the video image captured by sharpening, segmenting and smoothing and improved image clarity and quality; It fused genetic algorithms and Simulink platform to resolve and identify the optimal image. Results have shown that image grey scale values changes after the process of the fused optimal algorithm, high lightens the characteristics of maize diseases are high lightened, improves the recognition rate of maize disease video image remarkable and provides a valid basis for the identification and the diagnosis and treatment of maize disease.

Acknowledgments. This work was funded by the National High-Tech Research and Development Plan of China under Grants Nos. 2006AA10A309; The research and application of facilities for the safety of vegetables production technology based on Internet of Things (2011-Z20); Special funds for grain production development in Jilin province (an agricultural technology extension project) "research and application of vegetable security production technology based on facilities of Internet of things".

References

1. Russ, J.C.: The Image Processing Handbook. CRC Press (1995)
2. Cheng, H.D., Jiang, X.H., Sun, Y., et al.: Color image segmentation on: advances and prospect. Pattern Recogn. **34**(7), 2259–2281 (2001)
3. Zayas, I., Pomeranz, Y., Lai, F.S.: Discrimination between arthur and arkan wheat by mage analysis. Cereal Chem. **62**(2), 478–480 (1985)
4. Sasaki, Y., Okamoto, T., Imou, K., Tor, T.: Automatic diagnosis of plant disease. J. JSAM **61**(2), 119–126 (1999)
5. Sasaki, Y., Suzuki, M.: Construction of the automatic diagnosis system of plant disease using genetic programming which paid its attention to variety. ASAE Meeting Presentation (paper No. 031049) (2003)

6. El-Helly, M., El-Beltagy, S., Rafea, A.: Image analysis based interface for diagnostic expert systems. In: Proceedings of the Winter International Synposium on Information and Communication Technologies, Trinity College Dublin, pp. 1–6 (2004)
7. Sammany, M., El-Beltagy, M.: Optimizing neural networks architecture and parameters using genetic algorithms for diagnosing plant diseases. In: Proceeding of International Computer Engineering Conference, IEEE (Egypt Section) (2006)
8. Chen, J., Wen, J.: Such as automatic determination of the extent of the use of computer vision in cotton pests. Agric. Eng. **17**(2), 157–160 (2001)
9. Huang, H., Cai, J.: Computer vision research to improve detection of Nosema disease. Jiangsu Univ. (Nat. Sci.) **24**(2), 43–46 (2003)
10. Bai, J., Zhao, X.: Strong tin-rich conifer seedlings and other computer vision feature extraction. Northeast Forestry Univ. **28**(5), 94–96 (2000)
11. Wu, J., Xu, L.: Color Image Segmentation Techniques. China Image Graph. **10**(1), 1–10 (2005)
12. Xu, L., Wu, J.: Advances in computer vision technology in crop growth monitoring. Agric. Eng. **3**(2), 279–283 (2004)
13. Rui, T.: Grapes based on machine vision automatic identification technology. Northeast Forestry Univ. **36**(1), 95–97 (2008)
14. Mao, H.P., Xu, G., Li, P.: Recognition tomato nutrient deficiency of computer vision. Agric. Mach. **34**(2), 73–75 (2003)
15. Mao, H.P., Wu, X., Li, P.: Based on computer vision tomato nutrient deficiency neural network. Agric. Eng. **21**(8), 106–109 (2005)
16. Xu, G., Mao, H.P., Li, P.: Leaf color image color feature extraction. Agric. Eng. **18**(4), 150–154 (2002)
17. Tian, Y.W., Zhang, C., Li, C.: Application of SVM in plant disease spot shape recognition. Supp. Agric. Eng. **20**(3), 134–136 (2004)

The Design and Implementation of Online Identification of CAPTCHA Based on the Knowledge Base

Yu'e Song[1,2], Chengguo Wang[1], Ling Zhu[3], Xiaofeng Chen[1], and Qiyu Zhang[1(✉)]

[1] Yantai Academy, China Agricultural University,
No. 2006, Coastal Middle Road, Gaoxin District,
Yantai 264670, Shandong Province, China
aeaeae623@163.com, rcraingo@163.com,
wangcg@126.com, cxfeng1979@126.com
[2] School of Electrical and Information Engineering,
Beijing Polytechnic College, Beijing 100042, China
[3] Shandong Institute of Business and Technology,
College of Statistics, Yantai 264005, China
oklab@qq.com

Abstract. The Completely Automated Public Turing Test to Tell Computers and Humans Apart (CAPTCHA) identification is designed to distinguish between computers and humans and it prevents the web application programs from malicious attacks, so it has been applied widely. However, great challenges must be faced with the development of CAPTCHA identification. In order to improve the safety of the professional system, the CAPTCHA online identification based on the knowledge base, which has high security and bases on semantic questions and the professionalization of professional system, is put forward combining with the recessive CAPTCHA. The specific implementation course of the new online identification method is worked out according to the example of animal identification. The application of the verification code is suitable for people who have the corresponding professional knowledge. Because the computer has great difficulty to answer semantic information questions, which are also professional issues, so the new online identification method based on the verification of knowledge has very high security.

Keywords: CAPTCHA · Online identification · Knowledge base · Animal

1 Introduction

With the rapid development of internetwork, security problem of the web application becomes an extremely important issue for us. The HTTP attack based on the form automatically submission is a common way of network attack. According to the HTTP protocol, the attacker can write program to simulate the method of form submission, and submit the abnormal data to site service automatically and rapidly. This constitutes the basic HTTP attacks. An attacker can repeat logging to break a user's password and

D. Li and Z. Li (Eds.): CCTA 2015, Part II, IFIP AICT 479, pp. 92–99, 2016.
DOI: 10.1007/978-3-319-48354-2_9

this will lead to a leakage of users' privacy information. In order to prevent the attacker using program automatic login, Completely Automated Public Turing test to tell Computers and Humans Apart (CAPTCHA) technology has been widely used [1].

The CAPTCHA is a kind of program algorithm to distinguish between computers and humans, so the procedure must be able to generate and evaluate computer test which human can easily pass but not for computers [2, 3]. Because the computer cannot solve CAPTCHA question, the user who answer the question can be considered human [4].

In order to protect the network, CAPTCHA has been applied widely, such as preventing spam ads in the blog post, protecting website registration and the E-mail address, online polls, preventing dictionary attacks, the search engine robots, worms and spam, etc.

Since CAPTCHA has been proposed, different research institutions and scholars have developed a variety of CAPTCHA. CAPTCHA has different ways of classification [5]. According to the type of information, CAPTCHA can be divided into text CAPTCHA, image CAPTCHA, graphics CAPTCHA, audio CAPTCHA and video CAPTCHA. According to the way of recognition, CAPTCHA can be divided into dominant CAPTCHA and implicit CAPTCHA. According to the interaction, CAPTCHA can be divided into static CAPTCHA and dynamic CAPTCHA. Along with the development of the CAPTCHA, CAPTCHA recognition technology is also developing and some methods have been put forward, such as the matching shape context [6], template matching [8] and neural network identification methods [7]. This makes the security of the CAPTCHA has a huge challenge. Dynamic CAPTCHA and recessive CAPTCHA have a good security and is the research direction in the future.

The hidden CAPTCHA [5] refers to answering the question of the CAPTCHA expressing according to the semantic of CAPTCHA, for example, CAPTCHA system first randomly generates an expression $(5 + 3)*9/4$ and requires the user to answer the expression values; CAPTCHA system picks up a few images from the graphics library and users need to rotate the graphics to the right direction. Though artificial intelligence has a rapid development, the computer has much difficulty to answer semantic information questions, so the hidden CAPTCHA is safe.

In this paper, the CAPTCHA technology is studied deeply. Based on the implicit CAPTCHA and combining with the characteristics of professional system, a new kind of CAPTCHA is proposed based on the knowledge base and the security of the system can be effectively improved using the new kind of CAPTCHA.

2   Knowledge Representation

In the knowledge base, knowledge representation methods are logical notation, production representation, frame representation and object-oriented representation, semantic representation and the XML representation and representation of ontology [9], etc. According to the characteristics of the CAPTCHA, we choose production knowledge representation description.

Shortliffe firstly introduced the concept of production in the famous expert system MYCIN. The structure IF (E1 & E2 & … & En) THEN A is called the rule. It means that if the logical expression of E1 & E2 & … & En established, the conclusion A is right. The expression E1 & E2 & ... & En is called former part of the rule and is any legal logical expressions. It is the prerequisite for reasoning by using the rule. A is called later part of the rule and is the result of reasoning using the rule. [10]. The rule knowledge representation has many advantages, such as simple and clear reasoning, the reasoning machine design and implementation is simple and has a good characteristics in some specific application environment, etc.

3 The Design of CAPTCHA Based on the Knowledge Base

For some professional systems, CAPTCHA can be structured based on knowledge base. Because users have the corresponding knowledge and can reason the related results according to the precondition. Let us use a simple animal identification as an example to illustrate how to construct CAPTCHA.

We give the following rules about animal identification:

IF the animal has hair THEN the animals are mammals.

IF the animal has milk THEN the animals are mammals.

IF the animal has feathers THEN the animal is a bird.

IF the animal can fly AND lay eggs THEN the animal is a bird.

IF the animal eats meat THEN the animal is a carnivorous animal.

IF the animal has a canine tooth AND claw AND eyes staring at front THEN the animal is a carnivorous animal.

IF the animal is mammals and has claw THEN the animal is a hoof animal.

3.1 The Design of the Database and Table for Knowledge Base

According to the rules of reasoning above, we designed the rules table, inferences table and synonym table. Rules table save the atomic conditions of precondition, which are the minimum condition of premise condition. The above animal identification rules are in the rules table as shown in Table 1.

Table 1. Animal identification rules

Serial number	Rules
1	Have hair
2	Have milk
3	Have feathers
4	Can fly
5	Can lay eggs
6	Eat the meat
7	Have canine tooth
8	Have claws
9	Eye star at the front
10	Have hoof

The result of reasoning is text messages. There are different representations for the same text messages and the computer can't recognize it very well, therefore automatic word segmentation can be used for the results and CAPTCHA. In this process, the word which not be used can be removed and the keywords will be extracted, then we can match the keyword. For Chinese word segmentation, IK Analyzer 2012 can be used. The IK Analyzer is an open source lightweight Chinese word segmentation toolkit based on Java language. In the 2012 version, we support configuring IKAnalyzer. CFG.XML file to expand proprietary dictionary and stop using dictionary and dictionary format is utf-8 without BOM in Chinese text files [11]. Stop using words are not really meaning of function words in both English and Chinese [10] and can be ignored because they does not affect the understanding of sentence meaning. The stop using dictionaries are built on the basis of the literature [10, 11]. In order to assist CAPTCHA judgment, two options are increased which must be contained keywords and must not contained keywords. Meanwhile, in order to reduce the complexity of the system reasoning, the result is made as easy as possible. Inferences table is shown in Table 2.

Table 2. Inference table data

Premise condition	Results	Whether the word is segmented	Must contained keywords	Must not contained keywords
1	Mammals	no	no	no
2	Mammals	no	no	no
3	Birds	no	no	no
4, 5	Birds	no	no	no
6	Predators	no	no	no
7, 8, 9	Predators	no	no	no
1, 10	Hoofed animals	no	no	no
2, 10	Hoofed animals	no	no	no

Synonym of the word in the results is stored synonym table, including Chinese, English and acronyms.

In the MySQL database we design different table structures, which are shown in Tables 3, 4 and 5.

Table 3. Rule table

Field	Data type	Note
Id	int	Automatic numbering, primary key
Rule	varchar(100)	

Table 4. Inferences table

Field	Data type	Note
Id	int	Automatic numbering, primary key
Condition	varchar(100)	
Result	varchar(100)	
Segmentation	char(1)	
Key	varchar(200)	
Antonym	varchar(200)	

Table 5. Synonym table

Field	Data type	Note
Id	int	Automatic numbering, primary key
Key	varchar(100)	
Synonym	varchar(100)	

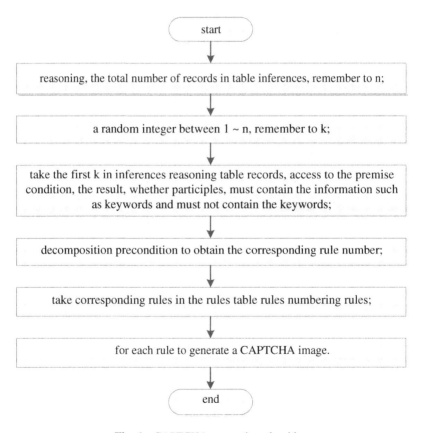

Fig. 1. CAPTCHA generation algorithm

3.2 CAPTCHA Generation Algorithm

(1) Reason the total number of records in inferences table and remember to n;
(2) Randomly select the integer between 1–n, remember to k;
(3) Take the kth records in the inferences table and access the premise condition, the result, whether participles, the keywords which must be contained and which must not be contained;
(4) Decompose precondition to obtain the corresponding rule number;
(5) Take corresponding rules in the rules table rules numbering rules;
(6) Generate a CAPTCHA image for each rule.

The algorithm flow chart is shown in Fig. 1.

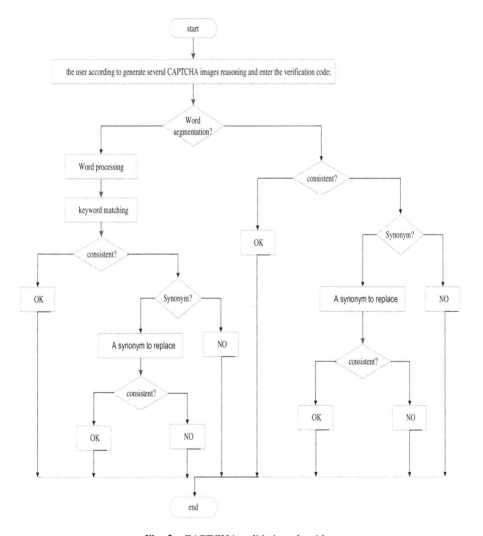

Fig. 2. CAPTCHA validation algorithm

3.3   CAPTCHA Validation Algorithm

(1) The user reasons according to generated CAPTCHA images and enter the CAPTCHA;

(2) Word segmentation? If no, compare the CAPTCHA entered by the user and the results and judge whether they are consistent. If consistent, agree on. If inconsistent, judge whether there is a synonym and whether consistent after replacement. If unanimity, agree on. If inconsistent, not through;

(3) If the words need segmentation, do words segmentation to the CAPTCHA entered by the user and results and match the keyword. If they are consistent, agree on. If inconsistent, judge whether there is a synonym and whether consistent after replacement. If unanimity, agree on. If inconsistent, not through. The algorithm flow chart is shown in Fig. 2.

3.4   CAPTCHA Implementation

The realization of the CAPTCHA is shown in Fig. 3.

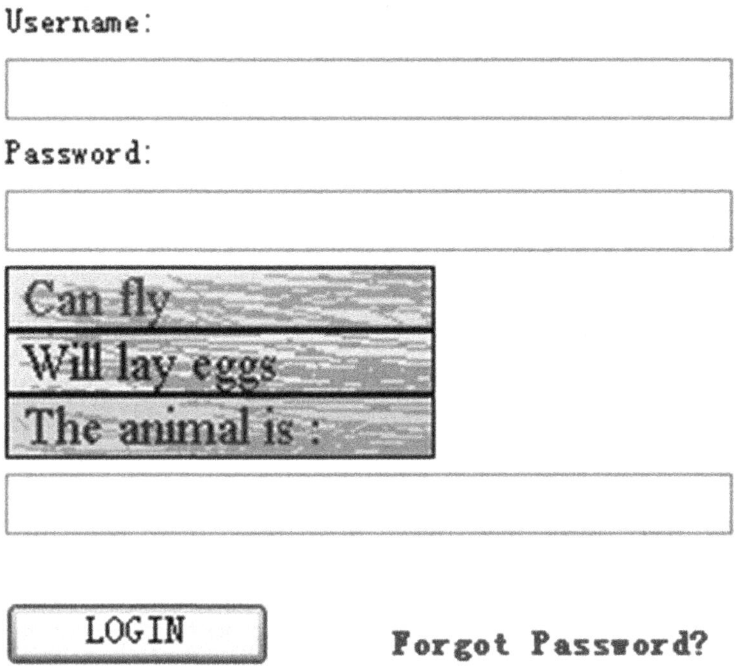

Fig. 3. Authentication code implementation

4 Conclusion

The CAPTCHA has a variety of forms, but the development of CAPTCHA recognition technology causes a hidden danger for the security of the CAPTCHA. In order to improve the security of the CAPTCHA, a new kind of CAPTCHA based on knowledge base is put forward combining the implicit CAPTCHA, which is based on semantic information question and answer and the professional system. This new CAPTCHA can significantly improve the security of the professional system. The CAPTCHA designed in this paper is suitable for professional system but not for general system, such as E-mail.

Acknowledgments. This work was supported by the scientific research fund project of China Agricultural University Yantai academy (YT201311, 201201Ja), the science and technology plan project of Beijing education committee (Grant no. KM201510853006) and key scientific research project of Beijing Polytechnic College (Grant no. bgzykyz201502, bgzykyz201503).

References

1. Ji, Z.: Principles and prevention of HTTP attacks based on identifying code recognization. Comput. Eng. **32**(20), 170–172 (2006)
2. Ying, X.: The research on user modelling for internet personalized services. Ph.D. thesis, National University of Defense Technology (2003)
3. von Ahn, L., Blum, M., Langford, J.: Telling humans and computers apart automatically. Commun. ACM **47**(2), 57–60 (2004)
4. Tao, R., Song, Y.E., Wang, Z.J.: Ambiguity function based on the linear canonical transform. IET Signal Process. **6**(6), 568–576 (2012)
5. Wang, B., Wang, J., Du, K., et al.: Research on attach and strategy of CAPTCHA technology. Appl. Res. Comput. **30**(9), 2776–2779 (2013)
6. Mori, G., Malik, J.: Recognizing objects in adversarial clutter: breaking a visual CAPTCHA. In: IEEE Conference on Computer Vision and Pattern Recognition (CVPR), vol. 1, pp. 124–141 (2003)
7. Zuo, B., Shi, X., Xie, F., et al.: A neural network based approach to recognizing the verification code. Comput. Eng. Sci. **31**(12), 20–22 (2009)
8. Huang, S., Xu, M.: Recognition and improvement of identifying code. J. Nanjing Normal Univ. (Eng. Technol. Ed.) **9**(2), 84–88 (2009)
9. Liu, J.-W., Yan, L.-F.: Comparative study of knowledge representation. Comput. Syst. Appl. **20**(3), 242–246 (2010)
10. Zhang, X., Gao, H., Zhao, Z.: The rule representation for knowledge in database style. Comput. Eng. Appl. **38**(1), 200–202 (2002)
11. Zhang, Q.: Research and design of spam email filter system based on bayesian algorithm spam. M.S. Thesis, Qufu Normal University (2006)

Research and Application of Monitoring and Simulating System of Soil Moisture Based on Three-Dimensional GIS

Guifen Chen[✉], Jian Lu, Ying Meng, Liying Cao, and Li Ma

School of Information Technology, Jilin Agricultural University,
Changchun 130118, China
guifchen@163.com

Abstract. Taking maize precise assignments section as the research object, using arcgis10.2 as the visual drive tool, comprehensive application of 3D modeling, database and GIS technology, this paper proposed a method of rapid establishment of agricultural areas of 3D virtual scene, developed soil moisture monitoring 3D simulation system based on GIS. The system had the 3D model search module, the dynamic scene simulation module, the dynamic monitoring and spatial information management function module and so on. Using Jilin Province Nong'an County town of Helong corn precision operation as an example, the system were preliminary application, the application results showed that the system can realize the effective management of agriculture areas, soil moisture spatial information and virtual display. According to the 3D model of soil moisture dynamic monitoring module, for the analysis and application of the regional soil moisture laid a solid foundation, and provided effective digital management platform for precision agriculture technique.

Keywords: Three-dimensional GIS · Simulating system · Soil moisture · Black soil

1 Introduction

In the "precision agriculture" techniques in GIS application development system is the implementation of "precision agriculture" practices of key technologies, namely the use of acquired farmland in the crop yield and environmental factors that affect crop growth difference information, the establishment of agricultural information database, data after conversion, processing and spatial analysis in GIS generated maps differentially information by analyzing the reasons for the differences affecting production, the development of economic, rational decision-making program production, crop management prescription map form, guidance farmland positioning job. GIS As for storage, analysis, processing and presentation of geospatial information computer software platform, the technology has matured. It is in the "precision agriculture" spatial information database technology systems used to establish agricultural land management, soil data, natural conditions, crop growth of the seedlings, the development trend of pests, crop yields space distribution of spatial information geographical statistics such as processing, pattern conversion and expression, to analyze the differences and

© IFIP International Federation for Information Processing 2016
Published by Springer International Publishing AG 2016. All Rights Reserved
D. Li and Z. Li (Eds.): CCTA 2015, Part II, IFIP AICT 479, pp. 100–110, 2016.
DOI: 10.1007/978-3-319-48354-2_10

the implementation of the regulation provides prescription information. But the two-dimensional plane GIS information cannot meet the growing application requirements, in terms of the relative two-dimensional GIS, three-dimensional GIS to be more comprehensive and accurate expression of the decision-making, production and management of modern agriculture. Three-dimensional GIS will be integrated into crop cultivation and management of decision support systems, and crop production management and Growth forecasting simulation model input-output analysis and intelligent agriculture expert system together, and with the participation of decision makers according to the spatial differences in yield, analyze the reasons, to make a diagnosis, provide scientific prescription, and to the field crop management prescription map GIS support of the formation of the regulatory guidance of scientific operations.

Global development status, the United States, Canada, Britain used in the fertilization of the most mature, basic access to commercial use. Japan mainly focused on gathering information related to the sensor and agricultural machinery automatic control and other aspects involved. Precision agriculture is also widely used in forestry production, mainly related to the relevant areas of fertilization, precision seeding, pest control, harvesting operations and water management. At present, except for a few countries, precision agriculture has not been large-scale deployments in the world, mainly because of its development of key technologies yet to produce a breakthrough in the practical value, information gathering technology and the cost is more expensive and other reasons, but for precision agriculture technology international the development potential and applications with a broad consensus, which is an important part of the development of agriculture as a high-tech applications, referred to as an important approach to sustainable agricultural development.

China is a large agricultural country; precision agriculture is an emerging concept, until the 1990s, began a study of precision agriculture. Establish a certain scale test area in Beijing, Shaanxi, Heilongjiang, Xinjiang, Inner Mongolia and other places, but overall is still at an experimental demonstration and nurture the development stage. In terms of technology, management and economic efficiency, our precision agriculture compared with the developed countries there is still a big gap, but also faced with inadequate technical support, information collection system was incomplete, expert systems are imperfect, precision is not high, the application conditions are not ripe and other conditions. Especially in high-precision agricultural machinery precision control system products, long-term dependence on imported goods, has seriously hampered the development of precision agriculture.

Although there are many difficulties in the development of precision agriculture, but agricultural machinery involving precise control of the relevant technology matured, where as RTK technology, wireless data transmission technology, navigation path planning technology, hydraulic control technology has been widely used in related industries. In 1998, the Ministry of Agriculture in Beijing Shunyi to establish a northern precision agriculture demonstration zone; in 2000, precision agriculture has been included in the national 863 high-tech research program and tested explore precision agriculture in Shanghai, Beijing and other places. Up to now, Chinese Academy of Sciences, Chinese Academy of Agricultural Sciences, China Agricultural University, Beijing Forestry Sciences, Shanghai Academy of Agricultural Sciences, the Shanghai Meteorological Bureau and other units have carried out research on precision agriculture, has been in Beijing, Hebei, Shandong, Shanghai and Xinjiang, We established a number of precision agriculture experiment and demonstration area. Heilongjiang as China's agriculture developed areas, first to the construction of agricultural information, the end of 2012, has been established in a number of farms Precision Agriculture Demonstration Zone. After several years of development, China's agricultural engineering and technical staff made great achievements in the technical system in precision agriculture.

2   Regional and Methods

2.1   Regional

Was chosen for the national "863" project "maize crop in Precise Operation System and Application" demonstration bases close Nong'an is located in the Song Liao Plain, it is one of the important commodity grain base, Jilin Province, located 60 km northwest of Changchun City, longitude 124° 32′–125° 45′, latitude 43° 54′–44° 56′; the average annual rainfall 507.7 mm, mainly fertility rating of chernozem soil, new soil in the wind sand, saline, alkaline earth, swamp soil, peat soil, paddy soil. Black soil is one of the most fertile Farmland County, concentrated in the eastern and southern counties Bao, Lung, pot roast, patron and three hillock township (town) loess sediments situation plateau, in the province black belt edge, close to the semi - arid region climate.

2.2   Research Program

The system uses GPS, wireless sensor data on soil moisture test area to acquire, construct spatial database in the data set, based on a relational database, dynamic database, database and graphic three-dimensional image library. Spatial query and spatial interpolation operation on the basis of the database, and the establishment of three-dimensional dynamic simulation model. Its technical route shown in Fig. 1.

2.3   Data Collection and Knowledge Acquisition

Use GPS to collect topography, spatial data sampling points, establishment of spatial database; according to the sampling point information, using GIS to draw the sampling grid maps, soil nutrients and other attribute data collection, build a relational database; via wireless sensors collect soil moisture, soil temperature, etc. soil moisture data to create dynamic database; based on artificial intelligence technology, the establishment of corn precision fertilization model library; according to corn growth stages, shooting its growing field, precise image and video work processes, create video library.

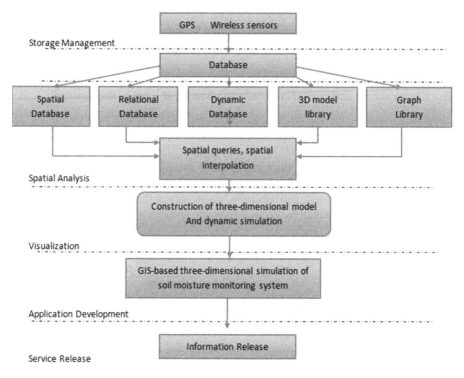

Fig. 1. Technology roadmap

2.4 Data Analysis and Processing

ArcGIS based platform, using 3DAnalyst tools fertility levels, topography, soil moisture and soil temperature, soil moisture data, spatial interpolation, spatial processing framework by two-dimensional and three-dimensional analysis of a variety of combinations, the analysis of the factors of soil moisture management Relations between.

(1) Production of spatial interpolation diagram dimensional GIS-based soil moisture data

According to the production of corn precise job objective laws, the use of Kinging interpolation method topography, soil moisture and other discrete sampling data interpolated into regular volume data to generate spatial interpolation diagram for three-dimensional reconstruction; image data on a regular grid display needs to be different depths of soil moisture, soil temperature and other data to interpolate interpolation applications including light and shade effect, in order to meet each pixel display area spatial variation diagram drawing generation.

(2) Analysis of three-dimensional interpolation maps

The use of three-dimensional spatial processing technique, terrain and soil moisture on different depths, spatial variability of soil temperature diagram variety of combinations, analyze the relationship between soil moisture, topography, soil temperature between the factors to solve complex soil moisture monitoring spatial analysis problems.

2.5   Construction and Scenario Simulation Model of Three-Dimensional GIS

(1) Three-dimensional GIS Model

Use ArcGIS and 3Ds Max platform to build three-dimensional GIS model soil moisture monitoring semantic layer, scale layers and geometric layer, Visual scenes based on soil moisture needs, soil moisture monitoring simulation system is divided into the terrain surface, underground two-level three-dimensional space to build and description:

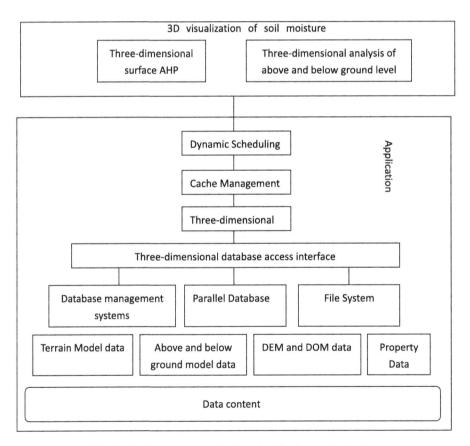

Fig. 2. Soil moisture monitoring visualization analysis chart

① Construction of the terrain surface level digital elevation model

Using digital elevation model surface modeling technology, through points, triangles and three grid surface modeling, surface soil for the complete spatial information, level surface topography of the terrain, such as establishing a digital elevation model, Ensure reasonable space division and regional identification on the core area of the terrain surface precision work space level;

② Construct three-dimensional models of above and below ground level perspective

Up and down the three-dimensional level is based on the distribution of soil moisture and fertility grade level description of three-dimensional space, mainly to solve the three-dimensional object is generated in a two-dimensional abstract representation and down overlapping problems and meet soil moisture and fertility grade level object precise three-dimensional space Expression and analysis requirements; When the traditional two-dimensional GIS approach to ecological issues, there are many more environmental factors limitations, three-dimensional and three-dimensional visualization of GIS technology integration, the establishment of three-dimensional GIS database, developed three-dimensional GIS visualization management system for three-dimensional visualization analysis, realization soil moisture Dynamic simulation monitoring scenes. Fig. 2:

3 Development of GIS-Based Three-Dimensional Simulation of Soil Moisture Monitoring System

ArcGIS Engine application systems use the .NET Framework, developed soil moisture monitoring system is based on simulation of three-dimensional GIS. System module is divided into: a three-dimensional model query module (three-dimensional map queries, database queries), dynamic scene simulation module (scene management, scene viewing), dynamic monitoring module (spatial analysis, trend analysis) and spatial information management module and other functions to dynamically simulation and dynamic monitoring module as an example to illustrate the system function.

3.1 Dynamic Scene Generations

Because of the complexity and fidelity three-dimensional virtual scene simulation results directly affect the system, so ensure the scene fidelity based on the use arcgis, combined 3DsMax platform provides plug-in function, according to the database generation parameters scenes programming precision agriculture three-dimensional virtual scene generation area, soil moisture simulation scenarios to achieve a three-dimensional visualization, this process has high reusability, greatly enhances the flexibility of the scene. The whole process including terrain generation, import and scene effect feature set.

(1) Establish a two-dimensional layer

Terrain model is the basis for generating a three-dimensional virtual scene, with its other models as a carrier able to determine the actual position in three-dimensional space.

The system generates a terrain model mainly refers to building on the ground subsurface model. First, arc map of agriculture and fertility level security boundary demarcation drawn, as shown in Fig. 3.

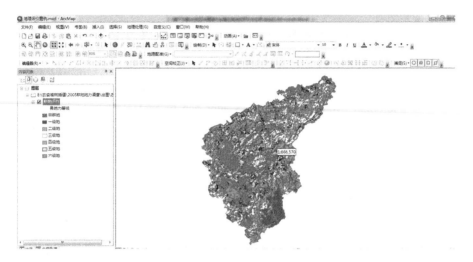

Fig. 3. Nong'an fertility rating dimensional distribution map

(2) Establishing a three-dimensional model

We will draw a good two-dimensional distribution of shp file into arcscene by setting the vector layer elevation, vectorStretched layer operations such as underground model simulation, the establishment of Nong'an fertility levels and underground vertical distribution of three-dimensional model diagram (Fig. 4).

3.2   Dynamic Monitoring Module

Choose different depths maize precision work together core demonstration areas Nong'anjiulongzhen (0–20 cm, 20–40 cm, 40–60 cm, 60–80 cm) soil moisture dynamic monitoring data to establish the soil moisture dynamic monitoring module.

(1) The vertical distribution of soil moisture and temperature analysis

Four models of the subsurface soil moisture, and soil temperature dynamic monitoring results shown in Figs. 5 and 6.

As can be seen in Fig. 5, soil moisture content with increasing soil depth increases; changes in soil moisture content at different times is not the same, around May 11 soil

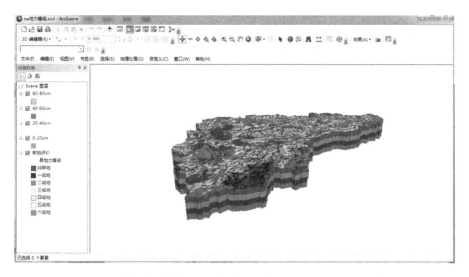

Fig. 4. Nong'an fertility grade three-dimensional vertical profile

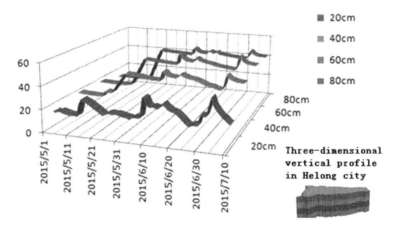

Fig. 5. Vertical distribution of soil moisture content

moisture content in the least, around July 3 maximum moisture content of soil; changes in soil moisture at different depths same trend.

As can be seen from Fig. 6: Soil temperature content with increasing soil depth is reduced; changes in soil temperature at different times is not the same, around May 13 the lowest soil temperature, soil temperature around July 10 highest; different soil depths water is substantially the same tendency.

(2) Soil moisture corn growing season temperature distribution analysis

The dynamic monitoring of 0–20 cm, 20–40 cm, 40–60 cm, 60–80 cm soil mois-ture four levels, soil temperature, soil moisture, temperature trends Fig. 7 through 11.

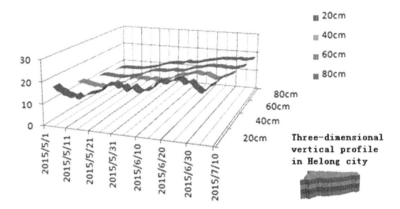

Fig. 6. Vertical distribution of soil temperature Content

We can see from Figs. 7, 8, 9 and 10, in the growth period, the change of soil temperature, and soil moisture change is obvious; Soil moisture content is the lowest, at the end of June or so most abundant on the 4th of July; Soil temperature around on June 10th lowest, around July 10th, the highest temperature.

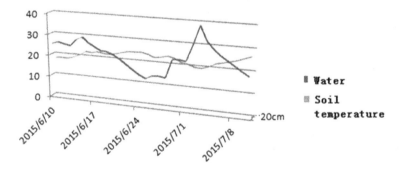

Fig. 7. 0–20 cm soil temperature and crop growth trend humidity

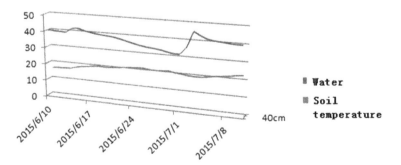

Fig. 8. 20-40 cm soil temperature and crop growth trend humidity

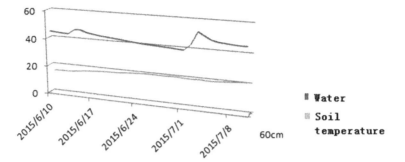

Fig. 9. 40–60 cm soil temperature and crop growth trend humidity

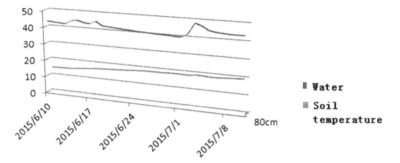

Fig. 10. Crop growing period of 60–80 cm soil temperature and humidity change trend chart

4 Discussion

For the realization of maize precision operation area soil moisture monitoring spatial information management standardization and 3D virtual display, in order to improve the visual simulation technology of reusability as the goal, the study through the ArcGIS platform, combined with 3dsmax software, spatial database and 3D model, established the soil moisture monitoring and simulation system based on 3D GIS.

(1) the building in our study of soil moisture monitoring simulation system is no longer confined to the two-dimensional map spatial information expression for agriculture, but with vivid features to express 3D virtual scene, through this way to simulate the cognitive process of corn precision work, realize the interaction between a user and a system mode transformation from traditional 2D space to multi-dimensional space, effectively play the 3D scene realistic and 2D map overall advantage, perfect interpretation of the essence of agricultural spatial information

(2) This system generate the region soil fertility level and vertical distribution of soil moisture of the underground three dimensional model diagram by using ArcMap to Nong'an boundary and soil fertility grade dividing rendering, taking the 3D

model diagram reflect different depths (0–20 cm, 20–40 cm and 40–60 cm, 60–80 cm) soil moisture conditions and dynamic monitoring data.

(3) According to the dynamic monitoring of soil moisture module to establish the three-dimensional model, the research laid the foundation for the application of the analysis of regional soil moisture.

Acknowledgments. This work was funded by the National High-Tech Research and Development Plan of China under Grants Nos. 2006AA10A309; The research and application of facilities for the safety of vegetables production technology based on Internet of Things (2011-Z20); Special funds for grain production development in Jilin province (an agricultural technology extension project) "research and application of vegetable security production technology based on facilities of Internet of things".

References

1. Shi, G.: Application research GPS and GIS technology in precision agriculture monitoring system. Hubei Agric. Sci. **10**(50), 1948–1950 (2011)
2. Zhu, G., Ma, C.T., Sun, L.C., et al.: Fast construction method D geographic information system three-dimensional model. Geogr. Geogr. Inf. Sci. **23**(4), 29–32, 40 (2007)
3. Zhang, Y., Liu, D., Jiang, C.: Urban feature modeling method-a case study of Heilongjiang Institute campus. For. Sci. Technol. Inf. **45**(3), 60–62 (2013)
4. Zhang, H.R.: In the field of mobile GIS technology. Surveying Spat. Inf. **37**(4), 137–138, 141 (2014)
5. Alkevli, T., Ercanoglu, M.: Assessment of ASTER satellite images in landslide inventory mapping: Yenice-Gok5ebey (Western Black Sea Region, Turkey). Bull. Eng. Geol. Environ. **70**(4), 607–617 (2011)
6. Zuo, X., Yu, Z., Xu, Y.: On the construction of mine 3D GIS. Geol. Explor. **37**(4), 63–67 (2001)
7. Huang, H.: Three-Dimensional Reconstruction Method of Stereo Vision Research. Bo ten dissertations Shanghai Jiaotong University. Shanghai Jiaotong University, Shanghai (2001)
8. Zhang, Li, X.: ArcGIS formation model of a tunnel. China Water Transp. **13**(3), 115–116 (2013)
9. Yao, X., Wang, C.: Advocated the establishment of deposit space-based three-dimensional model of ArcScene and Visualization. Eng. Geol. Comput. Appl. **1**, 125–126 (2006)
10. Cheng, P., Liu, S., Wang, W., Chen, H.: Research and application of three-dimensional geological model construction method. Jilin Univ. Earth Sci. **4**(2), 309–313 (2004)
11. Wang, M.: Three-dimensional geological modeling of clouds present situation and development trend. Geotext. Base **20**(4), 68–70 (2006)
12. Zhang, K., Wu, W., Bai, Y., et al.: Three-dimensional geological ArcGIS-based visualization. Liaoning Tech. Univ. **26**(3), 345–347 (2007)
13. Shoufeng, H., Li, Y., Yang, X.: Method based on 3D geological constructs. In: Fourteenth East China Provinces and One City of Surveying and Mapping Institute Symposium, pp. 170–173 (2012)
14. Chen, Y.: Bell ear research mine geological exploration and visualization management system modeling. Min. Res. Dev. **24**(1), 37–40 (2004)

Colorimetric Detection of Mercury in Aqueous Media Based on Reaction with Dithizone

Zihan Wu[1,2,3], Ming Sun[1,2,3(✉)], and Ling Zou[1,2,3]

[1] College of Information and Electrical Engineering,
China Agricultural University, Beijing 100083, China
wuzihan_cau@163.com, sunming@cau.edu.cn,
1475126006@qq.com
[2] Key Laboratory of Modern Precision Agriculture System Integration Research,
Ministry of Education, Beijing, China
[3] Key Laboratory of Agricultural Information Acquisition Technology,
Ministry of Agriculture, Beijing, China

Abstract. This study investigates the colorimetric reaction between dithizone and mercury in aqueous media which generate the orange Hg-dithizone complexes extracted by chloroform. Then combining with spectral analysis, the UV-vis spectral data of the complexes are obtained to build a forecast model. By means of the multiple linear regression model with SG smoothing method, RPD (Residual Predictive Deviation) of 3.2461 is reached with the detection limit of 0.1129 ug/L. This colorimetric method was found to be rapid, simple and sensitive for the detection of mercury in aqueous medium.

Keywords: Mercury · Dithizone · Colorimetry · Spectroscopy

1 Introduction

There are two forms of mercury in nature, organic mercury complexes and inorganic mercury, which can change to each other forms dependent on the ambient conditions [1]. Pollution caused by heavy metals has negative effects on human health and the environment, which is from rock weathering and industrial emissions, volcanic movement [2, 3]. Mercury can be passed by aquatic plants and animals through food chain to human bodies with long term bio-accumulation, which show great toxicity with little content. Mercury ions have a strong affinity with thiol groups in the body so that mercury ions could interact with substances which contains most thiol groups such as proteins and enzymes involved in the body's metabolism [4–6]. Therefore, finding a rapid and efficient method for the detection is of great realistic importance. For its detection, atomic emission spectrometry, high performance liquid chromatography and stripping voltammetry have been developed for detection of mercury in soil and aqueous samples, which need expensive instruments and usually need professional operators [7]. Colorimetric detection is simple and rapid, which can read out with the naked eyes though most methods have been proposed [8]. Reagent colorimetric method refers to comparing or measuring the colored substance solution or to measure the concentration of contents, based on chromogenic reactions that generate colored compounds.

© IFIP International Federation for Information Processing 2016
Published by Springer International Publishing AG 2016. All Rights Reserved
D. Li and Z. Li (Eds.): CCTA 2015, Part II, IFIP AICT 479, pp. 111–116, 2016.
DOI: 10.1007/978-3-319-48354-2_11

2 Materials and Instrumentation

2.1 Materials

$HgCl_2$, Dithizone, carbon tetrachloride, sodium sulfite, nitric acid, distilled water were purchased from Yixiubogu (Beijing, China). All the chemicals used for the study were following the standard procedures.

35 ml mercury solutions with various concentrations are put into centrifuge tubes, separately. 2 ml sodium sulfite solution with 20 % concentration is added to prevent dithizone from being oxisized. The pH value of the reaction solution is adjusted to 1.0 by using nitric acid and at last 10 ml dithizone solution with 2.5 mg/L concentration is dissolved in carbon tetrachloride. The reaction solution in the tubes is drastically Oscillated for 1 min, and Layered after being laid aside, extract the lower solution which is the complex (Hg-dithizone) generated by mercury and dithizone.

2.2 Instrumentation

RP1003H electronic balance and 78-1 magnetic heated stirrer were needed in the solution preparation process. UV-vis spectra of the complex are recorded on UV-2450 Shimadzu UV spectrophotometer.

3 Results and Discussion

3.1 Data Preprocessing

The UV-visible absorption spectra of reagents in the colorimetric experiment are shown in Fig. 1. It can be noted that the band for Hg-dithizone complex is at about 450 nm, which does not coincode with that of dithizone, indicating that dithizone has no effect on the absorbance of Hg-dithizone complex, even it over-doses. Other reagents used in the experiment also do not affect the absorbance of complex, and $HgCl_2$ solution has no significant absorption peak. Consequently, the mercury concentration can be measured by the absorbance of complex. The UV-visible spectra of Hg-dithizone complexes with different concentrations are shown in Fig. 2.

In the paper, varieties of pretreatment methods on original spectral data are employed, such as Savitzky-Golay convolution smoothing (SG), derivative algorithm, multiplication scatter correction (MSC) and standard normal variate (SNV), after the models are set up. Least squares fit is applied as the digital filter after SG smoothing of the data, which makes the least squares polynomial fit to the data in a moving window and no longer a simple average operation. Derivative algorithm is used for obtaining the spectral intensity, whose basic idea is to select the original spectrum at several points to constitute a window and derivative spectrum can be obtained by deriving data in the window. MSC aims to obtain an ideal spectrum by correcting the scatter of samples. Firstly, assuming the contribution of scatter associated with the wavelength to spectrum to the composition is different; therefore, the spectral data can be divided into two parts. Since the absorbance value of each wavelength point should meet certain

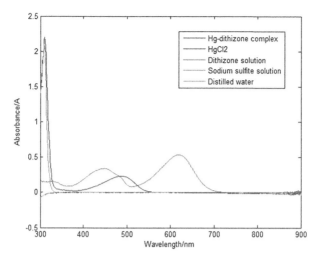

Fig. 1. UV-visible absorption spectra of reagents in the colorimetric experiment

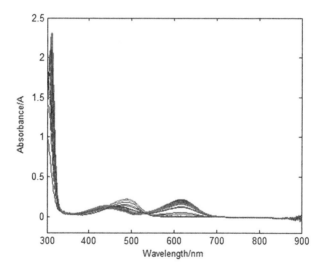

Fig. 2. UV-visible absorption spectra of Hg-dithizone complexes with different concentration

distribution (e.g. normal distribution) in every spectrum, the original spectrum subtracts the average of the spectrum, which is divided by the standard deviation of the spectral data, therefore, SNV is obtained.

3.2   The Selection of Relevant Bands

Since the spectral data are from 300 nm to 900 nm, which include large amount of data to be processed, correlation coefficient is employed to select relevant bands. Therefore, the amount of computation is reduced without loss of information and the purpose of fast and efficient detection is achieved. Figure 3 shows the correlation coefficient between absorbance and the concentration of original data. It can be observed directly that among 460 nm–520 nm absorbance and wavelength has high correlation, and at 484 nm correlation coefficient reaches the maximum where the spectral data is used to build up a single linear regression. In addition, the spectral data at 470 nm, 480 nm, and 490 nm is used for multiple linear regression modeling and PLS modeling.

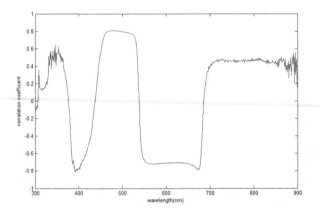

Fig. 3. Correlation coefficient between absorbance and the concentration of original data

3.3   Modeling Results

This study employs three models for data analysis, that are single linear regression (SLR), multiple linear regression (MLR), partial least squares regression (PLS). Table 1 summarizes the results of different models combined with different data pre-processing methods. Therefore, the best model for prediction based on evaluation indexes is chosen. The paper adopts 5 evaluation indexes, that are multiple correlation coefficients for modeling set (Rc), multiple correlation coefficients for validation set (Rv), root mean square error of calibration for modeling set (RMSEC), root mean square error of prediction for validation set (RMSEP), ratio of the standard deviation to the RMSEP. The one of the most important indexes is Residual Predictive Deviation (RPD). If RPD values is greater than 3, the model for the detection of reagents is ideal and can be used for quantitative analysis. If RPD is from 2.5 to 3, the prediction accuracy should be improved. If RPD is less than or equal to 2.5, the model cannot be used for quantitative analysis.

For the experimental data in the Table 1, SG smoothing method has the best smoothing effect, by which noise is effectively eliminated to improve the signal-to-noise ratio. However both of SNV and MSC destroy the original spectral data, for both of RPD are smaller than 3. Noise is introduced when first-order difference and second-order difference are employed, which results in the signal-to-noise ratio reduced

Table 1. Results of different models combined with different data preprocessing methods

Model	Pre-process methods	Model evaluation indexes				
		Rc	Rv	RMSEC	RMSEP	RPD
Single linear regression	Original data	0.9442	0.9880	0.1301	0.3448	3.1006
	MSC	0.9401	0.9114	0.1347	0.3315	2.9963
	SNV	0.9367	0.8834	0.1383	0.3417	2.9175
	SG	0.9459	0.9888	0.1282	0.3445	3.1487
	First-order differential	0.5587	0.7158	0.3276	0.2469	1.2317
	Second-order differential	0.2779	0.2668	0.3794	0.2096	1.0634
Multiple linear regression	Original data	0.9526	0.9864	0.1201	0.3610	3.3590
	MSC	0.9407	0.9214	0.1340	0.3306	3.0111
	SNV	0.9388	0.9260	0.1360	0.3270	2.9664
	SG	0.9492	0.9905	0.1243	0.3185	3.2461
	First-order differential	0.7949	0.8307	0.2397	0.3520	1.6834
	Second-order differential	0.3192	0.3313	0.3743	0.2625	1.0779
Partial least squares regression	Original data	0.9458	0.9895	0.1282	0.3460	3.1454
	MSC	0.9396	0.8992	0. 1352	0. 3386	2.9845
	SNV	0.9376	0.9049	0.1373	0.3340	2.9381
	SG	0.9461	0.9900	0.1279	0.3460	3.1531
	First-order differential	0.7614	0.8944	0.2560	0.3072	1.5758
	Second-order differential	0.3186	0.3253	0.3744	0.2593	1.0776

and a great deal of detail of the experimental data lost. When the differential width is too large, it will result in over-smoothing of the data.

High-precision Models can be obtained by SLR, MLR and PLS, but the accuracy of MLR model is better than both of SLR and PLS models, which is more suitable to process this data set. However, by PLS model, multicolinearity can be eliminated and RPD and other indexes can be improved. The results show that reagent colorimetric method is feasible to detect mercury in UV-vis spectrum.

4 Conclusions

In summary, a rapid, useful and sensitive colorimetric method for detection of Hg^{2+} in aqueous solution was developed. Based on the spectral data, different methods of preprocessing combining with SLR, MLR and PLS models is employed to get the best forecast model. By means of the multiple linear regression model combined with SG smoothing method, RPD of 3.2461 is reached with the detection limit of 0.1129 ug/L, which suggests that combination of reagent colorimetry and spectroscopy may be the one of the best approaches for detection of mercury in aqueous media, which can also be expanded to for the qualitative and quantitative estimation of other heavy metal.

Acknowledgment. Fund for this research is provided by the Public Service Sectors (Agriculture) Research and Special Funds for Modern Fishing Digital and Physical Networking Technology Integration and Demonstration (No. 201203017).

References

1. Selid, P.D., Xu, H., Collins, E.M., Collins, M.S.F., Zhao, J.X.: Sensing mercury for biomedical and environmental monitoring. Sensors **9**, 5446–5459 (2009)
2. Puiso, J., Jonkuviene, D., Macioniene, I., Salomskiene, J., Jasutiene, I., Kondrotas, R.: Biosynthesis of silver nanoparticles using lingonberry and cranberry juices and their antimicrobial activity. Colloid Surf. B **121**, 214–221 (2014)
3. Huang, C., Chang, H.: Selective gold-nanoparticle-based turn-on fluorescent sensors for detection of mercury(II) in aqueous solution. Anal. Chem. **78**, 8332–8338 (2006)
4. Bernard, A.: Health, pp. 80–805 (2011)
5. Gao, S., Jia, X., Chen, Y.: Old tree with new shoots: silver nanoparticles for labelfree and colorimetric mercury ions detection. J. Nanopart. Res. **15**, 1385 (2013)
6. Zargoosh, K., Babadi, F.F.: Highly selective and sensitive optical sensor for determination of Pb2+ and Hg2+ ions based on the covalent immobilization of dithizone on agarose membrane. Spectrochim. Acta Part A Mol. Biomol. Spectrosc. **137**, 105–110 (2015)
7. Maity, D., Kumar, A., Gunupuru, R., et al.: Colorimetric detection of mercury(II) in aqueous media with high selectivity using calixarene functionalized gold nanoparticles. Colloids Surf. A Physicochem. Eng. Aspects **455**(30), 122–128 (2014)
8. Zhang, N., Xu, J., Xue, C.: Core–shell structured mesoporous silica nanoparticles equipped with pyrene-based chemosensor: synthesis, characterization, and sensing activity towards Hg (II). J. Lumin. **131**, 2021–2025 (2011)

Study on the Prediction Model Based
on a Portable Soil TN Detector

Xiaofei An[1,2(✉)], Guangwei Wu[1,2], Jianjun Dong[1,2], Jianhua Guo[1,2],
and Zhijun Meng[1,2]

[1] Beijing Research Center for Intelligent Agricultural Equipment,
Beijing 100097, China
anxf@nercita.org.cn
[2] National Research Center of Intelligent Equipment for Agriculture,
Beijing 100097, China

Abstract. As the development of precision agriculture, it is necessary to obtain soil total nitrogen (TN) content and other element parameters. With the NIRS technology, a soil detector for soil total nitrogen content was developed. It included two part: optical part and control part. The detector took each lamp-house connected with the incidence of Y type optical fiber in turn by a manual rotation, The different wavelength lamp-house signal was transferred to the surface of soil by the input fibre. The reflected signal would be converted by photoelectric sensor, the optical signal was converted to electrical signal. After the power circuit, amplier circuit, and AD convert circuit, the electrical signal was processed by MCU. Finally, the result of soil total nitrogen content could be displayed on LCD. With the forty-eight apple orchard soil samples of Beijing surburb, the predicted models were established by seven different methods (MLR, PLSR, SVM, BPNN, GA + BPNN, GA + SVM and PSO + SVM). The model established by genetic algorithm (GA) optimizing BP neural network was optimal, with R_C of 0.94, R_V of 0.78, RMSEC and RMSEP of 0.037 and 0.067. The results showed that the soil total nitrogen content detector had a stable performance. The established model had perfect accuracy and strong robustness.

Keywords: NIR · Soil parameter · Modeling method · Soil sensor

1 Introduction

Soil macro and micro parameters such as TN, P, K, and OM could demonstrate the fertilization directly. It was the basis to carry out precision agriculture. At present, a lot of scholars [2–12] have carried out the farmland soil characteristics from spatial variability. The research content mainly concentrated on the soil nutrient (N, P, K, organic matter, et al.), soil moisture, elements such as electricity conductivity, pH value. Soil fertility, the physical and chemical parameters could be analyzed by spectral analysis methods, which could short analysis time, reduce cost and improve the analysis efficiency in the laboratory. With nearly 30 years' development, it has made remarkable progress and achievements under laboratory conditions by near infrared spectral method for the soil spectral data preprocessing.

© IFIP International Federation for Information Processing 2016
Published by Springer International Publishing AG 2016. All Rights Reserved
D. Li and Z. Li (Eds.): CCTA 2015, Part II, IFIP AICT 479, pp. 117–126, 2016.
DOI: 10.1007/978-3-319-48354-2_12

In 1999, Shibusawa has studied on the parameters of soil (nitrogen, phosphorus and potassium, organic matter, electricity conductivity, et al.) by near infrared spectral analysis technology. The sensitive wavebands have been selected and the parameters corresponding prediction models were also established. Huan-junliu, et al. selected 675 soil samples of Heilongjiang province to build VIS NIR waveband and linear model of soil organic matter that the prediction accuracy was as high as $0.936(R^2)$. Gao Hongzhi, JiWenJun et al. have used the maximum continuous projection algorithm to eliminate the effect between multivariate linear, extracted the modeling characteristics of soil total nitrogen and organic matter wavelength, the use of random forest, SVM, and ANN, partial least squares method (PLS) high precision models were obtained. Dong-jian et al. has used 900-1700 nm wavelength range of diffuse reflection spectrum to identify and eliminate abnormal samples. The accuracy of the model became excellent with the continuous projection algorithm (SPA, Successive Projections Algorithm) to the optimal modeling variable wavelength selection, and then through the three linear modeling methods affection the result of the organic matter content prediction was analyzed. PLS method and RBF neural network prediction model were also established. Yan shanshan introduced the continuum removal method to extract the sensitive wavebands of SOM, analysis when the change of soil organic matter content, the change rule of the spectrogram to extract the sensitive wavelength of 600 nm, 900 nm and 2210 nm as the center wavelength, the BP neural network to establish the model of SOM. With the prediction model data, root mean square error MSE was 0.286, the correlation coefficient was 0.979 and the model was superior to the whole band.

Since the 1990's, some soil scientists and agricultural information technology scholars have researched on soil sample spectrum analysis of research status in real time based on the NIR spectroscopy. And several kinds of prototypes have been developed. [10–20] (Adamchuk, 1999; Shibusawa, 2000; Mouazen, 2005; Maleki, 2007; Christy, 2008; XingZhen, 2008; Gao Hongzhi, 2011). Sudduth and Hummel have developed a rapid detection sensor which was used to identify the soil texture and soil organic matter in the 1990's based on NIRS. The accuracy of the detector could reach 90 %. With the development of NIRS technology, the NIR lamphouse could be used in the field condition which made it easy to collect data in the outside. Several large companies have also product specificity lamp-house for the portable soil sensor and on-the-go soil sensor. Several foreign scholars have also obtained excellent result of soil moisture, soil organic matter and so on. Different scholars from studying the spectral characteristics of the soil, combined with chemometrics methods, established the soil moisture, soil organic matter, soil total nitrogen and soil parameters prediction model. foreign scholars also started to research on the effect of factor under the field condition to improve the model accuracy.

Hence, we decided to develop a kind of portable soil TN detector based on NIRS technology. And then it would make the basis for the field condition to detect soil TN.

2 Materials and Methods

2.1 Soil TN Detector Design

(1) Selection of sensitive wavelengths

It was necessary to determine the used NIR wavelength lamp-house for the soil TN sensor, firstly. Several scholars had studied on the determination of NIR wavelength lamp-house. Zheng Lihua et al. had tried four kinds of methods to estimate soil TN content, such as BPNN, MLR, SVM and wavelet analysis. With the different methods, she also proposed the selected sensitive wavelength, including 24 suggestion wavelengths. The selected wavebands were 844 nm,859 nm, 923 nm,931 nm, 972 nm, 984 nm, 1028 nm,1064 nm,1092 nm,1124 nm,1187 nm,1208 nm, 1215 nm,1286 nm, 1311 nm,1389 nm,1394 nm,1536 nm,1559 nm,1673 nm,1684 nm,1895 nm,1833 nm, 1991 nm, 2150 nm and 2234 nm.

According to the mentioned results, An et al. suggested a new group of wavelengths to build the BPNN estimation model, 1550 nm, 1300 nm, 1200 nm, 1100 nm, 1050 nm, 940 nm. The R^2 (calibration and validation) were 0.85 and 0.77,respectively. Under the laboratory condition and field condition, the models have different affect factor, as a result, it would be necessary to determine new model parameters after finishing the development.

(2) Structure design of soil TN sensor

With the NIRS technology, a soil detector for soil total nitrogen content was developed. It was consisted of optical part and control part. The detector took each lamp-house connected with the incidence of Y type optical fiber in turn by a manual rotation, The different wavelength lamp-house signal was transferred to the surface of soil by the input fibre. Once the signal was converted by the photoelectric sensor, the optical signal was converted to electrical signal. After the power circuit, amplier circuit, and AD convert circuit, the electrical signal was processed by MCU. Finally, the result of

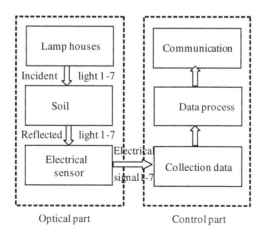

Fig. 1. Structure

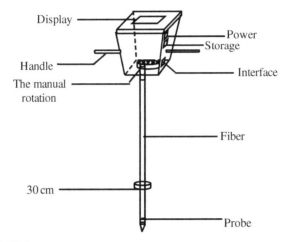

Fig. 2. Soil TN content detector structure

soil total nitrogen content could be displayed on LCD. Figure 1 shows the structure, and Fig. 2 shows Soil TN content detector structure.

(3) Optical part

In order to develop the portable detector, appropriate lamp-houses and transmission channel should be selected correctly. In this study, Lighting Emitting Diode (LED) was selected as active lamp-house. According to the result of 2.1 section, six wavelength of LED were selected. All the selected lamphouses had the advantagement of low cost, narrow bandwidth and portable. They were at the wavelength of 1550, 1300, 1200, 1100, 1050, and 940 nm. In order to analyze the effect of soil moisture, another special LED was also selected, which was at the wavelength of 1450 nm. When the soil TN sensor began to work, all the LEDs would be open by turn, and the optical signal could be transmitted by the input fiber. When the LED optical signal arrival at the surface of soil, it would be divide into two part. One part would be absorpted by the soil and the other part would be reflected by the soil. According to the different content of soil properties, the reflectance value would be different. For the soil moisture content, as the soil moisture content increased, the reflectance value would reduce. When the locating dowel was pulled up, the rotary table was back to original place by the clockwork spring.

(4) Control part

The soil TN detector was controlled by the MCU. Figure 3 shows the MCU process chart. Once the optical signal was converted into electrical signal, it was processed by the IU convert circuit, amplier circuit, AD convert circuit and LCD display circuit by turn. All the weak electrical signal would become standard signal for the MCU. In the MCU, it could be display stored. Both the reflectance value and absorbance value would be stored at the same time. With the help of soil TN predicted model, the soil TN content could be displayed on LCD.

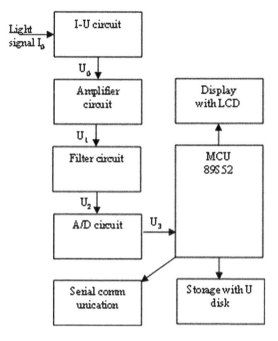

Fig. 3. MCU process chart

2.2   Experimental Methods

The soil coverage rate was 82 % in Beijing suburb, the area was about 137.8 hectare. Among them, the brown soil area was 89.1 hectare, about 65 % of Beijing soil area. Soil type was typical brown earth soil which has a strong representation. They were collected from Beijing suburb with the ranges of N 40.14395738° ~ 40.1434362° and E 116.2154066° ~ 116.2150152°. The depth was 1-20 cm under the soil surface. In order to reduce the effect of operation error, all the soil samples were taken as uniform number, crushing and drying. And then they were divided into two parts, one part was measured by the traditional chemical methods; the other part was used to detect by the developed portable sensor.

All the data were measured for twice by the soil TN detector. For the first time, the absorbance of standard whiteboard was measured. And for the second time, the probe output signal was also obtained. With the help of equation of (1) and (2), the absorbance of the soil sample could be obtained. According to the Eqs. (1) and (2), the absorbance would be obtained.

$$r_i = V'_i/V_i \times 100\% \tag{1}$$

$$A_i = \lg(1/r_i) \tag{2}$$

Where: i is 1550 nm, 1450 nm, 1300 nm, 1200 nm, 1100 nm, 1050 nm, and 940 nm; V_i is the output value at the wavelength of i by putting probe on the standard whiteboard; V_i' is the output at the wavelength of i by pushing probe into the soil sample; r_i is the reflectance value at the wavelength of i; A_i is absorbance of the wavelength of i.

2.3   The Evaluation of Soil TN Predicted Model

In order to evaluate the portable soil TN detector accuracy and performance, three parameters were chose to evaluate them, include RMSEC, RMSEP and RPD value. Equations (3), (4) and (5) shows the calculation formulas.

$$RMSEC = \sqrt{\frac{\sum_{i=1}^{n} (y_i - y_{ci})^2}{n_c}} \tag{3}$$

$$RMSEP = \sqrt{\frac{\sum_{i=1}^{n} (y_i - y_{pi})^2}{n_p}} \tag{4}$$

$$RPD = \frac{\sqrt{\sum_{i=1}^{n} (y_i - \overline{y})^2}}{\sqrt{\sum_{i=1}^{n} [(y_i - y_{pi})^2 - \frac{\sum_{i=1}^{n} (y_i - y_{pi})^2}{n_c}]^2}} \tag{5}$$

Where: i was soil sample serial; y_i was soil TN measured by FOSS Kjeltec^{TM}2300; y_{ci} is calibration group soil TN predicted value; y_{pi} is validation group soil TN predicted value; $\overline{y}$ is soil total nitrogen content average valuemeasured by FOSS Kjeltec^{TM}2300; n_c calibration group number; n_p is validation group number.

3   Result and Discussion

3.1   Soil Spectral Data Pretreatment

Table 1 showed the statistics data of soil parameters. The content range of that was 0.007 % ~ 0.286 %, average content of soil TN was 0.16 % and the standard deviation was 0.007. From the distribution coverage, soil total nitrogen variation range was across crop growth cycle, it was suitable for modeling analysis.

The data obtained from the detector has been processed by the average filtering software and hardware of the first-order RC low-pass filter. According to the formula (1) and (2), soil reflectance and absorbance values at the six different wavelengths were obtained. Relevant relations showed that the correlation coefficient between absorbance and soil TN content were 0.19, 0.31, 0.10, 0.21, 0.10 and 0.11, respectively.

This paper proposed a new soil spectral data pretreatment method, the differential absorbance was selected as a new independent variable factor. Differential absorbance data was obtained by formula (6). Figure 4 showed the result of absorbance and soil TN content with the new data pretreatment method. The correlation coefficient between

Table 1. Statistics data of soil parameters

Number	Parameter	Soil sample	Max value	Min value	Average value	Median value	Standard deviation
1	SM (%)	48	20.9	10.5	16.4	17.2	0.004
2	TN (%)	48	0.286	0.007	0.16	0.17	0.007
3	pH	48	8.34	7.71	7.99	8.00	0.02
4	OM (%)	48	1.31	0.69	0.95	0.93	0.02
5	P (mg/L)	48	454.2	45.90	118.15	107.20	9.59
6	K (mg/L)	48	351.90	81.73	165.20	152.95	9.37

differential absorbance ($A_{940,1100}$, $A_{1050,1550}$, $A_{1100,1200}$, $A_{1200,1550}$, $A_{1300,1550}$, $A_{1550,1450}$) and TN content with the new data pretreatment method were 0.26, 0.35, 0.28, 0.35, 0.28 and 0.21, respectively. Although the correlation coefficient was still low, it was improved by 36.8 %, 12.9 %, 180 %, 28.6 %, 180 % and 90.9 %, comparing with the original absorbance.

$$A_{i,j} = \frac{A_i - A_j}{A_i + A_j} \qquad (6)$$

Where: i and j are 940,1050,1100,1200,1300,1450 and 1550 nm; $A_{i,j}$ is differential absorbance value; A_i is original absorbance value.

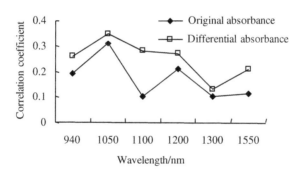

Fig. 4. Data pretreatment result

3.2 Establishment of Soil TN Predicted Models

Both the calibration and validation set were established by the 48 soil samples. The original absorbance A_i and the differential absorbance $A_{i,j}$ were as input factor. The soil TN predicted model was established by the MLR, PLSR, BPNN method, SVM method, GA + BPNN, improved support vector machine (GA + SVM) method and the improved support vector machine method (PSO + SVM) respectively. Table 2 showed the results of different soil TN predicted models.

Table 2. Comparison of different soil TN predicted models

Num	Method	Variable	Calibration set	R_C	Validation set	R_V	RMSEC	RMSEP	RPD
1	MLR	A_i	32	0.50	16	0.51	0.174	0.165	0.367
2	PLSR	A_i	32	0.83	16	0.65	0.076	0.106	1.446
3	BPNN	A_i	36	0.90	12	0.71	0.045	0.026	1.042
4	GA + BPNN	A_i	36	0.89	12	0.79	0.051	0.064	2.554
5	SVM	A_i	36	0.81	12	0.68	0.065	0.135	1.179
6	GA + SVM	A_i	36	0.90	12	0.77	0.044	0.090	1.941
7	PSO + SVM	A_i	36	0.96	12	0.70	0.135	0.070	2.514
8	MLR	$A_{i,j}$	32	0.58	16	0.44	0.161	0.137	0.521
9	PLSR	$A_{i,j}$	32	0.89	16	0.65	0.069	0.059	2.916
10	BPNN	$A_{i,j}$	36	0.86	12	0.74	0.054	0.040	1.064
11	GA + BPNN	$A_{i,j}$	36	0.94	12	0.78	0.037	0.067	2.577
12	SVM	$A_{i,j}$	36	0.69	12	0.72	0.097	0.106	1.638
13	GA + SVM	$A_{i,j}$	36	0.99	12	0.61	0.006	0.063	3.183
14	PSO + SVM	$A_{i,j}$	36	0.99	12	0.67	0.006	0.053	4.191

3.3 Discussions

The model with differential absorbance was obvious better than that with original absorbance for all the seven different methods from Table 2. For the RMSEC, it reduced from 0.085 to 0.061 and improved about 27.4 %. For the RMSEP, it reduced from 0.094 to 0.075, improved about 3.1 %. The model accuracy become more better.

For the linear method, PLSR model had the Rc of 0.83, Rv of 0.65, which were both higher than that of MLR model. RMESC and RMSEP were also lower than that of MLS model. For the nonlinear methods, BPNN, SVM, GA + BPNN, GA + SVM and PSO + SVM had the Rc of 0.69, 0.86, 0.94, 0.86 and 0.94, respectively. It had the increased trend gradually. The correlation coefficients of validation were 0.72, 0.74, 0.78, 0.74 and 0.78, respectively. Among them, the GA + BPNN method had the highest value. All the RMSEC and RMSEP satisfied the requirement except SVM method. According to the RPD value, GA + BPNN, GA + SVM and PSO + SVM model had excellent performance.

As a result, the GA + BPNN model had high Rc and Rv, and the RMSEC, RMSEP, RPD value satisfied the requirement. It could be as the soil TN forecasting model.

4   Conclusions

The soil TN forecasting model was also established by seven different methods (MLR, PLSR, SVM, BPNN, GA+BPNN, GA+SVM, PSO+SVM). The model was evaluated by Rc, Rv, RMSEC, RMSEP and RPD.

(1) A kind of new portable soil TN detector was developed. LED was selected as the lamp house, and optimization amplification and filter circuit were designed.

(2) A new soil spectral data pretreatment method suitable for portable detector was proposed. According to the method, the correlation coefficient between spectral absorbance and soil TN was improved, obviously.

(3) The soil TN forecasting model was established by seven different methods (MLR, PLSR, SVM, BPNN, GA + BPNN, GA + SVM and PSO + SVM). The GA + BPNN model had high Rc of 0.94 and Rv of 0.78, and the RMSEC, RMSEP, RPD satisfied the requirement. Especially, the RPD value reached 2.577. It could be as the soil TN forecasting model.

Although the soil parameter predicting model were established, the robustness and universality still had disadvantage. In the next step, the effect of soil other properties should be considered in the model.

Acknowledgment. This study was supported by 863 program (2012AA101901), and NSFC (6132009)(31301237).

References

1. Sinfield, J.V., Fagerman, D., Colic, O.: Evaluation of sensing technologies for on-the-go detection of macro-nutrients in cultivated soils. Comput. Electron. Agric. **70**(1), 1–18 (2010)
2. Zhang, J., Tian, Y., et al.: Estimating model of soil total nitrogen content based on near-infrared spectroscopy analysis. Trans. CSAE **28**(12), 183–188 (2012)
3. Shenk, J.S., Workman Jr., J.J., Westerhaus, M.O.: Application of NIR spectroscopy to agricultural products. In: Handbook of Near-infrared Analysis, Practical Spectroscopy Series, pp. 383–431 (1992)
4. Li, M., Sasao, A., Shibusawa, S., Sakai, K.: Soil parameters estimation with NIR spectroscopy. J. Japan. Soc. Agric. Mach. **62**(3), 111–120 (2000)
5. Chang, C.W., Laird, D.A.: Near-infrared reflectance spectroscopic analysis of soil C and N. Soil Sci. **167**(2), 110–116 (2002)
6. Gao, H.Z., Luo, Q.P.: Near infrared spectral analysis and measuring system for primary nutrient of soil. Spectrosc. Spectral Anal. **31**(5), 1245–1249 (2011)
7. Ji, W., Li, X., Li, C., et al.: Using different data mining algorithms to predict soil organicmatter based on visible near infrared spectroscopy. Spectrosc. Spectral Anal. **32**(9), 2393–2398 (2012)
8. He, D., Chen, X.: Research on soil organic matter content measurement by spectral analysis. J. Chin. Agric. Mech. **46**(2), 1–5 (2015)
9. Yan, S., Cheng, X., Song, H.: Study on the extraction of sensitive band of soil organic matter near infrared spectrum basd on continum removal. Shanxi Agric. Univ. J. **36**(1), 72–76 (2016)
10. Adamchuk, V.I., Lund, E.D., Sethuramasamyraja, B.: Direct measurement of soil chemical properties on-the-go using ion-selective electrodes. Comput. Electron. Agric. **48**(2), 272–294 (2005)
11. Reeves III, J.B., McCarty, G.W., Meisinger, J.J.: Near infrared reflectance spectroscopy for the analysis of agricultural soils. J. Near Infrared Spectrosc. **7**, 179–193 (1999)
12. Reeves III, J.B., McCarty, G.W.: Quantitative analysis of agricultural soils using near infrared reflectance spectroscopy and fibre-optic probe. J. Near Infrared Spectrosc. **9**, 25–34 (2001)

13. Bowers, S.A., Hanks, R.J.: Reflection of radiant energy from soils. Soil Sci. **100**(2), 130–138 (1965)

14. Bogrekci, I., Lee, W.S.: Improving phosphorus sensing by eliminating soil particle size effect in spectral measurement. Trans. ASABE **48**(5), 1971–1978 (2005)

15. Bogrekci, I., Lee, W.S.: Effects of soil moisture content on absorbance spectra of sandy soils in sensing phosphorus concentrations using UV–vis–nir spectroscopy. Trans. ASABE **49**, 1175–1180 (2006)

16. Bernard, G.B., Didier, B., Edmond, H.: Deteriming the distributions of soil carbon and nitrogen in particle size fractions using near-infrared reflectance spectrum of bulk soil samples. Soil Biol. Biochem. **40**, 1533–1537 (2008)

17. Christy, C.D.: Real-time measurement of soil attributes using on-the-go near infrared reflectance spectroscopy. Comput. Electron. Agric. **61**, 10–19 (2008)

18. Xing, Z., Yang, Q., Yang, S., et al.: Design of seed pelleter and experimental study on pelleting technology. Agric. Mech. Res. **3**, 98–101 (2008)

19. Sudduth, K.A., Hummel, J.W.: Soil organic matter, CEC, and moisture sensing with a portable NIR spectrophotometer. Trans. ASAE **36**(6), 1571–1582 (1993)

20. Hummel, J.W., Sudduth, K.A., Hollinger, S.E.: Soil moisture and organic matter prediction of surface and subsurface soils using an NIR soil sensor. Comput. Electron. Agric. **32**(2), 149–165 (2001)

A Research on the Task Expression in Pomology Information Retrieval

Dingfeng Wu, Jian Wang$^{(\boxtimes)}$, Guomin Zhou, and Hua Zhao

Agricultural Information Institute of CAAS, Beijing 100081, China
wangjian02@caas.cn

Abstract. By analyzing the progresses of pomology information retrieval which are driven by tasks, the author put forward an idea of expressing the tasks in pomology information retrieval. An interview test was used to validate the idea and based on analyzing the data of the test, a method of expressing the tasks in pomology information retrieval, in which the difference of searchers' acknowledgement is considered, was put forward.

Keywords: Pomology information retrieval · Task · Task expression

1 Introduction

With the rapid rise of agricultural websites and databases, the research of agricultural information retrieval is becoming more and more important to agricultural industry and agricultural scientific research. The study on information retrieval is now undergoing a great change from simply focusing on the statistical model and the mathematical algorithm to considering about the situation factors, such as user needs, job tasks, social environment and so on [1]. Within those circumstance variables, the job task is very important [2]. The researchers announced that the users' information retrieval behavior originated from their information needs, and the information needs caused by the job tasks, so the job tasks became the driving factors of the information retrieval behavior [3]. Apperceiving the users' job tasks can help the IR system to understand the users' intention and improve the precision of information retrieval system. To achieve this, the users' abstract tasks must be described in a specific way.

2 The Idea of the Task Expression in Pomology Information Retrieval

By observing and analyzing the pomology information retrieval process, we found that in pomology information retrieval, a user have to master some pomology knowledge which is necessary to complete his job task [2]. By taking that pomology knowledge, a user upgraded his pomology knowledge structure to which we called target pomology knowledge structure of the task. When there is a gap between the target pomology knowledge structure and the inherent pomology knowledge structure, the information retrieval behavior will be triggered to make up the gap. We believe that the pomology

D. Li and Z. Li (Eds.): CCTA 2015, Part II, IFIP AICT 479, pp. 127–132, 2016.
DOI: 10.1007/978-3-319-48354-2_13

knowledge gap often consists of some sub pomology knowledge gaps and that's why a series of interrelated information retrieval behavior often be observed together. Because every information retrieval behavior corresponds to a retrieval request query, so in pomology information retrieval, tasks can be expressed in to a series of retrieval queries. Figure 1 shows a typical pomology information retrieval process driven by job task. The user input two related retrieval request to make up the knowledge gap and upgraded the inherent pomology knowledge structure to the target pomology knowledge structure. In this process, task eventually expressed as a series of information retrieval queries.

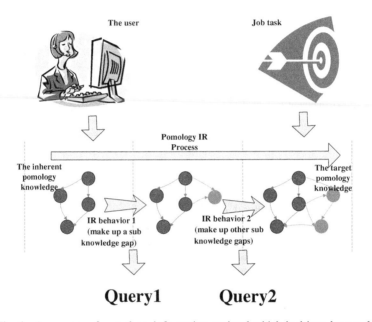

Fig. 1. A progress of pomology information retrieval which is driven by a task

3   Materials and Methods

In order to verify the assumption mentioned above, we interviewed the users of the pomology information retrieval system. In the interview process a large number of simulated job tasks were arranged and at information retrieval process were recorded and analyzed.

The participants including pomology scientific researchers, agricultural experts, scholars and orchard workers, all of whom had the experience of using pomology information retrieval system. The experiments took place at two places, one was the reality experiment site which was at the 101 room of office building of Pomology institute of CAAS, the other one was the Internet remote site, which used the instant messaging platforms, screen recording software, microphones and other video and audio signal capture tools to ensure the real-time communication between the participants and the experiment assistants and record the details of test results and test information behavior.

The setting of the simulation tasks is one of the key problems to be solved of the experiment design. Simulated tasks need to satisfy two conditions: for the first, It must be closely related to the pomology production and scientific research, and it is better to be derived from the practice of production and scientific research in order to be accepted by the participants. For the second, Must have some complexity, in order to identify the change of the ideas of the participants in the IR process, so the simulated tasks should not be single information search tasks but be tasks with practical background and realistic objectives. The source of the work task is provided by the participants themselves, that's because that a lot of typical and impression tasks were accumulated by the participants in their past pomology production and scientific research. The benefit of using the experienced tasks is that the task can be accepted by the participants easily and the retrieval behavior will be simple and intuitive to analyze since the participants are familiar with the tasks. Before the experiment, serval tasks should be prepared in case of the participants can provide few tasks.

In order to extract the intention of every information retrieval behavior smoothly, during the experiment, the participants were asked and guided by the experiment assistants to help them to expose their own aim of every search behavior. The search intention can be exposed in the two part of the experiment. The one is when the retrieval query was typed in, the query directly reflects the information retrieval intention. It should be noticed that sometimes it is difficult to find out a suitable query which can reflect the intention of participants accurately, so a lot of queries are ambiguous, incomplete, or too general. In response to the above situation, the experiment assistants have to ask questions to clarify the intent of the participants, sometimes even help the participants to get queries which are more suitable. The other one is when the participants do the relevance judgment after they get the result list. Usually, the result list is a superset of information and knowledge which the participants need, so the participants can be lead to express the judgement results and evidences which can further exposure the search intentions when the relevance judgments are being carried out.

4   Data and Analysis

There were eleven participants in this experiment who completed twenty-six virtual tasks. During the experiment, seventy-nine information retrieval behaviors are observed.

The retrieval intention can be obtained when analyzing the queries and the relevance judgment behaviors. The users intention can be divided into the following three classes: get the Knowledge points which directly related to tasks (intention 1), get the facts related to the tasks (intention 2) and get the Information that has nothing to do with the tasks (intention 3). If the results of the analysis of the query and the analysis of the relevance judgment behaviors are the same, it is concluded that the conclusion is drawn, but if they are different with each other, the intention obtained from the analysis of the relevance judgment behaviors will be accepted firstly. If there are multiple intentions, It is suspected that multiple information retrieval intents may be merged into one information retrieval behavior and that behavior should be departed into many sub-behaviors to be analyzed. On the basis of analyzing every information retrieval behavior, the participants' intentions can be counted out as shown in Table 1.

Table 1. The statistical results of information retrieval intention

Tasks	Total	Intention 1	Intention 2	Intention 3	Intention 1 ratio (%)	Intention 2 ratio (%)	Intention 3 ratio (%)
Task1	3	3	0	0	100	0	0
Task2	4	3	1	0	75	25	0
Task3	2	2	0	0	100	0	0
Task4	4	4	0	0	100	0	0
Task5	4	2	1	1	50	25	25
Task6	4	3	0	1	75	0	25
Task7	1	1	0	0	100	0	0
Task8	2	2	0	0	100	0	0
Task9	4	3	1	0	75	25	0
Task10	2	2	0	0	100	0	0
Task11	2	2	0	0	100	0	0
Task12	2	2	0	0	100	0	0
Task13	2	1	1	0	50	50	0
Task14	3	3	0	0	100	0	0
Task15	7	3	4	0	42.86	57.14	0
Task16	2	2	0	0	100	0	0
Task17	3	2	1	0	66.67	33.33	0
Task18	2	2	0	0	100	0	0
Task19	4	1	1	2	25	25	50
Task20	2	1	1	0	50	50	0
Task21	2	2	0	0	100	0	0
Task22	4	4	0	0	100	0	0
Task23	4	3	1	0	75	25	0
Task24	4	4	0	0	100	0	0
Task25	3	2	1	0	66.67	33.33	0
Task26	3	3	0	0	100	0	0
Grand total	**79**	**62**	**13**	**4**	**78.48**	**16.46**	**5.06**

Experimental data shows that during the pomology information retrieval process, 80 % of the retrieval behavior is focus on the Knowledge points which directly related to tasks, 16 % of the retrieval behavior is around the facts related to the tasks which can suppose the users to deduce out the knowledge points they need, the remaining 5 % are unrelated to the tasks. Considering about the reason of the information retrieval behavior, we found that job tasks are the most important driving factor which leads to several times more information retrieval behavior than the other factors. Only the task 15 is an exception, that's because in task 15, the participants were asked to find out the treatment method of the apple bagging black spot disease which is rare and lack of research, so the participants have to look for the potentially useful information by searching other similar diseases.

Twenty-six simulation tasks in the experiment include five categories: pomology disease information, pest information, variety information, breeding information and pomology cultivation techniques [4]. We found that different participants will look for different knowledge points when facing the same tasks and this phenomenon might because that their inherent knowledge structures are different. For example, when facing a task which asks the participants to find out the treatment method of a pomology disease, the plant protection professional participants are more inclined to search methods and new drugs, but the other professional participants usually look for the Mechanism and symptoms of the disease firstly. A possible explanation of this phenomenon is that the plant protection professional participants only had few knowledge gaps to make up, so they only need to look for the information which most urgent need to confirm. Relatively speaking, the other professional participants who are not so familiar with pomology diseases have more gaps to fill, so they need to get the basic information first. For the same reason, we noticed that when facing some tasks which are in the same category, a participant might use different Search strategy.

5 Conclusion

Through the above analysis, it is proved that a job task can be expressed as a series of information retrieval intention and further expressed as a collection of queries. If we do not consider the relevant knowledge structure, the job task can be expressed as a collection of queries for all the knowledge points needed for the task. A task can be represented by the following formula:

$$Task \rightarrow \{Query_1, Query_2, \ldots, Query_n\}$$

If we consider about the participant's level of familiarity of the task, the weights should be added to the elements in the collection, if $F_1, F_2, \ldots, F_n$ represents the level of familiarity with the knowledge points and $P_1, P_2, \ldots, P_n$ represents the weights of queries in the collection, the task can be expressed as the following formula:

$$Task \rightarrow \{P_1Query_1, P_2Query_2, \ldots, P_nQuery_n\}$$
$$P_k = 1/F_k \ k \in [1, n]$$

If the user is particularly familiar with the knowledge point, F_k trend approaches infinity, P_K is approaching zero, the expression of the task will not contain $Query_k$. On the contrary, if the user is not familiar with the knowledge point, the weight of $Query_k$ will increase.

Acknowledgements. Funding for this research was provided by "Agricultural system intelligent control and virtual technology team of Science and technology innovation project of CAAS" (**CAAS-ASTIP-2015-AII-03**).

References

1. Ingwersen, P., Jarvelin, K.: The Turn Integration of Information Seeking and Retrieval in Context. Springer, Dordrecht (2005)
2. Kelly, D., Cool, C.: The effects of topic familiarity on information search behavior. In: Hersh, W., Marchionini, G. (eds.) Proceedings of the Second ACM/IEEE Joint Conference on Digital Libraries (JCDL 2002), pp. 74–75. ACM, New York (2002)
3. Ingwersen, P.: Information Retrieval Interaction. Taylor Graham, London (1992)
4. 王中英, 果树学概论(北方本), 农业出版社 (1994)

Prediction of the Natural Environmental High Temperature Influences on Mid-Season Rice Seed Setting Rate in the Middle-Lower Yangtze River Valley

Shouli Xuan[2], Chunlin Shi[2(✉)], Yang Liu[2], Yanhua Zhao[3], Wenyu Zhang[2],
Hongxin Cao[2], and Changying Xue[1]

[1] CMA Henan Key Laboratory of Agrometeorological Support and Applied Technique,
Zhengzhou, 450003, China
xuecy9@163.com
[2] Institute of Agricultural Economics and Information,
Jiangsu Academy of Agricultural Sciences, Nanjing, 210014, China
shirleyxuan2008@hotmail.com, shicl@jaas.ac.cn,
senslover@yeah.net, research@wwery.cn,
caohongxin2003@yahoo.com.cn
[3] Suqian Meteorological Bureau, Suqian, 223800, China
664552404@qq.com

Abstract. The impact of high temperature on rice yield has been evaluated by using simulation models or conducting experiments with controlled high temperature and sowing times. In this paper, observed daily maximum temperature data at 6 representative stations in the Middle-Lower Yangtze River Valley (MLYRV) from 1984 to 2013 was analyzed to examine the daily relationship between rice seed setting rate (RSSR) and the maximum temperature by using data from the mid-season late-maturing indica rice variety regional experiments conducted in field conditions at the same representative stations from 2004 to 2011 (totally using 615 samples and 69 rice varieties). The results indicated that RSSR appears to be sensitive to high temperature from 36 days before full heading to 4 days before full heading (with the significance of the negative correlation between RSSR and maximum temperature in this period above 99 % confidence level) and the most sensitive at about 14 days before full heading (near the meiosis phase), indicating that for the mid-season late-maturing indica rice variety in the MLYRV, more attention should be paid to the high temperature damage at the meiosis stage. According to the extracted high temperature sensitive period, statistical forecast models were established to predict the regional rice high temperature damage in the MLYRV by using atmospheric circulation indices in preceding 12 months with a correlation coefficient between the predicted and observed heat stress index of 0.95 and a normalized root mean square error (*NRMSE*) of 28.4 %. In addition, a high temperature-induced rice sterility simulation model was also used to quantitatively forecast the meiosis phase high temperature influence on rice at the site scale. The *NRMSE* of the simulated and forecasted relative seed setting rate was 4.74 % and 2.84 %, respectively. In conclusion, the presented prediction models were useful to improve the rice high

© IFIP International Federation for Information Processing 2016
Published by Springer International Publishing AG 2016. All Rights Reserved
D. Li and Z. Li (Eds.): CCTA 2015, Part II, IFIP AICT 479, pp. 133–142, 2016.
DOI: 10.1007/978-3-319-48354-2_14

temperature damage forecast and were expected to be helpful to rice high temperature disaster prevention and reduction in the MLYRV.

Keywords: Middle-Lower Yangtze River Valley · Mid-season rice · High temperature · Seed setting rate · Statistical forecast model

1   Introduction

Global average temperature has increased by 0.74 °C in 1906–2005 and is projected to continue to increase [1]. The projected climate change is recognized to increase extremely high temperatures events [2–5]. China is one of the world's most productive rice regions, contributing ~27.5 % of the world rice production [6]. The Middle-Lower Yangtze River Valley (MLYRV) is an important rice producing areas in China. Under the background of climate warming, high temperature stress imposes an increasing risk to rice production and becomes one of the major constrains in increasing productivity of rice in the MLYRV [7–11].

High temperature influence on rice has been studied in many researches [12–15]. It has been pointed out that the heat stress damage mainly occurs in the reproductive period of rice. In previous studies, high temperature effect on rice yields were mainly investigated by using model simulations or control experiments [16–21]. Several studies have paid attention to effects of observed climate on rice [22–24]. For example, Zhao et al. analyzed the effects of observed mean maximum temperature at the stages of flowering and pre-milk on rice components in 1981–2003 [25]. These studies mainly concentrated on the growing-season mean or development stage mean (mostly flowering stage and filling stage) influence of high temperature on rice. Actually, the impact of high temperature largely differed with the local climate, land suitability, cultivated rice variety and time of sowing and harvesting et al. However, the critical period of rice damage caused by natural environmental high temperature in actual rice production is still not clear.

In this study, we tried to identify the most sensitive stage of mid-season RSSR to high temperature in the MLYRV in field conditions. Based on the extracted key sensitive period, the influence of high temperature on rice in the MLYRV was predicted by using the statistical forecast models and the high temperature-induced rice sterility simulation model.

2   Data and Methods

2.1   Data

This study was based on the rice data (including growth period and seed setting rate) at 6 stations (Yichang, Yueyang, Hefei, Nanchang, Hangzhou, Yangzhou) in the MLYRV from 2004 to 2011, which were obtained from the national regional yield trial of southern China (totally using 615 samples and 69 mid-season late-maturing indica rice varieties). The daily maximum temperature data and the monthly 74 circulation indices in 1984–2013 from the Chinese Meteorological Administration were also used.

2.2 Methods

2.2.1 Definition of the Heat Stress Index

In previous studies, the high temperature effects on rice were often investigated by using meteorological indices which were usually defined as the days of daily mean temperature >30 °C or days of daily maximum temperature >35 °C [26–28]. In the present study, the heat stress index (I_{hs}) calculated for the high temperature influence assessment was based on the work of Teixeira et al. [29]. It was assumed that (1) rice crop are only sensitive to high temperature during thermal sensitive period (*TSP*, days); (2) heat damage occurs when maximum temperature (T_{max}) exceeds the critical temperature threshold (T_{crit}); and maximum effect occurs when T_{max} exceeds the limit temperature threshold (T_{lim}). To calculate I_{hs}, "daily" high temperature stress intensity (I_{hsd}) is firstly estimated as a function of T_{max} (Eq. (1)).

$$Ihsd = \begin{cases} 0.0 & T_{max} < T_{crit} \\ \dfrac{T_{max} - T_{crit}}{T_{lim} - T_{crit}} & T_{crit} \le T_{max} < T_{lim} \\ 1.0 & T_{max} \ge T_{lim} \end{cases} \tag{1}$$

The value of T_{crit} and T_{lim} used in this study is 35 °C and 45 °C respectively. The daily values of I_{hsd} are then accumulated and averaged throughout the *TSP* (Eq. (2)).

$$I_{hs} = \frac{\sum_{i=1}^{TSP} I_{hsd}}{TSP} \tag{2}$$

The period from July 11 to August 12 is chosen as the thermal-sensitive period in this paper.

2.2.2 Validation of the Forecast Model

The normalized root mean square error (*NRMSE*) was used to evaluate the predictive ability of the statistical forecast models.

$$NRMSE = \frac{\sqrt{\sum_{i=1}^{n} (Y_i - X_i)^2 / n}}{\overline{X}} \times 100\% \tag{3}$$

where Y_i is the forecasted value, X_i the observed value, n the number of observations, and $\overline{X}$ is the mean of all observed values.

3   Results

3.1   Daily Variation Characteristics of the Maximum Temperature

In order to investigate the daily variation characteristic of the maximum temperature in significant growth stages of rice, averaged daily maximum temperatures (mean of 615 samples) were calculated near the full heading phase. For example, for the full heading day, the averaged maximum temperature was calculated as the mean of maximum temperatures in all the full heading days in 615 samples. As shown in Fig. 1, for the mid-season rice variety in MLYRV, persistent high temperature mainly occurred in the month before full heading (according to the standard of 33 °C). After the full heading stage, there was a consistent decrease in the averaged maximum temperature. Thus, the persistent high temperature occurred concomitantly with the booting and heading stages, resulting in serious loss and harm to rice production in this region.

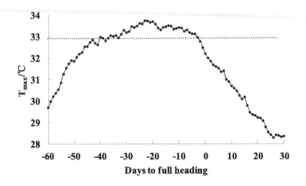

Fig. 1.   Daily variation of the T_{max}

3.2   High Temperature Effects on RSSR

To indentify the sensitive stage of RSSR to high temperature in the MLYRV, daily relationships between maximum temperature (5-day running means) and RSSR before and after full heading day were calculated. As shown in Fig. 2, the RSSR is sensitive to T_{max} in the period of 4–36 days before full heading, corresponding to the averaged date from July 11 to August 12. Obviously, the high temperature at about 14 days before full heading had the most significant impact on RSSR. However, the RSSR seemed to be not closely related to the high temperature at the full heading day. A possible reason might be due to the lower maximum temperature at the full heading day (32.2 °C, see Fig. 1). After the full heading day, the maximum temperature did not have any influence on RSSR. Therefore, for the mid-season rice variety in MLYRV, more attention should be paid to the heat damage around 2 weeks before full heading.

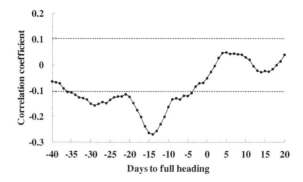

Fig. 2. Daily correlation between RSSR and T_{max} (5-day running means) (dashed indicate significance at 99 % confidence level)

The relationship between RSSR and averaged maximum temperature in 4–36 days before full heading in 2004–2011 was further investigated in Fig. 3. The results indicated that RSSR was decreased significantly with increasing maximum temperature in the sensitive period (4–36 days before full heading). The correlation coefficient between regional averaged RSSR and averaged maximum temperature in 4–36 days before full heading during 2004–2011 was −0.9 (significance above 99.9 % confidence level). It was found that all the years with the averaged T_{max} larger (less) 33 °C exhibited below (above) 80 % RSSR in the MLYRV. Especially for the year 2004, the MLYRV suffered from a severe heat wave in summer. In this year, the averaged maximum temperature in 4–36 days before full heading day reached above 34 °C with the RSSR of 76 %.

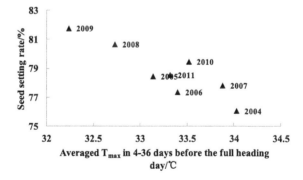

Fig. 3. The relationship between RSSR and averaged T_{max} in 4–36 days before full heading

3.3   Statistical Forecast Models of I_{hs}

As discussed above, the prediction of high temperature influence on rice with several months lead time is valuable to the rice production in the MLYRV, where the higher than normal summertime temperature will directly cause decreases in rice yield. To further predict the impacts of high temperature on mid-season rice in MLYRV, a rice heat stress index (I_{hs}) was calculated by using *TSP* as the extracted sensitive period (from July 11 to August 12) and statistical models were presented for predicting the high temperature stress on rice in this section.

Presented in Fig. 4 was the inter-annual variation of the I_{hs} in the extracted sensitive period of the mid-season rice in the MLYRV (solid line). It can be seen that the I_{hs} showed a significant ability to describe the rice heat-stress events occurred in the MLYRV (such as the year 2003 and 2013). As is known, local climate variability mainly depends on the large scale atmospheric circulations. Therefore, the correlation coefficients between the rice heat stress index in the MLYRV and 74 circulation indices in preceding 12 months (January to June in current year and July to December in last year) was examined to identify key preceding predictors of the I_{hs}. Statistical prediction models for the I_{hs} in the MLYRV were then constructed by using a multi-linear stepwise regression method based on the extracted predictors. We applied year-by-year validation to further verify the prediction models. The year-by-year validation was applied by establishing prediction models using circulation indices in preceding 20 years to forecast the I_{hs} in the following year (for example, using circulation indices in 1984–2003 to establish the forecast model and predict the I_{hs} in 2004; using circulation indices in 1985–2004 to establish the forecast model and predict the I_{hs} in 2005; ······; using circulation indices in 1993–2012 to establish the forecast model and predict the I_{hs} in 2013). The year-by-year validation method was used because it was closer to the approach in actual forecasting. Results of the year-by-year validation test during 2004–2013 (Fig. 4) demonstrated that the constructed prediction models shown good predictive abilities for forecasting the I_{hs} in the MLYRV with the correlation between the predicted and observed I_{hs} of 0.95 (significance above 99.9 % confidence level) and the *NRMSE* of 28.4 %.

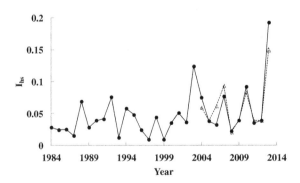

Fig. 4. The observed (solid) and forecasted (dashed) I_{hs} values in the MLYRV

3.4 Quantitative Forecast Model of High Temperature Effects on Rice

The heat stress in the MLYRV could be well forecasted with several months lead time by using the presented statistical models. However, the application of large area averaged prediction information was quite limited in production decisions. Therefore, a simulation model of high temperature induced rice sterility at a meiosis phase was used to predict the relative RSSR in MLYRV at the site scale.

According to the research of Shi et al. [30], the daily RSSR could be expressed in terms of temperature in meiosis phase with a quadratic equation:

$$Y_i = 1 - c \cdot (T_i - T_C)^2 \tag{4}$$

where Y_i is the daily relative RSSR, T_i is the daily T_{max}, c and T_C are variety parameters. The relative RSSR decreases when the maximum temperature in the meiosis stage is higher than T_C. Because of the varied daily maximum temperature in actual production, the total influence of high temperature on relative RSSR is expressed in the product of daily relative RSSR:

$$Y = \prod_{i=t0}^{t1} Y_i \tag{5}$$

where Y is the total influence of high temperature during meiosis phase on relative RSSR, $t0$ and $t1$ is the starting and ending time of the meiosis phase, respectively.

In the present study, the year 2009 without heat stress was set as the CK year. The observed rice relative RSSR was defined as rRSSR(i)/rRSSR(2009) × 100 %, where rRSSR(i) and rRSSR(2009) is the relative RSSR in forecast year and in 2009, respectively. As shown in Fig. 3, the MLYRV experienced significantly high temperature stress in the year 2004 and 2007. Therefore, the rice data in 2004 and 2007 was used to calibrate and validate the model, respectively. The daily T_{max} in crucial high temperature sensitive stage (7–19 days before full heading, see Fig. 2) were used to run the high temperature-induced rice sterility simulation model and trail-and-error method was adopted to calibrate the parameters. The result of calibration showed that for the mid-season rice in the MLYRV, the value of parameter c and T_C is 0.000246 and 30.6°C, respectively. Figure 5 presented the simulation and prediction results of the high temperature-induced rice sterility simulation model (observed RSSR is the average of 69 rice varieties). Because of the missing data in Nanchang (2004 and 2009) and Yangzhou (2007), we actually used 5 samples in 2004 and 4 samples in 2007. The results shown that the high temperature-induced rice sterility simulation model accurately simulated the relative RSSR in 2004 with a NRMSE of 4.74 % and forecasted the relative RSSR in 2007 with a NRMSE of 2.84 % (the correlation coefficient between the predicted and observed relative RSSR in 2007 reached 0.96 with the significance above 95 % confidence level), indicating the good predictive skill of the high temperature-induced rice sterility simulation model in the MLYRV.

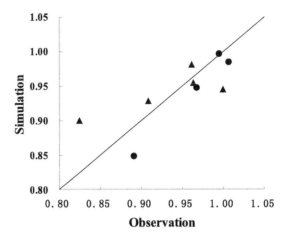

Fig. 5. The simulated (2004, triangle) and forecasted (2007, circle) results of the relative RSSR

4 Conclusions

Previous studies have found that rice plant was the most sensitive to high temperature at heading time. In this study, daily relationship between RSSR and temperature was investigated by using observed daily maximum temperature data at 6 representative stations in the MLYRV. Our results indicated that for the mid-season rice in the MLYRV, years with the maximum temperature in meiosis stage of rice above than normal may experience significant yield losses in the MLYRV. The results in the present study reinforced the need to plan adaptive strategies for rice production with regard to heat stress in the meiosis phase of mid-season rice in the MLYRV. Based on the revealed high temperature sensitive period, high temperature-induced rice sterility simulation model and statistical models were established to forecast rice heat stress in MLYRV. The results of validation tests further illustrated the good prediction ability of the constructed forecast models.

Acknowledgment. Funds for this study was provided by the National Natural Science Fund (31401279), the CMA/Henan Key Laboratory of Agro-meteorological Support and Applied Technique (AMF201201), the Jiangsu Agriculture Science and Technology Innovation Fund (CX(12)5059), the China Meteorological Administration Special Public Welfare Research Fund (GYHY201306035), the China Agricultural Administration Special Public Welfare Research Fund (201203032), and the Jiangsu Province Science and Technology Support Program (BE2012391).

References

1. IPCC. Climate Change 2007: Impacts, adaptation and vulnerability. Contribution of Working Group II to the Fourth Assessment Report of the Intergovernmental Panel on Climate Change. Cambridge University Press, Cambridge (2007)

2. Wheeler, T.R., Craufurd, P.Q., Ellis, R.H., et al.: Temperature variability and the yield of annual crops. Agric. Ecosyst. Environ. **82**, 159–167 (2000)
3. Tao, F., Hayashi, Y., Zhang, Z., et al.: Global warming, rice production, and water use in China: developing a probabilistic assessment. Agric. For. Meteorol. **148**(1), 94–110 (2008)
4. Lu, P., Yu, Q., Wang, E., et al.: Effects of climatic variation and warming on rice development across South China. Clim. Res. **36**, 79–88 (2008)
5. Zhang, T., Zhu, J., Reiner, W.: Responses of rice yields to recent climate change in China: An empirical assessment based on long-term observations at different spatial scales (1981–2005). Agric. For. Meteorol. **150**, 1128–1137 (2010)
6. FAO. Food and Agriculture Organization of the United Nations (2013). http://faostat.fao.org Last visited: 07.07.2015
7. Zheng, J., Sheng, J., Tang, R., et al.: Regularity of high temperature and its effects on pollen vigor and seed setting rate of rice in Nanjing and Anqing. Jiangsu J. Agric. Sci. **23**(1), 1–4 (2007). (in Chinese)
8. Ma, X., Xu, Y., Zhao, H.: Impacts of maximum or minimum temperature on yield and yield components of single-season Indica rice in Yangtze-Huaihe area. Geogr. Res. **27**(3), 603–612 (2008). (in Chinese)
9. Tian, X., Luo, H., Zhou, H., et al.: Research on heat stress of rice in China: progress and prospect. Chin. Agric. Sci. Bull. **25**(22), 166–168 (2009). (in Chinese)
10. Jiang, M., Jin, Z., Shi, C., et al.: Occurrence patterns of high temperature at booting and flowering stages of rice in the middle and lower reaches of Yangtze River and their impacts on rice yield. Chin. J. Ecol. **29**(4), 649–656 (2010). (in Chinese)
11. Xie, Z., Du, Y., Gao, P., et al.: Impact of high-temperature on single cropping rice over Yangtze-Huaihe river valley and response measures. Meteorol. Monthly **39**(6), 774–781 (2013). (in Chinese)
12. Satake, T., Yoshida, S.: High temperature-induced sterility in Indica Rices at flowering. Japan. J. Crop Sci. **47**(1), 6–17 (1978)
13. Matsui, T., Omasa, K., Horie, T.: High temperature-induced spikelet sterility of japonica rice at flowering in relation to air temperature, humidity and wind velocity conditions. Japan. J. Crop Sci. **66**(3), 449–455 (1997)
14. Matsui, T., Omasa, K., Horie, T.: The difference in sterility due to high temperatures during the flowering period among Janonica-rice varieties. Plant Prod. Sci. **4**(2), 90–93 (2001)
15. Kim, J., Shon, J., Lee, C.K., et al.: Relationship between grain filling duration and leaf senescence of temperate rice under high temperature. Field Crops Res. **122**(3), 207–213 (2011)
16. Erda, L., Wei, X., Hui, J., et al.: Climate change impacts on crop yield and quality with CO_2 fertilization in China. Philos. Trans. R. Soc. Lond. B **360**(1463), 2149–2154 (2005)
17. Krishnan, P., Swain, D.K., Bhaskar, B.C., et al.: Impact of elevated CO_2 and temperature on rice yield and methods of adaptation as evaluated by crop simulation studies. Agric. Ecosyst. Environ. **122**(2), 233–242 (2007)
18. Liu, W., Zhang, X., Yu, W., et al.: Assessment method study on high temperature damage to rice. Meteorol. Environ. Sci. **32**(1), 33–38 (2009). (in Chinese)
19. Zhang, Q., Zhao, Y., Wang, C.: Study on the impact of high temperature damage to rice in the lower and middle reaches of the Yangtze river. J. Catastrophol. **26**(4), 57–62 (2011). (in Chinese)
20. Jagadish, S.V.K., Craufurd, P.Q., Wheeler, T.R.: High temperature stress and spikelet fertility in rice. J. Exp. Bot. **58**(7), 1627–1635 (2007)
21. Shi, C., Jin, Z., Tang, R., et al.: A model to simulate high temperature-induced sterility of rice. Chin. J. Rice Sci. **21**(2), 220–222 (2007). (in Chinese)

22. Tao, L., Tan, H., Wang, X., et al.: Effects of high temperature stress on super hybrid rice Guodao 6 during flowering and filling phases. Chin. J. Rice Sci. **21**(5), 518–524 (2007). (in Chinese)
23. Tong, Z., Li, S., Duan, W., et al.: Temperature-driven climatic factors and their impact on the fertility of hybrid rice at anthesis. Chin. J. Eco-Agric. **16**(5), 1163–1166 (2008). (in Chinese)
24. Peng, S., Huang, J., Sheehy, J.E., et al.: Rice yields decline with higher night temperature from global warming. Proc. Natl. Acad. Sci. U.S.A. **101**(27), 9971–9975 (2004)
25. Zhao, H., Yao, F., Zhang, Y., et al.: Correlation analysis of rice seed setting rate and weight of 1000-Grain and agro-meteorology over the middle and lower reaches of the Yangtze river. Sci. Agric. Sin. **39**(9), 1765–1771 (2006). (in Chinese)
26. Yao, F., Zhang, J.: Change of relative extreme high temperature events and climate risk in rice growing period in China from 1981 to 2000. J. Nat. Disasters **18**(4), 37–42 (2009). (in Chinese)
27. Ren, Y., Gao, P., Wang, C.: High temperature damage to paddy rice in Jiangsu Province and its cause analysis. J. Nat. Disasters **19**(5), 101–107 (2010). (in Chinese)
28. Xie, X., Li, B., Wang, L., et al.: Spatial and temporal distribution of high temperature and strategies to rice florescence harm in the lower-middle reaches of Yangtze river. Chin. J. Agrometeorol. **31**(1), 144–150 (2010). (in Chinese)
29. Teixeira, E.I., Fischer, G., van Velthuizen, H., et al.: Global hot-spots of heat stress on agricultural crops due to climate change. Agric. For. Meteorol. **170**, 206–215 (2013)
30. Shi, C., Jin, Z., Zheng, J., et al.: Quantitative analysis on the effects of high temperature at meiosis stage on seed-setting rate of rice florets. Acta Agron. Sin. **34**(4), 627–631 (2008). (in Chinese)

Study on the Mutton Freshness Using Multivariate Analysis Based on Texture Characteristics

Xiaojing Tian[1,2], Jun Wang[3], Jutian Yang[2], Shien Chen[2], and Zhongren Ma[1(✉)]

[1] The Key Bio-engineering and Technology Laboratory of National Nationality Commission, Northwest University for Nationalities, Lanzhou 730030, China
smile_tian@yeah.net, mzr@xbmu.edu.cn
[2] College of Life Science and Engineering, Northwest University for Nationalities, Lanzhou 730024, China
jutianyang988@163.com, chshien@163.com
[3] Department of Biosystems Engineering, Zhejiang University, 388 Yuhangtang Road, Hangzhou 310058, China
Jwang@zju.edu.cn

Abstract. Aiming at discrimination and prediction of mutton freshness by texture profile, the texture parameters of mutton stored at 1°C, 4°C and Room temperature were analyzed. The analysis methods of Canonical Discriminant Analysis (CDA) and Principal component analysis (PCA) were used to analyze texture parameters of mutton. The results of PCA showed that mutton sample stored at three temperatures clustered into groups according to their freshness, changing along the direction of PC1. Better classification results were found by CDA. The changing trends of mutton freshness were described by Multiple Linear Regression (MLR) and Partial Least Square analysis (PLS), and effective predictive models were found for indices of days stored, TVB-N and pH using texture parameters. With optimum analysis methods, texture parameters could classify and predict freshness of mutton stored at three temperatures. Texture profiles were proved to be a fast and objective tool for the prediction of mutton freshness.

Keywords: Mutton · Freshness · Texture parameters · Multivariate analysis

1 Introduction

It is well known that a country's meat consumption and demands for healthier diets increase reflect its degree of development. Trends for healthier meat could result in increasing demands for meat with high protein, low fat and good taste, such as mutton, which could lead to a diet with better nutritional (lower fat and cholesterol) [1, 2] and sensorial features (flavor, juiciness, tenderness). The consumption of mutton has a significant increase in China in recent years. However, the shelf life of mutton sold as whole carcasses or as bone-in cubes at retail store is subject to variation of storage temperatures, contamination of microorganisms, enzymes and the processing

© IFIP International Federation for Information Processing 2016
Published by Springer International Publishing AG 2016. All Rights Reserved
D. Li and Z. Li (Eds.): CCTA 2015, Part II, IFIP AICT 479, pp. 143–154, 2016.
DOI: 10.1007/978-3-319-48354-2_15

conditions, which damage the functional, sensorial and nutritional quality of mutton [3]. It is of great importance to monitor the changes of mutton freshness during storage.

Several techniques are used to monitor the deterioration of meat, including the use of microbiological techniques, detection of metabolite concentrations, spectroscopy methods, electronic tongue and electronic nose, and sensorial panels. The microbiological techniques were used to evaluate consequences of bacteria-bacteria interactions [4, 5], monitor the spoilage and freshness of meat, and study on the initial microflora and total number of bacteria changes storaged at different temperatures [6]. For spectroscopy methods, near infrared reflectance spectroscopy (NIR) [7], fourier transform near infrared spectroscopy [8] and fluorescence spectroscopy [9] were shown to be effective for meat freshness assessment. The metabolic concentrations of protein, lipids, fat acid, carbohydrates, such as ATP [9], glucose and derived compounds or biogenic amines [10], detected by spectroscopy methods and physicochemical analysis, were used to express meat freshness. The electronic tongue and electronic nose showed their ability in freshness evaluation of pork [11, 12] and beef [13]. The senses of taste, smell, touch, sight and hearing were used to describe food properties by sensorial panels. The juiciness, tenderness, flavor and color of meat were determined by sensorial panels [14, 15]. However, there were few works done on the texture changes along with freshness of meat.

For the physical properties, the texture profiles, as combination of sensory perception, mechanical properties and geometric features, could directly describe the organization status of food texture characteristics in the form of numerical description. Texture Profile Analysis (TPA) [16] is a very useful method for texture analysis. The texture parameters of hardness, springiness, chewiness and cohesiveness were obtained by TPA, and these parameters were used to describe the eating quality of food and freshness assessment of fruits, fishes and meat. The relationships between TPA parameters of meat and sensory characteristics were studied aiming at finding the correlation of texture parameters and sensory analysis [17, 18]. TPA was used to monitor food quality and control of food quality, food shelf life [19] and consumer preference studies, modification formulation and processing process [20].

The studies of meat texture were mostly on the differentiation and substitution of sensory perception of raw meat and processed meat products. However, little detailed information is available on quantitative determination of physicochemical parameters of meat freshness, and correlation of texture parameters with physicochemical indexes [21, 22], except for the relationship between freshness and elasticity of fish [23] and chicken [24].

The potential use of Texture Profile Analysis for detection of mutton freshness was conducted by texture analyzer in this study. The objectives were: (1) investigation the use of a texture analyzer to monitor the freshness of mutton stored at temperatures of 1° C, 4°C and room temperature, with the help of multivariate analysis methods, (2) build models for the prediction indicators of mutton freshness, (3) develop an objective method for detection of mutton freshness.

2   Experiments and Methods

2.1   Experimental Samples

All the meat samples used were obtained from local market. For pretreatment, the fat and connective tissues were removed before detection, and the mutton samples were cut into cubes of 20 mm × 20 mm × 15 mm. Cuboids of mutton were stored at 1°C, 4°C and room temperature of 20 ± 2°C, respectively. The samples were brought to room temperature before analyzed by TPA and physichemical methods for freshness indicators. Samples detected at the day they were taken to the lab were named d0, samples which were stored for 1 day were named d1, samples which were stored for 2 day were named d2, and so on d3, d4, d5, d6, d7, d8, d9, d10, and d11.

2.2   Experimental Methods

The texture analyzer of TMS-Pro (Food Technology Corporation, USA), equipped with a 1000 N load cell, was used for texture measurements. The software of V1.13-002 Texture Lab Pro (Food Technology Corporation, Virginia, USA) was used to acquire and analyze data. TPA analysis was conducted according to the method of Tian [19] with some modifications. All the measurements were carried out with 2 compression cycle test, with cylinder probe of 15 mm diameter, 40 % compression of the original sample height. 5 s was left between the two compression cycles. The force-time deformation curves of mutton sample were gained with 1 N trigger force at speed of 60 mm min^{-1}. The hardness, chewiness, cohesiveness, springiness and gumminess of mutton were calculated from the force-displacement curve for each sample. Eight replicates were measured for TPA parameters of mutton for each treatment.

For tenderness, a digital muscle tenderness meter (C - LM3, Tenovo international co., Limited, China) was used to measure the tenderness of treated mutton samples. Tenderness measurement was evaluated with three replicates by cutting through the sample perpendicular to fibre direction, and expressed in form of shear force. Shear force was calculated as the average shear force of three samples.

As freshness indicator of mutton, content of total volatile basic nitrogen (TVB-N) was detected according to the method of semimicro-Kjeldahl determination in GB_T 5009.44 [25] and pH was detected according to the method in GB/T 9695.5 [26]. 3 duplicates were conducted for each sample, and the average of 3 duplicates were used.

For data processing, the texture characteristic parameters of mutton were analyzed by multivariate data analysis methods to discriminate the samples according to the freshness of mutton. Principal Component Analysis and Canonical Discriminant Analysis were used for visualization of mutton with different freshness using texture parameters. The changing trends of mutton freshness were described by Multiple Linear Regression and Partial Least Square analysis, and the predictive results were compared. For data processing, the software program of the SAS (V8, SAS Institute Inc., Gary, USA) was used.

3 Results and Discussion

3.1 Changes of Mutton Freshness

The deterioration of mutton samples was monitored by TVB-N, which is the typical indicator of meat freshness. pH of the sample was detected as a reference indicator for mutton freshness. The changing trends of TVB-N and pH of mutton samples stored at 1°C, 4°C and room temperature were shown in Figs. 1 and 2. The value of TVB-N exceeded the limit level of 25 mg/100 g at the 2nd day for samples stored at room temperature, the 8th day for samples stored at 4°C and 1°C. The pH exceeded over the limit level of 6.7 at the 2nd day for samples stored at room temperature, the 7th day for samples stored at 4°C, and the 8th day for samples stored at 1°C. The TVB-N and pH of mutton stored at three different temperatures increased with storage time. Faster changes were observed in higher storage temperature, which is in accordance with the changes of meat freshness by Nan [27].

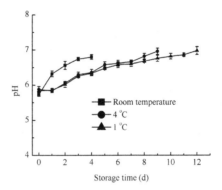

Fig. 1. Trend for pH versus storage time (day) at three temperatures

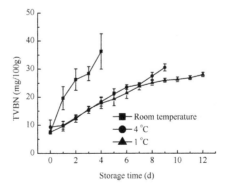

Fig. 2. Trend for TVB-N versus storage time (day) at three temperatures

3.2   Texture Characteristic Parameters of Mutton Changed with Freshness

The texture of mutton changed according to the deterioration of mutton when stored, and the changes of texture parameters and shear force were presented in Figs. 3 and 4.

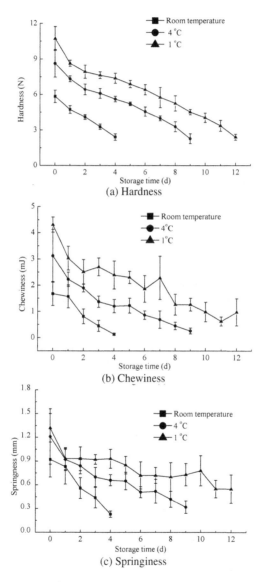

(a) Hardness

(b) Chewiness

(c) Springiness

Fig. 3. Trend for texture profiles of mutton versus storage time (days) for three storage temperatures

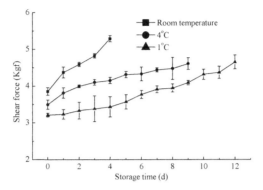

Fig. 4. Trend for shear force of mutton versus storage time (days) at three storage temperatures

Of all the variables measured, the values of hardness, chewiness, springiness all decreased over days at three storage temperatures. The shear force increased as the freshness of mutton decreased, indicating the tenderness decreased with the increasing of days stored. The higher the storage temperature was, the faster these parameters changed. Same decreasing trends of hardness, chewiness and springiness to storage time were observed for grass carp [28] and chicken [19].

3.3   Discrimination of Mutton Freshness by PCA and CDA

For discrimination of mutton freshness changed along storage, PCA and CDA were used to cluster mutton samples. Hardness, chewiness, springiness and cohesiveness, obtained by texture analysis, were used as independent variables to perform the PCA and CDA analysis by SAS V8. The results of PCA for mutton samples stored at three different storage temperatures were shown in Fig. 5. PC1 and PC2 accumulated 90.50 %, 94.58 % and 92.14 % of the total variance for mutton stored at room temperature, 4°C and 1°C, respectively. As shown in Fig. 5a, with PCA method, mutton samples stored at room temperature clustered into groups according to the days stored, although samples of d2 overlapped with that of d3. The freshness of mutton was clearly discriminated into three groups, with d0 as fresh group, d1 as secondary fresh group, d2, d3 and d4 as deteriorated meat, which were same with freshness level indicated by TVB-N. As shown in Fig. 5b, with PCA method, the mutton samples stored at 4°C also clustered into three groups according to the freshness of mutton, with samples of d0–d2 in the fresh group, samples of d3–d7 in the secondary fresh group, and samples of d8–d9 into the group of deteriorated meat, although samples of two adjacent days overlapped with each other. The PCA results of mutton samples stored at 1°C were shown in Fig. 5c, an obvious deteriorating trend was found for clusters of mutton with different freshness, although samples stored at 1°C overlapped for the adjacent 2 to 3 days.

All the mutton samples were grouped into distinct clusters according to their freshness by the method of PCA, except that some samples of two adjacent days overlapped with each other. What's more, in the decreasing direction of PC1, an obvious deteriorating trend was found for clusters of mutton with different freshness for three storage temperatures. Mutton with different freshness could be discriminated by PCA.

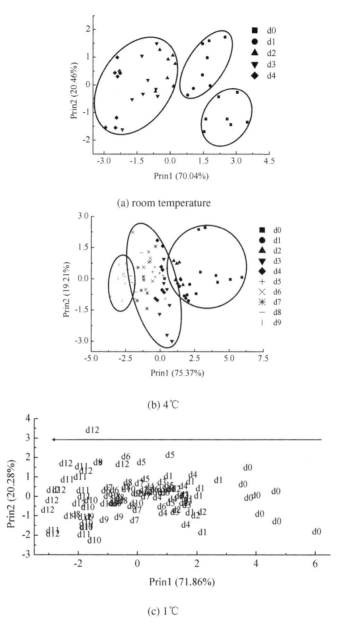

(a) room temperature

(b) 4℃

(c) 1℃

Fig. 5. Qualitative identification of mutton freshness by PCA based on the texture profiles for three storage temperatures

The results of CDA for mutton samples stored at three different storage temperatures were shown in Fig. 6. The first two CANs explained 99.18 %, 95.94 % and 97.77 % of the total variance of 100 % for mutton stored at room temperature, 4°C and 1°C, respectively. As shown in Fig. 6a, samples stored at room temperature were grouped into 5 clusters according to the days stored, with d0 as the fresh grade, d1 as the secondary freshness, d2, d3 and d4 as the deteriorated meat. In Fig. 6b and c, similar results were obtained. Mutton samples were grouped into 3 groups according to the freshness indicated by TVB-N value, except that few samples located into the wrong groups. With CDA, better discrimination results were found. In the descending direction of CAN1, a decreasing trend was found for freshness of mutton for three storage temperatures.

Mutton samples stored at three storage temperatures were classified according to their freshness, indicated by TVB-N, by methods of PCA and CDA, and better classification results were found by CDA.

3.4   Prediction of Mutton Freshness Using Texture Characteristic Parameters

Aiming at studying the relationship between texture parameters and freshness of mutton, the method of multiple linear regressions (MLR) and partial least square regression (PLS) were used, with texture parameters as input data. The predictive results were compared to find better prediction model. The correlation coefficient of R^2 and root mean square error of RMSE were used to estimate the performance of predictive models. Lower RMSE and larger R^2 always lead to better calibration models.

For data sets containing mutton samples, 72 (6 × 12) samples for calibration and 24 (2 × 12) for validation for mutton stored at 1°C, 54 (6 × 9) for calibration and 18 (2 × 9) for validation for mutton stored at 4°C, 24 (6 × 4) for calibration and 8 (2 × 4) for validation for mutton stored at room temperature, were used to build the prediction model for mutton freshness, and the predictive model were validated.

With texture parameters as independent variables, and days stored, TVB-N and pH as dependent variables, the multivariate projection method of PLS was conducted with leave-one-out technique. The accuracy was estimated using R^2 and RMSE. As shown in Table 1, good correlations for calibration data set were found between texture parameters and freshness indicators with $R^2 \geqslant 0.890$ for TVB-N, $R^2 \geqslant 0.891$ for pH, $R^2 \geqslant 0.941$ for days stored at three storage temperatures. The values of R^2 increased with the rising of storage temperature from 1°C to 4°C and then to room temperature. When these models were used to predict mutton freshness for samples in the test set, the lowest R^2 for pH, TVB-N and days stored (0.821, 0.772 and 0.831) were found for samples stored at 4°C. In the former studies, chemical parameters were successful predicted by method of PLS [29, 30], When the PLS method was used with texture characteristics, similar results were found that the freshness of mutton stored at different temperatures could be obtained by PLS using texture parameters.

The relationship between texture characteristics parameters and freshness indicators of mutton were established by MLR algorithm. Larger R^2 and lower RMSE indicate adequate fits. As shown in Table 1, a linear correlation between texture parameters and

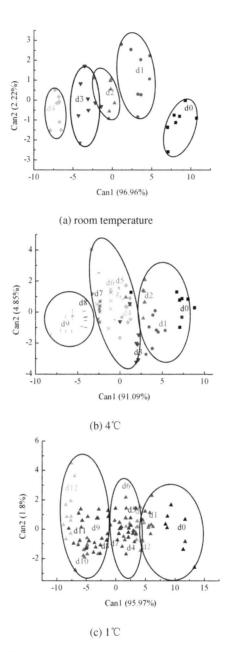

(a) room temperature

(b) 4 ℃

(c) 1 ℃

Fig. 6. Qualitative identification of mutton freshness by CDA based on the texture profiles for three storage temperatures

Table 1. Comparison of two predictive methods (PLS and MLR) for TVB-N, pH and days stored for mutton stored at three storage temperatures

Methods	Storage temperature	Parameters	Training set		Test set	
			R^2	RMSEC	R^2	RMSEP
PLS	Room temperature	TVB-N	0.953	2.05	0.977	2.56
		pH	0.926	0.1	0.913	0.15
		Days stored	0.968	0.25	0.94	0.43
	4°C	TVB-N	0.93	1.85	0.821	2.91
		pH	0.912	0.11	0.772	0.17
		Days stored	0.949	0.66	0.831	1.16
	1°C	TVB-N	0.89	2.22	0.87	2.36
		pH	0.891	0.12	0.853	0.14
		Days stored	0.941	0.94	0.923	1.03
MLR	Room temperature	TVB-N	0.949	1.96	0.972	2.39
		pH	0.917	0.1	0.923	0.14
		Days stored	0.962	0.25	0.94	0.43
	4°C	TVB-N	0.925	1.85	0.821	2.91
		pH	0.905	0.11	0.772	0.17
		Days stored	0.945	0.66	0.832	1.24
	1°C	TVB-N	0.885	2.21	0.87	2.37
		pH	0.888	0.12	0.853	0.14
		Days stored	0.938	0.94	0.923	1.03

freshness indicators was illustrated by MLR. Models were obtained with high correlation as $R^2 \geqslant 0.885$ for TVB-N, $R^2 \geqslant 0.888$ for pH, $R^2 \geqslant 0.938$ for days stored at three temperatures. For MLR, the values of R^2 increased with the rising of storage temperature. When the MLR models were used to predict samples in test set, the R^2 were all higher than 0.8, except for pH at 4°C (0.772). The MLR model was proved to be effective in the prediction of TVB-N, pH and days stored for mutton samples.

Models built by PLS and MLR showed high ability to predict the content of TVB-N, pH and days stored for mutton. Compared by values of R^2 and RMSE, better results were found by PLS with high correlation for the training and test sets. Freshness indicators, days stored were best predicted, than comes the TVB-N value. The prediction of pH was not so good.

4   Conclusions

The potential use of texture analyzer to monitor the freshness of mutton stored at different temperatures was evaluated in this study. It was found that texture parameters obtained by TPA analysis could be successfully applied for the monitor of meat deterioration and prediction of freshness indicator (TVB-N, pH, and days stored). Both PCA and CDA showed high ability in the discrimination of mutton samples with different freshness, and the results of CDA were better than that of PCA. The predictive models built by PLS and MLR showed high capacity (R^2 higher than 0.88 for all parameters in calibration data set, higher than 0.77 in validation data set) in mutton freshness prediction based on the texture parameters. The days stored were best predicted, and then comes TVB-N, and pH. What's more, PLS was more effective than MLR when they were used to study the freshness of mutton. The texture analyzer could provide the meat industry with a fast, objective and useful tool for the determination of mutton freshness.

Acknowledgments. Funds for this research was provided by the Changjiang Scholars and Innovative Research Team in University (IRT13091), the Chinese National Foundation of Nature and Science through Project 31560477 and 31160440, the Research Funds for introduced talents of Northwest University for Nationalities through Project xbmuyjrc201408.

References

1. Park, Y.W., Kouassi, M.A., Chin, K.B.: Moisture, total fat and cholesterol in goat organ and muscle meat. J. Food Sci. **56**(5), 1191–1193 (1991)
2. Schweigert, B., Payne, B.J.: A summary of the nutrient content of meat. American Meat Institute Foundation (1956)
3. Ventanas, S., Estevez, M., Tejeda, J.F., et al.: Protein and lipid oxidation in Longissimus dorsi and dry cured loin from Iberian pigs as affected by crossbreeding and diet. Meat Sci. **72**(4), 647–655 (2006)
4. Borch, E., Kant-Muermans, M.-L., Blixt, Y.: Bacterial spoilage of meat and cured meat products. Int. J. Food Microbiol. **33**(1), 103–120 (1996)
5. Gill, C.O.: Meat spoilage and evaluation of the potential storage life of fresh meat. J. Food Prot. **46**, 444–452 (1983)
6. Fu, P., Li, P.: Study on the initial microflora and its changes of chilling pork meat. Food Sci. **27**(11), 119–124 (2006). (In Chinese)
7. Grau, R., Sánchez, A.J., Girón, J., et al.: Nondestructive assessment of freshness in packaged sliced chicken breasts using SW-NIR spectroscopy. Food Res. Int. **44**(1), 331–337 (2011)
8. Chen, Q., Cai, J., Wan, X., et al.: Application of linear/non-linear classification algorithms in discrimination of pork storage time using Fourier transform near infrared (FT-NIR) spectroscopy. LWT-Food Sci. Technol. **44**(10), 2053–2058 (2011)
9. Oto, N., Oshita, S., Makino, Y., et al.: Non-destructive evaluation of ATP content and plate count on pork meat surface by fluorescence spectroscopy. Meat Sci. **93**(3), 579–585 (2013)
10. Luo, A., Zhu, Q., Zheng, H., et al.: Study on TVB-N assay for freshness index of chilled beef by compound preserving technique. Food Sci. **25**(2), 174–179 (2004)
11. Gil, L., Barat, J.M., Baigts, D., et al.: Monitoring of physical–chemical and microbiological changes in fresh pork meat under cold storage by means of a potentiometric electronic tongue. Food Chem. **126**(3), 1261–1268 (2011)
12. Musatov, V.Y., Sysoev, V.V., Sommer, M., et al.: Assessment of meat freshness with metal oxide sensor microarray electronic nose: A practical approach. Sens. Actuators B Chem. **144**(1), 99–103 (2010)
13. Hong, X., Wang, J., Hai, Z.: Discrimination and prediction of multiple beef freshness indexes based on electronic nose. Sens. Actuators B Chem. **161**(1), 381–389 (2012)

14. Jayasingh, P., Cornforth, D.P., Brennand, C.P., et al.: Sensory evaluation of ground beef stored in high-oxygen modified atmosphere packaging. J. Food Sci. **67**(9), 3493–3496 (2002)

15. Salinas, Y., Ros-Lis, J.V., Vivancos, J.L., et al.: Monitoring of chicken meat freshness by means of a colorimetric sensor array. Analyst **137**(16), 3635–3643 (2012)

16. Mochizuki, Y.: Texture profile analysis. Current Protocols in Food Analytical Chemistry (2001)

17. Caine, W.R., Aalhus, J.L., Best, D.R., et al.: Relationship of texture profile analysis and Warner-Bratzler shear force with sensory characteristics of beef rib steaks. Meat Sci. **64**(4), 333–339 (2003)

18. Ruiz de Huidobro, F., Miguel, E., Blazquez, B., et al.: A comparison between two methods (Warner–Bratzler and texture profile analysis) for testing either raw meat or cooked meat. Meat Sci. **69**(3), 527–536 (2005)

19. Tian, X., Wen, S., Shen, X., et al.: Study on freshness of chicken based on texture characteristics. Sci. Technol. Food Ind. **33**(17), 63–66 (2012). (In Chinese)

20. Martinez, O., Salmeron, J., Guillen, M.D., et al.: Texture profile analysis of meat products treated with commercial liquid smoke flavourings. Food Control **15**(6), 457–461 (2004)

21. Herrero, A.M., de la Hoz, L., Ordóñez, J.A., et al.: Tensile properties of cooked meat sausages and their correlation with texture profile analysis (TPA) parameters and physico-chemical characteristics. Meat Sci. **80**(3), 690–696 (2008)

22. Węsierska, E., Palka, K., Bogdańska, J., et al.: Sensory quality of selected raw ripened meat products. Acta Scientiarum Polonorum Technologia Alimentaria **12**(1), 41–50 (2013)

23. Li, T.: Relationship between the freshness and elasticity of chicken. Jilin, 50–59 (2008). (In Chinese)

24. Li, Z., Liu, Z.: The relationship between freshness and elasticity of fish. J. Zhengzhou Inst. Technol. **24**(4), 37–39 (2003). (In Chinese)

25. GB/T9695.5: National Standard of the People's Republic of China, Method for analysis of hygienic standard of meat and meat products (2003)

26. GB/T9695.5: National Standard of the People's Republic of China, Meat and meat products-Measurement of pH (2008)

27. Nan, Q.: Manual for Meat Industry, 1st edn, pp. 72–74. China Light Industry Press, Beijing (2003)

28. Zheng, R., Cao, C., Bao, J., et al.: Study on texture changes of grass carp frozen at different temperaturea with different treatment. Sci. Technol. Food Ind. **33**(01), 344–347 (2012). (In chinese)

29. Apetrei, C., Apetrei, I.M., Nevares, I., et al.: Using an e-tongue based on voltammetric electrodes to discriminate among red wines aged in oak barrels or aged using alternative methods: correlation between electrochemical signals and analytical parameters. Electrochim. Acta **52**(7), 2588–2594 (2007)

30. Rudnitskaya, A., Polshin, E., Kirsanov, D., et al.: Instrumental measurement of beer taste attributes using an electronic tongue. Anal. Chim. Acta **646**(1–2), 111–118 (2009)

Research and Application on Protected Vegetables Early Warning and Control of Mobile Client System

Guogang Zhao[1,2], Haiye Yu[1,2(✉)], Lianjun Yu[4], Guowei Wang[1,2,3], Yuanyuan Sui[1,2], Lei Zhang[1,2], Linlin Wang[1,2], and Jiao Yang[3,5]

[1] College of Biological and Agricultural Engineering, Jilin University, Changchun 130022, China
zhaoguogang2000@qq.com, 41422306@qq.com, haiye@jlu.edu.cn, suiyuan0115@126.com, z_lei@jlu.edu.cn, wanglll1211@foxmail.com
[2] Key Laboratory of Bionic Engineering, Ministry of Education, Changchun 130022, China
[3] School of Information Technology, Jilin Agricultural University, Changchun 130118, China
61516131@qq.com
[4] Changchun City Academy of Agricultural Sciences, Changchun 130111, China
120142901@qq.com
[5] Sixth Middle School in Changchun, Changchun 130000, China

Abstract. Experimental vegetable production is greatly influenced by many environmental factors. There is urgent need to supervise its air temperature and humidity, soil temperature and humidity in the facilities information such as real-time monitoring, and timely early warning and control, in case of irreparable damage. This paper adopts the IOT technology to design and develop mobile client system which is used to make warnings and protect vegetables in order to realize real-time monitoring in the process of vegetable production in the aspects of timely warning of environmental information and control.

Keywords: Protected vegetables · The internet of things · Early warning · Control system · Mobile client terminal

1 Introduction

In protected vegetables production, when the light is few, light intensity weak, day and night temperature differ, relative humidity of air is high and air flow is not good(slow speed), or soil temperature and humidity is too high or too low, they have an effect on the normal growth and development of vegetable crops. At present, in the regulation and control of greenhouse, there are man-made management which means poor regulatory capacity, awful management, bad proportion of the input-output to cause final low economic rewards. If we want to get the normal condition for growth and development of protected vegetables, it is necessary to regulate environmental factor timely and scientifically which may possibly affect the growth of vegetables in the greenhouse to meet the need for the development of vegetable.

© IFIP International Federation for Information Processing 2016
Published by Springer International Publishing AG 2016. All Rights Reserved
D. Li and Z. Li (Eds.): CCTA 2015, Part II, IFIP AICT 479, pp. 155–162, 2016.
DOI: 10.1007/978-3-319-48354-2_16

Internet of thing has a perception as its premise to realize fully interconnected networks between man and man, man and material or substances [1–4]. Agricultural internet of thing makes use of the sensor to obtain environmental date through a variety of instruments to monitor real-time display or control the parameters which are involved in to ensure a great harvest in the aspects of the crops, livestock and poultry. At presence, foreign counterparts have widely adopted sensor of agricultural things of internet in the seeding, production, transportaion and storage, etc. Moreover, they have accomplished the successful monitoring in the aspects of packaging and irrigation and Chinese scholars have made initial achievement at home [5–10].

When we set up the system to monitor vegetable for early warning and control, most of people could not keep the computer next all the time which often caused the huge damage. However, mobile phones are often carried with you every second. The problem is solved by the establishment of a monitoring and control system based on the mobile client's facilities effectively.

2   System Design

2.1   System Framework

The system makes use of c/s model to achieve the service mode of multi-clients sharing data. Taking advantage of WCF, receiving and storing, sending data can be realized. The main framework of the system [11–16] is shown in Fig. 1.

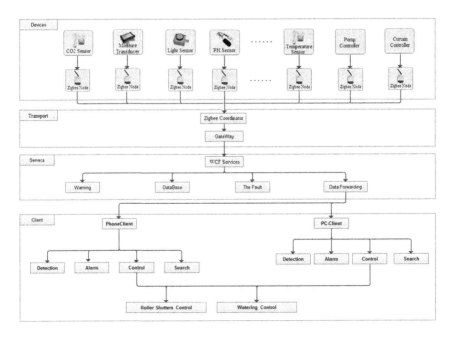

Fig. 1. The overall architecture of the system

2.2 Technology Adopted

2.2.1 Storage for Remote Data

In the system, the data transmission of the sensor is not regular. It may send a message to the server at any time. If the server acts as a connection to the remote and meanwhile needs to do a lot of I/O storage, the process will has a greater impact on database performance so as to influence the functional performance of the system. Therefore, this system tries to establish a specific database server which can alleviate slow performance effects caused by I/O storage.

2.2.2 Server Balanced Loading

Considering the future promotion of the system, if the data acquisition and equipment control were done in thousands of greenhouses, a steep pressure phenomenon may bring the slow response of the server, denial of server or system crashes and other issues. In order to prevent the occurrence of the problems, this system uses the NAT load balancing method, which means an external IP address is mapped into a plurality of internal IP addresses for each connection to convert every connection requirement to an address of an internal server dynamically, and meanwhile the external connection request is introduced to the server which is converted to the address to achieve the purpose of loading balancing.

2.2.3 WCF Server

Windows communication foundation (WCF) is developed by Microsoft company as application interfaces of a set of data communication. WCF communication mode mainly has three kinds of simple request reply: simple request-reply mode; no replay model and double-replies mode (that is data communication happening between clients and server sensors who can send the message to the clients at the same time.) this system presents the last model to realize the data communication between client terminals and servers, between transmission terminals and severs and between control equipment and servers.

2.2.4 Automatic Detection of Server

This gateway system with a remote hardware client and the sensor is used in TCP/IP protocol. This is a long chain with each communication at least maintained by a connection. They are not the system resources to be saved. When the client is using a proxy, even if the client communication with the server channel breaks, the servers can not capture the information so that the client has been out of connection with the server while the server still helps client to maintain this communication, which caused the huge waste of resources. Therefore, the system uses the heartbeat program that is server sends a data package to every communication client from time to time. If packets fail, it will continuously try to send three times. If they all failed then the system will recover resources greatly to improve utilization rate of resource.

2.3 System Control Process

The sensor transmits the data to the control center through the LTE gateway and after that, control center after processing sent the warning data to the mobile phone user. Mobile phone user get early warning data and send control information to the control center. The control center will convert control information into commands to send to the LTE node in order to achieve the equipment control.

2.4 Data Communication Protocol

Data communication is important for the communication between the two parties in the network. Actual line in the internet needs to realize the communication with a hardware device or the server and the clients terminals to ensure the smooth running of the whole system. There are two kinds of main data communication format in this paper, a kind of is the data collected by the sensor, the collected data is sent to the server; another is the control data, the server receives from mobile client's instructions, and instructions are sent to the need to control the device. The data communication protocol is shown in Table 1.

Table 1. Data communication

Type	Equipment	Data communication
Monitor	Collect	num-time-airt1-airt2-airh-soilt1-soilt2-soilh
Control	Water pump	num-priority-state
	Rolling machine	num-priority-state

3 System Realization

3.1 Data Collection

System can receive data from sensors and analyze received data by data protocol parsing as text displayed in cell phone client window. As shown in Fig. 2.

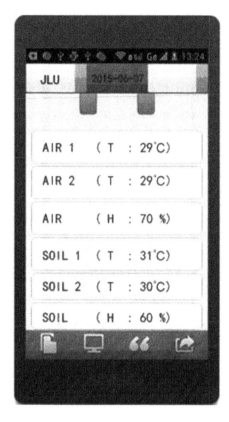

Fig. 2. Date collecting

3.2 Early Warning

When the various aspects of environment for growing crop changed over a given threshold value (the threshold's set up shown in Fig. 3) the system will automatically alarm and notify the user in the form of sound and vibration to show temperature and humidity greater than warning value data information and hence prompt the user to grasp the real-time to control the environment of the greenhouse condition and control the corresponding equipment to protect crops in better environment.

3.3 Device Control

When the environmental information exceeds the threshold, the system can manually realize the straight control to the equipment (pump roller shutter) to achieve the timely control of the environment. As shown in Fig. 4 of the device control interface:

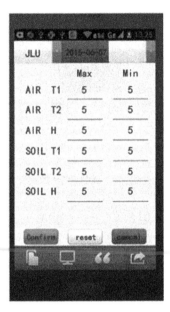

Fig. 3. Threshold settings

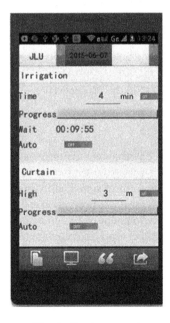

Fig. 4. Device control

4   Conclusions

The paper adopts Web Services technique to explain the service of data collection, storage, early warning, controlled and provide the interface for clients and date communication; it mainly makes use of the IOT technology through the wireless network to achieve the communication between hardware and hardware for smooth data communication; also, it applies language and database to achieve facilities and vegetables and remote control of the mobile phone client software. Thus this paper solves the time and space limit of data acquisition and control. After testing, if the sensor sends a data every ten minutes, environmental data can be received within zero point one seconds. When the phone after sending a control command, equipment can work within zero point one.

The system has been running for nearly a year in Jilin Province, and it has played an effective role in regulating the quality and yield of greenhouse crops.

Acknowledgment.   Funds for this research was provided by National 863 subjects (2012AA10A506-4, 2013AA103005-04), Jilin province science and technology development projects (20110217), China Postdoctoral Science Foundation the 54th surface funded (2013M541308), Jilin University Young Teachers Innovation Project (450060491471).

References

1. Chen, H.-M., Cui, L., Xie, K.-B.: Chin. J. Comput. Inst. Comput. Technol., Chin. Acad. Sci. Beijing **36**(1), 168–188 (2013)
2. Sun, Q.-B., Liu, J., Li, S., et al.: Internet of things: summarize on concepts, architecture and key technology problem. J. Beijing Univ. Posts Telecommun. **33**(3), 1–9 (2010)
3. Hu, Y.-L., Sunm Y.-F., Yin, B.-C.: Information sensing and interaction technology in internet of thing. Chin. J. Comput. **35**(6), 1147–1162 (2012)
4. Cheng, M., Wang, R.-H.: Advance in technical research and application of internet of things. Geomatics World, 22–27 (2010)
5. Li, Z., Wang, T., Gong, Z., et al.: Forewarning technology and application for monitoring low temperature disaster in solar greenhouses based on internet of things. Trans. Chin. Soc. Agric. Eng. (Trans. CSAE) **29**(4), 229–236 (2013)
6. Sheng, P., Guo, Y., Li, P.: Intelligent measurement and control system of facility agriculture based on zigbee and 3G. Trans. Chin. Soc. Agric. Mach. **43**(12), 229–233 (2012)
7. Yan, X., Wang, W., Liang, J.: Application mode construction of internet of things (IOT) for facility agriculture in Beijing. Trans. CSAE **28**(4), 149–154 (2012)
8. Wang, M.-H.: The Internet of things in agriculture application development demand for modern scientific instruments. Modern Sci. Instrum. **3**, 5–6 (2010)
9. Zhou, X.-B.: Facility agriculture online monitoring system based on internet of things. J. Taiyuan Univ. Sci. Technol. **32**(3), 182–184 (2011)
10. Ge, W., Zhao, C.: State-of-the-art and developing strategies of agricultural internet of things. Trans. Chin. Soc. Agric. Mach. **45**(7), 222–230 (2014)
11. Han, X., Wang, H.-B., Liu, K.-J.: Service-oriented SOA middleware design based on .NETFramework WCF component. J. Chin. Comput. Syst. **31**(12), 2359–2364 (2010)
12. Tan, Q.: Web application research based on WCF services framework and silverlight. Comput. Modernization **1**, 79–81 (2011)

13. Qian, T.: Design and implementation of computer lab management system based on WCF. Comput. Technol. Automat. **29**(4), 135–137 (2010)
14. Zhao, M., Zhang, X., Xiong, J.: Research of vivil aviation reservation system based on .NET and WCF. **33**(4), 1653–1659 (2012)
15. Yan, L., Wang, G., Wu, S.: Facilities vegetables monitoring and control system design based on WCF and Internet of things. Adv. Mat. Res., 2637–2640 (2014)
16. Yan, L., Hao, D., Zhang, H.: The facilities vegetables warning and control system based on mobile phone SMS. Appl. Mech. Mat., 576–579 (2014)

The Study of Winter Wheat Biomass Estimation Model
Based on Hyperspectral Remote Sensing

Xiaowei Teng[1,2,3,4,5], Yansheng Dong[1,2,3,4(✉)], and Lumin Meng[5]

[1] Beijing Research Center for Information Technology in Agriculture, Beijing 100097, China
tengxw1990@163.com
[2] National Engineering Research Center for Information Technology in Agriculture,
Beijing 100097, China
dongys@nercita.org.cn
[3] Key Laboratory of Agri-informatics, Ministry of Agriculture, Beijing 100097, China
[4] Beijing Engineering Research Center of Agricultural Internet of Things, Beijing 100097, China
[5] College of Geomatics, Xi'an University of Science and Technology, Xi'an 710054, China
1193824953@qq.com

Abstract. Biomass plays an important role in crop growth and yield formation. The study of biomass has been expanded to remote sensing sphere, which provides more ways to the obtainment of crop biomass. To carry out the study of winter wheat biomass estimation model, the field experiments were conducted at Rougu test area and Wugong test area, Shanxi Province in the cropping season 2013–2014. The biomass estimation model was based on the Time-Integrated Value of NDVI (TINDVI) and Leaf Water Content Index (LWCI), which was used to predict the winter wheat biomass. And the model was validated with the ground measured biomass. The results showed that the determination coefficient (R^2) and root mean square error (RMSE) between the measured and the estimated biomass were 0.7949 and 2.689 t/ha, respectively. The estimated biomass was exactly similar to the field measured biomass, therefore this model had a good application prospect.

Keywords: Winter wheat · Biomass · Hyperspectral remote sensing · TINDVI · LWCI

1 Introduction

The information of crop growing and health condition is essential to the optimization of crop production [1]. Biomass is an important indicator of crop condition monitoring [2]. The quantity of biomass affects the grain yield directly [3]. Many methods can be used to monitor crop biomass, including the estimation of remote sensing information, the prediction of crop model based on remote sensing and so on.

In recent years, a large number of scholars build numerous regression models of biomass estimation with the correlativity between remote sensing information and measured biomass, which have proved that the agronomy parameters just like Biomass

D. Li and Z. Li (Eds.): CCTA 2015, Part II, IFIP AICT 479, pp. 163–169, 2016.
DOI: 10.1007/978-3-319-48354-2_17

and Leaf N could be evaluated. He Cheng used three kinds of data, the Thematic Mapper data, the 30 meters Digital Elevation Model data and field observation data, to get the functional relation which was possibly applied between vegetation biomass and remote sensing image information [4]. Liu Ming used ten spectral vegetation indices to estimate the LAI and biomass, and the consequences indicated that the correlations between ten vegetation indices and LAI, aboveground biomass were significant, thus using vegetation indexes to inverse LAI and aboveground biomass was feasible [5]. Gao Shuai designed the microwave and optical remote sensing integrated vegetation indexes with the RADARSAT-2 and HJ-1 data, and the results showed that compared with original methods, these vegetation indexes had a better evaluating performance with the structure parameters just like maize LAI, height and biomass [6]. Calera et al. monitored the growth of corn and barley by applied remote sensing data, and the results showed that the value of TINDVI was linearly related to dry biomass [7]. With the purpose of establishing a spatial-temporal model for future TINDVI, Andreas Westergaard-Nielsen combined TINDVI and observed temperatures with a downscaled regional climate model (HIRHAM5) [8]. Hunt mentioned a leaf water content index (LWCI), which was calculated by Landsat near infrared band reflectance and shortwave infrared band reflectance, and proved that the leaf water content index had a good correlation with the relative water content [9]. Daeha Kim mentioned a biomass estimation model, which was based on the studies of Calera and Hunt [10].

Research of remote sensing technology for biomass estimation has been undertaken by many predecessors, and they have made corresponding progress, which has made a significant contribution. However, most of the estimation models are not universal, and are only suitable for local area, which goes against the popularization of these models. Daeha Kim mentioned a biomass model based on the value of NDVI and crop information data, but the model was not been validated by the field measured biomass. The paper adopts the biomass model of Daeha Kim, and to estimate the biomass collected at Rougu test area and Wugong test area with the hyperspectral remote sensing, hoping that the model can be applied in different areas.

2   Site Description

Field experiments were implemented in the Rougu test region and Wugong test region, Shanxi Province, China (Fig. 1). The test sites are located in central of Shanxi GuanZhong Plain, which is a typical arid and semi-arid region. The average temperature of summer is 26.1°C, and the average temperature of winter is −1.2°C. The average rainfall of the test sites is 635.1 mm. The main crops of the two areas are winter wheat and summer maize. Most of the winter wheat is sowed in the middle of October, and is harvested in early June. Five times field experiments were carried out and the measured agronomy parameters (Table 1) were obtained which included wheat canopy spectra, aboveground biomass and yield. The wheat canopy spectra was measured by the America ASD FieldSpec FR 2500 field spectral radiometer. The aboveground biomass and yield were measured by wheat samples obtained from each plot.

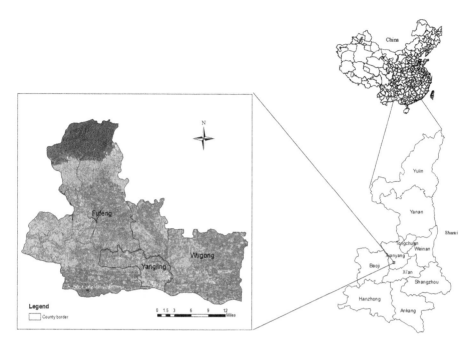

Fig. 1. The study area

Table 1. The testing time and growth period of winter wheat in the cropping season 2013–2014

Testing time	Mar 05–06	Mar 28–29	Apr 27	May 16–17	Jun 10
Growth period	Returning green stage	Jointing stage	Heading period	Pustulation period	Mature period

3   Methods

3.1   Model Description

Combining the previous researches [7, 9], Daeha Kim [10] invented the biomass model which was based on TM, ETM and OLI images, and involved the growth of TINDVI and the influence of water stress. The core principle of this model is that it translates the value of TINDVI into biomass under comprehensively considering the water stress coefficient. The equation for calculating biomass production is as follows:

$$B_{ASD} = m \times W \times TINDVI \tag{1}$$

Where m is conversion coefficient from the value of TINDVI to biomass. W is the water pressure coefficient or water pressure factor. B_{ASD} is biomass estimated from hyperspectral remote sensing. NDVI is the Normal Differential Vegetation Index, which is computed from reflectances of red band and near infrared band [11]. TINDVI is the Time Integrated Value of NDVI. And it is calculated as:

$$TINDVI = \int_{t_0}^{t_1} \left(NDVI - NDVI_{soil} \right) dt \tag{2}$$

$$NDVI_{ASD} = \frac{NIR - RED}{NIR + RED} = \frac{R_{890} - R_{670}}{R_{890} + R_{670}} \tag{3}$$

Where t_0 is the start time of returning green stage, and t_1 is the time when biomass is estimated. $NDVI_{soil}$ is NDVI of bare soil. RED and NIR are red band reflectances and near infrared band reflectances, respectively. R_{890} and R_{670} are the reflectances of 890th band and 670th band, respectively.

For dry leaves, the reflectance of shortwave infrared band is almost equal to the reflectance of near infrared band [12–14]. For green leaves, near infrared band has the maximum reflectance of the six Thematic Mapper band, whereas the shortwave infrared band reflectance is reduced because of absorption by water [14, 15]. So that the difference between the shortwave infrared band reflectance and the near infrared band reflectance for the green leaves should be equivalent to the water absorptance in the green leaves [9]. Absorbance is usually calculated by $-\log(1 - a)$, where a is the absorptance [16]. So that the biomass model regards the LWCI as the water pressure coefficient, and the water pressure coefficient [9] is defined by NIR and SWIR as:

$$W = LWCI = \frac{-\log\left[1 - (NIR - SWIR)\right]}{-\log\left[1 - (NIR - SWIR)_{FT}\right]} \tag{4}$$

Where SWIR is the reflectance of shortwave infrared band. This study uses the 890th and 1610th band reflectances of hyperspectral remote sensing instead of NIR and SWIR. The subscript means that the reflectances of these leaves are in the state without water stress.

The biomass conversion coefficient is associated with the region and the cultivar of winter wheat, which is determined by the value of TINDVI and the conservative biomass at mature. The equation of biomass conversion coefficient m is as follows:

$$\hat{m} = \frac{B_h}{E[W \times TINDVI]} \tag{5}$$

Where B_h is the conservative biomass at mature, and $E[W \times TINDVI]$ is the average of $W \times TINDVI$ in the region.

3.2   Model Verification

The R^2 and RMSE were applied to evaluate the biomass estimation results. The more the R^2 close to 1, the more the consistency well. And the more the RMSE close to 0, the more the error small. RMSE is calculated based on Eq. (6).

$$RMSE = \sqrt{\frac{\sum_{i=1}^{n}(E_i - M_I)^2}{n}} \qquad (6)$$

4   Results and Discussions

The estimated winter wheat biomass in the cropping season 2013–2014 was validated with the field measured biomass. In the analysis, the R^2 between estimated biomass and measured biomass in the period of Mar 05–06, Mar 28–29, Apr 27 and May 16–17 were 0.2836, 0.2042, 0.3412 and 0.1976, respectively (Fig. 2). The estimated biomass precision was slightly lower in single period. The maximum and minimum correlation coefficients were 0.3412 and 0.1976, respectively. The biomass estimation precisions were in the range of acceptable precision.

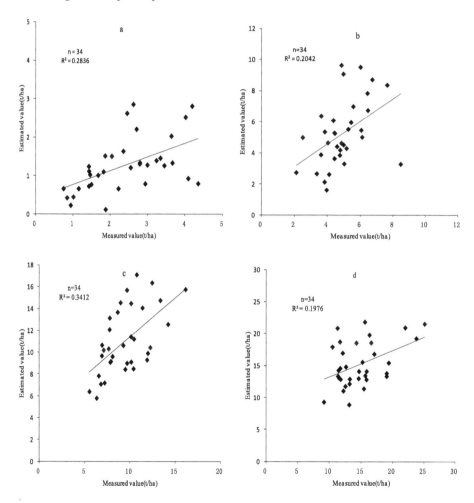

Fig. 2. The relationship between estimated biomass and measured biomass in the period of Mar 05–06 (a), Mar 28–29 (b), Apr 27 (c) and May 16–17 (d)

The R^2 and RMSE between the estimated and measured biomass in the whole period of winter wheat were 0.7949 and 2.689 t/ha, respectively (Fig. 3). The whole period of winter wheat estimated biomass had a higher correlation with the measured biomass.

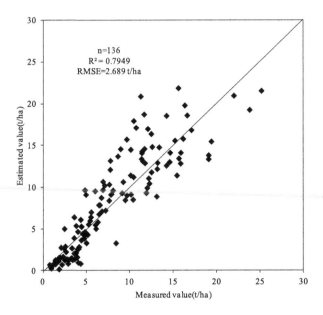

Fig. 3. The relational graph between estimated biomass and measured biomass in the whole period of winter wheat

Thus, the correlations between estimated biomass and measured biomass suggested that this biomass estimation model was viable. In this study, the TINDVI and LWCI were applied to estimate the biomass of winter wheat, which has more advantages than the biomass regression models. Because it can be applied to different areas. Moreover, it combined the former studies by considering both the TINDVI growth and the effect of water stress on biomass growth [10]. But this model had little consideration on rainfall data and irrigation data. Only four biomass tests were carried out in the cropping season 2013–2014, which lead to a deviation from the NDVI real situation curve. So that the each times single predicted precisions were slightly lower. This model is a new biomass model, thus the further research is required to verify and perfect the model.

5   Conclusion

This study carried out winter wheat biomass estimation based on a new biomass model through hyperspectral remote sensing, and it was verified by the field experimental data from Rougu test area and Wugong test area. The estimated wheat biomass had a higher correlation with the field measured biomass. Certainly, more field experiments should be carried out. In order to furtherly verify and perfect the biomass model, the meteorological data and field management data should also be considered. Finally, the model should be applied to remote sensing images. This paper may be valuable in guiding further study about crop biomass estimation.

Acknowledgments. The work is supported by the National Science Foundation (NNSF) of China (41401476).

References

1. Kross, A., Mcnairn, H., Lapen, D., Sunohara, M., Champagne, C.: Assessment of RapidEye vegetation indices for estimation of leaf area index and biomass in corn and soybean crops. Int. J. Appl. Earth Obs. Geoinf. **34**, 235–248 (2015)
2. Du, X., Meng, J.H., Wu, B.F.: Overview on monitoring crop biomass with remote sensing. Spectro. Spectral Anal. **11**, 3098–3102 (2010)
3. Ma, Q.R., Li, S.Z., Zhao, H.Q., Yang, G.X., Wu, D.Y., Dong, W.H.: A study on accumulation and increment distribution of biomass of summer maize in Zhengzhou. Chin. J. Agrometeorology **28**(4), 430–432 (2007)
4. He, C., Feng, Z.K., Han, X., Sun, M.Y., Gong, Y.X., Gao, Y., Dong, B.: The inversion processing of vegetation biomass along Yongding river based on multispectral information. Spectrosc. Spectral Anal. **32**(12), 3353–3357 (2012)
5. Liu, M., Feng, R., Ji, R.P., Wu, J.W., Wang, H.B., Yu, W.Y.: Estimation of leaf area index and aboveground biomass of spring maize by MODIS-NDVI. Chin. Agric. Sci. Bull. **31**(6), 80–87 (2015)
6. Gao, S., Niu, Z., Huang, N., Hou, X.: Estimating the leaf area index, height and biomass of maize using HJ-1 and RADARSAT-2. Int. J. Appl. Earth Obs. Geoinf. **24**, 1–8 (2013)
7. Calera, A., González-Piqueras, J., Melia, J.: Monitoring barley and corn growth from remote sensing data at field scale. Int. J. Remote Sens. **25**(1), 97–109 (2004)
8. Westergaard-Nielsen, A., Bjørnsson, A.B., Jepsen, M.R., Stendel, M., Hansen, B.U., Elberling, B.: Greenlandic sheep farming controlled by vegetation response today and at the end of the 21st Century. Sci. Total Environ. **512**, 672–681 (2015)
9. Hunt, E.R., Rock, B.N., Nobel, P.S.: Measurement of leaf relative water content by infrared reflectance. Remote Sens. Environ. **22**(3), 429–435 (1987)
10. Kim, D., Kaluarachchi, J.: Validating FAO AquaCrop using landsat images and regional crop information. Agric. Water Manage. **149**, 143–155 (2015)
11. Rouse, J.W., Haas, R.H., Schell, J.A., Deering, D.W., Harlan, J.C.: Monitoring the vernal advancement and retrogradation (greenwave effect) of natural vegetation. Texas A & M University, Remote Sensing Center (1974)
12. Thomas, J.R., Namken, L.N., Oerther, G.F., Brown, R.G.: Estimating leaf water content by reflectance measurements. Agron. J. **63**(6), 845–847 (1971)
13. Rock, B.N., Williams, D.L., Vogelmann, J.E.: In Machine Processing of Remotely Sensed Data Symposium, pp. 71–81. Purdue University, West Lafayette (1985)
14. Knipling, E.B.: Physical and physiological basis for the reflectance of visible and near-infrared radiation from vegetation. Remote Sens. Environ. **1**(3), 155–159 (1970)
15. Tucker, C.J.: Remote Sensing of Leaf Water Content in the Near Infrared. US National Aeronautics and Space Administration, Goddard Space Flight Center, Greenbelt, Md. NASA-TM-80291 (1979)
16. Nobel, P.S., Jordan, P.W.: Transpiration stream of desert species: resistances and capacitances for a C3, a C4, and a CAM plant. J. Exp. Bot. **34**(10), 1379–1391 (1983)

Design and Implementation of TD-LTE-Based Real-Time Monitoring System for Greenhouse Environment Temperature

Xin Zhao[1,2], Yang Jiao[1,2], Lianjun Yu[3], and Chuanhong Zhang[4(✉)]

[1] College of Biological and Agricultural Engineering,
Jilin University, Changchun 130022, China
jlndzx@sina.com, 61516131@qq.com
[2] Sixth Middle School in Changchun, Changchun 130000, China
[3] Changchun City Academy of Agricultural Sciences,
Changchun 130111, China
120142901@qq.com
[4] Changchun University of Science and Technology, Changchun 130600, China
15844005115@163.com

Abstract. With the characteristics of fast transferring speed and wide coverage of TD-LTE network, the research designs and implements the TD-LTE-based Real-time Monitoring System for Greenhouse Environment Temperature. The TD-LTE network, which is connected with the data server of the Internet, can transmit the real-time temperature data that sensor collects to the data server. The B/S mode-based data server can realize data query, analysis and other functions. This monitoring system solves the problems that previous monitoring systems are slow and can not be transmitted at a distance, and it can also meet the need of the real-time data acquisition that precision agriculture needs.

Keywords: TD-LTE · Precision agriculture · Data collection · Long-distance monitoring

1 Introduction

Compared with the commonly used wired network, wireless network is a new type of network formation. It is no longer dependent on the network cables which the traditional wired network depends on, eliminating the need for network cabling, effectively saving the cost, being used more and more conveniently. In the past, most wireless networks construct network with the help of WIFI, ZigBee, Bluetooth and so on. This method can implement the wireless transmission, but the transmission distance is near. If you want to transfer to a far distance, you need get access to the Internet. TD-LTE fully Time Division Long Term Evolution, also simply LTE-TDD, is one of the duplex technologies used in mobile communication technology. It is also one of the international standards led by China. TD-LTE can directly get access to the Internet, which makes up for the shortcomings of the traditional wireless network and greatly improves the efficiency of the wireless network [1–4].

© IFIP International Federation for Information Processing 2016
Published by Springer International Publishing AG 2016. All Rights Reserved
D. Li and Z. Li (Eds.): CCTA 2015, Part II, IFIP AICT 479, pp. 170–177, 2016.
DOI: 10.1007/978-3-319-48354-2_18

The greenhouse, also called greenhouse, can keep heat. In the season that is not suitable for plant growth, it can ensure the normal growth of plants, and ensure the output. The greenhouse is an important part of modern agriculture [5]. In the modern precision agriculture, the control of the growth environment of crops in the greenhouse has been changed from the traditional personal management mode to the modern, centralized management mode. Highly centralized management can effectively reduce the management cost, can make timely and accurate judgment in the shortest time, ensure the normal operation of the greenhouse, and ensure the normal production of greenhouse crops [6–9]. Greenhouse temperature is the basis of greenhouse planting and the coverage of centralized management is large, and the wireless network, which the traditional greenhouse temperature collection system used, is ZigBee mode, while ZigBee itself does not have the ability to get access to Internet, so the traditional greenhouse temperature monitoring system can only work in a small range, which is a problem for centralized management [10–12]. Based on TD-LTE network and embedded system, the greenhouse environment temperature monitoring system, which is comprehensive covering network, provides the wireless transmission mode that is not limited by the distance, and the efficient and fast data transmission service.

2 Overall Design System

The system consists of a wireless remote temperature acquisition system and an online data server based on WEB. The core of the greenhouse environment temperature acquisition system is embedded system, which uses the temperature sensor to collect the real-time data of the greenhouse environment temperature, and transmits the collected

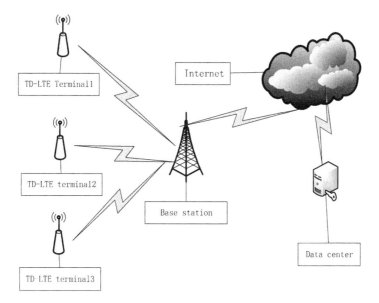

Fig. 1. Conventional and spectra after pretreatment

data to the data server on Internet through TD-LTE network. Database server is based on WEB technology, which achieve real-time monitoring, analysis and management of long-distance data. The overall structure of the system is shown in Fig. 1.

3   Temperature Acquisition Module Design

3.1   TD-LTE

LTE, which is based on the OFDMA technology, is the global standard by the 3GPP organization. TD-LTE, also LTE-TDD, fully Time Division Long Term Evolution, is one of the duplex technologies used in mobile communication technology. It is developed by the world's major companies and operators covered by the 3GPP organization. It is also one of the international standards led by China. It is a follow-up evolution technology scheme of TD-SCDMA, and its network transmission speed is dozens of times as fast as TD-SCDMA. TD-LTE can be directly connected to the Internet, and functions with the Internet serve, which greatly improving the efficiency of wireless network.

HUAWEI B593S is the production of TD-LTE produced by HUAWEI. The device supports TD-LTE through the built-in DHCP function, and also down is compatible with TD-SCDMA and EDGE networks, which can achieve a comprehensive coverage of the network. In the network access mode, the device supports 802.11 and RJ45, and 802.11 supports the connection of multi-users, so the system mainly used RJ45 mode.

3.2   Embedded

The hardware of this system includes the embedded microprocessor and peripheral equipment. The system mainly consists of CPU chip, FLASH chip, SDRAM chip, DM9000 network card, sensor and other hardware components. And the CPU, which is the core of the whole hardware system, uses the ARM series embedded processor $3C2440A by Sanxing. SDRAM and FLASH chips are the system storage module. FLASH stores boot code, kernel, and file of system (including the applications), and SDRAM provides the running space for the system and applications and stores temporary data when system working.

3.3   Operating System

Linux is an operating system that can run on a variety of different hardware platforms. The software platform of the smallest embedded Linux system consists of three parts—the boot process, the kernel image and the file system. When the system is powered, firstly the guidance program is implemented, then the kernel is called and booted by it in order to realize the detection and drive of hardware system, finally mount the file system and run the applications.

Linux kernel is well-structured and integrated with the most parts of the mainstream of the hardware driver. With the help of the Linux kernel, the whole system is more stable. And Linux has the perfect network function, which can make our system get good network compatibility. The Linux core of the current standard is too large to the resource-constrained embedded system. So due to the features that embedded software and hardware can be cut, we need to abandon our unused modules in the kernel, leaving only the functions needed by the system.

3.4 Data Communication

In order to guarantee the fast, safe and efficient data communications between the acquisition system and the server, it is necessary to carry on a series of conventions for the communication heat preservation format between each other. The system uses the data preservation format as "verification code ‖ node number ‖ date ‖ temperature 1 ‖ temperature 2 ‖ temperature 3 ##", and the "‖" is the data delimited identifier, and "##" is the data end identifier. In the acquisition frequency of temperature data, the system can be accurate to 1 s, but, in fact, such a high accuracy is not necessary, so the system's acquisition time is set to 1 min. Internet server intercepts properly the relevant data that the acquisition system sends and stores them in the database.

In order to guarantee the continuity and effectiveness of the data and prevent network outages and other emergencies, the system carries on the real-time transmission of data to the server, then backups them and stores the data to the embedded system. And at 0 o'clock everyday, the aggregated data will be sent to the server and be verified with server-side data so as to ensure the continuity and effectiveness of the data.

4 Server

The operating system of the server uses Linux operating system because Linux is the open source operating system. It has obvious advantages in terms of security. At the aspect of software cost, the windows system is charged and running software is also charged, so the cost is higher, while the Linux operating system is lower than the windows operating system and most running software are free software. At the aspect of centralized management, windows needs a large number of servers, while using Linux can effectively reduce the cost. Because Linux is selected as the operating system, Mysql is chosen to be the database. Mysql is also an open source database and it has the advantages, such as small size, fast speed and it can effectively guarantee the security of data storage.

4.1 Data Reception

Data transmission is through the Internet network, using the "standard" data communication format, so the server side intercepts the transmission data in terms of the calibrated data format. This program uses JAVA to write and the database uses Mysql to carry on the data storage. When the program is started, the process will automatically monitor the port to transmit data. Upon receipt of the data, the first data packet header data is verified to determine whether the data is sent to the client data, if verified on, then split the received data according to the agreed data format and stored them in the corresponding database.

4.2 Data Monitoring

The data monitoring platform features the real-time display for the greenhouse temperature data and analyzes the historical data. The monitoring platform is developed with the help of B/S structure model. The advantage of this model is that the application core of the system is focused on the server, which simplifies the development, maintenance and usage of the system. In the client side, there does not need carry out a separate application development, the client can interact with the server information through the browser. The client is also not limited to the windows operating system, but both Linux and MAC can be used.

The server uses Linux as the operating system. Database uses MySQL as the database, uses JAVA to carry out the relevant WEB page development and uses apache as the middleware. The network construction can effectively ensure the stable running of the system and the interaction between the user and the database. Through the browser to get access to the WEB page, the client can monitor the real-time temperature of the remote greenhouse and can view, analyze and export the historical data.

5 System Implementation

5.1 Data Acquisition

When the embedded system is running, it will automatically take 1 min to collect the temperature data in the greenhouse, as shown in Fig. 2.

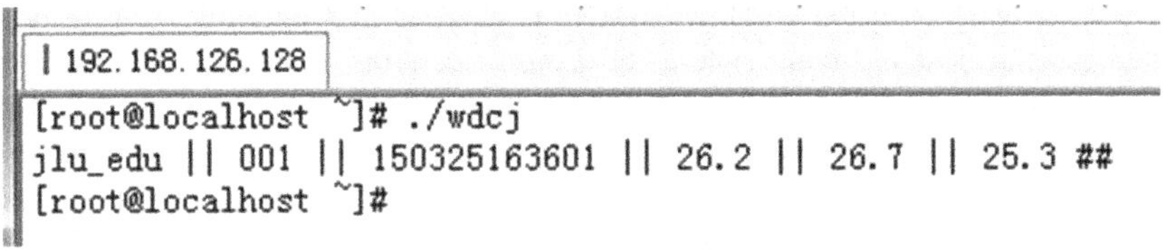

```
| 192.168.126.128 |
[root@localhost ~]# ./wdcj
jlu_edu || 001 || 150325163601 || 26.2 || 26.7 || 25.3 ##
[root@localhost ~]#
```

Fig. 2. Collect data

In addition to real-time transmission of data, the system also files historically data. At 0 o'clock daily, a full day's data will be uploaded to the server so as to ensure the continuity and effectiveness of data, as shown in Fig. 3.

```
| 192.168.126.128

[root@localhost ~]# cat h_time.log
jlu_edu || 001 || 150325163401 || 26.1 || 26.8 || 25.3 ##
jlu_edu || 001 || 150325163501 || 26.1 || 26.8 || 25.2 ##
jlu_edu || 001 || 150325163601 || 26.2 || 26.7 || 25.3 ##
jlu_edu || 001 || 150325163701 || 26.2 || 26.7 || 25.2 ##
jlu_edu || 001 || 150325163801 || 26.1 || 26.9 || 25.2 ##
jlu_edu || 001 || 150325163901 || 26.2 || 26.9 || 25.1 ##
jlu_edu || 001 || 150325164001 || 26.1 || 26.8 || 25.1 ##
jlu_edu || 001 || 150325164101 || 26.1 || 26.8 || 25.1 ##
jlu_edu || 001 || 150325164201 || 26.2 || 26.7 || 25.1 ##
jlu_edu || 001 || 150325164301 || 26.2 || 26.7 || 25.2 ##
jlu_edu || 001 || 150325164401 || 26.3 || 26.7 || 25.2 ##
jlu_edu || 001 || 150325164501 || 26.3 || 26.7 || 25.2 ##
jlu_edu || 001 || 150325164601 || 26.3 || 26.7 || 25.2 ##
jlu_edu || 001 || 150325164701 || 26.2 || 26.8 || 25.2 ##
jlu_edu || 001 || 150325164801 || 26.1 || 26.8 || 25.3 ##
jlu_edu || 001 || 150325164901 || 26.1 || 26.9 || 25.2 ##
jlu_edu || 001 || 150325165001 || 26.1 || 26.8 || 25.3 ##
jlu_edu || 001 || 150325165101 || 26.1 || 26.9 || 25.4 ##
[root@localhost ~]# █
```

Fig. 3. Collect data

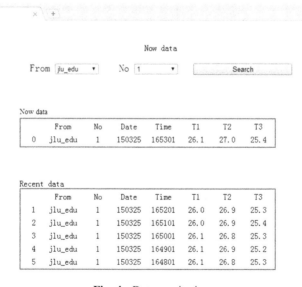

Fig. 4. Data monitoring

5.2   Data Monitoring

In the data monitoring, the system not only displays the current temperature, but also display the recent 5 temperature state below so as to facilitate the observation of data trends, as shown in Fig. 4.

5.3   Historical Data

The system has a historical data filtering function and meanwhile supporting to export to excel, as shown in Fig. 5.

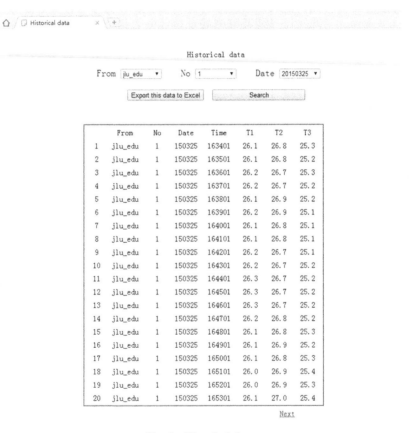

Fig. 5. Historical data

6   Conclusions

The system uses the embedded system and TD-LTE technology to implement the automatic collection of greenhouse environment temperature, and the collected data is transmitted to the data server side on Internet. Administrators can get access to the data server through the client's browser, and access contents information includes the current state of the greenhouse as well as browsing and analyzing the historical data. After testing, the system works stably and the transmission distance is more far, which provides a stable foundation for the modern agriculture centralized management. At the same time, the usage of the Linux operating platform has significantly reduced the cost of production, and produced certain economic benefits compared to the traditional windows platform.

Acknowledgment. Funds for this research was provided by National 863 subjects (2012AA10A506-4, 2013AA103005-04), Jilin province science and technology development projects (20110217), China Postdoctoral Science Foundation the 54th surface funded (2013M541308), Jilin University Young Teachers Innovation Project (450060491471).

References

1. Guan, W.: Analysis on key technologies and development trend of LTE. Guangxi Commun. Technol. **1**, 5–8 (2009)
2. Liao, J.: Discussion on TD-LTE: the only TDD international standard of LTE. Value Eng. **16**, 156 (2010)
3. Zhang, T.: An overview of LTE and its development. Telecom Eng. Tech. Stand. **11**, 1–6 (2010)
4. Hu, H., Dong, S., Jiang, Y., et al.: Analysis of fourth generation mobile telecommunications technology **32**(5), 1563–1567 (2011)
5. Ran, W.: The Data Acquisition System for the Greenhouse. Lanzhou University (2010)
6. Li, Z., Wang, G., Qi, F.: Current situation and thinking of development of protected agriculture in China. Chin. Agric. Mech. **1**, 7–10 (2012)
7. Yang, Q., Wei, L., Liu, W., et al.: Current situation and development strategy of agricultural research in China. Chin. Agric. Inf. **11**, 22–27 (2012)
8. Hu, J.: Analysis on the current situation and development of modern facilities agriculture. J. Agric. Mech. Res. **7**, 245–248 (2012)
9. Ma, Y.T., Mathieu, A., Wubs, A.M.: Parameter estimation and growth variation analysis in six capsicum cultivars with the functional-structural model GreenLab. In: Third International Symposium on Plant Growth Modeling, Simulation, Visualization and Applications, pp. 183–190 (2009)
10. Li, S.: The research on the intelligent greenhouse monitoring system based on wireless sensor network. J. Jiangxi Univ. Sci. Technol. **34**(1), 70–73 (2013)
11. Li, H., Lai, Z., Zhang, X.: Intelligent greenhouse irrigation subsystem design and implementation based on ZigBee Technology. J. Agric. Mech. Res. **1**, 95–98+107 (2014)
12. Li, L., Lin, T., Wu, Z.: Development of a wireless safety monitoring system for laboratories based on Zigbee. Exp. Technol. Manag. **1**, 108–112 (2011)

Research and Design of LVS Cluster Technology in Agricultural Environment Information Acquisition System

Guogang Zhao[1,2], Haiye Yu[1,2(✉)], Lianjun Yu[4], Guowei Wang[1,2,3], Yuanyuan Sui[1,2], and Lei Zhang[1,2]

[1] College of Biological and Agricultural Engineering, Jilin University, Changchun 130022, China
zhaoguogang2000@qq.com, {haiye,z_lei}@jlu.edu.cn,
suiyuan0115@126.com
[2] Key Laboratory of Bionic Engineering, Ministry of Education, Changchun 130022, China
[3] School of Information Technology, Jilin Agricultural University, Changchun 130118, China
41422306@qq.com
[4] Changchun City Academy of Agricultural Sciences, Changchun 130111, China
120142901@qq.com

Abstract. With the development of agricultural informatization, agricultural environment information acquisition platform needs to collect more and more data. And with the increase of the number of data acquisition terminal, a large number of concurrent data traffic is generated on the server. This will lead to the phenomenon of excessive load which directly affects the validity of the data. Therefore, based on LVS technology, this paper designed a solution for the agricultural environment information acquisition platform. Through the experiment we know that the LVS technology can guarantee the validity of the data acquisition, which lays the foundation for the development of precision agriculture.

Keywords: Agricultural information · IOT · Data acquisition · LVS · Precision agriculture

1 Introduction

The IOT is through the sensor technology, the acquisition of a variety of sensor information, through the network to connect with each other, information exchange and communication [1, 2]. The application of the same thing in precision agriculture, making precision agriculture has been rapid development [3–5]. By the traditional manual collection of environmental information, and gradually turned to the IOT technology to gather information, Improve the degree of information. Accurate and effective crop environment information data, in the greenhouse plant cultivation, can effectively ensure that the crop is in the best growth state, to ensure the yield and quality of agricultural products [6–10].

© IFIP International Federation for Information Processing 2016
Published by Springer International Publishing AG 2016. All Rights Reserved
D. Li and Z. Li (Eds.): CCTA 2015, Part II, IFIP AICT 479, pp. 178–184, 2016.
DOI: 10.1007/978-3-319-48354-2_19

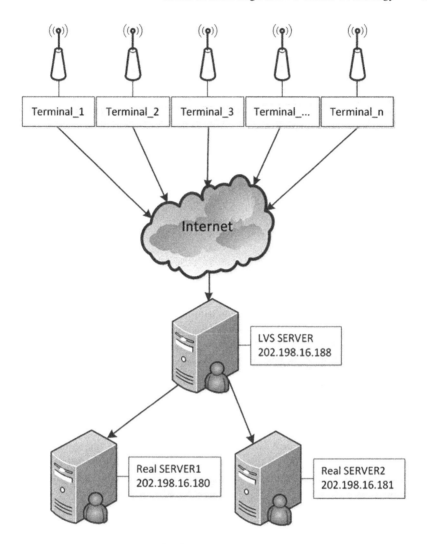

Fig. 1. System structure

With the popularity of the Internet, the IOT has been from the traditional mode of small scale, slowly into the Internet era. Agricultural environmental information collection mode based on the IOT is also a small range of environmental information collection, development into a large range, multi regional environmental information collection [11–13]. Agricultural environmental information acquisition system needs to deal with the network connection number of times also into explosive growth, if the network connection can not be processed in time, will lead to a decline in the quality of the system, especially in the greenhouse of agricultural facilities, if it can't effectively ensure the normal and stable operation of greenhouse, the loss will be unable to estimate. Based on LVS technology, the load balance of the agricultural environment information acquisition system is realized, and the service quality of the system is ensured, which lays the foundation for the development of precision agriculture.

2   System Structure

The system structure as shown in Fig. 1, based on LVS technology to construct the agricultural environment information acquisition system can according to each cluster server load, will come from the network terminal request reasonable allocation to each server implementation, can effectively solve the problem of high load of multi terminal and Fayin. To ensure the validity of the data terminal.

3   Key Technologies

3.1   Load Balancing

Load balancing (Balance Load): load balancing is a dynamic balancing technology in the network. Through the analysis of the data packets in the network, the server can handle the task in a reasonable dynamic distribution. Can ensure the network structure does not change, directly increase the network bandwidth and server, effectively improve the ability of processing data in the network, make the network flexibility and availability significantly improve the stability of the service operation [14, 15].

3.2   Cluster Profile

Cluster is usually the abbreviation of a system. This system is composed of many computers, as a whole, and provides services to the outside. At the same time, it can increase, modify or delete nodes in the cluster. Compared with the traditional single server, the cluster has the advantages of high stability, fast calculation speed and easy management [16, 17].

3.3   LVS

LVS full name is Virtual Server Linux, which is sponsored by China's Linux programmer, Dr. Zhang Wensong (who is responsible for the development of the core software) and the free software project, which is based on Linux system. LVS is a cluster based on load balancing technology. In the LVS, there must be a master server, and more than 2 true Real-server. Master server according to the scheduling algorithm to control the server Real-server, Real- is responsible for providing services. The client only needs to communicate with a IP address provided by the cluster. The cluster is just a single server, and the structure of the cluster is not an invisible to the client [18].

Server Real- can provide a variety of common services such as FTP, DNS, Telnet, SMTP, etc. Master control server is responsible for the control of server Real-. When the client sends a service request to the LVS, the server is specified by a server Real- to receive a request from the client and respond to its request, for the client to communicate with a fixed IP.

3.4 Keepalived

Keepalived is a software that works like a switch. It can detect the status of the server. When the server appears to restart, the keepalived will remove the server from the existing system. When the server is back to normal, it will resume to the system.

4 LVS Implementation

4.1 LVS Working Mode

LVS has three working modes, VS/NAT, VS/TUN, VS/DR. This paper chooses VS/TUN mode. The mode of the client to send the data packets, a two package, add a new destination IP address, and send it to the destination IP address corresponding to the real-server, it will be the results of the packet processing, directly returned to the client. The mode can effectively reduce the load of the pressure [19].

4.2 Firewall

In the network firewall usually refers to separate the internal network and external network and isolation, when two network for communication, according to certain rules, the communication of data conditional options, allowing data can freely in and out of the network, are not allowed to data, declined to its into the network, the maximum guarantee the safety of the network server. Linux operating system, also with a firewall, according to the different IP address and port to a certain degree of restriction. Because the system servers are used in the internal network, so close the Linux firewall can directly. The command:

service iptables stop

4.3 Yum

Yum (Yellow dog Updater, Modified), in Linux can from the specified server to download and install the RPM package, compared with traditional way of RPM installation, can automatic processing the RPM package dependencies, installed a can all depend on the package. The command:

yum -y install ipvsadm*

4.4   Configure Load Balancing Controller, the Command

ifconfig eth0:1 202.198.16.188 netmast 255.255.255.255 up
ipvsadm -A -t 202.198.16.188:80 -s wrr -p 20
ipvsadm -a -t 202.198.16.188:80 -r 202.198.16.180:8080 -i -w 1
ipvsadm -a -t 202.198.16.188:80 -r 202.198.16.181:8080 -i -w 1

4.5   Real-Server, the Command

modinfo tun   # test tun in Linux
modprobe tun   #load tun
ifconfig tunl0 202.198.16.188 netmask 255.255.255.255
broadcast 202.198.16.188   #add tunl0
ifconfig tunl0 up   #start tunl0
route add -host 202.198.16.188 dev tunl0
echo "2" >/proc/sys/net/ipv4/conf/tunl0/arp_announce
echo "1" >/proc/sys/net/ipv4/conf/tunl0/arp_ignore
echo "1" >/proc/sys/net/ipv4/conf/all/arp_ignore
echo "2" >/proc/sys/net/ipv4/conf/all/arp_announce

5   Test

The design goal of this paper is to realize the high load balance of agricultural environment information acquisition system. In the experiment, the client terminal is used to carry out concurrent access, and the data is sent to the server 500000, and the test data are as Table 1.

Table 1. Test data results

	Common mode			LVS mode		
	Valid data	Missing data	Loss rate‰	Valid data	Missing data	Loss rate‰
1	495234	4766	9.53	498548	1452	2.90
2	494656	5344	10.68	498731	1269	2.53
3	494531	5469	10.93	497913	2087	4.17
4	492967	7033	14.06	498355	1645	3.29
5	494789	5211	10.42	498367	1633	3.26
6	494159	5841	11.68	498212	1788	3.57
Average	494389	5610	11.22	498354	1645	3.29

The test results show that the agricultural environment information acquisition system based on LVS technology can effectively improve the service capability, and

Improve the service quality, which lays the foundation for the development of precision agriculture.

6 Conclusions

With the development of precision agriculture, the wide application of the IOT technology, the agricultural environment information acquisition system needs to provide uninterrupted service for 7×24 h. In this paper, the load balance of the agricultural environment information acquisition system is realized by using LVS technology, and the stable operation of the data acquisition system is effectively guaranteed.

Acknowledgment. Funds for this research was provided by National 863 subjects (2012AA10A506-4, 2013AA103005-04), Jilin province science and technology development projects(20110217), China Postdoctoral Science Foundation the 54th surface funded(2013M541308), Jilin University Young Teachers Innovation Project (450060491471).

References

1. Hu, Y., Sun, Y., Yin, B.: Information sensing and interaction technology in Internet of things. Chin. J. Comput. **36**(6), 1147–1163 (2012)
2. Liu, Q., Cui, L., Chen, H.: Key technologies and applications of Internet of things. Comput. Sci. **37**(6), 1–4, 10 (2010)
3. Yu, L., Lu, Y., Zhu, X.: Research advances on technology of Internet of things in medical domain. Appl. Res. Comput. **29**(1), 1–7 (2012)
4. Ma, X., Huang, Q., Shu, X., et al.: Study on the applications of internet of things in the field of public safety. China Saf. Sci. J. **20**(7), 170–176 (2010)
5. Li, F., Li, S.: Study on the development of smart library based on Internet of things technology. Libr. Inf. Serv. **3**(5), 66–70 (2013)
6. Li, D.: Internet of things and wisdom agriculture. Agric. Eng. **2**(1), 1–7 (2012)
7. Zhu, H., Wang, F., Suo, R.: The application of the internet of things in China modern agriculture. Chin. Agric. Sci. Bull. **27**(02), 310–314 (2011)
8. He, Y., Nie, P., Liu, F.: Advancement and trend of Internet of things in agriculture and sensing instrument. Trans. Chin. Soc. Agric. Mach. **44**(10), 216–226 (2013)
9. Yan, X., Wang, W., Liang, J.: Application mode construction of internet of things (IOT) for facility agriculture in Beijing. Trans. CSAE **28**(4), 149–154 (2012). (in Chinese with English abstract)
10. Liu, D., Zhou, J., Mo, L.: Applications of Internet of things in food and agri-food areas. Trans. Chin. Soc. Agric. Mach. **43**(1), 146–152 (2012)
11. Liu, G., Li, Y.: Research on differential GPS system of agricaltural machinery based on 3G. J. Agric. Mech. Res. **11**, 202–205 (2012)
12. Ping, S., Yangyang, G., Pingping, L.: Intelligent measurement and control system of facility agriculture based on Zigbee and 3G. Trans. Chin. Soc. Agric. Mach. **43**(12), 229–233 (2012)
13. Yang, C., Niu, L.: Design and implementation of Zigbee based environmental monitoring system for facility agriculture. J. South Central Univ. Nationalities (Nat. Sci. Edition) **31**(1), 88–92 (2012)

14. Zhou, Y., Liu, F.: Research on load balancing of web-server system. Comput. Digital Eng. **38**(4), 11–14, 35 (2010)
15. Chu, B., Liu, X., Liu, T.: Load balancing of system level application systems. J. Inf. Eng. Univ. **3**(4), 48–50 (2002)
16. Wang, Z., Jiang, X., Zhang, C.: Application of cluster in Internet web server. Comput. Eng. Des. **25**(3), 472–474 (2004)
17. Duan, G.: Application of cluster technology to the hospital informationization system integration. Comput. Syst. Appl. **21**(2), 38–41 (2012)
18. Zheng, L., Liu, J., Chen, H.: Performance evaluation and implementation analysis on Linux Virtual Server. J. Xiamen Univ. (Natural Science) **41**(6), 727–730 (2002)

Information Acquisition for Farmland Soil Carbon Sink Impact Factors Based on ZigBee Wireless Network

Bingbing Wang, Dekun Zhai, Lijuan Sun, Dandan Yang,
Zhihong Liu, and Qiulan Wu[✉]

School of Information Science and Engineering, Shandong Agricultural University,
Tai'an 271018, China
snowice310@163.com, 18763898271@163.com, wqlsdau@163.com,
qingjimao@126.com, 1240813159@qq.com, 1204992056@qq.com

Abstract. As the main part of terrestrial ecosystem, farmland ecological system's estimation of carbon sink not only can help to understand the soil organic carbon content in farmland and increase the crop yield, but also can have great significance to the research on farmland ecosystem's influence to global climate warming. Through analyzing soil organic carbon estimation model of carbon sink factors, this article provides a method using multi sensors technology to acquire impact factors of soil carbon sink estimation and upload the results to the host computers by ZigBee+GPRS technology to realize the estimation of regional soil organic carbon in farmland. This research sets its research area in wheat experimental field in Yanzhou, Shandong Province, and by laying sensor nodes in the experimental field to acquire impact factors, the research aims to build reliable transmission of acquired factors by the ZigBee wireless sensor network, combined with GPRS technology. The result shows that this method can realize the reliable transmission of impact factors of the soil carbon sink in farmland and is of great significance to improve the accuracy of the estimation of soil carbon sink.

Keywords: Farmland soil organic carbon · Carbon sink · ZigBee · GPRS

1   Introduction

In recent years, with the increasing of the greenhouse effect and the deepening of the global change research, the estimation of carbon sink has drawn more attention. Estimation of soil carbon sink in farmland, as a part of estimation of carbon sink in farmland, is the estimation of the organic carbon content in the farmland soil. As an important indicator of the stability of agricultural ecosystem, the estimation results of soil carbon sink in farmland have great significance in the field of maintaining the function of agricultural soil carbon sink, improving the levels of soil organic carbon, mitigating the climate warming tendency and ensuring food security. The estimation of soil carbon sink in farmland needs substantial amounts of impact factor data of carbon sink. The traditional way of gathering data comes from the analysis of the existing soil survey

© IFIP International Federation for Information Processing 2016
Published by Springer International Publishing AG 2016. All Rights Reserved
D. Li and Z. Li (Eds.): CCTA 2015, Part II, IFIP AICT 479, pp. 185–193, 2016.
DOI: 10.1007/978-3-319-48354-2_20

data, the previous research data and field test data and this kind of research method usually has many shortcomings, such as the amounts of data is small, data timeliness is not good and data cannot be continuously acquired [1–3].

In recent years, with the application of wireless sensor networks technology in China's precision farming, the technology provided the feasibility for obtaining accurate information of farmland ecosystem thanks to its low cost, low power consumption, easy networking and other advantages. With the help of ZigBee technology, Liu Hui et al. built the wireless sensor network to monitor the temperature and humidity of the farmland soil; Sheng Ping et al., by combining ZigBee and 3G technology, realized the automatic control, accurate control and remote real-time monitor to environment and video information of greenhouse crops; Sun Baoxia et al., using the wireless sensor network, made the real-time monitor to the rice field come true and this can realize the real-time acquisition, wireless transmission and remote monitor of the growth environmental parameters of the large area of rice field [4–9].

ZigBee technology is one of the wireless sensor network technology and it is a new short distance wireless transmission technology. The technology has the characteristics of high reliability of data transmission, low power consumption, short delay and easy network and provides the possibility of real-time, accurate and reliable transmission of soil carbon sink in farmland.

This article takes the experimental field in Yanzhou of long-term location experiment as the research object. Many terminal sensing nodes, router nodes and the corresponding coordinator nodes are set in the experimental field to build the ZigBee sensor network. Combining ZigBee with the GPRS technology, the research acquires and transmits efficiently the estimated impact factors of soil carbon sink in farmland and improves the accuracy of estimation of soil carbon sink.

2   Analysis on Impact Factors of Soil Carbon Sink in Farmland

At present, there are many kinds of estimation models of soil organic carbon in farmland, such as Model Soil-C, proposed by Huang Yao et al., Model CENTURY, built by Parton et al. from Colorado State University in the United States and Model RothC, built by Jenkinson et al. based on the long-term field experiment data collected from the famous Rothamsted station in UK. The model proposed by Huang Yao et al. is currently applicable to China's regional soil organic carbon estimation and this article applies this model to estimate the soil organic carbon in farmland in Yanzhou experimental field. There are four factors affecting the model proposed by Huang Yao et al.: soil temperature, soil humidity, soil pH and soil texture and the former three kinds of data can be collected by sensor technology [10–17].

3 Survey of the Research Area and Research Method

3.1 Survey of the Research Area

Yanzhou City is located in the southwest of Shandong Province, the Plains of Southwest of Shandong Province and is situated between east longitude 116°35′21″ ~ 116°51′36″, north latitude 35°23′31″ ~ 35°43′17″. Figure 1 shows the location of the research area.

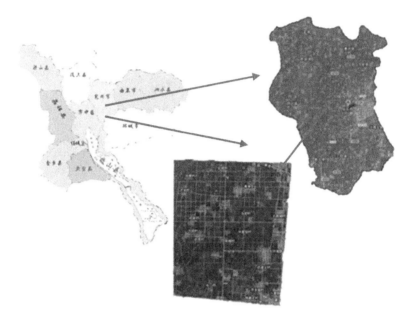

Fig. 1. Location of the research area

The research area has superior natural conditions and four distinct seasons. The average temperature over a year is 13.6°C. The average temperature of spring is 19°C, summer 22°C, autumn 20.5°C and winter −0.3°C. Soil in this area is mainly cinnamon soil and has good soil physico-chemical properties and permeability.

3.2 Reach Method

(1) The overall research plan
The system of collecting soil carbon sink in farmland, which based on ZigBee wireless sensor network, not only can collect automatic data, but also has function of accurate data transmission, data storage and management and data analysis and processing. Therefore, through three modules, data acquisition and pre-processing module of ZigBee carbon sink impact factors, data collection and storage module of carbon sink impact factors based on GPRS server side and management and estimation module of carbon sink impact factors based on Web, accurate collection and estimation of soil

carbon sink in farmland can be achieved and this can help make decisions about changes of China's regional soil organic carbon reserves. The overall structure is shown in Fig. 2:

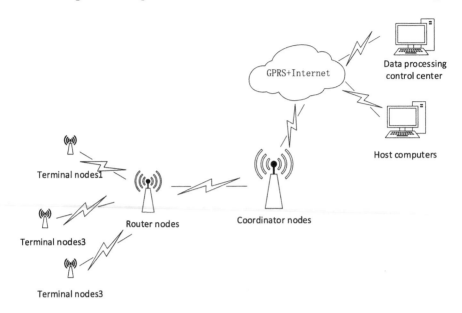

Fig. 2. The overall structure

Data acquisition and pre-processing module of ZigBee carbon sink impact factors employs sensors of soil temperature, sensors of soil humidity and sensors of soil pH as its terminal nodes to collect data and converges all these data to coordinator nodes, which have the function of data storage and pre-processing, through router nodes. Coordinator nodes, through core module, which has GPRS Modem, send data to the remote control center by GPRS network and Internet to store and apply.

Data collection and storage module of carbon sink impact factors in the sever side receives and stores data through SQL Sever database platform and data receiving and storage procedures running on the database. If the request to store the received data is legal, the sever will receive and store the collected data.

(2) Design of ZigBee nodes

ZigBee wireless sensor network nodes are composed by the terminal nodes, router nodes and coordinator nodes. The terminal nodes apply RFD (Reduced Function Devices), coordinator nodes FFD (Full Function Devices) and router nodes both RFD and FFD.

In this study, the terminal nodes in the experimental field are responsible for the data collection of carbon sink impact factors, therefore their main components, various related sensors (sensors of temperature and humidity, sensors of pH) can monitor the real-time soil carbon sink impact factors in the farmland; MCU (Micro Controller Unit) processes and transmits the collected data, which is mainly constructed by chip CC2530; the inverted F-style PCB antenna is responsible for receiving and sending signal; Flash memory chip is of low energy consumption; and the power module is supplied by lithium battery. Figure 3 shows its structure.

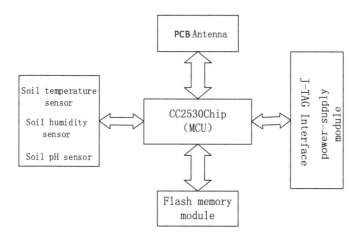

Fig. 3. Terminal node structure

In ZigBee wireless network, router nodes generally acts as a transit node. Collected data from the terminal nodes needs to be transmitted to the coordinator nodes. Usually distance between the two nodes shall not be so long, otherwise the data cannot be transmitted directly. In practical application, the increase of router nodes can keep the transmission between the terminal nodes and coordinator nodes to expand the network scope, so the router nodes are placed closely to the terminal nodes and far from the coordinator nodes.

The research applies the common communication interface RS-232 to realize the communication between the coordinator nodes and the host computers. The core module of the coordinator nodes transmits the collected data to the host computers by serial circuit RS-232. Serial circuit RS-232 is widely used in industrial standard [18].

(3) GPRS and Internet connection

In this article, the way to access to Internet is dependent on the GPRS network. After converged carbon sink impact factors, which are collected by remote ZigBee wireless sensor network, to the coordinator nodes, data are transmitted to the remote data center by GPRS DTU (Data Transfer Unit). Figure 4 shows the structure:

Data of carbon sink impact factors converged in the coordinator nodes, are transmitted transparently between data collection terminal and data center by GPRS network. The interfaces between the data terminals and GPRS DTU are generally RS-232, RS-485 or TTL and the interface used in this research is RS-232. The data control center of the host computers is the main equipment of the Internet center station. The center receives and processes the data transmitted by GPRS DTU, converts the protocol, and stores data to the database.

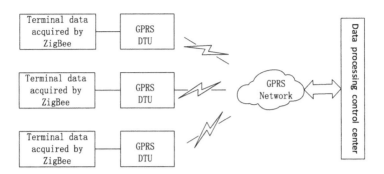

Fig. 4. Structure of GPRS and Internet connection

(4) Routing technology and routing algorithm
Accurate and reliable transmission of data in ZigBee wireless sensor network depends on not only the reasonable allocation and selection of routing nodes and routing protocols, but also the selection of data fusion processing technology.

In this research, topological structure of ZigBee network nodes is tree structure, including a coordinator node and a plurality of router nodes and terminal nodes. The equipment can only directly communicate to its own parent node or child node point to point, other communication can only be transmitted by tree structure routing. Therefore, it is very important to select the appropriate routing protocol for reliable transmission of data. The routing protocol is mainly responsible for the transmission of collected impact factors from terminal nodes to target nodes by network. It mainly includes two aspects: finding optimized path from the terminal nodes to target nodes and correctly forwarding data packets along the optimized path. On the basis of the existing ZigBee routing protocols, the article applies the cellular ant colony routing algorithm (CACO) proposed by Sun Yuwen et al. The algorithm combines the advantages of LEACH algorithm and the traditional ant colony algorithm and the network terminal nodes are mapped into cellular from cellular automaton to generate a search ants in the region. The randomly generated cluster heard node data are transmitted into the coordinator nodes in an effective and energy-saving way to fulfill the information routing process. The design of cellular node dormancy conversion mechanism and the pheromone updating rule can effectively reduce the blindness of the search path, balance network energy consumption and prolong the network lifetime [19, 20].

4   Comparison Analysis of Carbon Sink Impact Factors in the Experimental Area

The content of the research in this article has been tested in the experimental field in Yanzhou City. To verify the accuracy of carbon sink impact factors acquired by ZigBee wireless sensor network, the data is compared and calibrated with the corresponding data acquired by the portable instruments. Figures 5 and 6 show the comparison between data of soil temperature and soil humidity acquired by ZigBee and data acquired by the portable instruments from Group 3 No. 1 node for 24 h in a row on April 30, 2015 in the experimental area and the selected cycle is 30 min. Compared with the measured values, the average relative tolerance of soil temperature and soil temperature are 0.27 % and 0.68 % respectively. The way acquiring soil carbon sink impact factors based on multi sensors and ZigBee technology can not only obtain a large number of continuous data, but also meet the needs of soil carbon sink estimation in farmland.

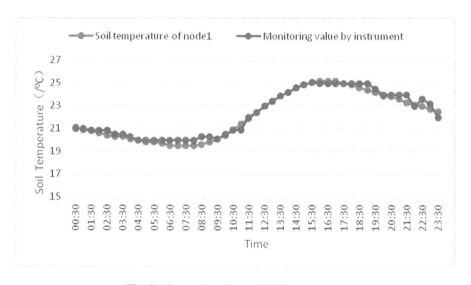

Fig. 5. Comparison figure of soil temperature

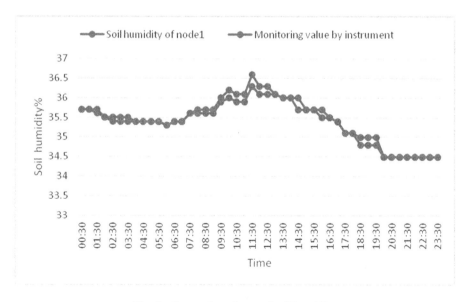

Fig. 6. Comparison figure of soil humidity

5 Conclusion and Summary

Aiming at the current situation of large data, poor continuity and low reliability about the regional soil carbon sink estimated impact factors in farmland and widely application of wireless sensor network technology in acquisition and detection of farmland basic information, the article proposed the way of obtaining soil carbon sink impact factors in farmland with the help of ZigBee wireless sensor technology.

Monitoring network scheme of clustered wireless sensor network is designed in the experimental area of the city of Yanzhou. Each cluster works as a monitoring site and there are four sites to collect carbon sink factors in the experimental area in a continuous and real-time way, combining with ZigBee+GPRS technology to achieve the reliable transmission of carbon sink factor data. By comparing and analyzing the collected data and the measured data, one conclusion can be achieved that this method can effectively obtain the impact factors of soil carbon sink.

Compared with the traditional way, the method to obtain the impact factors of farmland soil carbon sink in this article, can realize the real-time and continuous acquisition of carbon sink impact factor data, can effectively improve the soil organic carbon estimation, and can provide a theoretical basis for the influence of global carbon source/sink from farmland ecosystem.

Acknowledgment. Funds for this research was provided by the National High-tech R&D Program (863 Program) (2013AA102301), the innovation subject of Shandong Province Agricultural Application Technology, the youth project of Modern Agriculture Development Research Institute (14xsk3-03), and the youth project of National Natural Science Foundation of China (71503148).

References

1. Zheng, X., et al.: Research review on the estimation models and the applications of crop land carbon sequestration. J. Anhui Agric. Sci. **37**(35), 17649–17652, 17691 (2009)
2. Ling, E., et al.: Changes in soil organic carbon in croplands of China: II estimation of soil carbon sequestration potential. Soil Fertiler Sci. China **6**, 87–91 (2010)
3. Long, J., et al.: Advances of soil organic carbon model in farmland ecosystem. Chin. Agric. Sci. Bull. **28**(05), 232–239 (2012)
4. Liu, H., et al.: Development of farmland soil moisture and temperature monitoring system based on wireless sensor network. J. Jilin Univ. (Eng. Technol. Ed.) **38**(3), 604–608 (2008)
5. Sheng, P., et al.: Intelligent measurement and control system of facility agriculture based on ZigBee and 3G. Trans. Chin. Soc. Agric. Mach. **43**(12), 229–233 (2012)
6. Chen, Q., et al.: Design and implementation of the IOT greatway based on ZigBee/GPRS protocol. J. Comput. Res. Dev. **48**(9), 367–372 (2011)
7. Li, D.: Internet of things and wisdom agriculture. Agric. Eng. **2**(1), 3–4 (2012)
8. Sun, B., et al.: Real-time monitoring system for paddy environmental information based on wireless sensor network. Trans. Chin. Soc. Agric. Mach. **45**(9), 241–246 (2014)
9. Han, H., et al.: Design and application of ZigBee based telemonitoring system for greenhouse environment data acquisition. Trans. CSAE **25**(7), 158–162 (2009)
10. Yang, L.: Dynamics models of soil organic carbon. J. Forest. Res. **14**(4), 323–330 (2003)

11. Gao, L., et al.: Comparison of soil organic matter models. Chin. J. Appl. Ecol. **14**(10), 1804–1808 (2003)
12. Huang, Y., et al.: Model establishment for smiulating soil organic carbon dynamics. Sci. Agric. Sinica **34**(5), 465–468 (2001)
13. Zhou, T., et al.: Impacts of climat change and human activities on soil carbon storage in China. Acta Geogr. Sinica **58**(5), 727–734 (2003)
14. Ling, E., et al.: Changes in soil organic carbon in croplands of China: I analysis of driving forces. Soil Fertilier Sci. China **6**, 82 (2010)
15. Gao, L., et al.: Simulation of climate impact on soil organic carbon pool in black soil. J. Liaoning Tech. Univ. **24**(2), 288–291 (2005)
16. Xie, L., et al.: Review of influence factors on greenhouse gases emission from upland soils and relevant adjustment practices. Chin. J. Agrometeorology **32**(4), 481–487 (2011)
17. La, S.N., Bolonhezi, D., Pereira, G.T.: Short—term soil CO2 emission after conventional and reduced tillage of a no-till sugarcane area in southern Brazil. Soil Tillage Res. **91**, 244–248 (2006)
18. Wang, X., et al.: Design of ZigBee wireless temperature and humidity monitoring system based on CC2530. J. Chin. Agric. Mechanization **35**(3), 217–220 (2010)
19. Dai, Y.: Research of Monitoring of the FIELD Information Based on Wireless Sensor Network of ZigBee, pp. 9–15. Northwest A&F University (2010)
20. Sun, Y.: Research and Implementation of Field Environment Monitoring System Based on Wireless Sensor Networks, pp. 68–74. Nanjing Agricultural University (2013)

Penetration Depth of Near-Infrared Light in Small, Thin-Skin Watermelon

Man Qian[1,2,3,4,5], Qingyan Wang[2,3,4,5], Liping Chen[1,2,3,4,5(✉)], Wenqian Huang[2,3,4,5], Shuxiang Fan[1,2,3,4,5], and Baohua Zhang[2,3,4,5]

[1] Beijing Research Center of Intelligent Equipment for Agriculture, Beijing 100097, China
qianman101504@163.com, fanshuxiang8903@163.com

[2] National Research Center of Intelligent Equipment for Agriculture, Beijing 100097, China
{wangqy,chenlp,huangwq}@nercita.org.cn

[3] Key Laboratory of Agri-Informatics, Ministry of Agriculture, Beijing 100097, China

[4] Beijing Key Laboratory of Intelligent Equipment Technology for Agriculture, Beijing 100097, China

[5] College of Mechanical and Electronic Engineering, Northwest Agricultural and Forestry University, Yangling 712100, Shanxi, China
zhangbaohua@sjtu.edu.cn

Abstract. Non-destructive detection of internal quality in watermelon has very important significance for improving watermelon's production efficiency. Near-infrared (NIR) spectroscopy is one of the most popular non-destructive detection methods. However, it is challenging to collect spectra exactly due to the multiple scattering and absorbing by the skin and internal tissues. In order to obtain the interactions between light and watermelon tissues, the transportation feature of NIR light in small, thin-skin watermelon was studied in the range of 750–900 nm. For this purpose, the diffused transmission spectra were collected with removing the sample slices along the perpendicular bisector of the source-detector line. Based on the spectra in effective wavelength band, the penetration depth curves were fitted by least square method, and the results of different detecting positions (equator and top) were compared. It was shown that, light penetration depth on the equator was 8.3–9.5 mm, 8.7–17.8 mm and 18.9–38.5 mm with source-detector distance of 10 mm, 20 mm and 30 mm, respectively. The penetration depth on the top was less than the equator. And the penetration depth increased with source-detector distance increasing. With deeper penetration depth, more information about internal quality was carried by the diffused transmission spectra. However, the intensity of spectra was weaker. According to these results, a reasonable source-detector distance could be designed for collecting effective information about internal quality. This study is of potential significance for optimizing the handheld probe geometry for large fruit, and offers theoretical bases for non-destructive detection.

Keywords: Near-infrared spectroscopy · Penetration depth · Transportation features · Diffused transmission spectra · Small, thin-skin watermelon

© IFIP International Federation for Information Processing 2016
Published by Springer International Publishing AG 2016. All Rights Reserved
D. Li and Z. Li (Eds.): CCTA 2015, Part II, IFIP AICT 479, pp. 194–201, 2016.
DOI: 10.1007/978-3-319-48354-2_21

1   Introduction

Watermelon is one of the most popular fruits in summer, and its consumption is large. With the development of consumer's demand for fruit's quality, more attention have been paid to the internal quality, including soluble solid content, taste and nutritional value. And high internal quality can enhance the competition of domestic watermelon. However, due to the excessive watering and unreasonable fertilization, the quality of watermelon has fallen down with increasing planting area in recent years. Therefore, it is important to detect watermelon's internal quality at the postharvest handling and processing stage. However, the traditional detection method which mainly depends on destructive sampling is inefficient. It also cannot be applied widely. And it is urgent to develop a fast and non-destructive detection method to improve watermelon's internal quality.

Near-infrared spectroscopy has many advantages, including efficient analysis, fast execution, reproducibility and without sample pre-processing. And it plays a vital role in non-destructive and fast detection for fruits and vegetables. Near-infrared spectroscopy has already been used to detect watermelon's quality, such as firmness [1], soluble solids content [2–4] and maturity [5, 6]. And near-infrared spectroscopy with diffused transmission is more suitable for developing portable detecting system. Therefore, it is significant to acquire the transmission features and penetration depth of NIR light in watermelon's tissue. Many researches on light penetration depth in fruit have been reported by domestic and foreign scholars. The light penetration depth in fruit can be estimated by using the coefficient of absorption and reduced scattering [7]. The direct experimental measurement includes using a black knife to hinder light penetrating into the fruit [8], removing the sample slices gradually [9, 10] and piercing the fruit by optical probe to measure directly [11]. However, the research subjects are small fruits, such as apples, oranges, mandarins, tomatoes or pears. And researches on lager fruits have been rarely reported, such as watermelons.

To solve this problem, in this paper the diffused transmission spectra were collected, which were fitted by partial least square (PLS) method. Then the penetration depth on the equator and top were obtained with different source-detector distance (10 mm, 20 mm and 30 mm). And a suitable source-detector distance could be designed to reach effective transmission depth in watermelon. It provided technical basis for developing portable and non-destructive detection equipment for large fruits.

2   Materials and Methods

2.1   Fruit Samples

In this work, two small, thin-skin watermelons (Te Xiaofeng) were purchased from a fruit shop in Beijing. The vertical and horizontal diameters of watermelon are about 120 mm. And there are few defects on watermelon's surface. And they were placed in the room condition (20°C) for more than 24 h to avoid the influence of temperature on the detection results. The detective positions have no defects, and the light intensity can't destruct watermelon's tissue. The samples were labeled to detect the penetration depth for the equator and top, respectively.

2.2   The System Setup and Diffused Transmission Spectra Collection

Figure 1 illustrates the experimental setup. The system for collecting spectra included a halogen lamp (Illumination Technology, 50 w), portable Near-infrared spectrometer (USB 2000+, Ocean Optics, US, 500–1100 nm), probe of optical fiber (Ocean Optics, US), fruit holder and computer. The spectrometer's resolution is 1.7 nm and integral time is 2 s. In order to avoid stray light, the fiber probe, watermelon sample and fruit holder were placed in the black light box. Meanwhile, the source fiber probe and detector fiber probe aimed at the labeled detective position, tightly. Then the diffused transmission spectra could be collected.

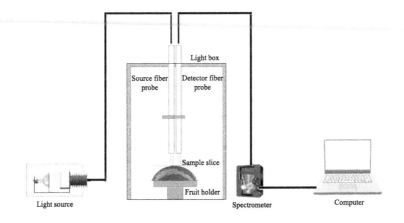

Fig. 1.   Illustration of spectra collection system

Figure 2a shows the illustration of spectra collection. The spectra were collected with removing the sample slices along the perpendicular bisector of the source-detector line on the basis that the light transmission path is "banana-shaped" between the source and detector [12]. Figure 2b shows the position of source fiber probe and detection fiber probe, including 1, 2, 3 source points and 1′, 2′, 3′ detection points. The distance between source and detector is 10 mm, 20 mm, 30 mm, respectively.

The half of watermelon was used as the original melon slice for spectra collection because of the light penetration depth less than half a watermelon. The spectra of each thickness slice were collected with different source-detector distance until the thickness is about 1 mm. The slice thickness values were 57.9, 45.8, 30.4, 23.3, 13.6, 11.3, 7.5, 5.6, 3.4, 1.2 mm. The spectra were collected for 5 times on the same position for each thickness slice, and the average spectrum of five times was used to analyze transportation features and penetration depth.

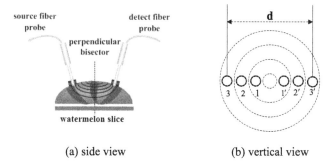

(a) side view   (b) vertical view

Fig. 2. Illustration of collecting spectra

2.3 Spectral Data Preprocessing and Analysis

2.3.1 Spectral Data Preprocessing

In this study, Savitzky-Golay was used to smooth the original spectra. The processing window was 7 and calculation time was 3. After Savitzky-Golay preprocessing, the feature information was strengthened and random noise was weakened.

2.3.2 Light Penetration Depth for Single Wavelength

For each wavelength (λ), the spectra of thickest slice are the reference spectra (I_1) to the spectra $(I_n, n = 1, 2, \ldots 10)$ of each thickness slice $(\delta_n, n = 1, 2, \ldots 10, \delta_1 > \delta_2 > \ldots > \delta_{10})$, called $g_n = \dfrac{I_1}{I_n}$, $n = 1, 2 \ldots, 10$ are compared.

Then a non-linear model of the form $g_n = 1 + a \cdot exp(b, \delta_n)$, with the parameter of a, b. And g_n (the I_1/I_n ratio) is the fitted value. δ_n (the thickness of watermelon slice) is fitted to measured points, using the least squares method. Then $\delta_1/0.1$ points of the fitted curve are taken, and the step length is 0.1 on the horizontal ordinate. An infinitesimal ε is given. A certain point $[\delta(x), g(x)]$ to neighboring point $[\delta(x + \delta_1/0.1), g(x + \delta_1/0.1)]$ is compared. If $[g(x + \delta_1/0.1) - g(x)]^2 < \varepsilon$, the value of x is approximately penetration depth of the wavelength (λ).

2.3.3 Light Penetration Depth in Effective Wavelength Range

Because the NIR light in 750–900 nm is low absorption and high scattered in biological tissues, the light can penetrate several centimeters in the tissues [13]. Due to the systematic error and random noise, the wavelength range in 750–900 nm is to analyze penetration depth, and the step length is 20 nm. For each wavelength, the penetration depth was fitted by lease square method as the single wavelength.

3   Results and Discussion

3.1   Features of NIR Spectra

Figure 3 shows the diffused transmission spectra with varying thickness on the equator when source-detector distance is 10 mm. Due to the high relative transmission and low reflectance, the spectra of the thinnest slice are at the bottom. The spectra of thicker slices are on the top. When the slice thickness is 7.5 mm, the intensity of spectra fell significantly. It indicates that the NIR light transmit the slice and the penetration depth is about 7.5 mm. When the slice thickness is less than or equal to the penetration depth, light will transmit the slice and thus reflect less. The signal to noise ratio is lower with further source-detector distance and the light transportation features make little difference on the equator and top.

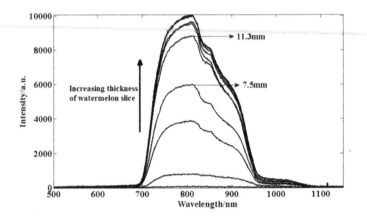

Fig. 3. The diffused transmission spectra with source-detector distance of 10 mm on the equator

3.2   The Penetration Depth for Single Wavelength

Figure 4 shows the fitted penetration depth curve for peak wavelength. When source-detector distance is 10 mm, the penetration depth is 8.4 mm on the equator. Table 1 shows the penetration depth of peak wavelength with different source-detector distance on the equator and top. As shown in Table 2, the maximum penetration depth is 29.8 mm and 21.1 mm on the equator and top with 30 mm source-detector distance, respectively. And the minimum penetration depth is 8.4 mm and 8.3 mm on the equator and top with 10 mm source-detector distance, respectively. It is noticed that the penetration depth is deeper with further source-detector distance on the equator and top. Meanwhile, the penetration depth on the top is less than the equator.

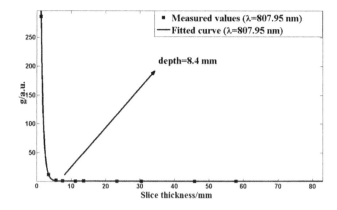

Fig. 4. The fitted curve of penetration depth for single wavelength

Table 1. Penetration depth for peak wavelength

Position	d (mm)	Peak wavelength (nm)	Penetration depth (mm)
Equator	10	807.95	8.4
	20	804.33	14.5
	30	793.46	29.8
Top	10	810.24	8.3
	20	804.00	13.2
	30	806.63	21.1

Table 2. Penetration depth in effective wavelength range

Position	d (mm)	Wavelength to largest penetration depth (nm)	Penetration depth range (mm)
Equator	10	755.26	8.3–9.5
	20	754.14	8.7–17.8
	30	770.21	18.9–38.5
Top	10	754.14	8.5–9.3
	20	757.50	7.2–15.6
	30	757.50	8.4–31.2

3.3   The Penetration Depth in Effective Wavelength Range

Figure 5 shows the penetration depth curves in effective wavelength band. On the equator, the penetration depth is nearly 10 mm with source-detector distance of 10 mm and the penetration depth curve is less volatile. It is because that the NIR light is mainly scattered due to the low penetration depth. The penetration depth is 8.7–17.8 mm and 18.9–38.5 mm with source-detector distance of 20 mm and 30 mm, respectively. Because the slice absorbs more and optical energy decreases with wavelength increasing, the penetration depth falls down. The transportation features make little difference on the equator and top. And the penetration depth on the top is less than the equator. It is probably due to less soluble solid content and more water in the top than the equator. The slice absorbs more and the penetration depth is less on the top.

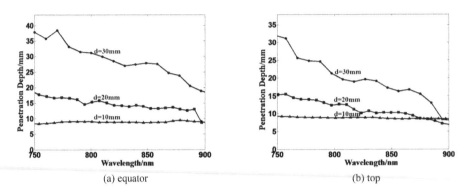

(a) equator   (b) top

Fig. 5. The penetration depth curves in effective wavelength range

Table 1 shows the penetration depth range and wavelength to largest penetration depth with different source-detector distance on the equator and top. The wavelengths to largest penetration depth are all in the range of 750–800 nm, indicating the NIR light in this range being low absorption feature. And the penetration depth increases with source-detector distance increasing on the equator and top, as is according to Monte Carlo model [14].

4   Conclusions

In this paper, the diffused transmission spectra in the range of 750–900 nm were collected with removing the sample slices. And the penetration depth curves of equator and top were fitted by partial least square method. Results showed that, the penetration depth increased with source-detector distance increasing, and more information about internal quality could be carried by spectra. However, the intensity of spectra was weaker. The transportation feature made little difference between equator and top, but the penetration depth on the top was less than the equator. The result is helpful to design a reasonable source-detector distance for achieving effective penetration depth. And it can provide reference for developing portable and non-destructive device for large fruit.

Acknowledgment.   The authors sincerely acknowledge the support by Beijing Municipal Natural Science Foundation (No. 6144024), and Beijing Academy of Agriculture and Forestry Sciences Foundation for Young Scholars (QNJJ201423).

References

1. Tian, H., Ying, Y., Lu, H., et al.: Study on predicting firmness of watermelon by Vis/NIR diffuse transmittance technique. Spectrosc. Spectral Anal. **27**(6), 1113–1117 (2007)
2. Zhang, F., Wang, Q., Ma, Z., et al.: Content determination of sugar and fiber in watermelon by near-infrared spectroscopy. Food Sci. **28**(1), 258–261 (2007)
3. Jie, D., Xie, L., Rao, X., et al.: Using visible and near infrared diffuse transmittance technique to predict soluble solids content of watermelon in an on-line detection system. Postharvest Biol. Technol. **90**, 1–6 (2014)
4. Tian, H.Q., Wang, C.G., Zhang, H.J., et al.: Measurement of soluble solids content in melon by transmittance spectroscopy. Sens. Lett. **10**(1–2), 570–573 (2012)
5. Abebe, A.T.: Total sugar and maturity evaluation of intact watermelon using near infrared spectroscopy. J. Near Infrared Spectrosc. **14**(1), 67–70 (2006)
6. Li, Y., Zhao, H., Chang, D., et al.: Maturity qualitative discrimination of small watermelon fruit. Spectrosc. Spectral Anal. **32**(6), 1526–1530 (2012)
7. Qin, J.W., Lu, R.F.: Monte Carlo simulation for quantification of light transport features in apples. Comput. Electron. Agric. **68**(1), 44–51 (2009)
8. Chen, P., Nattuvetty, V.R.: Light transmittance through a region of an intact fruit. Trans. ASAE **23**(2), 519–522 (1980)
9. Lammertyn, J., Peirs, A., De Baerdemaeker, J., et al.: Light penetration properties of NIR radiation in fruit with respect to non-destructive quality assessment. Postharvest Biol. Technol. **18**, 121–132 (2000)
10. Han, D., Chang, D., Song, S.: Information Collection of mini watermelon quality using near-infrared non-destructive detection. Trans. Chin. Soc. Agric. Mach. **44**(7), 174–178 (2013)
11. Fraser, D.G., Jordan, R.B., Künnemeyer, R., et al.: Light distribution inside mandarin fruit during internal quality assessment by NIR spectroscopy. Postharvest Biol. Technol. **27**(2), 185–196 (2003)
12. Feng, S.C., Zeng, F., Chance, B.: Monte Carlo simulations of photon migration path distributions in multiple scattering media. In: SPIE, vol. 1888, pp. 78–79 (1993)
13. Li, H., Xie, S., Lu, Z., et al.: Visible and near-infrared light scattering model of biological tissue. Acta Opt. Sin. **19**(12), 1661–1666 (1999)
14. Wang, L.H., Jacques, S.L., Zheng, L.: Monte Carlo modeling of light transport in multilayered tissues. Comput. Methods Programs Biomed. **47**(2), 131–146 (1995)

Design and Implementation of an Automatic Grading System of Diced Potatoes Based on Machine Vision

Chaopeng Wang[1,2,3,4,5], Wenqian Huang[2,3,4,5], Baohua Zhang[2,3,4,5], Jingjing Yang[2,3,4,5], Man Qian[1,2,3,4,5], Shuxiang Fan[1,2,3,4,5], and Liping Chen[1,2,3,4,5(✉)]

[1] College of Mechanical and Electronic Engineering, Northwest A&F University, Yangling, 712100, China
[2] Beijing Research Center of Intelligent Equipment for Agriculture, Beijing, 100097, China
[3] National Research Center of Intelligent Equipment for Agriculture, Beijing, 100097, China
[4] Key Laboratory of Agri-Informatics, Ministry of Agriculture, Beijing, 100097, China
[5] Beijing Key Laboratory of Intelligent Equipment Technology for Agriculture, Beijing, 100097, China
wangcp_nwsuaf@163.com, qianman101504@163.com,
fanshuxiang1989@163.com, huangwq@nercita.org.cn,
yangjj@nercita.org.cn, chenlp@nercita.org.cn, baobaopost@126.com

Abstract. Potato is one of the most important crops in the world. In recent years, potato and its processed products have gradually become important trade goods. As an important semi-manufactured product, diced potatoes need to be graded according to their three-dimensional (3D) size and shape before trading. 3D information inspection manually is a time-consuming and labor intensive work. A novel automatic grading system based on computer vision and near-infrared linear-array structured lighting was proposed in this paper. Two-dimensional size and shape information were extracted from RGB images, and height information was measured in NIR images combined with structured lighting. Then, a pair of pseudo-color and gray level height map images fusing with 3D size and shape information was constructed. Finally, diced potatoes were classified into either regular or irregular class according to their 3D information and criteria required by the industry. The grading system and proposed algorithm were testified by a total of 400 diced potatoes with different size and shapes. The test results showed that the detection error was in the range of about 1 mm, and the classification accuracy was 98 %. The results indicated that the system and algorithm was efficient and suitable for the 3D characteristic inspection of diced potatoes.

Keywords: Computer vision · Diced potatoes grading · Image processing · Structured light · Three-dimensional measurement

1 Introduction

Potato is the World's and China's fourth largest staple crop after rice, wheat and maize, it is also one of the most promising high-yield crops in China [1]. With the rapid development of global agriculture, potato and its processed products have great market

D. Li and Z. Li (Eds.): CCTA 2015, Part II, IFIP AICT 479, pp. 202–216, 2016.
DOI: 10.1007/978-3-319-48354-2_22

potential and become a primary part of the global agricultural trade as well. Consumers care more and more information about the products they bought, and well-informed high-quality products are easier to arouse consumers' interest and stimulate their purchasing desire [2]. For fruits and vegetables, appearance is one of the most important sensory quality attributes, it would not only influences the packaging and retail price, but also affects consumers' preferences and choice. Products with perfect appearance are always receive more favor of consumers favor and would have a better sales appeal [3, 4]. Diced potato as a kind of semi-manufacture in trade has strict requirements in size and shape. The irregular shape of diced potato not only deteriorates its appearance, but also has effects on the market value. Therefore, it is necessary to make classification for diced potatoes according to the 3D information before packaging or trading. In practice, diced potatoes are classified by human visual inspection, and 3D information inspection manually is a labor intensive work, it was also time-consuming. Manual processing pose added problems of maintaining the consistency and uniformity in grading [5, 6].

Currently, with the improvement of image processing, quality control with computer vision has been an important technology [7, 8]. For the requirement of speedy and real time, computer vision system has being developed as a significant part in quality detection and evaluation [9, 10]. Over the past several years, the on-line inspection system based on computer vision are widely used in realization of automatic detection of many different agricultural products, including the external and internal quality detection, but automatic classification for diced potato according to their 3D information is still not available. There is an increasing demand for on-line detection equipment based on computer vision that can mimic the human grading and realize accurate classification of diced potato with the unified standard to address the above issues with human visual inspection for food manufacturers.

With photo-electronics, image processing and computer technique rapid development, structured light vision technology has been widely used in computer vision system for 3D reconstruction and automatic measurement successfully [11]. There have been several techniques proposed with quite different characteristics for accurate measurement. In order to get the height of seedling, Feng et al. designed a automatic inspection system based on structured light vision, and result shows the height measurement error was less than 5 mm for the normally straight seedling [12]. However, the detection precision could not satisfy the requirement of diced potato inspection. Binocular visual was also extensive used for 3D size measurement. A binocular structure-light scanner was constructed to acquire the surface detail information of the work-pieces by Liu et al., and it was suitable for precision and efficiency demands of the large work-piece measurement [13]. However, 3D measurement based on binocular stereovision systems is not suitable used for online detection of diced potato just because it is time consume and the complexity of stereo matching. Coded structured light systems were commonly used for real-time acquisition of 3D surface data and it was considered as an important and widely used active shape acquisition technique [14, 15]. The encoded pattern plays a dominate role in the system, it could affect all the measurement performances, including the time consumption and accuracy. Xu et al. proposed one-shot pattern method for 3D shape measurement, result shows that the accuracy can achieve 0.18 mm. With the system, moving object could be inspected, and it can be implemented for automotive production lines [16]. A principle of uniquely color-encoded pattern projection was proposed by Chen et al. [17] to design a color matrix for improving the reconstruction efficiency based on coded structured light system. By using such a light pattern, it could realize the 3D vision reconstruction from a single image and accomplish the reliable and accurate measurement for scene objects. It could also be used in dynamic environment for real-time application. However, encoded structured light system is complex and the light pattern has important effect on measurement accuracy.

To realize automatic on-line classification of diced potatoes, a grading system based on computer vision with high accuracy and reliability performance is needed. For this purpose, intensive research works are being conducted to design and build a flexible, reliable and effective computer vision system by using a monocular stationary camera and a near-infrared linear-array structured lighting.

2   Objectives

In order to develop a real time grading system for diced potato classification by using computer vision and near-infrared linear-array structured lighting. To actualize this objective, several steps have to be completed: (1) RGB and NIR images synchronous acquisition through the same optical path of the camera for 3D information inspection. (2) 2D size measurement in RGB images for width and length inspection. (3) 2D shape feature (rectangle degree) extraction from the RGB images for contour shape inspection. (4) Height map (pseudo-color image and gray level images) construction according to the height information extracted from NIR images by using near-infrared light. (5) Evenness evaluation by using Gaussian distribution in height maps (gray level images). (6) Developing an efficient image processing algorithm based on computer vision system and near-infrared structured light to classify the diced potato into either regular or irregular class.

3   Materials and Methods

3.1   Diced Potato Samples

Fresh potatoes from a market in Beijing, were selected and processed into diced potatoes for study. The shape of some diced potatoes were influenced by the surface of potato, as a result, the contour shape of these diced potatoes would be triangle or other irregular shape, and the surface would be slant or rugged. Diced potatoes were classified into either regular or irregular class according to their 3D information and criteria required by the packinghouse. Namely, regular diced potatoes, which were traded at a higher price, can meet all the requirements of the geometric parameters, and irregular that doesn't satisfy at least one of the criteria. In our study, 17 potatoes were processed into 400 diced potatoes (length, width and height were about 15 mm for regulars, the geometries were uncertain for irregulars), including 270 regular and 130 irregular samples.

3.2   Computer Vision System

All the samples were inspected and classified by using the vision system as shown in Fig. 1(a). The vision system used in our study is same as the system described in [2]. It mainly consists of a 2CCD camera (JAI AD-080GE), a near-infrared linear-array structured lighting (800 nm, 200 mw), a lighting system (LED light), a computer, a conveyor belt driven by a stepper motor. The multi-spectral camera installed right above the conveyor belt was connected to network ports of computer with two RJ45 twisted-pairs as show in Fig. 1(b). It can acquire both NIR (800 nm) and RGB images through the same optical path simultaneously. The structured lighting was mounted on the upper left of conveyor belt and in the same horizontal plane with camera. The pair of two visible LED light source were distributed symmetrically at the both upper sides of conveyor for light supplement. The whole system was placed in a black box to prevent the interference from outside. In the process of 3D information detection, the image acquisition, image processing, final diced potatoes classification and conveyor belt control panel proposed in this paper was developed in MFC combined with OpenCV.

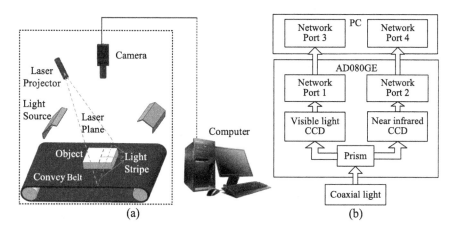

Fig. 1. Schematic illustration of computer vision system (a) Diagram of the computer vision system used in this research (b) Connection diagram of monocular camera and computer

The conveyor belt consists of inferior smooth material belt and stepper motor was used to transmit diced potatoes for the on-line detection. The specular reflection of projection light strip would be significantly reduced due to the inferior smooth surface of belt. And this would make it easy to extract the light strip and reduce the inspection error. The stepper motor was controlled by the driver and controller which were connected to computer through RS232 serial port to control the speed and other motion parameters of conveyor belt.

3.3   2D Shape Inspection Methods

Contour shape detection plays a domination role in 2D shape extraction. The majority of diced potatoes are regular, and the contour shape of qualified diced potato is regular rectangle. For others, they are influenced by the surface of potato, and the contour shape would be irregular. If the contour shape was not rectangle, the diced potato under detection would be judged as irregular directly and other shape information was not needed. Rectangle degree is a significant index to measure the contour shape of rectangular objects. Rectangle degree reflects the filling degree of an object to its external rectangle and rectangle factor can be used to distinguish whether the aim region is rectangle [18]. Therefore rectangle degree was used for the contour shape judgment with minimum circumscribed rectangle method in this paper.

$$R = S_0/S_{MER} \tag{1}$$

Where, R represents the rectangle factor, S_0 represents the area of diced potato's surface imaged by camera, S_{MER} represents the area of minimum circumscribed rectangle. For rectangular objects, the rectangle factor is infinite closed to 1, but for other objects, the rectangle factors vary from 0 to 1.

3.4   3D Size Inspection Methods

The inspection principle of height information based on triangulation theory is shown in Fig. 2. The conveyor belt is regarded as the reference plane, and it is configured to be parallel with the baseline of the camera and laser projector. The theory of height measurement can be explained by similar triangles $\triangle ABP$ and $\triangle CDP$. According to the triangle similarity, height of object can be calculated by:

$$\frac{h}{L-h} = \frac{d}{s} \tag{2}$$

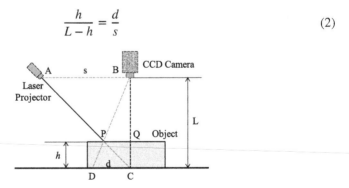

Fig. 2. The theory of the height measurement based on triangulation principle

Where, h represents the distance from point Q to C, and it also represents the height of object; d represents the distance between points C and D which can be extracted by image processing; L represents the vertical distance from the baseline of camera and laser projector to the conveyor belt; s represents the baseline distance from the laser projector to the CCD camera; L and s can be measured directly. Equation (2) can be equally transformed into the format as:

$$h = \frac{L - h}{s}d \tag{3}$$

Generally, the height of diced potato is lower than 20 mm, but the distance between baseline of camera and laser projector to the conveyor belt has exceeded 500 mm, large difference existed between them in magnitude. So Eq. (3) can be simplified as:

$$h = \frac{L}{s}d \tag{4}$$

For height measurement, the distortion distance d' caused by height of diced potato should be extracted for offset distance d detection. And it can be calculated by the distance between the projecting light stripes in inspected and reference images. The x-coordinate of point Q(x, y) is same of point C(x, y) due to the light strip is parallel to the y-axis of camera image plane, and it can be obtained before inspection. For points B, P, and D is collinear, and it is the same pixel for both reference plane point D and inspected part point P in the camera image plane, point P can be easily detected by image processing. A reference image with straight light strip should be acquired before inspection to obtain the reference coordinates Q(x, y), for which the coordinate position is fixed. In height measurement process, when diced potatoes were transmitted through the view of camera, NIR images with the distortional light strip should be acquired to get the information of detection coordinates P(x, y). The difference between x-coordinate of points P and Q could represent the distortion distance d' due to points P and Q were collinear, they had the same y-coordinate.

Both RGB and NIR images were used for 3D size detection. RGB images were processed for the 2D size measurement. With minimum circumscribed rectangle method, width and length of diced potato could be simply and conveniently detected in a small error range. NIR images were processed for height measurement. Figure 3 shows NIR images acquired by the monocular camera. When nothing is under detection, the light strip would be continuous without any distortion for the near-infrared structured light is just throwing lighting onto reference plane as shown in Fig. 3(a). But when diced potato is transmitted and through the vision system, parts of the light strip would project

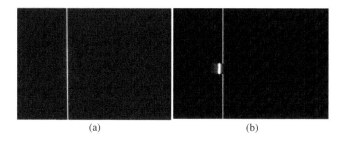

(a)   (b)

Fig. 3. NIR images acquired by AD080-GE (a) Reference image (b) Detection image

on the surface of it, and the light strip would be disconnected as shown in Fig. 3(b). The offset distance of light strip would change with the height of diced potato, so it is important to get the relationship between diced potato's actual height and the offset distance of light strip for height measurement.

Before the height measurement, threshold segmentation was used to carry on binary processing for NIR image. And then, light strip in binary images should be refined for pixel coordinate extraction. An applicable image thinning algorithm not only need to provide high inspection accuracy but also should has high efficient that could complete the image processing in the time interval between two images was acquired (the maximum frame rate of AD-080GE is 30 fps). In our research, the centerline extraction method was used for the image thinning. The center point coordinates $O(x_0, y_0)$ of light strip in each line of binary image can be calculated using Eq. (5):

$$x_0 = \frac{1}{A} \sum_{(x,y) \in R} x \, y_0 = \frac{1}{A} \sum_{(x,y) \in R} y \qquad (5)$$

Where, R represents the pixels set of light strip in each line of binary image; A represents the size of pixels set; x and y represents the x-coordinate and y-coordinate of each pixel in light strip.

NIR images were detected line-by-line to get pixel coordinates of the centerline which could represent the position of light strip. After thinning of the binary images, the centerline coordinates could be extracted, and distortion distance d' could be calculated by:

$$d' = d_0 - d_1 \qquad (6)$$

Where, d_0 represents the centerline coordinates of reference image, d_1 represents the centerline coordinates of inspected image.

In our research, model materials (50 mm long, 30 mm wide and 2 mm high) were used layer by layer and formed different heights for the data sampling. Cubic spline interpolation was used to establish the relationship between offset distance d and distortion distance d' by using Matlab 2010. When d' were inspected by visual inspection system, the offset distance of light strip would be got by the fitting function, which could represents the relationship. Finally, the real height of diced potato h would be calculated with triangulation theory proposed above.

3.5   Surface Evenness Detection

For further research, height map, including pseudo-color and gray level images were constructed according to the distortion distance d' in the process of height measurement. The pseudo-color images could make it easier for human eyes to evaluate the performance of the height image, and the gray level images were used for the diced potato's surface evenness detection. The surface evenness was assessed based on the grey scale distribution of height map. The average and mean square error of d' could be calculated by the grey scale distribution, which could be extracted by the image processing of gray level image. Gaussian model was used for surface evenness detection, and the evenness was determined by the *ucl* (upper control limit) and *lcl* (lower control limit) of d' using Eqs. (7) and (8):

$$ucl = \mu + \sigma \tag{7}$$

$$lcl = \mu - \sigma \tag{8}$$

Where, μ represents the average of d', σ represents the mean square error of d'. Due to different distortion distance d' corresponds to different gray scale in gray level image, μ and σ can be determined using Eqs. (9) and (10):

$$\mu = \sum_{i=1}^{255} i * p_{(i)} \tag{9}$$

$$\sigma = \sqrt{\frac{1}{255} \sum_{i=1}^{255} (i - \mu)^2} \tag{10}$$

Where, i represents the gray scale of gray level image from 1 to 255, gray scale 0 is preclusive due to it is defined as the background (black) of the height image; $p_{(i)}$ represents the appearing probability of each gray scale appeared in height map image.

As show in Fig. 4, the shadow represents the gray scales in the scope of *lcl* and *ucl*. If the total probabilities of gray scales in the shadow area is higher than a specified threshold, the surface evenness would be judged as eligible.

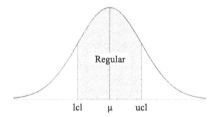

Fig. 4. Control limits for surface evenness detection

3.6 Whole Classification Algorithm for 3D Size and Shape Inspection

In this paper, 3D geometry characteristics of diced potatoes were extracted by using a combination of structured lighting method, triangulation theory and minimum circumscribed rectangle method. The image processing and shape classification algorithm mainly include the following steps: (1) Calibration: NIR image in which the light strip was uninterrupted was acquired before shape detection to get reference pixel coordinates; (2) 2D size measurement: RGB image was processed to get the length and width in pixels with minimum circumscribed rectangle method. Then the length and width of diced potatoes would be figured out with the parameters of camera and other parameters. (3) 2D shape feature extraction: RGB images were processed to get the rectangle degree of diced potatoes, according which to judge if the contour shape of diced potato was regular rectangle. (4) Height map construction: Pseudo-color images and gray level images which could represent the height information of diced potato were constructed simultaneous depend on the distortion distance d' between the projection light stripes in inspected and reference images. (5) Surface evenness evaluation: The grey-histogram of height map (gray level image) was created to get average of d' and mean square error of d', and then judge if the surface of diced potato was even with Gaussian model. (6) Grade judgment: All the shape parameters inspected through above steps were compared with the parameters of 3D characteristic information set according to the factory requirements for classification. If the geometrical information measured by computer vision system satisfied requirements, Flag would be defined as TRUE, and diced potato under detection would be regarded as regular. Otherwise, the Flag would be defined as FLASE and the diced potato would be regarded as irregular. Finally, diced potatoes were classified into either regular or irregular class. The flowchart of the whole classification algorithm is shown in Fig. 5.

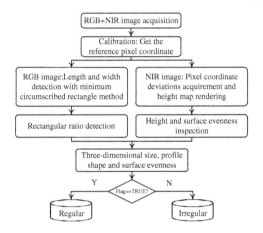

Fig. 5. Flowchart of the classification algorithm for diced potato

4   Results and Discussion

The automatic control software based on computer vision is shown in Fig. 6. The on-line inspection of diced potato would be conducted according to the classification algorithm mentioned above. Inspection results, including 3D size and classification information were shown in real time. Meanwhile, the height map images could be constructed with the moving of diced potatoes, and the pixels would be assigned to different color according to their height information and color bar.

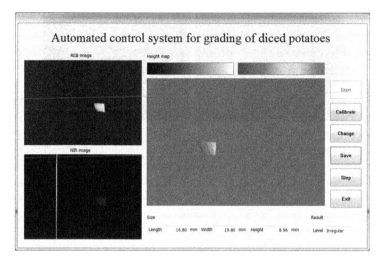

Fig. 6. The inspection system developed in VC++ 2010 for diced potato grading

4.1 Results of Light Strip Thinning and Smoothing

In the process of height measurement, it is important to extract the centerline of light strip in NIR image. Height information of each pixel in the light stripe could be calculated using the measurement principle of the triangulation method. The result of light strip thinning and smoothing was shown in Fig. 7. Figure 7(a) is the NIR image with light strip projecting on the surface of diced potato. As shown in Fig. 7(a), a light stripe with distortion was projected onto the scene and imaged by the camera, and it is obvious that the width of light strip on the surface of diced potato increases due to the light scattering. Figure 7(b) shows the results of light strip thinning. As shown in Fig. 7(b), the light strip is thinned as a very thin line with the width of only one pixel using the centerline extraction method. Figure 7(c) shows the distortion distance d' of each pixel in the centerline. The 3D profile shape of diced potatoes could be constructed by connecting all the distortion distance of the pixels in the centerline. Veining defects were often observed around light strip on the surface of reference plane or diced potato due to the slight scattered light of conveyor belt and instability of the laser projector. As shown in Fig. 7(c), the data after image thinning were not smooth and sometime it would influence the detection results. For this reason, median filtering was used to smooth the original data of distortion distance d' on account of the simple algorithm and better processing performance compared with other processing methods. Figure 7(d) shows the result of data after media filtering. As shown in Fig. 7(d), most of the burrs are eliminated, especially the data in area of reference plane.

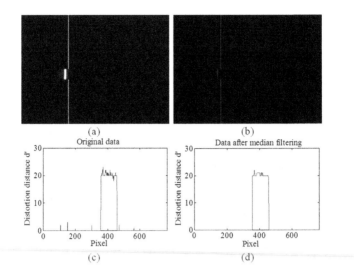

Fig. 7. The result of light strip thinning and data smoothing (a) Detection image (b) Result of image thinning (c) Original data of distortion distance d' (d) Results data after median filtering

4.2   Detection Result

Contour shape and surface evenness inspection were used to judge if the diced potato satisfied the dimensional requirements was regular pattern. In order to detect the surface evenness, pseudo-color images and gray level images were constructed as shown in Fig. 8. Figure 8(a) is the RGB image of regular diced potato, Fig. 8(b) and (c) are the pseudo-color image and gray level image constructed by the vision system according to

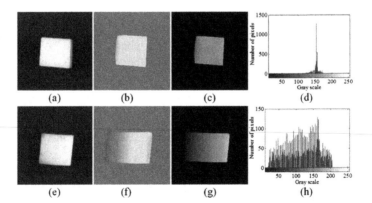

Fig. 8. Detection results of regular and irregular diced potato detection (a) RGB image of regular diced potato; (b) Pseudo-color image of regular diced potato; (c) Gray level image of regular diced potato; (d) Gray histogram of regular diced potato; (e) RGB image of irregular diced; (f) Pseudo-color image of irregular diced potato; (g) Gray level image of irregular diced potato; (h) Gray histogram of irregular diced potato

the 3D information, Fig. 8(d) is the gray histogram which could indicate the grey scale distribution of the gray level image. Images in the second row of Fig. 8 is the relevant images of an irregular diced potato. As shown in Fig. 8(f) and (g), the height map were distorted due to the slant or rugged surface, and the surface area of irregular diced potato was larger than the surface area of regular diced potato. Compared Fig. 8(d) with (h), it is obvious that, if diced potato under detection was regular, the grey scale distribution would be relatively concentrated and its' distribution trends is similar to Gaussian distribution. Otherwise, the grey scale distribution would be dispersive for irregular diced potato with the uneven surface.

In order to observe the classification effect, and testify detection algorithm proposed in this paper, the inspection results of different classes of diced potatoes were randomly selected. Measuring error was extracted by comparing the 3D size detected by the grading system with the real value measured by digital caliper. The measurement results, including length, width and height shows minor difference (in range of 1 mm) contrasted with the real value.

Figure 9 shows the detection results of geometric parameters. Figure 9(a), (b) and (c) shows the relationship of real values and measured values of length, width and height, respectively. The corresponding R^2 (coefficient of determination) is 0.9072, 0.8939 and 0.8968, this could meets requirements for the classification of diced potatoes. The rectangle degree and Gaussian probability were the other two important parameters to judge if diced potato was regular or not. For most irregular diced potatoes, these two parameters were influenced by the surface of potato. There has a relationship between the contour shape and surface evenness, it can be represented by the correlation between rectangle degree and Gaussian probability. As shown in Fig. 9(d), it is obvious that the rectangle degree of regular diced potato is higher than 0.9. By contrast, the rectangle degree of irregular diced potato is disperse because of the various contour shape, and for majority of them, the rectangle degree is lower than 0.9. The Gaussian probability of regular diced potato is higher than 0.68, and it is lower than 0.68 for major irregular diced potatoes due to their uneven surface. Therefore, 0.68 can be used as the decision condition for the surface evenness judgment. For minority irregular diced potatoes, the Gaussian probability is higher than 0.68, but 3D size and rectangle degree could not meet the requirements, it is greatly reduced the inspection error probability. In Fig. 9(d), there have many regular diced potatoes with the same rectangle and Gaussian probability, therefore some overlap was also observed.

The image processing mainly includes minimum circumscribed rectangle detection and height map construction aim at getting the 3D geometrical characteristics of diced potato. The intermediate process result images are shown in Fig. 10. Diced potatoes with different shape as shown in Row (a), all the geometrical information of sample 1 and 2 are regular and could satisfy the criteria, and other samples are irregular. Images in Row (b) are the detection results with minimum circumscribed rectangle method. Height map including pseudo-color and gray level image constructed in the process of height measurement are shown in Row (c) and Row (d), different height is denoted as different pixel values.

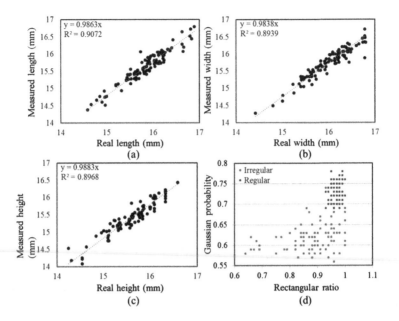

Fig. 9. Detection results of geometrical characteristics (a) Comparison of measured length and real length; (b) Comparison of measured width and real width; (c) Comparison of measured height and real height; (d) Distribution of Gaussian probability and rectangular ratio for regular and irregular diced potato

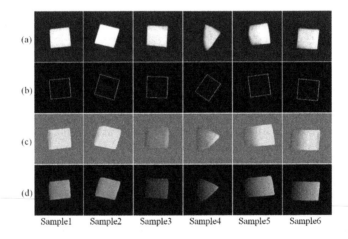

Fig. 10. Results of diced potatoes with deferent shape (a) RGB images; (b) Results of minimum circumscribed rectangle method; (c) Pseudo-color images; (d) Gray level image

The irregular diced potatoes mainly include 3 types: (1) Geometrical feature is regular, but 3D size would not satisfy the requirements as show in sample 3. Though the contour shape is regular rectangle and the surface is evenness, the height could not meet requirements. This is obvious when compared the height map of sample 3 with sample 1 or 2, especially in pseudo-color image. (2) Both of the contour shape and surface evenness could not satisfy the requirements as shown in sample 4, this would be classified as irregular directly with rectangle degree detection. (3) The contour shape of diced potato is regular rectangle, but the surface is slant or rugged, as shown in sample 5 or 6. It is obvious to observe the trend of the surface height change with the height map.

5 Conclusions

Real-time, on-line, low-cost computer vision system is urgent needed to replace the manual inspection for the classification of diced potato. An automatic grading system based on computer vision and near-infrared linear-array structured lighting was proposed in this paper. In order to reduce the interference of external conditions and make the inspection results more precise and stable, visible and near-infrared linear-array structured lighting were used respectively for contrast tests to select the projector with higher detection speed and accuracy. Meanwhile, convey belt with inferior smooth material was used for diced potatoes transmission. The matt surface would reduce the specular reflection of projection light strip significantly and this would make it easy to extract the light strip. In this research, RGB and NIR images acquired simultaneously by a monocular camera were processed for 3D geometrical characteristics extraction, including physical dimension, contour shape and surface evenness, which were used as the main decision conditions for the classification. Minimum circumscribed rectangle method and structured light measurement method were used for the 3D size inspection, meanwhile, height map include pseudo-color image and gray level image were constructed according to the height of diced potato for the surface evenness judgment with Gaussian module. The results demonstrate the detection error of the grading system was in range of about 1 mm, and the classification accuracy has reached 98 %. This preliminary research verified the possibility and feasibility of using computer vision combined with structure light measurement method for the classification of diced potato.

Acknowledgment. The authors gratefully acknowledge the financial support provided by National Key Technology R&D Program (No. 2014BAD21B01).

References

1. Qu, D.Y., Xie, K.Y., Jin, L.P., et al.: Development of potato industry and food security in China. Sci. Agric. Sin. **38**(2), 358–362 (2005)
2. Zhang, B.H., Huang, W.Q., Wang, C.P., et al.: Computer vision recognition of stem and calyx in apples using near-infrared linear-array structured light and 3Dreconstruction. Biosyst. Eng. **139**, 25–34 (2015)
3. El Masry, G., Cubero, S., Moltó, E., et al.: In-line sorting of irregular potatoes by using automated computer-based machine vision system. J. Food Eng. **112**(1–2), 60–68 (2012)
4. Zhang, B.H., Huang, W.Q., Gong, L., et al.: Computer vision detection of defective apples using automatic lightness correction and weighted RVM classifier. J. Food Eng. **146**, 143–151 (2014)

5. Al Ohali, Y.: Computer vision based date fruit grading system: design and implementation. J. King Saud. Univ. Comput. Inf. Sci. **23**(1), 29–36 (2011)
6. Blasco, J., Aleixos, N., Moltó, E.: Machine vision system for automatic quality grading of fruit. Biosyst. Eng. **85**(4), 415–423 (2003)
7. Razmjooy, N., Mousavi, B.S., Soleymani, F.: A real-time mathematical computer method for potato inspection using machine vision. Comput. Math Appl. **63**(1), 268–279 (2012)
8. Kong, Y.L., Gao, X.Y., Li, H.L., et al.: Potato grading method of mass and shapes based on machine vision. Trans. Chin. Soc. Agric. Eng. **28**(17), 143–148 (2012)
9. Gunasekaran, S.: Computer vision technology for food quality assurance. Trends Food Sci. Tech. **7**(8), 245–256 (1996)
10. Zhang, B.H., Huang, W.Q., Li, J.B., et al.: Principles, developments and applications of computer vision for external quality inspection of fruits and vegetables: a review. Food Res. Int. **62**, 326–343 (2014)
11. Zhou, F., Zhang, G.: Complete calibration of a structured light stripe vision sensor through planar target of unknown orientations. Image Vis. Comput. **23**(1), 59–67 (2005)
12. Feng, Q.C., Liu, X.N., Jiang, K., et al.: Development and experiment on system for tray-seedling on-line measurement based on line structured-light vision. Trans. Chin. Soc. Agric. Eng. **29**(21), 143–149 (2013)
13. Liu, J.W., Liang, J., Liang, X.H., et al.: Industrial vision measuring system for large dimension work-pieces. Opt. Precis. Eng. **1**, 126–134 (2010)
14. Chen, Y.J., Zuo, W.M., Wang, K.Q., et al.: Survey on Structured light pattern codification methods. J. Chin. Comput. Syst. **31**(9), 1856–1863 (2010)
15. Salvi, J., Fernandez, S., Pribanic, T., et al.: A state of the art in structured light patterns for surface profilometry. Pattern Recogn. **43**(8), 2666–2680 (2010)
16. Xu, J., Xi, N., Zhang, C., et al.: Real-time 3D shape inspection system of automotive parts based on structured light pattern. Opt. Laser Technol. **43**(1), 1–8 (2011)
17. Chen, S.Y., Li, Y.F., Zhang, J.: Vision processing for realtime 3-D data acquisition based on coded structured light. IEEE Trans. **17**(2), 167–176 (2008)
18. Rosin, P.L.: Measuring shape: ellipticity, rectangularity, and triangularity. Mach. Vis. Appl. **14**(3), 172–184 (2003)

A Soil Water Simulation Model for Wheat Field with Temporary Ditches

Chunlin Shi[(⊠)], Yang Liu, Shouli Xuan, and Zhiqing Jin

Institute of Agricultural Economics and Information/Key Laboratory
of Agricultural Environment in Lower Reaches of the Yangtze River,
Ministry of Agriculture of PRC, Jiangsu Academy of Agricultural Sciences,
Nanjing 210014, China
shicl@jaas.ac.cn, luisyang@126.com,
shirleyxuan2008@hotmail.com, jaasjzq@163.com

Abstract. Accurate soil water content simulation is the basis of disaster early-warning and evaluation about waterlogging and drought. In order to more accurately simulation the water movement in wheat field with temporary field ditches, a two-dimension soil water simulation model was developed in this study. The model included the water movement vertically(up and down) and horizontally(ribbing and ditch), and traditional runoff estimation was replaced by calculating the drainage water from ditches. The model could simulate the comprehensive effect of depth of plow layer, initial soil water content, precipitation intensity and infiltration rate of plow pan layer on runoff. The application of the model in Xinhua city, China showed good agreement between observation with simulation values.

Keywords: Water · Simulation model · Wheat field · Temporary field ditches

1 Introduction

Waterlogging is one of the major agrometeorological disasters in wheat production in middle and lower valley of Yangtze river due to consecutive rainfall and weak infiltration of the plow pan because of the prevailing rice-wheat rotation cropping system [1, 2]. The simulation on soil water content could provide support for waterlogging forecast and yield loss evaluation. Now there are many models to simulate the soil water content. In some models, such as CERES-wheat, VSMB, without considering the weak infiltration of plow pan, all excessive water will infiltrate to the next layer. So they could not simulate the waterlogging condition [3, 4]. Some models could simulate waterlogging, such as SWATRE [5], DRAINMOD [6], DHSVM [7], but they always were used to simulate the movement of soil solutions and water management at regional scales, so they ignored the diffusion of water. Lv (1998)simulated the dynamic change of soil water content at Zhejiang province with MARCOS model [8]. Recently, some models based on artificial neural networks have been developed to simulate the soil water content [9, 10]. Such models could simulate the water content changes, but their explanations in soil water movement is poor and their applicability in easy waterlogging regions have no enough evidences. With the development of drip irrigation in arid regions, some models could simulated the water movement in vertical and horizontal directions [11, 12].

© IFIP International Federation for Information Processing 2016
Published by Springer International Publishing AG 2016. All Rights Reserved
D. Li and Z. Li (Eds.): CCTA 2015, Part II, IFIP AICT 479, pp. 217–224, 2016.
DOI: 10.1007/978-3-319-48354-2_23

Most of above studies think water movement only at vertical direction. But many temporary field ditches were digged for drainage in wheat production in middle and lower valley of Yangtze river [13]. So above models maybe cannot accurately express water movement character in easy waterlogging regions. It is necessary to develop a model to simulate the soil water movement which can realize the role of the temporary filed ditches on drainage.

The objective of this paper aims to develop a 2-Dimension model to simulate water movement which will suit to wheat field with temporary ditches at rainy regions. The study will provide the basis for the evaluation of waterlogging disaster and optimization of water management.

2   Model Description

Rice-wheat rotation is a common cropping system in the middle and lower valley of the Yangtze river. Because of rice production, the infiltration rate of the plow pan layer is weak. When the rainfall is enough large in wheat production, the rainfall will be held up at the top of the plow pan and led to waterlogging damage to wheat plants, so in order to drain excess water, some temporary ditches were digged along the plant direction (Fig. 1).

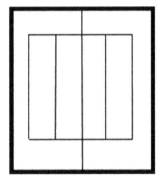

Fig. 1. Sketch map of temporary ditches in wheat field

2.1   Water Movement Character and Equation

Wheat field soil could be divided into three layers, plow layer, plow pan layer and underneath layer (Fig. 2). Rainfall will be absorbed by plow layer soil until its water content reach field capacity, the sparse and porous soil texture in plow layer led the excess rainfall can infiltrate easily to the bottom of the plow layer. But the tighten plow pan layer will led to weak infiltration, the rainfall will accumulate at the bottom of the plow layer, and water content at the bottom of plow layer will increase to saturation condition, thus waterlogging will happen. The saturation water will infiltrate though the plow pan layer or move to the temporary field ditches simultaneously.

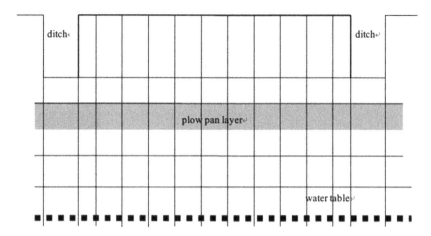

Fig. 2. Soil profile for the model

In order to better describe water movement, the soil profile was divided into six layers. The plow layer include two layers, which was determined as the depth of the ditch. Plow pan layer is look as one layer and the underneath layer include three layers. The depth of last layer is changed with the depth of water table.

The movement equation of soil water is listed as below:

$$
\begin{aligned}
&\text{Plow Layer 1}: D_1 \frac{dW(i,1)}{dt} = P - DR(i,1) - E_a - D(i,1) - E_V \cdot RR_1 - RO \\
&\text{Plow Layer 2}\;\; D_2 \frac{dW(i,2)}{dt} = DR(i,1) - DR(i,2) - D(i,2) - E_V \cdot RR_2 \\
&\text{Plow Pan Layer}\;\; D_3 \frac{dW(i,3)}{dt} = DR(i,2) - DR(i,3) - D(i,3) \\
&\text{Underneath Layer1}\;\; D_4 \frac{dW(i,4)}{dt} = DR(i,3) - DR(i,4) - D(i,4) \\
&\text{Underneath Layer2}\;\; D_5 \frac{dW(i,5)}{dt} = DR(i,4) - DR(i,5) - D(i,5) \\
&\text{Underneath Layer3}\;\; D_6 \frac{dW(i,6)}{dt} = DR(i,5) - DR(i,6) - D(i,6)
\end{aligned}
\tag{1}
$$

where D_i is the depth of the ith layer, $W(i,j)$ is the volume water content at grid (i,j). P is rainfall, E_a is soil evaporation, E_v is crop transpiration. $DR(i,j)$ is drainage amount at grid (i,j), RO is runoff and $D(i,j)$ is the water diffusion at grid (i,j). j is the layer number, and i is the horizontal position for grids.

2.2 Precipitation Infiltration

In plow layer, similar to other water balance models, we also assume that redundant water will all transfer to next layer when the water content is bigger than filed capacity until the top of the plow pan layer. When the excess precipitation led to soil saturation condition for soil at the bottom of the plow layer, water will move to temporary field ditches and infiltrate through to plow pan layer simultaneously. The infiltration amount (INF) is calculated using Eq. (2):

$$INF = \begin{cases} DT_0 & DT < DT_0 \\ a \cdot (DT - DT_0) + DT_0 & DT \geq DT_0 \end{cases} \qquad (2)$$

where DT is the depth of saturated layer at the bottom of the plow layer, DT_0 is maximum thickness of saturated layer for all infiltration in a day. a is coefficient, which relates to soil texture. The DTo in Eq. 2 was set to 2 mm in this study.

2.3 Water Diffusion

Water always moves from the positions with high hydraulic potential to those with low hydraulic potential. So when grid (i,j) has difference in water potential with nearby girds, including grids (i − 1,j), (i + 1,j), (i,j − 1) and (i,j − 1), water diffusion will happen. The diffusion amount can be expressed as Eq. (3):

$$D(i,j) = K_h \cdot (\psi_{i-1,j} - \psi_{i,j})/(Z_i - Z_{i-1}) + K_h \cdot (\psi_{i,j} - \psi_{i+1,j})/(Z_i - Z_{i+1}) \\ + K_V \cdot (\psi_{i,j-1} - \psi_{i,j})/(D_i - D_{i-1}) + K_V \cdot (\psi_{i,j} - \psi_{i,j+1})/(D_i - D_{i-1}) \qquad (3)$$

where Kv and Kh are the soil vertical and horizontal unsaturated hydraulic conductivity, respectively. Ψ is water potential and the subscribe expresses the position of grids. Di is depth of the ith layer and Zi was horizontal distance of grid i.

2.4 Runoff

Because the intensity of rainfall is small during the wheat growth season and the big roughness of wheat field, the surface runoff happens only when the plow layer was statured for the grids on the top of plow layer and the runoff values equal to the precipitation minus soil absorbable water. At the ditches, the water moved to ditches and precipitation minus infiltration amount and soil absorbable water will be leak out as runoff. The soil absorbable water equals to the saturation water content minus current soil water content.

2.5 Calculation of Evapotranspiration

The potential reference evapotranspiration is calculated as Priestley-Taylor model [4]:

$$ET_P = K_S \cdot SR \cdot (0.00488 - 0.00437 \cdot ALBEDO) \cdot (\bar{T} + 29) \qquad (4)$$

where, SR is daily solar radiation ($MJ/M^2/D$), ALBEDO was field reflection coefficient. T is mean air temperature, equals to 0.6TM + 0.4TN. TM and TN are the maximal and minimal temperatures respectively. K_S is temperature coefficient, calculated as below Eq. (5).

$$K_s = \begin{cases} 0.01\exp(0.18(\bar{T}+20)) & \bar{T} < 5 \\ 1.1 & 5 \le \bar{T} \le 24 \\ 0.05(\bar{T}-24)+1.1 & \bar{T} > 24 \end{cases} \tag{5}$$

ALBEDO is expressed as:

$$ALBEDO = \begin{cases} \gamma & sowing - emergence \\ 0.23 - (0.23-\gamma)\exp(-0.75L) & emergence - elongation \\ 0.23 + (L-4)^2/160 & elongation - maturity \end{cases} \tag{6}$$

where, γ is the reflection coefficient for bare soil, is set to 0.2.

The potential field evapotranspiration is calculated as Eq. (7) [14–16]:

$$ET_C = \begin{cases} ET_P & L \le 1.5 \\ \frac{(K_C-1)L + (5-1.5K_C)}{3.5}ET_P & 1.5 < L < 5 \\ K_C \cdot ET_P & L \ge 5 \end{cases} \tag{7}$$

where, K_C is maximal crop coefficient.

The potential soil transpiration is calculated as $E_P = ET_P \cdot \exp(-\delta L), \delta$ is extinction coefficient and set to 0.5 in this study. The crop evaporation is $ET_C - E_P$. The actual soil transpiration is:

$$E_a = \begin{cases} E_P & W > W_F \\ E_P \cdot \frac{W-W_P}{W_F-W_P} & W_P < W < W_F \\ 0 & W < W_P \end{cases} \tag{8}$$

where, W_P and W_F are soil water content for wilting point and field capacity.

2.6   Depth of Water Table

The depth changed as below Eq. (9):

$$\Delta WT = DR(i,6) + 2K_V \cdot (W(i,6) - WSD)/Z(6) \tag{9}$$

where ΔWT is change amount of water table, DR $(i,6)$ is infiltration amount of grid $(i,6)$, Kv is vertical diffuse coefficient, WSD is saturation soil water content for underneath layer. $Z(6)$ is the depth of 6th layer.

Considering the movement of underneath water, the change of water table could use the mean value of Δ WT at each point.

3   Model Application

3.1   Soil Parameter Description

Xinhua City located at Jiangsu province is the easy waterlogging area because the elevation is the lowest in Lixiahe plain. According to filed investigation, the depth and its attribution for each layer is list as Table 1.

Table 1. The soil parameters for each layer in Xinhua City

Layer	Thickness (cm)	Bulk density (g/cm³)	Field water capacity (%)	Wilting point (%)
1	13	1.21	37.9	8.9
2	7	1.42	36.6	8.7
3 plow pan	10	1.44	30.9	10.3
4	10	1.38	33.5	9.8
5	10	1.35	35.3	9.5
6	variable	1.35	35.3	9.5

The thickness of layer 6 change with the water table and the initial value was set to 40 cm, and parameters in layer 6 have no data, we assume their attributions equal to those in layer 5.

Water movement parameters are listed as belows

$$K_h = K_V = KS \cdot (\frac{\psi}{\psi_e})^{-2+3/\beta}$$

$$\psi = \psi_e \cdot (\frac{W}{W_S})^{\beta}$$

where, K_h and K_V are horizontal and vertical unsaturated hydraulic conductivity. KS is saturated hydraulic conductivity. Ψ and Ψ_e are hydraulic potential and saturated hydraulic potential.

3.2   Simulation

Using soil water content and water table observation data in 2012–2013 from Xinhua agrometeorology experiment station (32.95°N, 119.82°E), the model was validated. Because the wheat planted on Oct. 29, 2012, the model simulated from Oct. 20, 2012. The simulated water table change with observed values in good agreement. But the simulated soil water content was lower than observed values during before winter and after wheat heading. The possible reason for this phenomena maybe be the incorrect parameters input (Figs. 3 and 4).

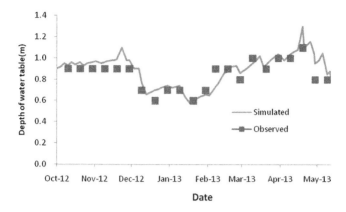

Fig. 3. The change of the depth of water table for 2012–2013 wheat production season

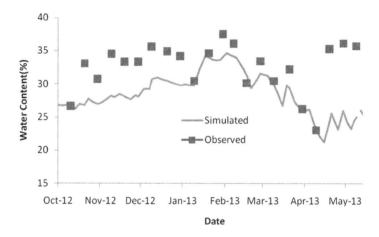

Fig. 4. The change of soil water content in layer 1 for 2012–2013 wheat production season

4   Discussion

In previous studies, the runoff always was estimated from the CN method proposed by department of agriculture, USA [4, 7]. Such method considered the effect of soil texture, initial soil water content on the runoff. But it could not describe the impact of precipitation intensity and depth of plow layer. In this study, we consider the all above factors, so it is more reasonable.

The water movement always were considered only in vertical direction in previous models [3, 4, 7]. This paper expanded it to 2-D space which could be suited to most area in wheat field. But at the both sides of wheat field, several connected ditches with different depth makes the simulation of soil water more complex [8]. So further study should be needed to explore water movement wholly in wheat field.

Although there are some other simple models or analytical models to describe the water movement [17, 18], they always could not suit to different ecological conditions.

Acknowledgment. This study was funded by the Special Fund for Agro-scientific Research in the Public Interest (201203032), Jiangsu Province Science and Technology Support Program (BE2012391), and the Fund for Independent Innovation of Agricultural Sciences in Jiangsu Province (CX(12)3055).

References

1. Li, C., Jiang, D., Wollenweber, B., Li, Y., Dai, T., Cao, W.: Waterlogging pretreatment during vegetative growth improves tolerance to waterlogging after anthesis in wheat. Plant Sci. **180**(5), 672–678 (2011)
2. Jin, Z., Shi, C.: An early warning system to predict waterlogging injuries for winter wheat in the Yangtze-Huai plain (WWWS). Acta Agron. Sin. **32**(10), 1458–1465 (2006)
3. Baier, W., Robertson, G.W.: Soil moisture modeling: conception and evaluation of the VSMB. Can. J. Soil Sci. **76**, 251–261 (1996)
4. Ritchie, J.T., Otter, S.: Description and performance of CERES-wheat: a user-oriented wheat yield model. USDA-ARS **38**, 159–170 (1985)
5. Belmans, C., Wesseling, J.G., Feddes, R.A.: Simulation model of the water balance of a cropped soil: SWATRE. J. Hydrol. **63**, 271–286 (1983)
6. Cox, J.W., MeFarlane, D.J., Skaggs, R.W.: Field evaluation of DRAINMOD for predicting waterlogging intensity and drain performance in south—western Australia. Aust. J. Soil Res. **32**(4), 653–671 (1994)
7. Wigmosta, M.S., Vail, L.W., Lettenmaier, D.P.: A distributed hydrology-vegetation model for complex terrain. Water Resour. Res. **30**(6), 1665–1679 (1994)
8. Jun, Lv: Simulation of water balance in crop growth field. Shuili Xuebao **1**, 45–50 (1998)
9. Elshorbagy, A., Parasuraman, K.: On the relevance of using artificial neural networks for estimating soil moisture content. J. Hydrol. **362**, 1–18 (2008)
10. Si, J., Feng, Q., Wen, X., Xi, H., Tengfei, Y., Li, W., Zhao, C.: Modeling soil water content in extreme arid area using an adaptive neuro-fuzzy inference system. J. Hydrol. **527**, 679–687 (2015)
11. Tian, F.Q., Gao, L., Hu, H.P.: A two-dimensional Richards equation solver based on CVODE for variably saturated soil water movement. Sci. Chin. Technol. Sci. **54**(12), 3251–3264 (2011)
12. El-Nesr, M.N., Alazba, A.A., Simunek, J.: HYDRUS simulations of the effects of dual-drip subsurface irrigation and a physical barrier on water movement and solute transport in soils. Irrig. Sci. **32**(2), 111–125 (2014)
13. Guo, S.Z., Peng, Y.X., Qian, W.P., Chen, Z.W.: Wheat Crop Science and Technology in Jiangsu Province. Jiangsu Science & technology publishing house, Nanjing (1994)
14. Hu, J.C., Cao, W.X., Luo, W.H.: A soil-water balance model under waterlogging condition in winter wheat. J. Appl. Meteorol. Sci. **15**(1), 41–50 (2004)
15. Brission, N., Seguin, B., Bertuzzi, P.: Agrometeorological soil water balance for crop simulation models. Agric. Meteorol. **59**, 267–287 (1992)
16. Paraskevas, C., Georgiou, P., Llias, A., Panoras, A., Babajimopoulos, C.: Evapotranspiration and simulation of soil water movement in small area vegetation. Int. Agrophys. **27**(4), 445–453 (2013)
17. Qian, L., Shufen, S.: Development of the universal and simplified soil model coupling heat and water transport. Sci. Chin. Ser. D Earth Sci. **51**(1), 88–102 (2008)
18. Shani, U., Ben-Gal, A., Tripler, E., Dudley, L.M.: Plant response to the soil environment: an analytical model integrating yield, waterm soil type and salinity. Water Resour. Res. **43**, W08418 (2008). doi:10.1029/2006WR005313

The Synchronized Updating Technology Research of Spatio-temporal Supervision Data Model About Organizing of Construction Landuse Data in Distributed Environment

Xiaolan Li[1,2], Bingbo Gao[1,2(✉)], Yuchun Pan[1,2,3,4], Yanbing Zhou[1,2], and Xingyao Hao[1,2]

[1] Beijing Research Center for Information Technology in Agriculture, Beijing 100097, China
[2] National Engineering Research Center for Information Technology in Agriculture, Beijing 100097, China
{lixl,gaobb,panyc,zyb,hxy}@nercita.org.cn
[3] Key Laboratory of Agri-Informatics, Ministry of Agricuture, Beijing 100097, China
[4] Beijing Engineering Research Center of Agricultural Internet of Things, Beijing 100097, China

Abstract. As China advances toward urbanization, the relation between supply and demand of land use is growing acutely. Effective supervision of various land use has become necessary to achieve reasonable and lawful use of land. The change of land use from farmland or unused land to construction land use in the process of urbanization usually undergoes the following stages: approval, expropriation, provision, application, and supplement. The change to construction land use similarly undergoes this process, and then the outcome before and after the change is compared to determine potential land use problems. The present process-oriented spatio-temporal data model for organization and supervision data management records the supervised spatio-temporal variation process and state by using new-added auxiliary tables. This model can efficiently organize and demonstrate multiple relationships among landuse stages and the evolution process in the life cycle of spatial entities. To achieve effective supervision from the upper to the lower in the three-level "province-city-county" distributed environment and to detect potential land use problems early, data from different places are collected and synthesized for analysis. To ensure that the database is up-to-date, real-time synchronization of spatio-temporal supervision data should be studied and the integrity of the spatio-temporal data model should be maintained. Therefore, this study introduces an updating mechanism based on the trigger technology for data synchronization. When importing basic data from distributed databases into the supervision database, simultaneous updating in related auxiliary tables is performed along with changes in the basic tables. That is, records are added automatically into auxiliary tables that reflect the spatio-temporal change of the process and state when basic data are imported. This method not only achieves and maintains consistency in the database, but also ensures integrity and currency in the auxiliary tables. Change information can be expressed duly in all supervision stages, and highly efficient supervision of construction landuse is guaranteed.

© IFIP International Federation for Information Processing 2016
Published by Springer International Publishing AG 2016. All Rights Reserved
D. Li and Z. Li (Eds.): CCTA 2015, Part II, IFIP AICT 479, pp. 225–236, 2016.
DOI: 10.1007/978-3-319-48354-2_24

Keywords: Distributed · Spatio-temporal supervision data model · Synchronized update · Consistency · Integrity

1 Introduction

In the urbanization process of China, the relation between land supply and demand is increasingly becoming distinct. With the aims of controlling construction land use growth rate, strengthening regulation to achieve economical and intensive land use, and preventing unreasonable or unlawful use of land use, the process of approval, supplement, application, expropriation, and investigation of construction land use must be supervised [1]. In the supervision data of construction land use in the urbanization process, a certain flow and life cycle are demonstrated in approval, expropriation, provision, application, supplement, and investigation these stages. Hence, vertical supervision, aside from horizontal analysis at these stages is necessary [2]. In the current supervision systems, land use data of related stages are scattered in thematic databases, and only the land use data of the current stage are supervised. Supervision data in different databases lack a dynamic connection, thereby resulting in difficulty to achieve full supervision of urbanization land use from the perspective of horizontal correlation [3, 4].

With this demand, Gao Bingbo et al. proposed a process-oriented spatio-temporal supervision data model for construction land use [5]. On one hand, this model introduces an improvement to the database structure to suit existing supervision data and constructs the incidence relationship among approval, supplement, application, expropriation, and investigation land use stages. On the other hand, auxiliary tables are employed to record the evolution process of related land use entities and determine multiple relationships among land use stages. This model realizes the supervision of urbanization development, including the recording of change processes to allow review of land use entities and reconstruction of land use projects, and improves the efficiency of supervision of construction land use.

In the process-oriented spatio-temporal supervision data model for construction land use, supervision data on land use are analyzed. Basic geographic data and thematic data are collected from different places to achieve efficient supervision of construction land use. In China, land use management databases and other related business systems are distributed vertically at three levels: "province-city-county" [6]. Data are obtained from different systems that are distributed in different network nodes. In the importation and updating of related supervision data in the "province-city-county" distributed environment, related change information should be reflected in the supervision database system duly. That is, when data in basic tables from the corresponding levels change, the auxiliary table should reflect such change to ensure consistent and up-to-date information in the supervision database. The problem of synchronized updating of the entire supervision database will be discussed in the succeeding section.

This study introduces a synchronized updating technology for the spatio-temporal supervision data organization model for construction land use in a distributed environment. The role of this technology in maintaining the consistency of database is analyzed.

2 Organization of Supervision Data in the Spatio-temporal Database

Supervision of construction land use refers to both vertical and horizontal supervision of the approval, supplement, application, expropriation, and investigation stages of the entire land use process. Land use starts from the application for approval of new construction land projects and ends with the completion of the construction project. To determine the construction land project and the land use entities involved at a certain time, the state of land use should be reconstructed at a specific moment. Gao Bingbo proposed a process-oriented spatio-temporal data organization and management model in construction land use supervision based on the requirement mentioned previously.

In the model, the basic data of each link are connected with each other in a certain manner. The evolution of land use entities in the approval, supplement, application, expropriation, and investigation stages is the foundation to construct the correlation among these stages. The storage structure of basic supervision data is shown in Fig. 1. In every two links, "consolidation, reclamation, and development" land use projects and supplementary cultivated land projects are connected by supplementary cultivated land use entities. Meanwhile, supplementary cultivated projects and approval projects are connected by the approval project serial number. The relationship between the approval project and expropriation project is established by expropriated land entities. The relationship between supplement and expropriation projects, which is much more complex, is established through the relationship between related supplement project and approval project and that between related approval project and expropriation project. The application project and supplement project are connected by related application land use entities. The vertical correlation between two links is established in the same manner, thereby allowing vertical supervision in the system by vertical association.

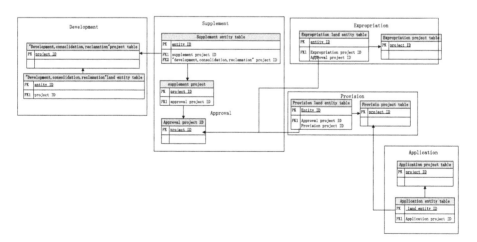

Fig. 1. Storage structure of the basic table data in supervision database

With regard to the basic data storage structure, four auxiliary tables, namely process, sequence, state, and evolution tables, are newly added in the supervision data organization model. The process table presents a list of different land use projects in the supervision database. The sequence table displays the relationship among land use projects. The state table contains information related to land use entities. Lastly, the evolution table exhibits the relationship among entities, such as cutting, merging, copying, and complementary. The organization of these tables and the storage structure are shown in Fig. 2. The four tables indicate the correlation of land use processes and the evolution information of land use entities in construction land supervision. In the basis of these auxiliary tables, we can immediately reconstruct the relationship among land use projects and recall land use entities. As a result, the historical state of construction land entities at a particular instance in a time series will be easy to extrapolate.

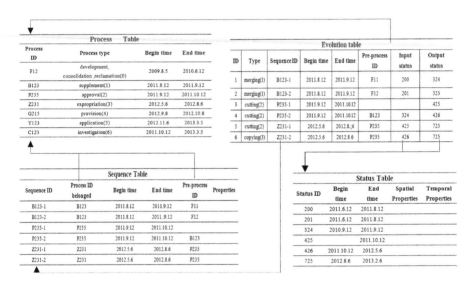

Process Table

Process ID	Process type	Begin time	End time
F12	development, consolidation reclamation(0)	2009.8.5	2010.6.12
B123	supplement(1)	2011.8.12	2011.9.12
P235	approval(2)	2011.9.12	2011.10.12
Z231	expropriation(3)	2012.5.6	2012.8.6
G215	provision(4)	2012.9.8	2012.10.8
Y123	application(5)	2012.11.6	2013.3.5
C123	investigation(6)	2011.10.12	2013.3.5

Evolution table

ID	Type	SequenceID	Begin time	End time	Pre-process ID	Input status	Output status
1	merging(1)	B123-1	2011.8.12	2011.9.12	F11	200	324
2	merging(1)	B123-2	2011.8.12	2011.9.12	F12	201	325
3	cutting(2)	P235-1	2011.9.12	2011.10.12			425
4	cutting(2)	P235-2	2011.9.12	2011.10.12	B123	324	426
5	cutting(2)	Z231-1	2012.5.6	2012.8.;6	P235	425	725
6	copying(3)	Z231-2	2012.5.6	2012.8.6	P235	426	725

Sequence Table

Sequence ID	Process ID belonged	Begin time	End time	Pre-process ID	Properties
B123-1	B123	2011.8.12	2011.9.12	F11	
B123-2	B123	2011.8.12	2011.9.12	F12	
P235-1	P235	2011.9.12	2011.10.12		
P235-2	P235	2011.9.12	2011.10.12	B123	
Z231-1	Z231	2012.5.6	2012.8.6	P235	
Z231-2	Z231	2012.5.6	2012.8.6	P235	

Status Table

Status ID	Begin time	End time	Spatial Properties	Temporal Properties
200	2011.6.12	2011.8.12		
201	2011.6.12	2011.8.12		
324	2010.9.12	2011.9.12		
425		2011.10.12		
426	2011.10.12	2012.5.6		
725	2012.8.6	2013.2.6		

Fig. 2. The storage structure of auxiliary table data in supervision database.

The traceback of supervision data mainly includes the spatial form or property of land use entities and the spatio-temporal process of approval, expropriation, provision, application, supplement, and investigation in land use two parts. The traceback of the evolution of land use entities can be achieved by examining the state and evolution tables. The traceback of the spatio-temporal process requires a step-by-step analysis of the evolution, sequence, and process tables to obtain land use project information at a certain time. In reconstructing the historical state of land use entities, the existence of land use projects and entities can be analyzed using the process and state tables, while the information of reconstructed land use projects and entities can be obtained using the sequence and evolution tables. Storing evolution information of spatio-temporal processes and entities by using auxiliary tables is more intuitive because these tables increase the operational and practical review and reconstruction work of supervision data and reduce the complexity of supervision operation that requires different supervision data at multiple time periods. Consequently, the efficiency of construction land supervision is improved.

3 Construction Land Use Supervision Database in a Distributed Environment

3.1 Supervision Data Flow in the "Province-city-county" Network

Supervision data mainly comprise basic geographic data, land use data at approval, expropriation, provision, application, supplement, and investigation stages, and thematic data. In the "province-city-county" network environment, three types of data are distributed in different network nodes and business systems. These data are extracted from other related databases or systems between two levels. The flow of supervision data in the distributed environment is shown in Fig. 3.

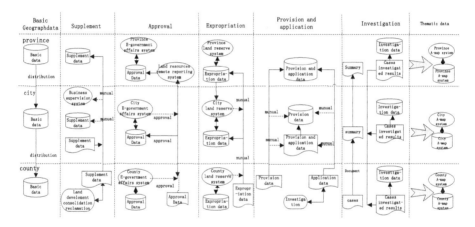

Fig. 3. The supervision data flow status in "province-city-county" distributed environment

In general, land use data at approval, expropriation, provision, application, supplement, and investigation stages in the "province-city-county" environment are complex. Databases at the top, bottom, or same levels must be accessed to extract related data at each corresponding stage. The supplement data in the supervision database at the county level contain consolidation, reclamation, and development data that are then imported into the supplement database at the city level. The supplement data of different cities are subsequently imported into the supplement database at the province level. The approval data are obtained from the electronic government affairs systems at the county level. City-level approval data can be extracted from the data at the county-level or from the electronic government affairs systems at the same city level. Province-level approval data are from the land resource remote reporting system or the provincial e-government affair system that gathered data from the subordinate level. The county-level expropriation data are extracted from the land reserve database, while the expropriation data at the city level is gathered from the related county data, and the provincial expropriation data is formed from city nodes with the same way. The provision data in county nodes are encoded manually, while the city- and province-level expropriation data are gathered from lower-level nodes. The application data at county level are obtained from the law enforcement and inspection processes. The city-level application data are extracted from subordinated county nodes, and the provincial application data are gathered from the same but lower county nodes.

The supervision database uses a series of indicators. Basic geographic data and thematic data are required as reference data, and land use data at approval, expropriation, provision, application, supplement, and investigation stages are checked if reasonable and legal to identify potential problems. The basic geographic data are propagated by the Ministry of Land and Resources to provinces, cities, and then to counties, forming the three-level, "province-city-county" basic geographic database. The thematic data in the supervision database are extracted from the province-level, city-level, and county-level a-map databases in the distributed environment.

In the upper analysis, related land use systems distributing scattered and system structures are found to vary in the "province-city-county " distributed environment. The data at different levels and in different business systems are obtained manually. Thus, the collection of full supervision data and supervision efficiency are affected. A high degree of automation of the data flow would clearly increase the satisfaction level of supervision business needs. Therefore, the current supervision data flow should be changed to a new and more adaptive data flow in such distributed environment.

3.2   Synchronization of Construction Land Supervision Data in a Distributed Environment

In the construction land supervision database, we use the model proposed by Gao Bingbo in storing supervision data. The supervision database is deployed in the distributed environment characterized by a three-level, "province-city-county" vertical management mode. We set provincial, city, and county supervision nodes, and the county and city nodes interact and share data with each other, as well as the city and provincial nodes, thereby achieving a top-to-bottom control and effective management. The interaction among nodes is shown in Fig. 4.

In the "province-city-county" network environment, we need to consider the content of two aspects in constructing the supervision database. One aspect is the corresponding relationship among different business data, such as basic geographic data and land use data, and the other aspect is the corresponding relationship between province and city nodes or between city and county nodes. The basic tables in the supervision database contain three types of data: geographic data, land use data, and thematic data. Geographic data could be obtained from superior level nodes of "province-city-county" network. Thematic data could be extracted from the a-map system at the same level. Land use data could be collected from subordinate level nodes by subscription and use of distribution technology on a certain interface. Lastly, data in auxiliary tables could be obtained from basic tables. When the data in basic tables are updated, related records in auxiliary tables are also updated. Figure 5 shows the new data flow of the construction land supervision database system in the distributed environment.

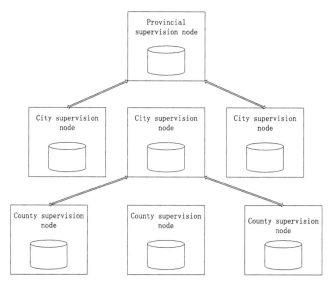

Fig. 4. Construction land use supervision database management system in "province-city-county" three-level network environment

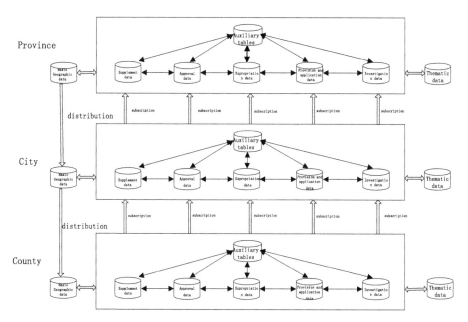

Fig. 5. New data flow of construction land supervision database system in distributed environment

In the new data flow, the superior database could gain data from subordinate databases by subscription and use of distribution technology based on the data service. Data service is used to package the data in each level and to provide access by deploying a responsive distribution service, registration service, and subscription service at a superior platform. Data service is equipped with the functions of service registration and real-time data push. Service information is registered to the registration service, and then the changed data is pushed to subscribers in real time. Subscription service uses the information registered by the data service. The subscription mission is registered in the registration service, and related data are relayed to the receiving interface. Distribution service receives the pushed data, charges the execution of subscribed data, and pushes changed data to the interface of the subscription service. Registration service maintains the registered information of the data service and the subscription mission and activates the subscription service to execute data push as the subscription mission. The concrete process of data acquisition is as follows: First, data are distributed by publishing the data as data service in a subordinate database server that requires superior synchronized updating. These data are then registered in the registration service. Second is the subscription of data. The subscription service at a superior-level platform creates subscription mission and registers on a registration server according to the current distributed land use data service. When data are changed, the data service will push changed data to the subscription service. The registration service will simultaneously activate the data distribution service according to subscription mission and send the changed data to the subscription service.

The previously described process can ensure the automatic updating of the database at the superior level as the data of the basic tables change at the subordinate level. Consistency is consequently guaranteed. However, the supervision database management system not only includes the basic tables, but also the auxiliary tables that contain data from the basic tables. As mentioned earlier, when the data of the basic tables are updated, the auxiliary tables must have a certain mechanism to create related records based on the basic supervision data and then form the topological relationship information of the spatio-temporal process. In this new data flow design, the integrity and up-to-date quality of the entire supervision database system are maintained, subsequently ensuring the accuracy and efficiency of supervision work.

4 Automatic Maintenance of Construction Land Use Supervision Data in Distributed Environment

The data in a distributed database are decentralized or stored in different physical location nodes. For the management system to be consistent, data in different nodes should remain consistent while mutual exchange or sharing between certain databases occurs. For these cases, the DBMS manufacturers have designed a mechanism, such as Oracle's advanced replication technology [7] and SQL Server's subscription replication technology [8], to achieve data synchronization and automatic maintenance. Some researchers have also studied synchronization and automatic maintenance from different perspectives and carried out many data synchronization technologies and methods [9], such as the Oracle technology [10], the multicast form in updating the database [11], the reduction method based on the SQL operation statement [12], and the data synchronization and updating technology for land use data [13, 14]. All these mechanisms can satisfy synchronization needs and automatic database maintenance. However, for the process-oriented, spatio-temporal supervision database management system, the relationship between basic and auxiliary tables is derived, and the traditional synchronization technology could not contain basic and auxiliary tables in the same database to allow synchronized updating and to maintain integrity. Therefore, further research on database synchronization technology is necessary.

In a distributed environment, the spatio-temporal supervision data model for construction lands includes not only the basic tables that contain land use information at each stage, but also the auxiliary tables that exhibit the association among land use projects and the evolution among land use entities. The model can reflect the evolution process of land use entities in their life cycle and backrtrace or reconstruct the land use stages at particular instances using this form. As shown in Fig. 6, the auxiliary tables have a close derivation relationship with the basic tables. The process table corresponds to different land use projects, the sequence table records the association between two different land use projects, the status table records land use entities and their duration, and the evolution table expresses the evolution process between entities. A corresponding relationship is present among basic tables, as well as a derivation relationship between basic and auxiliary tables. When basic table data are imported, data in auxiliary tables must be updated to maintain the integrity of the entire database.

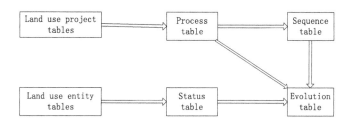

Fig. 6. The relationship of basic tables and auxiliary tables

Considering the characteristics of data in the approval, expropriation, provision, application, supplement, and investigation stages and the four auxiliary tables comprehensively, a spatio-temporal data synchronization and updating mechanism is developed in this study based on the combination of the trigger and the traditional land use database synchronization and updating technology. Figure 7 illustrates this developed mechanism and the specific description.

(a) In the importation of data on projects or programs and on the land use entities, related execution of maintenance operations is triggered.

(b) When data are imported into the basic tables that contain data of the projects or programs at the approval, expropriation, provision, application, and supplement stages, two triggers, namely, before and after triggers, will be generated. In the supervision database, "supplement" corresponds to the supplement program in the approval project, and the application project has the same series number as the provision project; this

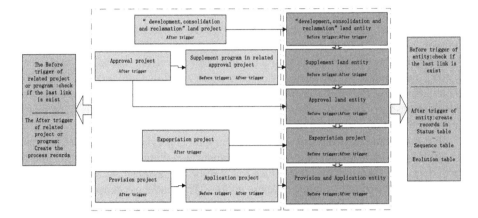

Fig. 7. The synchronized updating mechanism of spatio-temporal data based on trigger

series number is a unique key in the database. After the activation of the after trigger as data are imported into current land use basic tables, the process table creates related records. The before trigger is activated to check if the last stage of the project or program exists. The current stage data could be imported successfully based only on the existence of the data of the last stage. In the database, the importing operation of the provision program must be performed after the importing operation of its related approval project. For the application project, the provision project should be imported first and then the related land monitor information is obtained. The entire spatio-temporal supervision data of projects or programs should be imported into corresponding project or program tables in the following order: (1) consolidation, reclamation, and development project, (2) approval project, (3) supplement program in approval project, (4) expropriation project, (5) provision project, and (6) application project.

(c) When data are imported into the basic tables that record information on land use entities at the approval, expropriation, provision, application, and supplement land use stages, two triggers, namely, before and after triggers, are generated. The application land use entities are provision land use entities. The after trigger is used to add records into the status, sequence, and evolution tables, while the before trigger is used to check if the land use project, which is related to the current land use entity, exists. The importing operation of land use entity data should be followed by the importing operation of related land use project or program data. The entire spatio-temporal supervision data about land use entities should be imported into corresponding basic tables in the following order: (1) consolidation, reclamation, and development entities, (2) supplement entities, (3) approval project, (4) expropriation entities, and (5) provision entities.

In the distributed environment, to maintain the consistency and integrity of the database system, we can use the previously described synchronization and updating technology to import data using the principle "first land use projects, then land use entities."Based on the trigger technology, when land use project data are imported, the mechanism will synchronize and update the process table. When land use entity data are imported, the mechanism will synchronize and update the status, sequence, and evolution tables. Updates in basic table contents should also correspond to updates in auxiliary table records to maintain consistency and integrity of the construction land spatio-temporal supervision database and to ensure that all data are up to date. The entire process of tracing is performed quickly, thereby considerably enhancing vertical supervision efficiency.

The synchronization and updating technology ensures that the auxiliary tables are updated when the data are imported in the basic tables. This updating technology, which is based on the trigger mechanism of the database, is much more efficient than the synchronization and updating system proposed by Fu Zhongliang et al., which uses specific algorithms [15]. As the basic technology of the database, trigger is more effective than the extended program module in immediately responding to changes and in constructing the spatio-temporal topological relationship. Compared with the method that uses spatial data to check and record changes in the spatial database, which was carried out by Zhangjian [16], the technology reported in this paper does not require spatial operation and the use of log records. The auxiliary can reflect the spatial data change information indirectly, rendering the complex spatial operation unnecessary, and the trigger works on the auxiliary tables directly, thereby simplifying the updating process. Efficient supervision of spatio-temporal data of construction land use is also ensured.

5   Conclusions

In the current "province-city-county" distributed environment, the construction land use spatio – temporal supervision data model is used to execute vertical and horizontal supervision or to comprehend historical land use situation at a certain time by traceback and reconstruction. The supervision data are extracted from different places. Supervision databases should remain consistent for efficient construction of the construction land. Therefore, this study proposes a synchronization and updating technology based on the database trigger mechanism. The technology maintains land use project data and land use entity data that are imported into related tables at a certain order. The trigger mechanism will be activated as related records are added into auxiliary tables. When data from different nodes or levels are imported into the supervision database, the data are kept up to date and consistent in basic tables and auxiliary tables. Information on land use process evolution based on the land use entity life cycle is immediately generated in the database. While the synchronization and updating technology is based on the specific spatio-temporal supervision data organization model, suitable technologies for different and common situations may need further study.

Acknowledgment. Funds for this research was provided by the National Science and Technology Plan Projects (2013BAJ05B01).

References

1. Ministry of Land and Resources of the People's Republic of China. Circular on strengthening the supervision and administration of construction (National land resources (2008), No.92), 19–20 (2008)
2. Zonghua, L., Changling, L., Qiuying, T.: Realizaiton of Construction land dynamic supervision based on high-dimentional attribute intelligent fusion. Sci. Technol. Manag. Land Resour. **5**, 58–62 (2013)
3. Jiang, W., Wang, J., Yin, J., Chen, Y.: Research and construction of authorized supervision system for land expropriation and supplying. Geospatial Inf. **03**, 96–99 (2010)
4. Qiuying, T., Changlin, L., Rumin, W., et al.: The establishment and application on information system of construction land approval management of Wuhan. Urban Geotech. Invest. Surv. **05**, 9–12 (2011)
5. Gao, B., Zhou, Y., Pan, Y., et al.: Process-oriented spatio-temporal data model for organizing and management of supervision data of construction land use. Land Resour. Inf. **05**, 19–24 (2014)
6. Tian, D., Zhang, B.: Examination and approval mechanism for provincial, municipal and county database update of second land investigation. Land Resour. Inf. **03**, 39–42 (2010)
7. oracle8 Distributed database system. release2(8.1.6) [EB/OL]. http://technet.oracle.com/doc/oracle8!-816/server.816/a76960/2001,3,12
8. SQL Server Replication Revealed, Dell Power Solutions (2000)
9. Li, L.: Study on data consistency in distributed database system. Comput. Appl. Softw. **10**, 209–211 (2010)
10. Ding, K., Yan, H., Diao, X.: Research of data synchronization technology in distributed database. J. Naveal Univ. Eng. **05**, 100–104 (2004)
11. Wei, Z., Jiyun, L., Hui, J., et al.: Design and implementation of data synchronization based on distributed memory database. Mod. Electron. Technol. **37**(2), 77–83 (2014)
12. Haiming, Z.: The research and implementation of heterogeneous database synchronization technology based on SQL reduction method. Comput. Era **10**, 15–18 (2008)
13. Chen, W., Wen, X.: Design and implementation of distributed geographical database synchronization updating system. Bull. Surv. Mapp. **11**, 92–94 (2012)
14. Liu, S., Xu, S., Wang, L., et al.: Research on a trigger-based data synchronization system and the implementation of key techniques. Comput. Appl. Softw. **12**, 189–191 (2012)
15. Zhongliang, F.U., Gengqin, L.I.: Design and implementation of marinebase library synchronization update system. Geomatics World **22**(2), 93–96 (2015)
16. Jian, Z., Nan, L., Yu, L.: Research of spatial data real-time synchronization technology in PGIS. Urban Geotech. Invest. Surv. **1**, 48–50 (2013)

Comparison of Four Types of Raman Spectroscopy for Noninvasive Determination of Carotenoids in Agricultural Products

Chen Liu[5], Qingyan Wang[1,2,3,4], Wenqian Huang[1,2,3,4],
Liping Chen[1,2,3,4,5(✉)], Baohua Zhang[6], and Shuxiang Fan[5]

[1] Beijing Research Center of Intelligent Equipment for Agriculture,
Beijing 100097, China
[2] National Research Center of Intelligent Equipment for Agriculture,
Beijing 100097, China
{wangqy,huangwq,chenlp}@nercita.org.cn
[3] Key Laboratory of Agri-Informatics, Ministry of Agriculture,
Beijing 100097, China
[4] Beijing Key Laboratory of Intelligent Equipment Technology for Agriculture,
Beijing 100097, China
[5] College of Mechanical and Electronic Engineering,
Northwest A&F University, Yangling 712100, Shaanxi, China
xmyliuchen@126.com, fanshuxiang126@126.com
[6] State Key Laboratory of Mechanical System and Vibration,
Shanghai Jiaotong University, Shanghai 200240, China
zhangbaohua126@126.com

Abstract. Carotenoids are one class of naturally-occurring pigments with antioxidant properties. They can absorb light energy for use in photosynthesis for plants, and act as antioxidants to reduce risk of cancer for human. Carotenoids are confirmed to exist in agricultural products as the main source for human. Raman spectroscopy is a new technique for determination of carotenoids in agricultural products as it is both noninvasive and rapid. Four types of Raman spectroscopy could be used for contact measurement of carotenoids in fruits and vegetables: (1) Fourier transform Raman spectroscopy; (2) Resonance Raman spectroscopy; (3) Raman microspectroscopy; (4) Spatially Offset Raman spectroscopy. The experimental setups, advantages and applications of the above-mentioned Raman spectroscopies are discussed.

Keywords: Carotenoids · Fourier transform Raman spectroscopy · Resonance Raman spectroscopy · Raman microspectroscopy · Spatially offset Raman spectroscopy

1 Introduction

Carotenoids are one class of naturally-occurring pigments with antioxidant properties. They have positive effect on the immune system in animals and plants [1]. Most carotenoids cannot be synthesized by human beings directly. They widely exist in agricultural products such as tomatoes, watermelons, grapes and carrots. All carotenoids are

© IFIP International Federation for Information Processing 2016
Published by Springer International Publishing AG 2016. All Rights Reserved
D. Li and Z. Li (Eds.): CCTA 2015, Part II, IFIP AICT 479, pp. 237–247, 2016.
DOI: 10.1007/978-3-319-48354-2_25

produced from eight isoprene molecules and contain forty carbon atoms [2]. The best known carotenoids include lycopene and beta-carotene which is the vitamin A precursor. They all belong to the class of carotene which contains only carbon and hydrogen and no oxygen. There is another class called xanthophyll which contains oxygen such as lutein, zeaxanthin and astaxanthin. Carotenoids as natural pigments also contribute to the plants color when chlorophyll is not present [3]. That means carotenoids abound in ripe fruits and vegetables with the color of yellows, reds, or oranges [4].

At present, chromatography and spectroscopy are two common methods applied to determinate carotenoids in agricultural products [5–9]. High-performance liquid chromatography (HPLC) is the representative of chromatography. It is widely used for quantitative analysis due to its low detection limit and high accuracy [10]. However, this method may cost considerable amount of time for sample preparation because this method need to destroy sample. It also requires elaborately extracting the chromophore before experiment. However, it need not take any trouble about preprocessing in spectrum detection. Raman spectroscopy is a well-suited noninvasive method in this context. It is a non-destructive, non-contact analytical technique result in little sample preparation [11–13]. Based on that, many variations of Raman spectroscopies could be capable of enhancing the sensitivity, acquiring very specific information or improving the spatial resolution [14–17]. Raman spectroscopy has been proved to be a useful and powerful tool for carotenoids analysis.

2 Raman Spectroscopy for Carotenoids Analysis

A whole Raman spectroscopy instrument includes a laser excitation source, a sample chamber, a determination system and a data analysis system [18]. The lasers as the excitation sources provide directional monochromatic lights in vast majority of Raman measurements. Monochromaticity, directionality, and coherence are three unique properties of Raman excitation light source. The laser excitation sources are usually provided by lasers at 488, 514, 532, 633, 780/785 or 1064 nm. Compared with the requirement of lasers for fluorescence excitation, the linewidth of lasers for Raman excitation needs to be narrower. Another indispensable part of Raman system is sample chamber, which generally consists of lens, filters and sample holders. The role of lens is to obtain the most effective irradiation on the sample, and to maximize the collection of scattered lights. Filters, normally including notch filters and edge filters, are used to prevent Rayleigh scattering into the detected signals. The specifications of the filters are of great concern to the low wavenumber performance of Raman spectroscopy. A detection system always be divided into two parts: spectrometer and detector. Many technical parameters of spectrometer, such as resolution, spectral range, diffraction efficiency, aberrations, and stray lights, have crucial impact on the whole Raman system. However, the spectrometer would cause great loss of light intensity, thereby resulting in the decrease of sensitivity. It might be a way to improve sensitivity that obtains the required lights directly from narrowband filters without spectrometer. The detector could use a charge-coupled device (CCD) or a photomultiplier detector to collect Raman spectrum. The data analysis system is responsible for data pre-processing, data computing and data storage.

As we known, Raman spectroscopy is characterized by strong bands produced from stretching vibration of specific nonpolar groups. Based on conventional Raman determination, the C-C and C=C bonds in carotenoids just belong to nonpolar groups, and it is easy to observe strong bands within 1150–1170 cm^{-1} and 1500–1550 cm^{-1}, respectively. Additionally, a polyene chain with CH_3 groups attached could generate Raman shift in the 1000–1020 cm^{-1} region. However, there are some deficiencies in the process of in vivo conventional Raman determination: (1) fluorescence of samples is strong during the determination; (2) Raman spectrum is not abundant to distinguish various carotenoids; (3) it is unable to observe the structure of carotenoids in this resolution level; (4) it cannot detect internal Raman spectrum of solid samples. To solve above problems, Fourier transform Raman spectroscopy (FT-Raman), Resonance Raman spectroscopy (RRS), Raman microspectroscopy, and spatially offset Raman spectroscopy (SORS) are introduced into in vivo determination of carotenoids. These techniques will be discussed below. Besides, surface-enhanced Raman spectroscopy (SERS) is another rapid detection method. It depends on enhancement module contacting with samples to enhance the Raman scattering. This paper is mainly concentrated on the introduction of non-invasive and in vivo Raman methods, so SERS method is not discussed in this paper.

3 Fourier Transform Raman Spectroscopy

Raman spectroscopy relies on Raman scattering which is very weak excited by intense laser light [19]. Normally, the wavelength range of laser is in the visible, near infrared or near ultraviolet range [20]. When an intense laser light illuminating on carotenoids, the electrons in molecules of carotenoids would emit light with low energy, leading to both fluorescence emission and Raman scattering in a wide wavelength range. Under this background, FT-Raman is developed to reduce the interference of fluorescence. FT-Raman uses an Nd:YAG laser (1064 nm) to obtain the maximum information available, moreover, the problem of fluorescence background could be circumvented. In addition, a longer integration time and higher power lasers are required at the same time [21]. Figure 1 shows one type of FT-Raman experimental setup. The monochromatic light (1064 nm) emitting by the laser is collimated by the lens, reflected by the mirror, and irradiated on the sample. The Raman signal is collected by a Michelson interferometer after filtering by a dichroic filter. The Michelson interferometer could produce a relative improvement in signal-to-noise ratio. Then, the scattering light passed through a dielectric filter and finally focused into a liquid N_2 cooled detector.

The application of FT-Raman is determining special samples with strong fluorescence. FT-Raman eliminates the fluorescence background effectively by using excitation laser in near-infrared wavelength (1024 nm). It is an easy way to distinguish between similar isomers. Compare to Fourier transform infrared spectrum (FT-IR), FT-Raman shows good applicability on aqueous solution. For this reason, FT-Raman could determine carotenoids directly in vivo, such as plant epidermis, fruits, and human skin. H. Schulz et al. demonstrated the feasibility of FT-Raman spectroscopy applied on carotenoids determination in floristic kingdom. It presented an especial way to detect the distribution of several carotenoids and distinguish various quantity of double bonds

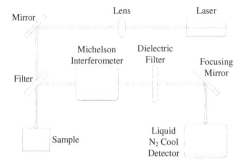

Fig. 1. Schematic diagram of FT-Raman spectrometer

conjugated carotenoids in the intact plant tissue [22]. Similarly, Vanessa E. de Oliveira studied on natural carotenoids in more than 50 samples of plant tissue by FT-Raman. They recorded the characteristic bands of C-C and C=C stretching and C-CH$_3$ bending in each sample. They also found that the characteristic key bands were arose from their own molecular interactions. [23]. Compared to ATR-IR and NIR reflection spectroscopy, FT-Raman spectrum of tomato products showed significant wavenumber shifts of targeted carotenoids, and it showed a good reliability on lycopene ($R^2 = 0.91$ and $SECV = 74.34$) and beta-carotene ($R^2 = 0.89$ and $SECV = 0.34$) [24]. Based on Raman images, the spatial distribution of carotenoids in intact carrot roots of different colors was evaluated by Malgorzata Baranska et al. [25]. They also defined tissue specific accumulation of several variants of carotenoids by Raman images in the carrot root cross section [26]. However, the laser device at 1064 nm wavelength is always large and expensive, and this is a limiting factor of FT-Raman to be portable.

4 Resonance Raman Spectroscopy

The feature of RRS is to define characteristic vibration bands of corresponding model compound. RRS requires a special wavelength of the laser which is used to obtain the selective spectra. The excitation frequency must be close in energy to the electronic transition of the sample. For carotenoids, the optional excitation wavelength range depends on the numbers of C=C in molecules from 410 to 520 nm. Carotenoids like lycopene, alpha-carotene, beta-carotene and their isomers, which own 11 conjugated carbon double-bonds, have approximate Raman shift at 488 nm. However, only lycopene owns special peak position at 514 nm with the maximal difference [27]. In other carotenoids, the maximum of absorption values are located at 515 and 490 nm for diadinoxanthin and diatoxanthin, respectively [28]. By comparing C=C band at different excitation wavelength for further study, the peak positions are very similar to those of lutein at 514, 496 or 476 nm. The peak position of violaxanthin is at 502 nm and another variant neoxanthin is at 488 or 458 nm [29]. But in practical applications, agricultural products contain more than one type of carotenoids in most instances [30]. Based on that, the optimal excitation wavelength for most carotenoids is determined at 488 or 514 nm. Figure 2 shows a typical RRS experimental setup applied to

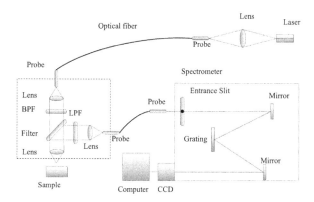

Fig. 2. Schematic diagram of RRS experimental setup

carotenoids. The constitutive requirement of RRS is similar to FT-Raman. The main distinction is the exciting beam filtered by a BPF (band pass filter) and focused onto the sample, then using a dichroic filter and a LPF (low pass filter) to eliminating the laser. The band of BPF and LPF are according to the sample under test.

The main advantage of RRS method is its high sensitivity. It makes the determination of a small quantity of carotenoid molecules in different agricultural products realized. Dane Bicanic et al. used RRS method combined with tristimulus colorimetry to rapidly detect beta-carotene in mango homogenates. At a low concentration of carotenoids, RRS featured the characteristic bands of beta-carotene at 1008, 1158 and 1523 cm^{-1}. The correlation between beta-carotene content and relative intensity of Raman scattering was linear with R^2 = 0.962 [31]. Bhosale P et al. also measured the levels of carotenoids contents in farm products especially fruit juice by RRS and compared to HPLC [32]. It shows complementation between the two methods: HPLC had a good performance in evaluating juices, while RRS was suitable for determining content of carotenoids in the same maturity of intact tomatoes. Based on resonance Raman effects, Ouyang S L, et al. obtained the content of lycopene and beta-carotene by measuring the second harmonic of stretching vibration of C=C directly in lycopene and beta-carotene, respectively [33]. Yan J, et al. also confirm that *9-cis*-beta-carotene exists in cytochrome *b6f* complex of spinach by HPLC and RRS [34]. Compared to chromatography, the measurement of RRS is quick, usually not exceeding 5 s [35]. However, in practical applications, the measurement may maintain for some time to decrease the influences of fluorescence background. In addition, the high universalizable of RRS methods could be another advantage [36].

5   Raman Microspectroscopy

Traditionally Raman microspectroscopy is used to measure Raman spectrum of a micro-scale region on a sample. The main application is to observe the molecular concentration profiles. The system is usually equipped with an optical microscope combined with an excitation laser, a sensitive detector and some other optical devices.

This method could focus on the distribution of carotenoids in tiny samples with a high spatial resolution of about 250 nm [37]. Since the size of laser beam is only several micrometers in diameter, it could reduce the fluorescence during the experiment. For this reason, Raman microspectroscopy is suitable for determining carotenoids in emulsions or homogenates [38]. To further increase the optical resolution, Confocal Raman microscopy is designed to investigate carotenoids, the principle of which is to eliminate out-of-focus lights by means of adding a spatial pinhole placed at the confocal plane of the lens. At the expense of long exposures, the optical resolution could confine in hundreds of nanometers [39]. A characteristic block diagram of Raman microspectroscopy experimental setup is shown in Fig. 3. A microscope objective is added to a common Raman spectrum measuring system. In the process of detection, the laser (785 or 514 nm) need to pass through the objective and focus on the sample. The scattered signal caused by laser is also collected by the objective. Therefore, a laser rejection filter need to attenuate the laser in a very low level. After that, the scattered light can be collected into the spectrometer. The spectrum is obtained from a CCD camera, analyzed and displayed on the computer similarly.

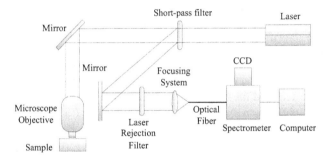

Fig. 3. Schematic diagram of Raman microspectroscopy experimental setup

The advantage of Raman microspectroscopy is the nano-scale resolution and high accuracy. Based on that, Pudney P D A, et al. showed the spatial locations of three main carotenoids within tomato cells, and also demonstrated the follow changes of carotenoids [40]. It is confirmed that the confocal Raman microscope is a powerful tool to measure the green tomatoes. It can offer a full-scale Raman spectrum of the organic composition in unripe tomatoes. This method was also applied to detect carotenoids in ripe tomatoes in which the major carotenoid is lycopene [41]. In addition, the influence factors of external environment were discussed. Cecilia A. Svelander et al. evaluated the microstructure of carotenes in tomato and carrot emulsions under varying degrees of high pressure homogenization. [42]. Paolo Camorani et al. proposed confocal micro-Raman spectroscopy as a spatially resolved method to evaluate the changes of carotenoid pattern after thermal treatment of carrots. They have used steaming, boiling and microwaving way to cook carrot. The difference before and after are measured by Raman spectroscopy and HPLC [43]. Combined with density functional theory, the molecular structure are also analyzed by comparing the calculated and determination results [44].

6   Spatially Offset Raman Spectroscopy

SORS is an unconventional method specialized in obtaining special information which beneath the surface of objects, such as drugs [45], explosives [46] and plastic [47]. Raman spectroscopies above-mentioned have pimping near-surface detection limits. In contrast, SORS has an effective retrieval of Raman spectra at deep subsurface layers. The basic principle of SORS is to obtain the Raman scattering information using two spatially offset points: one is the location of illumination and the other is the location of determination [48]. There are several millimeters separation distance between two places. The purpose is to avoid the Raman scattering at domination excitation region [49]. The experimental setup of SORS is unlike conventional Raman instruments. The main distinction is the mode of collecting Raman scattering. Figure 4 shows the distinctions of conventional Raman and SORS approach. As shown in Fig. 4, the laser excitation point and the sample collection point are at the same location in conventional Raman. In contrast, Raman spectra in standard SORS approach are collected from the spatial regions away from the place of illumination, and the direction of laser irradiation is opposite to the probe which collects the Raman scattering. When detecting transparent or liquid samples, an adapted SORS approach is applied for analyzing variety of samples [50]. This method enhances the scattering signal of the liquid or transparent sample directly under the Raman detection area. It realizes detecting bottled liquid directly cause of Raman scattering of liquid could not be easily lost within the noise. As the offset distances increasing, however, the Raman signal would be barely discernible. In this case, this method may require adding multivariate analysis.

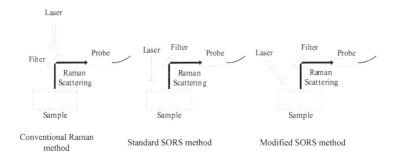

Fig. 4.   Schematic diagrams of conventional Raman and SORS

The advantage of SORS method is non-invasive determination of subsurface. Three Raman spectroscopies above can measure near-surface only with a few hundred micrometers depth. By contrast, the determination depth of SORS could reach several millimeters. Due to the complexity of SORS instrument, the application of SORS for carotenoids detection is not widespread. Qin et al. developed a benchtop point-scan Raman chemical imaging system to show the detail distribution of lycopene during postharvest ripening of tomatoes utilizing SORS technique. The tomato samples were selected and cut open from green stage to red stage. The results showed that the SORS technique was capable of obtain internal Raman chemical information in tomato [51]. Afseth N.K., et al. explored the possibility of determination carotenoids of salmon through the skin [52]. At an offsets around 5 mm with an 830 nm laser, the results clearly showed that information regarding carotenoids content could be extracting from both sides of the skin. These applications demonstrated the potential of SORS technology as a new method for subsurface determination of carotenoids.

7 Conclusion

Raman spectroscopy is a powerful tool to research on carotenoids due to its unique capability of fingerprint for identifying and characterizing the structure of carotenoids molecules. Raman spectroscopy also shows great potential in qualitative and quantitative analysis on carotenoids. In recent years, with advances in optical instruments, several variations of Raman spectroscopies is increasingly widespread such as FT-Raman, RRS, Micro-Raman and SORS. The contrast of these four methods is shown in Table 1. However, new technologies and ideas should be provided to improve insufficient. Many features such as background fluorescence, category distinction of carotenoids, interference and absorbance of reaction mixture still affect the detection results. For different agricultural products, the most crucial is to explore proper detection mode of non-invasive, non-destructive or in vivo. In addition, new perspectives such as femtosecond stimulated Raman spectroscopy and transmission resonance Raman spectroscopy are promising for improvement of the precision and applicability of carotenoids study. The combination of Raman spectroscopy and chromatography will be promising for overcoming many defects. It is expected that Raman spectroscopy will applied in on-line fast determination of carotenoids in the future.

Table 1. The contrast of four variations of Raman spectroscopy

Techniques	Principles	Advantages
Fourier Transform Raman Spectroscopy	Operation with FTIR spectrometer at 1064 nm	Fluorescence suppression
Resonance Raman Spectroscopy	Special excitation wavelength at 488 or 514 nm	Raman signal enhancement
Raman Microspectroscopy	Combination with an optical microscope	Higher spectral resolution
Spatially Offset Raman Spectroscopy	Uses two spatially offset points to obtain subsurface spectra	More subsurface information

Acknowledgment. This work is supported in part by Research Foundation for Young Scholar of BAAFS (No. QNJJ201423).

References

1. Blackburn, G.A.: Quantifying chlorophylls and Caroteniods at leaf and canopy scales: an evaluation of some hyperspectral approaches. Remote Sens. Environ. **66**(98), 273–285 (1998)
2. Viuda-Martos, M., Sanchez-Zapata, E., Sayas-Barberá, E., et al.: Tomato and tomato byproducts. human health benefits of Lycopene and its application to meat products: a review. Critical Rev. Food Sci. Nutr. **54**(8), 1032–1049 (2014)
3. Yang, D., Ying, Y.: Applications of Raman spectroscopy in agricultural products and food analysis: a review. Appl. Spectrosc. Rev. **46**(7), 539–560 (2011)
4. Zheng, J., He, L.: Surface-enhanced raman spectroscopy for the chemical analysis of food. Compr. Rev. Food Sci. Food Safety **13**(3), 317–328 (2014)
5. Pacheco, S.: Microscale extraction method for HPLC carotenoid analysis in vegetable matrices. Scientia Agricola **71**(5), 416–419 (2014)
6. Payyavula, R.S., Navarre, D.A., Kuhl, J.C., et al.: Differential effects of environment on potato phenylpropanoid and carotenoid expression. BMC Plant Biol. **12**(2), 39–57 (2012)
7. Fraser, P.D., Bramley, P.M.: The biosynthesis and nutritional uses of carotenoids. Prog. Lipid Res. **43**(3), 228–265 (2004)
8. Rodriguez-Amaya, D.B., Kimura, M., Godoy, H.T., et al.: Updated Brazilian database on food carotenoids: factors affecting carotenoid composition. J. Food Compos. Anal. **21**(6), 445–463 (2008)
9. Wright, S.W., Jeffrey, S.W., Mantoura, R.F.C., et al.: Improved HPLC method for the analysis of chlorophylls and carotenoids from marine phytoplankton. Marine Ecol. Prog. **77** (2–3), 183–196 (1991)
10. Stephanie, S., Brenda, C., Haiqun, L., et al.: Single v. multiple measures of skin carotenoids by resonance Raman spectroscopy as a biomarker of usual carotenoid status. Br. J. Nutr. **110** (5), 911–917 (2013)
11. Synytsya, A., Judexova, M., Hoskovec, D., et al.: Raman spectroscopy at different excitation wavelengths (1064, 785 and 532 nm) as a tool for diagnosis of colon cancer. J. Raman Spectrosc. **45**(10), 903–911 (2014)
12. Sselhausa, M.S., Sselhausb, G., Joriod, R.S.: Raman spectroscopy of carbon nanotubes. Phys. Rep. **409**(2), 47–99 (2005)
13. Moskovits, M.: Surface-enhanced Raman spectroscopy: a brief retrospective. J. Raman Spectrosc. **36**(6–7), 485–496 (2005)
14. Li, J.F., Huang, Y.F., Ding, Y., et al.: Shell-isolated nanoparticle-enhanced Raman spectroscopy. Nature **464**(7287), 392–395 (2010)
15. Macernis, M., Sulskus, J., Malickaja, S., et al.: Resonance Raman spectra and electronic transitions in Carotenoids: a DFT study. J. Phys. Chem. A **118**(10), 1817–1825 (2014)
16. He, L., Lamont, E., Veeregowda, B., et al.: Aptamer-based surface-enhanced Raman scattering detection of Ricin in liquid foods. Chem. Sci. **2**(8), 1579–1582 (2011)
17. Burgio, L., Clark, R.J.H.: Library of FT-Raman spectra of pigments, minerals, pigment media and varnishes, and supplement to existing library of Raman spectra of pigments with visible excitation. Spectrochimica Acta Part A Mol. Biomol. Spectrosc. **57**(7), 1491–1521 (2001)
18. Du, X., Chu, H., Huang, Y., et al.: Qualitative and quantitative determination of Melamine by surface-enhanced Raman spectroscopy using silver nanorod array substrates. Appl. Spectrosc. **64**(7), 781–785 (2010)
19. Ferrari, A.C., Meyer, J.C., Scardaci, V., et al.: Raman spectrum of graphene and graphene layers. Phys. Rev. Lett. **97**(18), 13831–13840 (2006)

20. Kelly, J.F., Blake, T.A., Bernacki, B.E., et al.: Design considerations for a portable Raman probe spectrometer for field forensics. Int. J. Spectrosc. **10**, 1–15 (2012)
21. Mangolim, C.S., Moriwaki, C., Nogueira, A.C., et al.: Curcumin–β-cyclodextrin inclusion complex: Stability, solubility, characterisation by FT-IR, FT-Raman, X-ray diffraction and photoacoustic spectroscopy, and food application. Food Chem. **153**, 361–370 (2014)
22. Schulz, H., Baranska, M., Baranski, R.: Potential of NIR-FT-Raman spectroscopy in natural carotenoid analysis. Biopolymers **77**(4), 212–221 (2005)
23. De Oliveira, V.E., Castro, H.V., Edwards, H.G.M., et al.: Carotenes and carotenoids in natural biological samples: a Raman spectroscopic analysis. J. Raman Spectrosc. **41**(6), 642–650 (2010)
24. Baranska, M., Schütze, W., Schulz, H.: Determination of lycopene and beta-carotene content in tomato fruits and related products: comparison of FT-Raman, ATR-IR, and NIR spectroscopy. Anal. Chem. **78**(24), 8456–8461 (2006)
25. Baranski, R., Baranska, M., Schulz, H.: Changes in carotenoid content and distribution in living plant tissue can be observed and mapped in situ using NIR-FT-Raman spectroscopy. Planta **222**(3), 448–457 (2005)
26. Baranska, M., Baranski, R., Schulz, H., et al.: Tissue-specific accumulation of carotenoids in carrot roots. Planta **224**, 1028–1037 (2006)
27. Darvin, M.E., Gersonde, I., Meinke, M., et al.: Non-invasive in vivo determination of the carotenoids beta-carotene and lycopene concentrations in the human skin using the Raman spectroscopic method. J. Phys. D Appl. Phys. **38**(15), 2696–2700 (2005)
28. Alexandre, M.T.A., Gundermann, K., Pascal, A.A., et al.: Probing the carotenoid content of intact Cyclotella cells by resonance Raman spectroscopy. Photosynth. Res. **119**(3), 273–281 (2014)
29. Andreeva, A., Velitchkova, M.: Resonance Raman studies of carotenoid molecules within photosystem I particles. Biotechnol. Biotechnol. Equip. **23**, 488–492 (2009)
30. Withnall, R., Chowdhry, B.Z., Silver, J., et al.: Raman spectra of carotenoids in natural products. Spectrochimica Acta Part A Mol. Spectrosc. **59**(10), 2207–2212 (2003)
31. Bicanic, D., Dimitrovski, D., Luterotti, S., et al.: Estimating rapidly and precisely the concentration of beta carotene in mango homogenates by measuring the amplitude of optothermal signals, chromaticity indices and the intensities of Raman peaks. Food Chem. **121**(3), 832–838 (2010)
32. Bhosale, P., Ermakov, I.V., Ermakova, M.R., et al.: Resonance Raman quantification of nutritionally important carotenoids in fruits, vegetables, and their juices in comparison to high-pressure liquid chromatography analysis. J. Agric. Food Chem. **52**(11), 3281–3285 (2004)
33. Ouyang, S.L., Zhou, M., Cao, B., et al.: Lycopene and β-Carotene content in tomato analyzed by the second harmonic. Spectrosc. Spectral Anal. **29**(3), 3362–3364 (2009)
34. Yan, J., Liu, Y., Mao, D., et al.: The presence of 9-cis-beta-carotene in cytochrome b(6)f complex from spinach. Biochim. Biophys. Acta **1506**(3), 182–188 (2001)
35. Tschirner, N., Schenderlein, M., Brose, K., et al.: Resonance Raman spectra of beta-carotene in solution and in photosystems revisited: an experimental and theoretical study. Phys. Chem. Chem. Phys. **11**(48), 11471–11478 (2009)
36. Efremov, E.V., Ariese, F., Gooijer, C.: Achievements in resonance Raman spectroscopy: Review of a technique with a distinct analytical chemistry potential. Anal. Chim. Acta **606**(2), 119–134 (2008)
37. Thygesen, L.G., Lokke, M.M., Micklander, E., et al.: Vibrational microspectroscopy of food. Raman vs. FT-IR. Trends Food Sci. Technol. **14**(1), 50–57 (2003)

38. Dane, B., Darko, D., Svjetlana, L., et al.: Correlation of trans-Lycopene measurements by the HPLC method with the Optothermal and Photoacustic signals and the color readings of fresh tomato homogenates. Food Biophys. **5**(1), 24–33 (2010)

39. Piot, O., Autran, J.C., Manfait, M.: Spatial distribution of protein and phenolic constituents in wheat grain as probed by confocal Raman microspectroscopy. J. Cereal Sci. **32**, 57–71 (2000)

40. Pudney, P.D.A., Gambelli, L., Gidley, M.J.: Confocal Raman microspectroscopic study of the molecular status of carotenoids in tomato fruits and foods. Appl. Spectro. **65**(2), 127–134 (2011)

41. Trebolazabala, J., Maguregui, M., Morillas, H., et al.: Use of portable devices and confocal Raman spectrometers at different wavelength to obtain the spectral information of the main organic components in tomato (Solanum lycopersicum) fruits. Spectrochimica Acta Part A Mol. Biomol. Spectro. **105**(6), 391–399 (2013)

42. Svelander, C.A., Lopez-Sanchez, P., Pudney, P.D.A., et al.: High pressure homogenization increases the in vitro bioaccessibility of alpha- and beta-Carotene in carrot emulsions but not of Lycopene in tomato emulsions. J. Food Sci. **76**(9), 215–225 (2011)

43. Camorani, P., Chiavaro, E., Cristofolini, L., et al.: Raman spectroscopy application in frozen carrot cooked in different ways and the relationship with carotenoids. J. Sci. Food Agric. **95** (11), 2185–2191 (2014)

44. Liu, W., Wang, Z., Zheng, Z., et al.: Density functional theoretical analysis of the molecular structural effects on Raman spectra of beta-Carotene and Lycopene. Chin. J. Chem. **30**, 2573–2580 (2012)

45. Olds, W.J., Jaatinen, E., Fredericks, P., et al.: Spatially offset Raman spectroscopy (SORS) for the analysis and detection of packaged pharmaceuticals and concealed drugs. Forensic Sci. Int. **212**(1–3), 69–77 (2011)

46. Loeffen, A.P.W., Maskall, G., Bonthron, S., et al.: Chemical and explosives point detection through opaque containers using spatially offset Raman spectroscopy (SORS). Proc. SPIE **8018**(1), 413–414 (2011)

47. Bloomfield, A.M., Loeffen, P.W., Matousek, P.: Detection of concealed substances in sealed opaque plastic and coloured glass containers using SORS. Proc. SPIE – Int. Soc. Optical Eng. **7838**(1), 125–131 (2010)

48. Matousek, P., Morris, M.D., Everall, N., et al.: Numerical simulations of subsurface probing in diffusely scattering media using spatially offset Raman spectroscopy. Appl. Spectrosc. **59** (12), 1485–1492 (2005)

49. Charlotte, E., Pavel, M.: Noninvasive authentication of pharmaceutical products through packaging using spatially offset Raman spectroscopy. Anal. Chem. **79**(4), 1696–1701 (2007)

50. Eliasson, C., Macleod, N.A., Matousek, P.: Noninvasive detection of concealed liquid explosives using Raman spectroscopy. Anal. Chem. **79**(21), 8185–8189 (2007)

51. Qin, J., Chao, K., Kim, M.S.: Investigation of Raman chemical imaging for detection of lycopene changes in tomatoes during postharvest ripening. J. Food Eng. **107**, 277–288 (2011)

52. Afseth, N.K., Bloomfield, M., Wold, J.P., et al.: A novel approach for subsurface through-skin analysis of salmon using spatially offset Raman spectroscopy (SORS). Appl. Spectrosc. **68**(2), 255–262 (2014)

The Molecular Detection of *Corynespora Cassiicola* on Cucumber by PCR Assay Using DNAman Software and NCBI

Weiqing Wang[✉]

Beijing Vocational College of Agriculture, Beijing, China
weiqingfine@163.com

Abstract. Objective: to establish a quick molecular detection method in Beijing, which can prevent the occurrence of cucumber target spot disease from the source. Methods and results: The DNA band of *Corynespora cassiicola* had been obtained by PCR and sequenced. Using DNAman software and NCBI database to analyze the sequence, the results showed that the obtained DNA was that of *Corynespora cassiicola* on Cucumber. A specific primers CCC1/CCC2 were obtained by DNAman software and NCBI database. It was also proved to can be used to distinguish *Corynespora cassiicola* from other pathogenic fungi using DNAman software and NCBI database.

Keywords: Molecular detection · *Corynespora cassiicola* · DNAman · NCBI

1 Introduction

A cucumber target spot disease caused by *Corynespora cassiicola* (Berk. & Curt.) occurs serious damage. The leaves are the main victims and in severe cases the pathogenic fungi can spread to the petioles and even to the vine. Both on the top and back of leaves, large and small necrotic lesions can be formed and there is one white bull's eye on the center and even cause leaves dry by the humidity lesion. With the development of modern agriculture, in 2012 the agricultural facilities are expected to develop 35 acres. Facilities agriculture has been positioned as a main direction in the development of agriculture of Beijing. Cucumber, as an important vegetable in Beijing has been more attention in pest and disease control work. Spot disease caused by *Corynespora cassiicola* in cucumber is very serious and rapid progression in recent years and has caused more and more damages in China. Its occurrence was the growing trend and is causing serious economic losses to farmers from 2005 to 2010.

The current study shows that seed infection is an important spread way of cucumber target spot disease (*Corynespora cassiicola*) and in seeds markets the parameters related to purity, germination rate, moisture content are often detected. But the seed infection is often ignored, which often results in the occurrence of target leaf spot disease caused by *Corynespora cassiicola* in cucumber and results in late serious impact on growth and yield.

Traditional detection methods are time-consuming, low sensitivity, susceptible to the interference of man-made and environmental factors, etc. so it is necessary to

---

The original version of this paper was revised: The affiliation of the author was corrected. The Erratum to this chapter is available at DOI: 10.1007/978-3-319-48354-2_62

D. Li and Z. Li (Eds.): CCTA 2015, Part II, IFIP AICT 479, pp. 248–258, 2016.
DOI: 10.1007/978-3-319-48354-2_26

establish a fast, easy and accurate detection method, provide the basis for early diagnosis and prevention of disease, which can prevent the occurrence of cucumber *Corynespora cassiicola* from the source. And if the seed can be resolved through physical measures, it can be effective technical measures to the comprehensive prevention and control of the target leaf spot in cucumber. This study aims to establish a molecular detection method using DNAMAN software and NCBI to detect the spot disease caused by *Corynespora cassiicola* in cucumber. The rDNA-ITS sequences of fungi which is the moderately repetitive sequence widely distributed in the genome. It has been reported to use the diversity of the rDNA-ITS sequences of fungi in the level of family, genus, and species to design specific primers to detect fungi.

2   Methodology

DNAMAN is the application of molecular biology in all software packages. The software package provides an integrated system with multiple features for efficient sequence analysis. This software can be used to do multiple sequence alignment for a restriction analysis, design primer, protein sequence analysis or graph. DNAMAN's speed, precision, and high quality versatility makes it a fundamental tool for each molecular biologist to rely on. It is also a sequence analysis software package to each university, research institution, laboratory and research scientist with affordable price. DNAMAN can be used in Microsoft Windows, MacOS and Linux. All three platforms of the DNAMAN file share the same format. DNAMAN common format of the system will help the communication between the PC, Macintosh and Linux, and make your work platform independent.

2.1   Materials and Methods

Procedures for the use of DNAman: we have got scar bacteria gene has been got, using DNAman software of sequence specific primers were designed, and the sequence structure analysis, to expect to get the sequence related biological information data, such as: open reading frame, amino acid sequence and protein translation simulation map, and he and other fungal homologous of distance analysis.

2.2   DNAMAN Software Sequence Alignment Method

Insertion sequence, multiple sequence alignment:

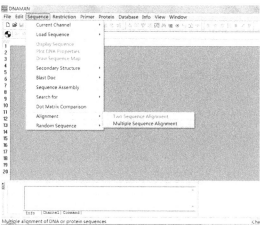

After analysis of the phylogenetic tree, you can export to Clustal format, you can use bioedit to view

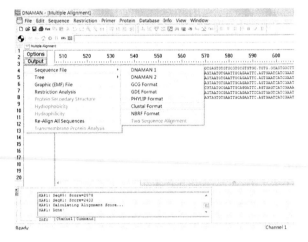

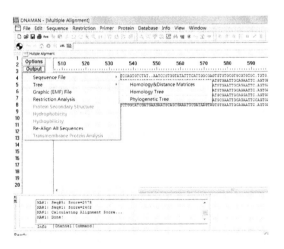

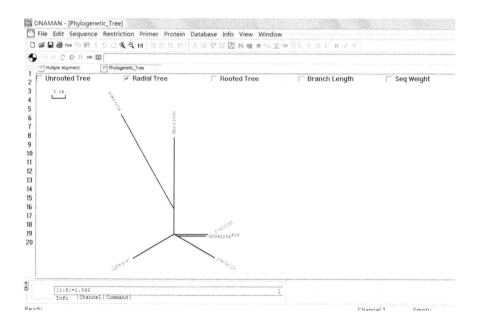

2.3   NCBI Database

Blast is a search programs based on sequence similarity developed by the U.S. Biotechnology Information (NCBI) database. Blast is the abbreviation of "partial similarity query tool" (Basic Local the Alignment Search Tool).

NCBI BLAST is a sequence similarity search program development, but also as to identify genes and genetic characteristics means. BLAST can be less than 15 s of time for the entire DNA sequence databases to search. NCBI provides additional software tools are: Finder open reading frame (ORF Finder), Electronic PCR, and sequencing submission tool, Sequin and BankIt. All of the NCBI database and software tools can be obtained from the WWW or FTP. NCBI and E-mail server, providing a text search or sequence similarity search an alternative way to access the database.

The blast software of NCBI Sequence alignment method

Input the sequence

Blast

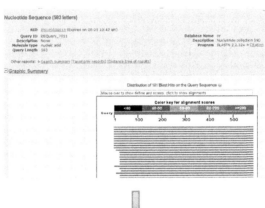

The results showed that the highest gene sequence was obtained.

By blast sequence analysis method of main program is: by gene sequence has received the scar bacteria directly in NCBI nucleic acid libraries for comparison, direct comparisons have been obtained and nucleic acid sequences base sequence homology. The highest number of genes that have been registered with the highest homology sequence homology was obtained by comparison. If the similarity reached 99 %, the obtained sequence should be the gene.

2.4   The Tested Strains Were Provided by the Plant Protection Station of Beijing

The tested strains were provided by the plant protection station of Beijing. The tested strains culture: the strains were inoculated into potato dextrose liquid medium, 28 °C, 145 r/m oscillation culture between 4 and 7 d, filtration collection mycelium, and then placed in dry heat sterilization box at the temperature of 60 °C drying – 20 °C frozen preservation reserve. The strain was inoculated onto the agar medium, 28 °C, and 4–7d.

2.5   Cucumber Seed Source

Tested cucumber seed source: Beijing 10 suburban counties (Fangshan, Miyun, Daxing, Changping, Pinggu, Yanqing, etc.); Cucumber (Zhongnong 16, Beijing 203, Beijing 204), fruit cucumber (Mini 2, Dai Duoxing).

2.6   DNA Extraction and rDNA-ITS Amplification and Sequencing

The method of the genomic DNA of the tested strains and the total DNA extraction was use by White T;: Total DNA was extracted from Trout C L's methods: the 0.25 g soil samples were milled into powder, and 0.5 mL 0.4 % of the milk powder solution vortex suspension was added. 12 000 R/m in centrifugal m 3, with 2 ml of supernatant was 0. 3 % SDS extraction buffer vortex mixed, then add other volume of phenol: chloroform: isoamyl alcohol (25: 24: 1, V/V) solution upside down mixing, 12000 r/m in centrifugal 25 min. In the supernatant phase to another centrifugal tube, the addition of 0.6 times the volume of cold ISO alcohol, 4 °C in 20 min, Centrifuge for 10 min in 12000r/m. In the supernatant, wash 2 times with 70 % alcohol. After drying, the precipitation was dissolved by TE with RNase A, and the reserve was kept at 20 °C.

rDNA-ITS region of cucumber scar blotch were amplified using ribosomal DNA in eukaryotes universal primers ITS1 and its4 [11] and the sequence as follows: ITS1: 5 '-TCCGTAGGTGAACCT2GCGG-3', ITS4: 5 '-TCCTCCGCTTATTGATATGC-3'. Reaction system (25 μL): ddH$_2$O17. 2 μL, 10 x buffer of 2.5 MμL, MgCl$_2$ 1. 5 μL, dNTP (10 mmol/L) of 0.5 μL, 10 mol/L of universal primers ITS1/ITS4 1 μL, 1 μL of template DNA, Taq polymerase 0. 3 μL. PCR reaction program was as follows: 94 °C 3 min as pre degeneration, 94 °C 1 min, 56 °C1 min, 72 °C 50 s as a cycle, and repeats 32 cycles; then 72 °C 10 min as a extension. The PCR products by 1.0 % agarose gel electrophoresis detection and recovery of connection and transformation, sequence by Beijing Shanghai Biological Engineering Co., Ltd.

2.7   Design of Specific Primers

The rDNA-ITS sequence of the *Cladosporium tenuissimum* (NO. FJ603350, FR822778, FR822800, FR822816, FR822843和FR822848) and the *Botryotinia fuckeliana* (NO. GU062311.1), *Trichoderma atroviride* (NO. JP665257), *Corynespora cassiicola* (NO. JQ595296.1) of cucumber were download by GenBank. Specific primers (CCC1/CCC2) designed and CCC1/CCC2 primers and the comparison between specific primers (CCC1/CCC2 and sequencing of the ITS sequence of the scars were got by using DNAMAN, DNAStar and Premer5 softwares.

2.8   Primer Specificity Verification

PCR was amplified by CCC1/CCC2 with the specific primers and the genomic DNA of all the strains tested as the template. Reaction system and procedure 1.5. PCR products were detected by 1 % agarose gel electrophoresis (Fig. 1).

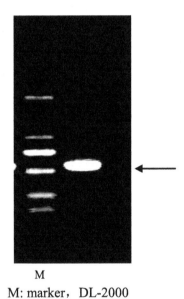

M
M: marker, DL-2000

Fig. 1.   Gene PCR electrophoresis of scar plaque

3   Results and Analysis

3.1   Obtaining the Sequence of *Corynespora Cassiicola*

According to GeneBank query using fungal ribosomal rDNA universal primers ITS1 and ITS4, then got the DNA of pathogenic fungi by PCR. After sequencing, the sequence was compared using the blast software of NCBI. The alignment results were: the similar rate of the DNA sequence of using ITS1 and ITS4 as primers and that of

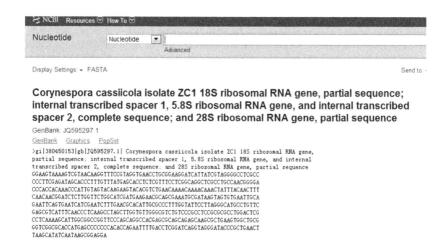

Fig. 2. The DNA sequence of GenBank accession number JQ595297

Fig. 3. Sequence producing significant alignments

scar fungus *Corynespora cassiicola* (GenBank accession number JQ595297 (Fig. 2)) can be reached 100 % (Fig. 3) in NCBI database. The results showed that DNA band obtained from PCR is the DNA fragment from fungus *Corynespora cassiicola*.

3.2   Homology Analysis

The homology analysis of the DNA sequences of *Corynespora cassiicola* (GenBank accession number JQ595297) and *Botryotinia fuckeliana* (GenBank accession number FJ903283, HM849615), *Cladosporium sp.* (GenBank accession number GU062286, JN974012), *Phytophthora capsici* (GenBank accession number HQ643180), *Pseudoperonospora cubensis* (GenBank accession number EU826114.1) were done by DNAman. The results showed that they had homology relationship between them (Fig. 4). The closest relationship with *Corynespora cassiicola* (GenBank accession number JQ595297) was *Botryotinia fuckeliana* (GenBank accession number FJ903283, HM849615), followed by *Cladosporium sp.* (GenBank accession number GU062286,

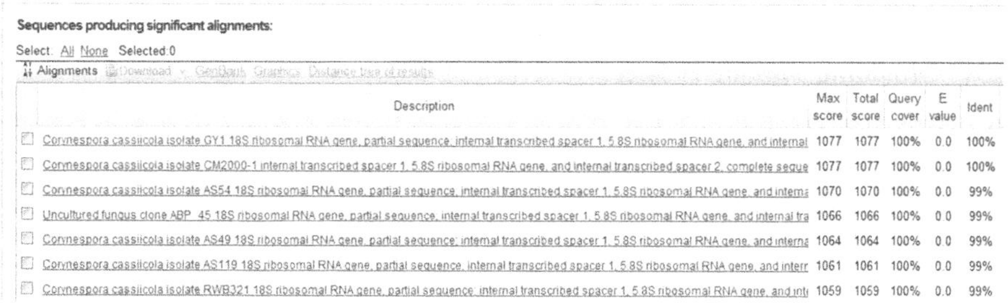

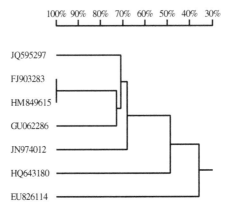

Fig. 4. The homology tree of *Corynespora cassiicola* and other Fungus pathogenic fungi

JN974012), and the most distant relative to a was *Pseudoperonospora cubensis* (GenBank accession number EU826114.1).

3.3 Obtaining Specific Primers

The DNA of *Corynespora cassiicola* was analyzed by the DNAman software (Fig. 4) and the specific primers were obtained from the Fig. 5.
    CCC1(TCGTAGGGGCCTCGCCCCCTTCGAGATAGCAC,
    CCC2(GAAGTGGCTGCGGGTCGGCGCACCATGAGC).

3.4 The Results of Primer Specificity Verification

The results showed that there was an band at about 600 bp in the PCR products that was product of genome DNA as the templates and the ITS1/ITS4 as primers, but there was no band in CK. This result showed that the extracted genomic DNA template was in accordance with the requirements of the amplification;

Using designed specific primers of ccc1/CCC2 as the primers and the genomic DNA extracted from all strains (*Trichoderma*, *Cladosporium sp.*, *Botryotinia fuckeliana* and *Corynespora cassiicola*) as the templates for PCR amplification, electrophoresis results (Fig. 6) found only in *Corynespora cassiicola* genomic DNA as the template amplified specific bands of about 600 bp, and expected results are consistent, other cucumber pathogenic fungi didn't amplify specific band. The results showed that the designed primers were relatively specific, and could be distinguished *Corynespora cassiicola* from other pathogenic fungi.

M:Marker, DL-2000; CK: ddH$_2$0; 1 ∼ 2: *Trichoderma*, *Cladosporium sp.* 3 ∼ 4 *Botryotinia fuckeliana*; 5 ∼ 6: *Cladosporium sp.*; 7 ∼ 10 *Corynespora cassiicola*

```
JQ595297    ......................GGAAGTAAAAGTCGTA       16
EU826114    TTTTGTATGGCGAATTGTAGTCTATAGAGGCGTGGTCAGC       40
FJ903283    ........TCTTGGTCAATTTAGAGGAAGTAAAAGTCGTA       32
GU062286    ........TCTTGGTCCATTTAGAGGAAGTAAAAGTCGTA       32
HM849615    ........................................        0
HQ643180    ........................................        0
JN974012    ......................GGAAGTAAAAGTCGTA       16
Consensus

JQ595297    ACAAGGTTTCCGTAGGTGAACCTGCGGAAGGATCATTATC       56
EU826114    GTGGGCGCTTGGGGTAAGTTCCTTGGAAGAGGACAGCATG       80
FJ903283    ACAAGGTTTCCGTAGGTGAACCTGCGGAAGGATCATTAC.       71
GU062286    ACAAGGTCTCCGTAGGTGAACCTGCGGAAGGGATCATTACA.      72
HM849615    ........TCCGTAGGTGAACCTGCGGAAGGATCATTAC.       31
HQ643180    ................CCACACCTAAAAAACTTTCCACG       23
JN974012    ACAAGGTTTCCGTAGGTGAACCTGGGGAAGGATCATTAT.       55
Consensus                   c         a            a

JQ595297    G.TAGGGGCCTCGCCCCCTTCGAGATAGCACC....CTTT       91
EU826114    GAGGGTGATACTCCCGTTCATCCCTGAGTGGCTCGTGCGT      120
FJ903283    ...AGAGTTCATGCCCGAAAGGGTAGACCTCCC.ACCCTT      107
GU062286    ...AGTGACCCCGGTCTAACCACCGGGATGTTC.ATAACC      108
HM849615    ...AGAGTTCATGCCCGAAAGGGTAGACCTCCC.ACCCTT       67
HQ643180    T.GAACCGTATCAACCCTTTTAGTTGGGGGTCTTGTACCC       62
JN974012    ...CGAGTTAGGGTCCCCAGGGCCCGAATCTCCCAACCCT       92
Consensus

JQ595297    GTTTATGAGCACCTCTC.GTTTCCTC..............      116
EU826114    ACGACCCGTTTTCTTTGAGTCGCGTTGTTTGGGAATGCAG      160
FJ903283    GTGTATTATTACTTT...GTTGCTTT..............      130
GU062286    CTTTGTTGTCCGACTCT.GTTGCCTCCGG...........      136
HM849615    GTGTATTATTACTTT...GTTGCTTT..............       90
HQ643180    TATCATGGCGAATGTTT.GG.ACTTCGGT...........       89
JN974012    TTTTTTTTCCAACCTCT.GTTGCTTCGGGGGGCCCGTCCT      131
Consensus                         g    c   t

JQ595297    .............GGCAGGCTC.........GCCTGCCA      133
EU826114    CGCAAAGTAGGTGGTAAATTCCATCTAAAGCTAAATATTG      200
FJ903283    ............GGCGAGCT.........GCCTTCGG      146
GU062286    ............GGCGACCCT.........GCCTTCGG      153
HM849615    ..........GGCGAGCT..........GCCTTCGG      106
HQ643180    ..........CCGGGCGAGTA..........GCTTTTTG      108
JN974012    TGATGGACCGCCGGGGGACCCCCCCTTGCGGTGTCCTCTG      171
Consensus               g

JQ595297    ACGGGGACCCACCACAAACCCATTGTAGTACAAGAAGTAC      173
EU826114    GTGCGAGACCGATAGCGAACAAGTACCGTGAGGGAAAGAT      240
FJ903283    GCCTTGTATGGTCGCCAGAGAATACCAAAACTCTTTTTAT      186
GU062286    GCGGGGGCT..CCGGGTGGACACTTCAAACTCTTGCGTAA      191
HM849615    GCCTTGTATGCTCGCCAGAGAATACCAAAACTCTTTTTAT      146
HQ643180    TTTTAAACCCATTTCACAATTCTGATTATACTGTGGGGAC      148
JN974012    GCCCGTGCCCGTCGATAGCCCACGTCTAAACTCTTGCTTA      211
Consensus

JQ595297    ACGTCTG.......AACAAAC..AAAACAAACTATT...      201
EU826114    GAAAAGAAC......TTTGAAAAGAGAGTTAAAGAGTA.C      273
FJ903283    TAATGTC.......GTCTGAGT..ACTATATAATAGT...      214
GU062286    CTTTGCA.......GTCTGAGT..AAACTTAATTAATAAA      222
HM849615    TAATGTC.......GTCTGAGT..ACTATATAATAGT...      174
HQ643180    GAAAGTC........TCTGCTT..TTAACTAGATAGC...      175
JN974012    AAACGTGTTTTTTTGCCTAAATTCATAACTAAAAAAAAC      251
Consensus                                       a
```

Fig. 5. The alignment of sequence of *Corynespora cassiicola* and DNA sequence of other fungus.

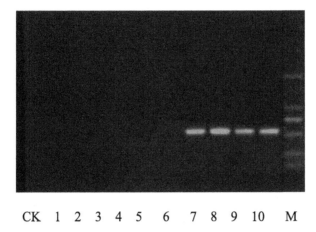

CK   1   2   3   4   5   6   7   8   9   10   M

Fig. 6. CCC1/CCC2 primer specific primers were used to detect PCR using different templates.

References

1. Dixon, L.J., Schlub, R.L., Pernezny, K., Datnoff, L.E.: Host specialization and phylogenetic diversity of Corynespora cassiicola. Phytopathol. **99**(9), 1015–1027 (2009)
2. Li, S., Hartman, G.L.: Molecular detection of fusarium solani f. sp. glycines in soybean roots and soil. Plant. Pathol. **52**, 74–83 (2003)
3. White, T., Bruns, J., Lee, S., et al.: Amplification and direct sequencing of fungal ribosomal RNA genes forphylogenetics. In: PCR Protocools: A Guide to Methods and Application, pp. 315–322. Academic Press, San Diego (1990)
4. Frederick, R.D., Snyder, C.L., Peterson, G.L., Bonde, M.R.: Polymerase chain reaction assay for the detection and discrimination of the soybean rust pathogens Phakopsora pachyrhizi and P. meibomiae. Phytopathology **92**, 217–227 (2002)
5. Trout, C.L., Ristaino, J.B., Madritch, M., Wangsomboondee, T.: Rapid detection of phytophthora infestans in late blight-infected potato and tomato using PCR. Plant Dis. **81**, 1042–1048 (1997)
6. Bonants, P.J.M., De Veerdt, M.H., Van Gent-Pelzer, M.P., Lacourt, I., Cooke, D.E.L., Duncan, J.M.: Detection and identification of Phytophthora fragariae Hickman by the polymerase chain reaction. Eur. J. Plant Pathol. **103**, 345–355 (1997)
7. Grote, D., Olmos, A., Kofoet, A., Tuset, J.J., Bertolini, E., Cambra, M.: Specific and sensitive detection of phytophthora nicotianan by simple and nested-PCR. Eur. J. Plant Pathol. **108**, 1997–2007 (2002)
8. Zhang, Z.G., Zhang, J.Y., Wang, Y.C., Zheng, X.B.: Molecular detectionof fusarium oxysporum f.sp. niveum and Mycosphaerella melonis ininfected plant tissues and soil. FEMS Microbiol. Lett. **249**, 39–47 (2005)

Simulation of Winter Wheat Phenology in Beijing Area with DSSAT-CERES Model

Haikuan Feng, Zhenhai Li, Peng He, Xiuliang Jin, Guijun Yang[(✉)],
Haiyang Yu, and Fuqin Yang

Beijing Research Center for Information Technology in Agriculture, Beijing
Academy of Agriculture and Forestry Sciences, Beijing 100097, China
{fenghaikuan123,yangfuqin0202}@163.com,
{lizh323,hepeng1009,jinxiuxiuliang}@126.com,
{yanggj,yuhy}@nercita.org.cn

Abstract. The Decision Support for Agrotechnology Transfer (DSSAT) model
was a worldwide crop model, and crop accurate simulation of phenology was
the premise to realize other functional simulations. The objective of this study
was to attempt to calibrate the parameters of wheat phenology coefficients,
including cultivar and ecotype coefficients, and develop the winter wheat phenology coefficients of Beijing area. To achieve this goal, field surveys of 7 years
in wheat growing seasons in Beijing were carried out during 2005–2012. The
trail-and-error method and GLUE method were used to calibrate the phenology
parameters with 4 growing seasons of 05/06, 06/07, 07/08 and 08/09. Three
growing seasons, 09/10, 10/11, and 11/12 were used for validation, and the
results showed good agreements between observed date and predicted date.
The RMSE of validation data for TS, BT, HD, AN, and MA were 1.63 d, 2.45 d,
3.16 d, 1.83 d, 3.56 d, respectively. Therefore, the calibrated parameters could
be used to monitor winter wheat phenology, and could be used for other research
as the basis phenology parameters of Beijing area.

Keywords: Phenology · DSSAT-CERES · Phenology coefficients · Winter
wheat

1 Introduction

Knowledge of crop phenology is essential for plant physiological indexes, crop production and crop managements [1, 2]. Plant growth phase could represent partitioning
of the assimilations into the plant organs [3]. Accurate prediction of crop production is
closely related with some critical phenology stage, for example anthesis [4]. Besides,
phenology is very important in the guidance of crop managements [3].

For most crops models, more than two phases can be used to describe their detail
phenological sub-routines in terms of temperature and crop development [5]. Phenology is described by the dimensionless state variable development stage in 'School of de
Wit' crop models, D. In these models, D is different numbers in different period, it is 0
during emergence, 1 at flowering, and change into 2 during maturity, and the development rate a function of photoperiod and environment temperature [6, 7]. The CERES

D. Li and Z. Li (Eds.): CCTA 2015, Part II, IFIP AICT 479, pp. 259–268, 2016.
DOI: 10.1007/978-3-319-48354-2_27

model gives a detailed description of phenology simulation. The growth stages of wheat in the CERES model are divided into 9 stages, and vernalization affect is considered as well [8–10]. Water and nutrient can have influence on the development of rate, and STICS considered these affect in phenology simulation [5, 11]. During 49 growing seasons, performance of eight crop growth simulation models of winter wheat are compared by Reimund et al., and those models are widely used, easily accessible and well-documented and nine models for crop during 44 growing seasons of spring barley [12, 13]. The application of rigorous statistical techniques can be used to calibrate sub model of APSIM-Oryza in phenology aspect, and the original method which is put forward by Sarath et al. makes it of great easy to do so [14].

The Decision Support for Agrotechnology Transfer (DSSAT) model was a worldwide crop model, and crop accurate simulation of phenology was the premise to realize other functional simulations. The DSSAT-CERES model shows a detailed description of phenology simulation. CERES-Wheat is used in the simulation of wheat anthesis and production in Southern Sardinia Italy, and Beijing, China, respectively by Dettori (2011) and Wang (2009) [15, 16]. Palosuo compared eight crop growth models including DSSAT for anthesis and maturity estimation [5], and the results showed that the phenological stages provided the most accurate estimates using DAISY and DSSAT. By considering terminal spikelet, booting, Xue predicted phenological development of winter wheat via using WE model and CERES-Wheat, in this model, heading, anthesis and maturity are also taken into account [17]. Cultivar coefficients in DSSAT-CERES, such as P1V, P1D, P5 of wheat, were mainly and generally considered to calibrate the phenology, while other phenology parameters, called ecotype coefficients including P1, P2, P3, P4, were set as default. However, default ecotype coefficients selecting resulted in deviations between simulated and observed phenology data, even though cultivar coefficients were adjusted. calibrate the parameters of wheat phenology coefficients, including cultivar and ecotype coefficients, and develop the winter wheat phenology coefficients of Beijing area. Calibration of the parameters of wheat phenology coefficients, including cultivar and ecotype coefficients, In this study, it is the objective to calibrate the parameters of wheat phenology coefficients, including cultivar and ecotype coefficients, and develop the winter wheat phenology coefficients of Beijing area.

2 Materials and Methods

2.1 Study Area and Phenology Investigation

A 7 year field observation was conducted at the Beijing District (40 °00'N ~ 40 °23'N, 116 °27'E ~ 116 °59'E), PR China, during the 2005–2012 growing seasons. The climate of the region is warm moderate semi-humid continental monsoon climate representative, in summer it is hot and rainy, cold and dry in winter, and spring and autumn is short. The mean temperature of the whole year is 10–12 °C, and the mean rainfall is 600 mm.

According to Zadoks [18] and Tottman [19], the data of main phenological periods were recorded during 2005–2012 growing seasons. The phenological periods included

Table 1. Phenological data of Beijing district for 7 growing seasons

Stage	2006	2007	2008	2009	2010	2011	2012
ST	Sep 25	Sep 28	Sep 25	Sep 28	Sep 25	Sep 27	Sep 25
EM	Oct 2	Oct 5	Oct 2	Oct 5	Oct 2	Oct 4	Oct 2
TS	Mar 12	Mar 6	Mar 6	Mar 9	Mar 22	Mar 13	Mar 22
BT	May 4	Apr 28	May 1	Apr 28	May 6	Apr 30	May 2
HD	May 10	May 6	May 7	May 6	May 14	May 8	May 8
AN	May 15	May 11	May 12	May 12	May 19	May 14	May 13
MA	Jun 17	Jun 15	Jun 18	Jun 15	Jun 20	Jun 19	Jun 18

sowing time (ST), emergence (EM), terminal spikelet initiation (TS), booting (BT), heading (HD), anthesis (AN) and maturity (MA), and detailed data were shown in Table 1.

2.2   Model Description and Input Data Set

The DSSAT model successfully used 25 years before by worldwide investigators for various uses [10, 20] (Jones et al. 2003; He et al. 2012). The DSSAT model simulates the physiologically ecology process of crops vegetative growth and reproductive growth, crop photosynthesis, respiration, dry-matter distribution and plant growth and aging [5, 10, 21]. The version 4.5 of DSSAT can simulate more than 29 different kinds of crops, including maize, peanut, soybean, rice, wheat, et al. [22]. Notably, accurate forecasting of phenological development is significant in agroecosystem, and is the premise to realize other functional simulations [1, 17, 23] (Xue et al. 2004; Sakamoto et al. 2005; Xu et al. 2009). In DSSAT-CERES-Wheat [3], the growth stages of wheat are divided into 9 stages, while stages 1 to 5 are the main wheat growing stages [23]. The thermal time unit is the kernel of the most phenological development in DSSAT, and photoperiod and vernalization affect are considered the most limiting factor in the crop growth stage between emergence and terminal spikelet initiation as well.

For the sake of run a crop model and appraise a simulation, meteorological data, soil data, crop management information, and experiment data are required [24, 25]. Daily weather data, including minimum and maximum air temperature, and precipitation were acquired from China Meteorological Data Sharing Service System (CMDSSS). The website is http://cdc.cma.gov.cn. While solar radiation is obtained from sunshine hours of CMDSSS with the Angstrom formula as used in Allen et al. [26].

Each soil horizon of soil data included lower limited volume water content (VWC), soil texture, upper limited VWC at saturation, field capacity, saturated hydraulic conductivity, soil organic carbon, inorganic nitrogen, PH, and bulk density (Table 2). These parameters are obtained from field measurements before sowing time.

Cultivar parameters were optimized with the GLUE methods [27, 28], a brief review of GLUE method is given in He et al. [28, 29]. The parameters contains vernalization sensitivity coefficient (P1V), photoperiod sensitivity coefficient (P1D), grain filling phase duration (P5), kernel number (G1), kernel size (G2), single tiller weight (G3), and phyllochron interval (PHINT). ecotype coefficients includes duration

Table 2. Soil profile characteristics of the experiment field

Depth (cm)	Sand %	Silt %	Clay %	pH	Org. C %	Total N %	LL %	DUL %	Sat. %	BD g/cm^3
0–10	22.6	53.9	23.5	8.00	1.04	0.11	8.8	27.3	51.1	1.66
10–20	22.6	53.9	23.5	8.03	1.04	0.11	8.8	27.3	51.1	1.60
20–40	22.5	54.1	23.4	8.08	1.01	0.10	8.7	27.3	51.3	1.35
40–60	14.9	47.8	37.3	7.94	0.68	0.08	12.3	34.8	54.7	1.16
60–80	14.9	47.8	37.3	7.98	0.66	0.08	12.3	34.8	54.7	1.13
80–100	16.7	43.0	40.3	8.03	0.59	0.07	12.3	34.8	54.7	0.99

Note: Org. C, LL, DUL, SAT, BD represent soil organic carbon, lower limited VWC, field capacity, upper limited VWC at saturation, and bulk density, respectively.

of end juvenile to terminal spikelet stage (P1), duration of terminal spikelet to end leaf growth stage (P2), duration of end leaf growth to end spike growth stage (P3), and duration of end spike growth to end grain fill lag stage (P4). These parameters were calibrated by the trail-and-error method one by one.

2.3 Statistical Analysis

Three statistical indices were used to appraise performance of the model, comparing simulated results with measured data. The first is the mean error (E):

$$E = \frac{1}{n} \sum_{i=1}^{n} (S_i - M_i)$$

The value of E could show the deviation between analogic and practical measured data. Positive values indicate the simulated data is larger, and vice versa. Meanwhile the lower the absolute value was, the higher the accuracy and precision of the model simulation was considered to be. The second is root mean square errors (RMSE):

$$RMSE = \sqrt{\frac{1}{n} \sum_{i=1}^{n} (S_i - M_i)^2}$$

where S_i and M_i are the analogic and practical measured data, respectively, and n is the number of treatments. Generally, the RMSE shows a close agreement between measured values and predicted values. The last is the index of agreement (d) of Willmott [29]:

$$d = 1 - \frac{\sum_{i=1}^{n} (S_i - M_i)^2}{\sum_{i=1}^{n} (|S_i - \overline{M}| + |M_i - \overline{M}|)^2}$$

where $\overline{M}$ is the mean of the n measured data. The value of d ranges from $-\infty$ to 1.0, and the closer the index value is to one, the better the agreement between the simulated and measured data and vice versa.

3   Results

3.1   Thermal Unit of Phenology Stage

The thermal unit (TU) of each phenology stage from 2005 to 2012 was showed in Fig. 1. The TU of each phenology stage is the sum of the TU per day of the growth period. As seen in Fig. 1, there was a large difference in the 7 growing seasons result from environmental conditions. The average TU of ST-TS, TS-BT, BT-HD, HD-AN, AN-MA period were 820, 602, 147, 103, 802 degree-days, respectively. The total TU of the 09/10 growing season was the lowest among the 7 seasons, in which the TU of each period except BT-HD period were lower than the average TU of each period of 7 seasons. The TU of each growing stage varied in each year, which represented that thermal unit was one of influence factor to phenology stage.

3.2   DSSAT-CERES Model Calibration

Four growing seasons, 05/06, 06/07, 07/08, and 08/09, were used to calibrate the DSSAT-CERES phenology parameters. The calibrated parameters of P1, P2, P3, P4, P1V, P1D, P5, were 267 degree-days, 600 degree-days, 175 degree-days, 300 degree-days, 39.99 d, 87.40, 635 degree-days, respectively.

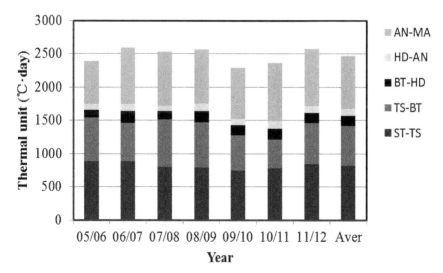

Fig. 1. The thermal unit of each phenology stage from 2005 to 2012 (Note: Aver represent the average of thermal unit of each phenology stage.)

Observed and predicted day after planting (DAP) for TS, BT, HD, AN, MA were showed in Table 3 and Fig. 2. The results showed that phenology simulation using DSSAT-CERES achieved a good simulation. The DSSAT-CERES model can predict HD, AN, MA better than TS and BT. The best prediction was for MA, which the E, RMSE, and d were 0 d, 1.00 d, and 0.97, respectively. The differences of observed and predicted DAP were only 1 d. The E, RMSE, and d for HD were −0.75 d, 1.66 d, 0.90, respectively, and for AN were −2.25 d, 2.50 d and 0.83, respectively. The deviations of observed and predicted DAP were no more than 3 d and 4 d, respectively. The E, RMSE, and d for TS were +0.50 d, 4.18 d and 0.73, respectively, and for BT were −3.00 d, 4.24 d, and 0.53, respectively. As the reason, there were large differences of observed and predicted DAP, which were +7 d for TS in 08/09, +6 d for BT in 05/06, and −7 d for BT in 07/08. In brief summary, the DSSAT-CERES model can simulate the phenology after phenology parameters calibrated, and the later periods simulating were better than earlier stage.

Table 3. Statistical indices of phenology parameters calibration with observed (O) and predicted (P) day after planting (DAP) using the DSSAT-CERES model for TS, BT, HD, AN, MA in 4 growing seasons, 05/06, 06/07, 07/08, 08/09 growing seasons

Indices	TS	BT	HD	AN	MA
E	0.50	−3.00	−0.75	−2.25	0
RMSE	4.18	4.24	1.66	2.50	1.00
d	0.73	0.53	0.90	0.83	0.97

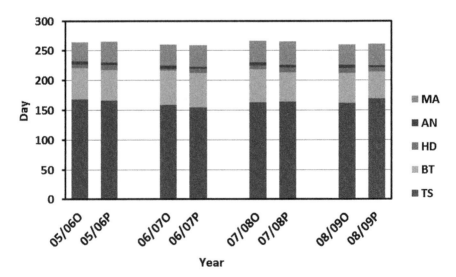

Fig. 2. Phenology parameters calibration with observed (O) and predicted (P) day after planting (DAP) using the DSSAT-CERES model for TS, BT, HD, AN, MA in 4 growing seasons, 05/06, 06/07, 07/08, 08/09 growing seasons (Note: D represented the deviation of predicted DAP and observed DAP.)

3.3 Validation of Phenology Stage Simulation with DSSAT-CERES

Three growing seasons, 09/10, 10/11, and 11/12, were used to test the reliability of the DSSAT-CERES model with the calibrated phenology parameters. Validation results were listed in Table 4 and Fig. 3. The simulation of TS, BT, HD, and AN were in accordance with the observed stage. The E, RMSE, and d values for TS, were 1.33 d, 1.63 d, and 0.98, respectively, for BT were 1.33 d, 2.45 d, and 0.90, respectively, for HD were 2.67 d, 3.16 d, 0.86, respectively, and for AN were 0.67 d, 1.83 d, 0.94, respectively. There were large differences between observed and predicted MA dates, which the E, RMSE, and d were 1.33 d, 3.56 d, 0.60, respectively. The difference between observed and predicted MA dates for 09/10 growing season was 6 d. The validation results showed that there were good agreement between observed and predicted DAP using the DSSAT-CERES model for each phenology stage.

Table 4. Statistical indices of observed (O) and predicted (P) day after planting (DAP) using the DSSAT-CERES model for TS, BT, HD, AN, MA in 3 growing seasons, 09/10, 10/11, 11/12 growing seasons.

Indices	TS	BT	HD	AN	MA
E	1.33	1.33	2.67	0.67	1.33
RMSE	1.63	2.45	3.16	1.83	3.56
D	0.98	0.90	0.86	0.94	0.60

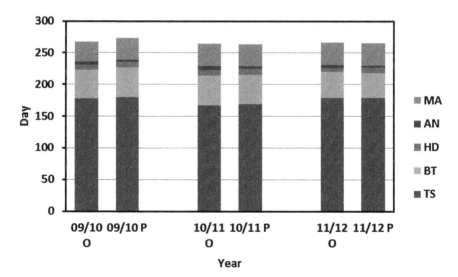

Fig. 3. Validation of observed (O) and predicted (P) day after planting (DAP) using the DSSAT-CERES model for TS, BT, HD, AN, MA in 3 growing seasons, 09/10, 10/11, 11/12 growing seasons (Note: D represented the deviation of predicted DAP and observed DAP.).

4 Discussion

Phenology stage was influenced by external environment factor, including temperature, light, water, and nutrient [30], and temperature and day length were the main driving factor. Figure 1 showed that the TU of each growing stage varied in each year, which indicated TU was nonlinear with phenology stage [3]. The 09/10 growing season was cooler as compared to the other growing seasons (Table 1), and the TU of 09/10 growing season was much lower than other growing stage.

The research purpose was to attempt to calibrate the parameters of wheat phenology coefficients, including cultivar and ecotype coefficients. The calibrated parameter of P2, 600 degree-days, was much higher than the P2 of CAWH01, USWH01, and UKWH01, which were given parameters in DSSAT 4.5. However, the calibrated P2 was reasonable in consideration of the P2 definition, stage terminal spikelet to end leaf growth period. There were about 6 leaves from terminal spikelet stage to end leaf growth for the winter wheat in North China [19], and a phyllochron, defined as the interval of time between leaf tip appearance PHINT (degree-days), is about 100 degree-days.

The RMSE of each phenology stage were ranged from 1.0 d to 4.24 d for calibrated data, and from 1.60 d to 3.56 d for validation data. They were acceptable results, because there were little differences in determining phenology dates during the actual operation. There were large differences between predicted dates and observed dates for 09/10 growing season, because the 09/10 growing season was cooler than the others. Overall, the calibration and validation results showed that there were good agreement between observed and predicted DAP using the DSSAT-CERES model for each phenology stage.

The study tried to calibrate the phenology parameters, including cultivar and ecotype coefficients. Therein, comparison of calibrated ecotype coefficients and default ecotype coefficients provided by DSSAT should be focused on further investigation.

5 Conclusion

This study was to attempt to calibrate the parameters of wheat phenology coefficients, including cultivar and ecotype coefficients, and develop the winter wheat phenology coefficients of Beijing area. The calibrated parameter of P2 was larger than the default value provided by DSSAT, while it was reasonable in consideration of actual growth progress of winter wheat in Beijing area.

Four growing seasons, 05/06, 06/07, 07/08, 08/09 season, were used as parameters calibration, and three growing seasons, 09/10, 10/11, 11/12 season, were used for validation. The results showed fine agreements between observed and predicted DAP using the DSSAT-CERES model for each phenology stage. The calibrated parameters, as the basis phenology parameters of Beijing area, could be used for other research, such as total above-ground biomass simulation, yield prediction, water and nitrogen balance study.

References

1. Sakamoto, T., Yokozawa, M., Toritani, H., et al.: A crop phenology detection method using time-series MODIS data. Remote Sens. Environ. **96**(3), 366–374 (2005)
2. Xu, S.J., Lin, M.Y., Xu, Z.W.: Advance on dynamic simulation model for crop development. J. Inner Mongolia Univ. Natl. **24**(2), 167–171 (2009)
3. Ritchie, J.T.: Wheat phasic development. Model. Plant Soil Syst. (Modelingplantan) **31**, 31–54 (1991)
4. Haboudane, D., Miller, J.R., Tremblay, N., et al.: Integrated narrow-band vegetation indices for prediction of crop chlorophyll content for application to precision agriculture. Remote Sens. Environ. **81**(2), 416–426 (2002)
5. Palosuo, T., Kersebaum, K.C., Angulo, C., et al.: Simulation of winter wheat yield and its variability in different climates of Europe: a comparison of eight crop growth models. Eur. J. Agron. **35**(3), 103–114 (2011)
6. Bouman, B.A.M., Van Keulen, H., Van Laar, H.H., et al.: The 'School of de Wit'crop growth simulation models: a pedigree and historical overview. Agri. Syst. **52**(2), 171–198 (1996)
7. Boogaard, H.L., Van Diepen, C.V., Rotter, R.P.: User's guide for the WOFOST Control Center 1.8 and WOFOST 7.1. 3 crop growth simulation model. Alterra Wageningen University (2011)
8. Jones, C.: CERES-Maize: a stimulation model of maize growth and development. In: NTIS, p. 195. Springfield, VA, USA (1985)
9. Ritchie, J.T., Singh, U., Godwin, D., Hunt, L.: A user's guide to CERES Maize, V2. 10. International Fertilizer Development Center (1992)
10. Jones, J.W., Hoogenboom, G., Porter, C.H., et al.: The DSSAT cropping system model. Eur. J. Agron. **18**(3), 235–265 (2003)
11. Brisson, N., Gary, C., Justes, E., et al.: An overview of the crop model STICS. Eur. J. Agron. **18**(3), 309–332 (2003)
12. Palosuo, T., Kersebaum, K.C., Angulo, C., et al.: Simulation of winter wheat yield and its variability in different climates of Europe: a comparison of eight crop growth models. Eur. J. Agron. **35**(3), 103–114 (2011)
13. Rötter, R.P., Palosuo, T., Kersebaum, K.C., et al.: Simulation of spring barley yield in different climatic zones of Northern and Central Europe: a comparison of nine crop models. Field Crops Res. **133**, 23–36 (2012)
14. Nissanka, S.P., Karunaratne, A.S., Perera, R., et al.: Calibration of the phenology sub-model of APSIM-Oryza: Going beyond goodness of fit. Environ. Model Softw. **70**, 128–137 (2015)
15. Dettori, M., Cesaraccio, C., Motroni, A., et al.: Using CERES-Wheat to simulate durum wheat production and phenology in Southern Sardinia, Italy. Field Crops Res. **120**(1), 179–188 (2011)
16. Wang, X., Zhao, C., Li, C., et al.: Use of ceres-wheat model for wheat yield forecast in Beijing. In: Li, D., Zhao, C. (eds.) Computer and Computing Technologies in Agriculture II, vol. 1, pp. 29–37. Springer, Heidelberg (2009)
17. Xue, Q., Weiss, A., Baenziger, P.S.: Predicting phenological development in winter wheat. Clim. Res. **25**(3), 243–252 (2004)
18. Zadoks, J.C., Chang, T.T., Konzak, C.F.: A decimal code for the growth stages of cereals. Weed Res. **14**(6), 415–421 (1974)
19. Tottman, D.R.: The decimal code for the growth stages of cereals, with illustrations. Ann. Appl. Biol. **110**(2), 441–454 (1987)

20. He, J.Q., Dukes, M.D., Hochmuth, G.J., Jones, J.W., Graham, W.D.: Identifying irrigation and nitrogen best management practices for sweet corn production on sandy soils using CERES-Maize model. Agri. Water Manage. **109**, 61–70 (2012)
21. Thorp, K.R., DeJonge, K.C., Kaleita, A.L., et al.: Methodology for the use of DSSAT models for precision agriculture decision support. Comput. Electron. Agri. **64**(2), 276–285 (2008)
22. Liu, H.L., Yang, J.Y., Drury, C.F., et al.: Using the DSSAT-CERES-Maize model to simulate crop yield and nitrogen cycling in fields under long-term continuous maize production. Nutr. Cycl. Agroecosyst. **89**(3), 313–328 (2011)
23. Xu, S.J., Lin, M.Y., Xu, Z.W.: Advance on dynamic simulation model for crop development. J. Inner Mongolia Univ. Natl. **24**(2), 167–171 (2009)
24. Gao, L.Z.: Foundation of Agricultural Modeling Science. Tianma Book Limited Company, Hong Kong (2004)
25. Hoogenboom, G., Jones, J.W., Traore, P.C., Boote, K.J.: Experiments and data for model evaluation and application. Improving Soil Fertility Recommendations in Africa using the Decision Support System for Agrotechnology Transfer (DSSAT), pp. 9–18. Springer, Netherlands (2012)
26. Allen, R.G., Pereira, L.S., Raes, D., Smith, M.: Crop evapotranspiration guidelines for computing crop water requirements-FAO irrigation and drainage paper 56. In: FAO, Rome, vol. 300, p. 6541 (1998)
27. Jones, J.W., He, J., Boote, K.J., Wilkens, P., Porter, C.H., Hu, Z.: Estimating DSSAT cropping system cultivar-specific parameters using Bayesian techniques. In: Methods of Introducing System Models into Agricultural Research, (Methodsofintrod), pp. 365–394 (2011)
28. He, J.Q., Dukes, M.D., Jones, J.W., Graham, W.D., Judge, J.: Applying GLUE for estimating CERES-Maize genetic and soil parameters for sweet corn production. Trans. ASABE **52**(6), 1907–1921 (2009)
29. He, J., Jones, J.W., Graham, W.D., et al.: Influence of likelihood function choice for estimating crop model parameters using the generalized likelihood uncertainty estimation method. Agri. Syst. **103**(5), 256–264 (2010)
30. Willmott, C.J.: Some comments on the evaluation of model performance. Bull. Am. Meteorol. Soc. **63**, 1309–1369 (1982)
31. Han, X.M., Shen, S.H.: Research progress on phenological models. Chin. J. Ecol. **27**(1), 89–95 (2008)

Design of Monitoring System
for Aquaculture Environment

Hua Liu[1], Liangbing Sa[1,2], Yong Wei[1(✉)], Wuji Huang[1,2],
and Binjie Shi[1,2]

[1] College of Engineering and Technology, Tianjin Agricultural University,
Tianjin 300384, China
{41599386,810874355,595183963,578982393}@qq.com,
hwjl544565913@163.com
[2] Students' Innovate Center, Tianjin Agricultural University,
Tianjin 300384, China

Abstract. In order to strengthen aquaculture monitoring and management, real-time to improve the breeding environment, improve farm income reduce labor intensity, proposed aquaculture environmental monitoring system for wireless networks. The system consists of sensor nodes, base stations, the main controller, routing, consisting of PC and mobile phone users. Through field testing, the system can complete the environment and the waste discharge, and display and comparison of real time data. At the same time, through the replacement of different sensors applied to different environments.

Keywords: Sensor node · Wireless network · Monitoring · Environment

1 Introduction

In recent years, with the development of large-scale farming, large-scale farming, is becoming more intensive, modern management techniques of information and automation become increasingly important. A intelligent sensing, remote control monitoring system of the bottom sensors and video surveillance equipment integrated together, will complete "comprehensive perception, wireless transmission, intelligent processing" of information management. At the same time, Users use human machine interface configuration observation of environmental information, the growth status of the animal farms anywhere and anytime. In addition, whenever unusual circumstances scene occurs, the system automatically sends an alarm message [1, 2].

By various types of sensor nodes placed in breeding farms, real-time acquisition of farm's environmental parameters, wireless networking technology parameters are transmitted to human Kingview PC interface or the clouds, while the data stored in the SD card. Camera monitoring specific breeding farms, the user can actually get the field of animal growth conditions. Master chip for data analysis, processing, while controlling the venue of the controller, when the index is unreasonable to make the environment when working inside the controller parameter adjustment to a reasonable value. SMS module while the system was added, is allowing users to monitor more blossoms [3–5].

© IFIP International Federation for Information Processing 2016
Published by Springer International Publishing AG 2016. All Rights Reserved
D. Li and Z. Li (Eds.): CCTA 2015, Part II, IFIP AICT 479, pp. 269–276, 2016.
DOI: 10.1007/978-3-319-48354-2_28

2 Hardware Design

2.1 System Structure

The system is divided into three parts: monitor control section (nodes), data transmission section (wifi), terminal control platform (PC, SMS terminal). Wireless technology will detect and control parts connected together to achieve the purpose of real-time transmission of data, and the data stored in the SD card. Kingview interface makes PC more humanly, real-time data to generate reports, web publishing, automatic alarm [6]. SMS module for remote users can be monitored and controlled via mobile terminals, real-time observation inside and precise regulation of the venue environment at any time, to achieve two-way monitoring and control services. Frame is shown in Fig. 1.

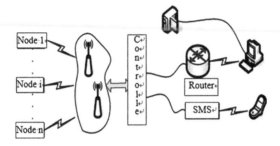

Fig. 1. System topology

2.2 Functions Achievement

(1) The combination of control system for the farm and internet of things, and the operating mode remote Kingview interface, allows the operator to control environmental indicators breeding farms;

(2) Record growth environment section animal breeding farms in each period to ensure the quality in farm;

(3) realize the remote control, as long as the control system is in the same network, it can be controlled at any place worldwide;

(4) the use of serial communication node can reduce the number of nodes, and the wireless communication between nodes and nodes;

(5) the use of solar power each node, so that the green agriculture. Enough during the day solar energy into electrical energy for all-weather use system;

(6) Monitoring livestock and poultry excrement, thus increasing the warning and other functions.

Main module functions is shown in Fig. 2.

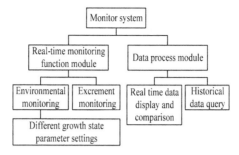

Fig. 2. Monitoring system module

2.3 Master Chip

Atmega 2560 is a simple system controller chips of the minimum system. Arduino Mega is a core with ATmega 2560 microcontroller board, the group itself has 54 digital I/O input/output terminal (14 PWM outputs), 16 sets of analog inputs, 4 UART (hardware serial ports), using a 16 MHz crystal oscillator. Since Atmega 2560 microcontroller chip is rich in external resources, pins and more simple. Temporarily idle pin can do the follow-up development community. Atmega 2560 smallest single-chip system includes a power supply module, MCU, download the module, reset module, the principle is shown in Fig. 3.

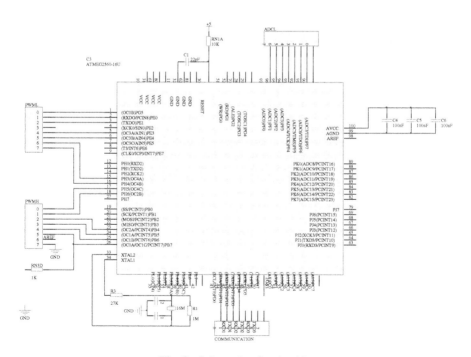

Fig. 3. Schematic of main chip

2.4   2.4G Wireless Communication

2.4G Wireless module implements the data transmission between nodes. The module with low power consumption, transmission speed and other little choice wireless serial communication module, communication distance up, meets the design requirements of the system.

WLC_24L01 is a wireless transceiver module based on the Nordic Semiconductor of Nrf24L01. The module integrates all RF related functions and devices, users need only a simple configuration registers through the SPI interface, which can realize communication, reducing the user of the wireless product development cycle. Figure 4 is the interface circuit between the module and MCU.

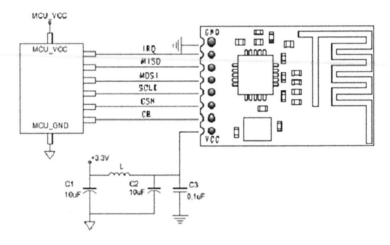

Fig. 4.   Interface circuit

2.5   Sensor Nodes

According to the detection system environment, different sensors may be suitably selected nodes, such as temperature sensors, humidity sensors, light sensors, ammonia sensors, CO_2 sensors, H_2S sensors. Wireless sensor node system block diagram [7, 8] is shown in Fig. 5.

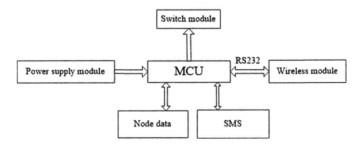

Fig. 5.   Block diagram of wireless sensor node

3 Software Design

3.1 Serial to Ethernet Configuration

By entering RM04 configuration interface: work mode selection Serial to Ethernet, network protocol select TCP server, remote IP does not work as a server, the server to open port TCP listening port serial parameters set here according to the needs of our users set the baud rate 9600, 8 data bits, parity bit NONE, 1 stop bit. We do not enable network parameters DHCP configured static IP: 196.168.11.254, submit configured so that Ethernet has been constructed [9].

Open Configuration software, create a new project in the "Device" option, select the communications port (free to choose a serial number as long as the actual serial interface and can not), and communication (MODBUS (Ethernet via the "Device Configuration Wizard" NIC)). Then enter the port address 196.168.11.254, it should be noted that if the virtual IP Ethernet module and PC must be in the same LAN, if the WAN must be fixed IP. New data variables, the same variables and attributes required to address the corresponding element. Communication in the picture elements and data dictionary variable corresponds to good when using different devices and data variables whenever a different port mappings on the line.

3.2 System Algorithms

The system uses incremental PID algorithm. Read node data back to the data processing whether the comparison satisfies a predetermined value, if it satisfies the use of PID algorithm for automatic control, the controller compares the acquisition value and

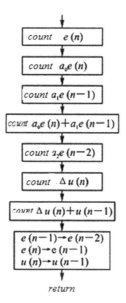

Fig. 6. Algorithm process

the setting value and issue the command, and then sent to the node allows the controller work. The principle is the following formula (1):

$$\Delta u(n) = u(n) - u(n-1)$$
$$= K_P[e(n) - e(n-1)] + K_P \frac{T}{T_I} e(n) + K_P \frac{T_D}{T}[e(n) - 2e(n-1) + e(n-2)]$$

$$(1)$$

Algorithm process is shown in Fig. 6.

4 Results and Discussion

4.1 Test Results

In the system debugging process we determined the drinking water tank, light intensity, the concentration of ammonia, carbon dioxide concentration, room temperature, the solution temperature. The following data is testing laboratory simulation data made by the various indicators line chart: Fig. 7 shows raw and spectra after pretreatment respectively.

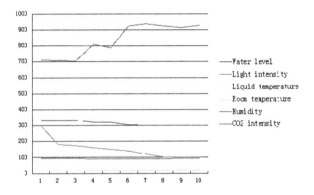

Fig. 7. Conventional and spectra after pretreatment

4.2 Discussion

In the system, monitoring device and execution unit in the entire operation of the system to achieve a friendly communication with users of wireless data transmission. At the same time, by detecting the data analysis, the system can detect changes in the environment in real time, and timely feedback to the user. Since the control system uses a PID algorithm in the software design process, so it has the following characteristics.

(1) use incremental PID control algorithm such that a given value is not mutated, but there is a certain inertia delay of slow variables, so the system has a certain adaptability and stability;

(2) incremental algorithms do not need to accumulate, accuracy problems after the calculation error, the calculation of the amount of control less affected;
(3) incremental algorithm is derived from the incremental control, it will not affect the overall operation of the system.

5   Conclusions

By testing system is stable, and another set of system test data, test data accurate rate of 98 %, to achieve the desired results. While highlighting the control principle in control process (PID incremental control algorithm), such as: regulation, adjust the water level of light intensity and so on. The system is stable, easy to control, simple. System is small, low cost choice in favor of SMEs. The sample source of catering waste oil in the research is limited and cannot completely represent diversity and complexity of catering waste oil. In addition, the law breakers usually add catering waste oil to qualified edible vegetable oil according to a certain proportion, and then sell the fake oil, therefore, it needs to further collect representative adulterated samples in the future.

(1) twenty-four hours of uninterrupted real-time acquisition, display various environmental parameters within breeding sites, including: light intensity, temperature, humidity, carbon dioxide concentration and ammonia concentration;
(2) camera monitoring breeding farms circumstances;
(3) PC interface configuration king collected real-time display of environmental parameters, making statements;
(4) When any abnormal when an alarm, the system automatically send control commands;
(5) automatic timing system for the phone to send environmental parameters, and you can use SMS to send control commands to the system.

Acknowledgment. Funds for this research was provided by Innovation and Entrepreneurship Training Program of Tianjin (201410061074) and College students of science and technology innovation projects of Tianjin Agricultural University.

References

1. Diao, Z., Yin, J., Zhu, X.: Electronic, model design of breeding environment monitoring system with multi sensor data fusion. Technol. Softw. Eng.
2. Huang, H.: Study on System of Livestock Farming Environment Control Base on PIC18F2580, pp. 23–30. Huazhong Agricultural University, Wu Han, China (2009)
3. Zhou, Y., Sun, B., Li, J.: Environmental monitoring system of livestock and poultry breeding based on STM32F103R6. Jiangsu Agric. Sci. **41**(1), 375–377 (2013)
4. Baonong, H., Wei, N.: Design of pig farms culture environmental monitor system base on WSN. J. Chin. Agric. Mech. **35**(1), 260–263, 269 (2014)

5. Chen, N., Zhou, Y., Xu, H., et al.: Design of wireless monitoring and control system for aquaculture environment based on ZigBee and GPRS. Ttansducer Microsyst. Technol. **30**(3), 1000–9787 (2011)
6. Wang, R., Xu, B., Wei, R., et al.: Design and implementation of an intelligent environmental monitoring system for animal house based on wireless sensor net (WSN). Jiangsu J. Agric. Sci. **26**(2), 562–566 (2010)
7. Zhang, W., He, Y., Liu, F., et al.: The environmental control system base on IOT for scale livestock and poultry breeding. J. Agric. Mech. Res. **2**, 245–248 (2015)
8. Sun, L.: Design of Livestock and Poultry Farming Monitoring and Control System, vol. 12, pp. 4–17. Nanjing Agricultural University, Nanjing, China (2011)
9. Jiang, R.: The Design and Implementation of the Barn Aquaculture Environmental Monitoring System, vol. 6, pp. 11–33. Northeast Agricultural University, Harbin, China (2013)

Research on the Agricultural Skills Training Based on the Motion-Sensing Technology of the Leap Motion

Peng-fei Zhao[1,2,3,4], Tian-en Chen[1,2,3,4(✉)], Wei Wang[1,2,3,4], and Fang-yi Chen[1,2,3,4]

[1] Beijing Research Center for Information Technology in Agriculture, Beijing, 100097, China
[2] National Engineering Research Center for Information Technology in Agriculture, Beijing, 100097, China
[3] Key Laboratory of Agri-Informatics, Ministry of Agriculture, Beijing, 100097, China
[4] Beijing Engineering Research Center of Agricultural Internet of Things, Beijing, 100097, China
{zhaopf,chente,wangw,chenfy}@nercita.org.cn

Abstract. With the increasing development of virtual reality technology, the motion-sensing technology used in agricultural skills training more widely, it plays a great role in promoting the agricultural production, scientific research and teaching. To break the space-time constraints, in order to train the farmers to know well the high precision agricultural skills, and increase the user's immersive and interactive, we propose a training method based on motion-sensing technology. This article in view of the grape vines binding technology, puts forward a kind of agricultural skills training methods based on the leap motion technology. Through maya bone modeling technology to realize 3-D simulation of grape vines, the system completes the interactive simulation of grape vines binding based on leap motion technology.

The experimental results show that the system can be a very good simulation of the grape vines binding process. System is stable, reliable and strong commonality, it can be used for simulating different plants vine binding, and the system innovative interactive, it can increases the user experience.

Keywords: Leap motion · Motion-sensing · Agricultural · Skills training

1 Introduction

Without using complex control equipment, the motion-sensing technology allow people use the body movements and digital equipment interact with the environment, and in accordance with people's movements to complete a variety of commands. Motion-sensing technology used in agricultural skills training, can solve many problems, for example, the experimental equipment that is insufficient, long time consuming problem. So, the technology can improve teaching efficiency and quality [1]. Training system based on the motion-sensing technology can demonstrate farming activities of seeding simulation, crops such as fruit picking, crop cultivation simulation. This article presents an interactive design training method for virtual agriculture grape park based leap motion motion-sensing technology. First, build a three-dimensional grapes park foundation

D. Li and Z. Li (Eds.): CCTA 2015, Part II, IFIP AICT 479, pp. 277–286, 2016.
DOI: 10.1007/978-3-319-48354-2_29

template library and grape dimensional model, and then track user gestures, and identify the user crawling vines, enhance vines, binding vines gestures, etc. When a series of binding vine action is completed, by setting a good UI interface, the user is prompted to re-tie the vines or success to bind vines.

2 Virtual 3D Model

Traditional agricultural skills training let the farmers learn the technology in indoor classroom, then the farmers go to work in the filed, the process is often greatly limited by the space-time, weather factors, material, etc. Using the virtual farm training, the farmers can avoid damaging the crops because of mistake of farming. In agricultural training using a virtual farms, farmers can change the farmland environment and cultivation measures, directly observe changes in crop growth and yield, quality, so as to deepen understanding of the new technology, new varieties and master degree. The traditional agricultural science and technology promotion mode can not achieve this effect.

Virtual farm plant modeling refers to the use of physical or mathematical method, the mathematical model obtained to describe the system needs to simulation, it is essential for the digital simulation steps [2]. Three-dimensional modeling technology is mainly divided into computer modeling and software modeling. Due to computer modeling is composed of surface, the technology require higher performance computer, and many difficulties exist in the process of model building. So, the system adopts the Maya software modeling.

2.1 Maya Modeling Technology

Maya software is a leader in the field of three-dimensional digital animation and visual effects. The Maya function is flexible operation, highly realistic rendering, and maya integrates the animation with digital effects technology. Moreover, it integrates 3D modeling, animation, special effects with the rendering plate that provides a variety of tools for making model and animation need [3]. From the actual physical simulation to character animation, and the particle system complex, maya can almost making all of CG (Computer Graphics) forms.

Types of modeling in Maya software is divided into NURBS modeling, Polygons modeling, Subdivision modeling. The establishment of bone is the process of using joint and bone building level connection structure. The bones are the joint structural level, it can be animated and positioned the deformable objects. The user can make objects become joints and bone sub objects, and use the skeleton to control the movement of objects.

2.2 Grapes Vines Modeling

In this system, according to the laws of the grape phyllotaxis, we build the model of grape leaves through the Maya software. In this method, through key nodes and nurbs curve, we build the structure of branches [4]. The construction of the grapes 3D model steps are as follows:

First, through the digital imaging equipment to capture static grape appearance, and the photo processing for texture mapping real.

Second, consult the grape vine growth morphology, in accordance with the law, construct three-dimensional grape contour model.

Third, according to the laws of the grape phyllotaxis, build a virtual 3-D model of grape leaves.

Fourth, according to the grape inflorescence rule and the growth cycle of the fruit, build relevant three-dimensional form in different periods the grapes.

Fifth, in the virtual model of grape vines, to join the bone invasion, simulate the grapes and vine deformation.

As shown in Fig. 1, Fig. 1a according to the growth rule of nature, extract the grape contour skeleton. Figure 1b through the optimization design of pipelines, the creation of fruit and three dimensional profile. Figure 1c through texture mapping and the optimized model, 3D virtual simulation of grape.

(a) Outline the skeleton   (b) The fruit profile   (c) 3-D Simulation model

Fig. 1. The construction of the grapes 3D model

3   Motion-Sensing Technology

3.1   Kinect Motion-Sensing Technology

Motion-sensing interaction devices provide the user with high accuracy and low cost requirements, the user can easily distinguish different gestures, and the user can define a gesture to grasp virtual objects, or make a pull or folding finger movements.

Kinect is the use of peripheral body sense device most widely, it can complete real-time capture and tracking of human body in real-time. But kinect recognition accuracy is about 4 mm, which requires the user need to have larger movement range [5]. And 30 frames per second capture ability make equipment cannot distinguish the rapid movements of players, delay is very bad. Because the weakness of precision and delay temporarily, the kinect is unable to solve on the wrist precision farming activity.

3.2 Leap Motion Motion-Sensing Technology

The sensing orange of Leap Motion accuracy can reach one percent mm, so it can let a person directly control computer through the fingers, including image scaling, moving, rotating and instruction operation, precise control, etc. [6]. In this space that your 10 finger movements can be tracking in real-time, error in the 1/100 mm. The leap motion is superior to Kinect in the interactive speed and accuracy that can guarantee the user successfully completed more accurate, more precise farming activities. As shown in Fig. 2, Leap motion captured five fingers outstretched palm movement [7].

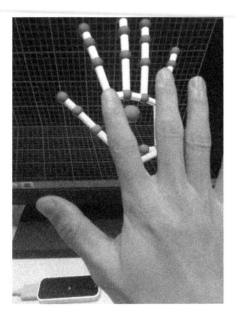

Fig. 2. Hand simulation

3.3 Custom Gesture Model

Through the Maya modeling software, we can create the 3-D hands model, as shown in Fig. 3. In the system, the model is imported into the system, and we create the respond events of the hands [8]. In response to an event trigger, the system can complete the simulation of gestures.

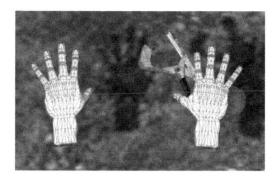

Fig. 3. Hands gesture model

As shown in Fig. 4, Fig. 4a, in the real world, the system capture the left hand stretched out, and in the system, the response effect of virtual left. As shown in Fig. 4b, in the system, the effect of virtual left hand clenched.

(a) Left hand open   (b) Left hand clench

Fig. 4. Left hand status

As shown in Fig. 5, Fig. 5a in the real world, the system capture the right hand stretched out, and in the system, the response effect of virtual right hand. As shown in Fig. 5b, in the system, the effect of virtual right hand clenched.

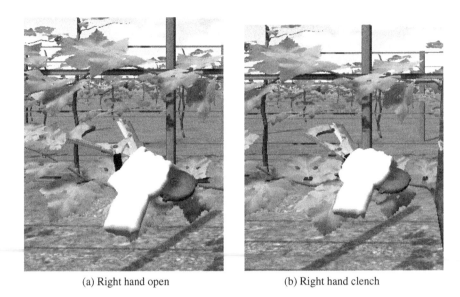

(a) Right hand open   (b) Right hand clench

Fig. 5. Right hand status

3.4   Inverse Kinematics

Inverse dynamics is a method to calculate the root nodes information through the modeling nodes information. In a nutshell, IK (Inverse Kinematics) algorithm confirms the location information of end, according to the position information of the end nodes derivate the position information of skeletal class N chain of the father [9]. Through this method specifies the role of "position", namely the first rendered roles into bone joint

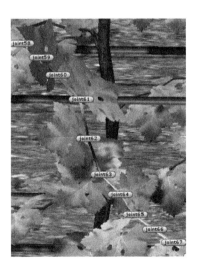

Fig. 6. Joint point graphic

structure, the so-called position expressed as the angle formed therefrom, including the state vector θ and the direction vector and the position of the node. For the IK algorithm, the basic formula is:

$$\theta = f^{-1}(X)$$

Among them, X is usually the end location information, and use this algorithm to calculate to meet the conditions of the θ [10]. As shown in Fig. 6, we add the joint points in the vine branches and use IK inverse dynamics, the users can hold the vines users and the vines around the key points for spatial arbitrary direction of movement.

3.5 System Working Process

After the user login, the system will automatically peripheral initialization, and check whether the normal peripheral, if not connected system or the driver is not installed, then the system prompt error and exit the training system. If the connection is normal, the system will pop up a simple introduction to the gesture operation, as shown in Fig. 7. Next, the user hands could be detected in the Leap Motion space, with the left hand grabbing vine, right hand to pick up tying vine machine, and then the vines binding in the right position. The system will be based on the vine binding position to judge whether the user is tied to vines success. Tied cane is successful, then the system exit; If bind cane not successful, he system will prompt the user for training again. System work process is shown in figure.

Fig. 7. Operation method

4   The Experimental Results

User login, such as the system peripherals are connected properly, then start tying the rattan movement simulation. As shown in Fig. 8, the user's left hand appears in the Leap Motion detection range, there will be left hand in the virtual system, and prompts the user use the left hand grab the vine. As shown in Fig. 9, users grab the vines, according to the arrow tips and improve near to a grape vines upward.

Fig. 8. Virtual left hand   **Fig. 9.** Grasp the vine

After users use the left hand complete a series of actions and keep the state invariant. When users right hand appear in the Leap Motion range of visibility, the system appears virtual right hand with tied rattan machine. At a specific location, the vines machine will be fixed on the grape vine, as shown in Fig. 10. When bound nodes meet the requirements, system prompts tied rattan successfully, and exit the system, as shown in Fig. 11.

Fig. 10. Hand tied the vine   **Fig. 11.** Complete grasp vine

As shown in Fig. 12, it is the system sketch. The frame rate of the system is 54 frames/sec. As shown in Fig. 13, it is the sketch that how to operate the system.

Fig. 12. System sketch   **Fig. 13.** Operate the system

According to the above results, compared with the reference [5], Leap Motion can capture users movements, such as, finger movements. Moreover, the system runs smoothly and the FPS can reached 54 frames/sec, so that the system consumes less IT resources. Finally, the cost of the system hardware is low, and the operation is simple, the method is suitable for all kinds of agricultural training. So, it is feasible to carry on the agricultural training that using Leap Motion to capture the finger movements.

5   Conclusions

Compared with the teaching system of motion-sensing technology based on current, the system aimed at the agricultural field demand, the method can meet high precision farming gesture operation, and achieve fast response, low delay capture effect. Through the teaching system, whenever and wherever, the farmers can be better familiar with the technology of tying vines, and break the limit of space-time to grasp the essentials of tying vines. Achieve the demand of science and technology, science and technology education peasants farming, realize the modernization of agriculture.

The system can make full use of human gestures convenience, make the interactive design and experience of agricultural park is more natural, more real, and can meet the agriculture park aided design, virtual experience farming operations and science educa-tion application requirements.

Acknowledgment. Funds for this research was provided by the Technology Innovation Ability Construction Projects of Beijing Academy of Agriculture and Forestry Sciences (KJCX20140416), and The Beijing Municipal Science and Technology Projects

(D141100004914003). The research work also has been done under the help of the team of the Information engineering.

References

1. Xu, M., Sun, S., Pan, Y.: The research of motion control in virtual human. J. Syst. Simul. **15**(3), 338–340 (2003)
2. Li, H., Luo, C., Tang, J.: Application of virtual agriculture technology. Comput. Eng. Des. **8**(29), 2059–2061 (2008)
3. Qiu, Y.: Design of characters walking movement based on Maya technology. Comput. Mod. **3**(4), 22–25 (2010)
4. Liu, C., Feng, J., Jiang, J.: Cluster analysis of Chinese wild grape species based on morphological characters. J. Plant Genet. Resour. **2**(6), 847–852 (2011)
5. Fan, J., Zhou, G.: Kinect based agricultural virtual teaching. J. Anhui Agric. Sci. **42**(12), 3706–3709 (2014)
6. LeapMotion SDK. https://developer.leapmotion.com
7. Leap Motion Product. https://www.leapmotion.com
8. Chen, H., Ma, Q., Zhu, D.: Research of interactive virtual agriculture simulation platform based on Unity3d. J. Agric. Mech. Res. **34**(3), 181–186 (2012)
9. Li, Z., Peng, Y.: The research based on the 3D inverse dynamics of IK algorithm. Image Process. Multimedia Technol. **32**(24), 33–36 (2013)
10. Tolani, D., Goswanmi, A., Balder, N.I.: Real time inverse kinematics techniques for anthropomorphic limbs. Graph. Models **62**(5), 353–358 (2000)

Study of Spatio-temporal Variation of Soil Nutrients in Paddy Rice Planting Farm

Cong Wang[1,2,3,4,5], Tianen Chen[1,2,3,4,5](✉), Jing Dong[1,2,3,4,5], Shuwen Jiang[1,2,3,4,5], and Chao Li[2,5]

[1] Beijing Research Center for Information Technology in Agriculture,
Beijing 100097, China
[2] National Engineering Research Center for Information Technology
in Agriculture, Beijing 100097, China
{wangcong,chente,dongj,jiangsw}@nercita.org.cn,
3961695@qq.com
[3] Key Laboratory of Agri-Infomatics, Ministry of Argiculture,
Beijing 100097, China
[4] Beijing Engineering Research Center of Argicultural Internet of Things,
Beijing 100097, China
[5] Erdaohe Farm, Jiamusi 156330, Heilongjiang, China

Abstract. It is significant to analyze the spatial and temporal variation of soil nutrients for precision agriculture especially in large-scale farms. For the data size of testing results is growing every time after sampling mostly by the frequency of once a year or several months, in order to discover the variation trends of specific nutrient which would be instructive for the fertilization in the future. In this study, theories of GIS and geostatistics were used to characterize the spatial and temporal variability of soil nutrients in paddy rice fields in the Erdaohe farm of Heilongjiang Province, China, which located in the north of Daxing'an Mountains, has an area of nearly 36.1 million hectares for paddy rice planting. The soil samples, collected from 2009 to 2013 once a year, were sampled based on the spatial distribution of paddy rice fields, counting as 651 in 2009, 1488 in 2010, 954 in 2011, 483 in 2012, and 471 in 2013. These samples were analyzed for pH, soil organic matter (SOM), available nitrogen (AN), available phosphorus (AP), and available potassium (AK). In this study, we calculated and compared the spatial and temporal variation in whole farm area, using methods of exploratory statistical and geostatistical analysis. Conclusion acquired is that from 2009 to 2013, the spatio-temporal variations decreased in soil pH, AN, AP, AK, and increased in SOM. Moreover, according to the comparison of interpolation results, these five soil properties in Erdaohe farm remained not very stable in the past five years, which could implicate important significance in future research for consideration of correlation amongst fertilization, rice yield and other factors especially in large-scale farms.

Keywords: Spatio-temporal variation · Soil nutrition · Geostatistical analysis · Semivariogram model

© IFIP International Federation for Information Processing 2016
Published by Springer International Publishing AG 2016. All Rights Reserved
D. Li and Z. Li (Eds.): CCTA 2015, Part II, IFIP AICT 479, pp. 287–299, 2016.
DOI: 10.1007/978-3-319-48354-2_30

1 Introduction

Soil nutrients provide a scientific accordance in fertilizer applications, especially in paddy rice planting farms. However, soil properties not only have spatial variability, but also oscillate with the time changing. It's meaningful to analyze the extent of temporal and spatial variation of soil nutrient contents for more reasonable fertilizer applications.

Lots of works have been actualized on soils by measuring and analyzing the spatial dependencies on soil fertility [1–4]. For example, Weijun Fu et al. studied the spatial variation of soil nutrients in a dairy farm in southeastern Ireland [5], Kelin Hu et al. studied patterns of spatial and temporal variation of SOM in Beijing's urban–rural transition zone [6], and Zhang Xing-Yi studied the spatial variability of nutrient contents in northeast China where has black soils [7]. However, most of the researches learned temporal variation in period of time with long intervals, neglecting the continuous changing year after year [8]. In this paper, methods of statistics and geostatistics were applied to study the spatio-temporal variation for data of soil pH, available nitrogen (AN), soil organic matter (SOM), available potassium (AK), and available phosphorus (AP) collected from 2009 to 2013 once a year in a paddy rice planting farm in northeast China.

2 Materials and Methods

2.1 Study Area

The farm of Erdaohe located in northeast boundary of China, closed to Russia across the Ussuri River in the east, and the Heilongjiang in the north (Fig. 1). This area belongs to the Sanjiang Plain, and has a humid or semi-humid continental monsoon climate of the North Temperate Zone, which is suitable for agriculture production,

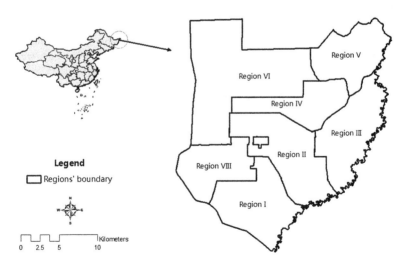

Fig. 1. Location of the study area

especial for paddy rice planting. The total area of this farm is 534.2 km, with an area of 362 km for cultivated land, including 360.7 km for paddy rice planting.

2.2   Soil Sampling

The soil samples were planed to collect at the depths of 0-20 cm from 2009 to 2013 once year (Fig. 2). The sample time were mostly between autumn harvests and fertilizers. Generally, this work was mostly done by experienced technicians, who has a quantity knowledge of agriculture production in the sampling region. The count of sampling points are 651 in 2009, 1488 in 2010, 954 in 2011, 483 in 2012, and 471 in 2013. On the other hand, location of these points were selected according to space distribution of land parcels, soil types, land use types and experience of technicians. After sampling, soil samples were naturally dried at ventilation place and then sieved to pass a 2-mm mesh after crushed. In this article, soil test results of pH, soil organic matter (SOM), available potassium (AK), available nitrogen (AN), and available phosphorus (AP) are used for spatio-temporal variation analysis, the soil test methods are revealed in Table 1.

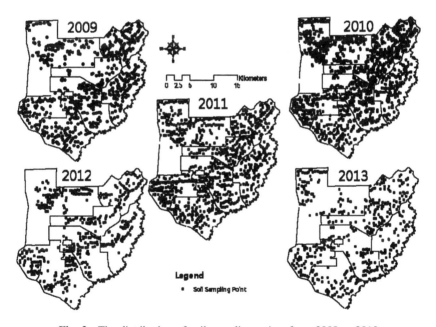

Fig. 2. The distribution of soil sampling points from 2009 to 2013

Table 1. Soil test methods used in study area

Test item	Method	Unit of test result
Soil pH	Potentiometry method (water soil ratio is 2.5:1)	——
SOM	Potassium dichromate sulfuric acid heating method	g/kg
AN	NaOH hydrolyzation diffusion method	mg/kg
AP	NaHCO3 - Molybdenumblue method	mg/kg
AK	Atomic absorption spectrophotometric method	mg/kg

3 Analytical Methods

3.1 Exploratory Statistical Analysis

In this paper, methods of exploratory statistical and Geostatistical analysis are both chosen to study the temporal and spatial variation of soil nutrient contents in this paper's study area. First of all, descriptive analysis indexes such as maximum (max), minimum (min), median, mean, skewness and kurtosis, coefficient of variation (C.V.) and standard deviation (S.D.) were chosen to achieve the summary information of soil nutrients distribution. These indexes can be divided into three different types: location, spread, and shape, which provide diverse descriptions of soil nutrients. Index of mean is the arithmetic average of data measured, which shows the center of the distribution of the original data.

Besides indexes described above, Normal Q–Q plots (quantile–quantile plots) were created to identifying the probability and some distinct outliers (same as extreme values). Usually on the x-axis marked the observed values, and for a normal distribution values expected were marked on the y-axis. In general, samples which have a normal distribution cluster would follow a diagonal straight line [9]. Meanwhile, on the normal Q–Q plots, it's easy to observe the low or high value outliers, because these points will be away from the calculated normal Q–Q line.

3.2 Geostatistical Analysis

In this study, the spatial variation of each soil nutrient content was measured by geostatistics. Experimental variogram evaluator is approximately uninfluenced for any inherent random function, whatever it's really sensitive to external values since it is on account of squared distinctions among calculated data. The semivariogram model was established for each microbiological parameter for the sake of characterizing the level of spatial variability between neighboring samples. Meanwhile, the proper model function was suitable to the semivariogram model. The value of semivariogram **Y**(h) was calculated using equation below

$$\gamma(h) = \frac{1}{2N(h)} \sum_{i=1}^{N(h)} (Z(x_i) - Z(x_i + h)) \tag{1}$$

In the equation, h represents the demarcation distance from the locations of x_i to locations of $x_i + h$. $Z(x_i)$ and $Z(x_i + h)$ represents the values which are calculated for the regionalized variables at locations x_i or $x_i + h$. The last one N(h) represents the quantity of two sets at any demarcation distance of h [10, 11].

Including spherical, Gaussian, exponential, linear and power models, there are several models available to adjust the experimental semivariogram [12–14]. On the other hand, a semivariogram includes three primary parameters which define the spatial structure of original data as: $\mathbf{Y}(h) = C_0 + C$. Where C_0 delegates the nugget effect which means the local variation coming at scales smaller than the samples' interval, like sampling error, measurement error and fine-scale spatial variability. The sum of C_0 and C is the sill which represents total variance in the equation. The distance is called the range at which the semivariogram levels away from the sill. Furthermore, sampling points are not spatially connected whenever the numerical value of separation distances is larger than the range [10].

The equation of different models are described below. Model of spherical aniso-tropic was adjusted to the empirical semivariance, which is defined as:

$$\begin{cases} \gamma(h) = C_0 + C_1\left[\dfrac{3h}{2a} - (h/a)^3/2\right] & 0 < h < a \\ \gamma(h) = C_0 + C_1 & h > = a \\ \gamma(0) = 0 & h = 0 \end{cases} \tag{2}$$

In the equation, C_0 represents the nugget value which means the spatial variability produced by the random components such as micro-scale processes and measured error. C_1 is the structural variance which means the spatial heterogeneity produced by spatial autocorrelation. $C_1 + C_0$ represents the sill, while A represents the range (or spatial correlation distance).

Other stationary models such as Gaussian (Eq. (3)), exponential (Eq. (4)) and linear (Eq. (5)) equations are defined as:

$$\gamma(h) = C_0 + C_1\left[1 - \exp(-h/a)^2\right] \tag{3}$$

$$\gamma(h) = C_0 + C_1[1 - \exp(-h/a)] \tag{4}$$

$$\gamma(h) = C_0 + bh \tag{5}$$

Where C_0, C_1, h and a represent the same meanings as spherical anisotropic model, while b is slope of the semivariance line in Eq. (5).

4   Results and Discussion

4.1   Variation of Soil Properties in Past Five Years

According to exploratory statistical analysis, Table 2 shows the soil nutrients determined values of minimum, maximum, median, mean, coefficient of variation (C.V.), standard deviation (S.D.), kurtosis and skewness from 2009 to 2013. First of all, considering the values of different soil nutrients described by min, max, median and mean, in the past five years, test results of pH shows a relatively inflexible constant, while the other four properties appears more variable. Among these nutrient types, except for SOM, the variation degree tested by C.V. (%) all decreased with small fluctuations in the past five year. Furthermore, for pH data the C.V. value was relatively small while that for AK data was relatively large.

Table 2. The statistical values of soil properties.

	Year	min	max	mean	Median	S.D.	C.V. (%)	Skewness	Kurtosis
pH	2009	4.60	6.40	5.50	5.60	0.26	4.77	-0.22	3.79
	2010	5.10	6.50	5.55	5.50	0.21	3.73	0.45	3.36
	2011	4.98	6.40	5.57	5.58	0.20	3.60	0.20	3.17
	2012	4.88	6.52	5.52	5.52	0.24	4.42	0.52	4.64
	2013	4.94	6.21	5.54	5.54	0.17	3.11	-0.19	3.10
SOM	2009	11.20	69.56	40.44	40.66	8.40	20.77	0.00	3.62
	2010	17.79	59.19	39.55	44.66	6.51	16.45	-0.18	2.69
	2011	12.50	74.20	39.25	39.08	8.69	22.13	0.34	3.83
	2012	23.20	154.00	43.11	41.80	10.56	24.49	3.37	31.05
	2013	12.60	379.80	43.29	41.00	24.43	56.44	10.67	143.39
AN	2009	111.20	507.50	231.85	229.30	50.18	21.64	0.79	5.34
	2010	83.76	376.70	243.00	275.70	56.94	23.43	0.22	2.75
	2011	86.90	485.59	236.77	260.71	41.73	17.63	0.48	4.59
	2012	135.35	371.63	220.44	216.16	32.80	14.88	0.82	4.88
	2013	109.90	351.20	203.67	197.20	41.08	20.17	1.09	4.67
AP	2009	3.60	79.90	27.70	27.20	9.67	34.91	0.40	4.48
	2010	3.93	63.40	27.88	34.70	9.39	33.67	-0.20	2.72
	2011	3.57	52.09	28.44	29.39	8.58	30.16	-0.50	3.63
	2012	3.30	54.14	30.90	31.36	9.29	30.07	-0.18	3.02
	2013	3.70	66.40	30.03	30.60	9.28	30.91	-0.29	3.53
AK	2009	25.00	531.00	149.82	132.00	73.95	49.36	1.67	7.23
	2010	27.00	569.00	157.95	141.00	65.84	41.69	1.44	6.06
	2011	53.54	609.30	163.47	143.23	73.26	44.82	2.21	10.80
	2012	38.00	609.00	176.91	163.42	67.53	38.17	1.90	11.26
	2013	71.00	699.00	192.71	170.00	86.59	44.93	2.07	9.96

On the other hand, the degree of dispersion tested by S.D. shows that the sequence from high to low is AK, AN, AP, SOM, AP and pH, besides, the degree of dispersion of SOM increased significantly in recent two years. At last, kurtosis and skewness indicated the shape of distribution of soil raw data compared with normal distribution. The results showed that except for AK and several years' data of SOM and AN, the others were closest to normal distribution.

4.2 Normal QQ-Plots Analysis and Data Transformation

Since parts of soil values did not fit the normal distribution, data Fig. 3 shows the normal QQ-plots of five different soil nutrients in 2009. Besides AK values displayed a concave shape, the pH, SOM and AP values obeyed a shape of nearly straight line, which means a near normal distribution. While the AN data obeyed a straight diagonal line with data points nearby except for several points deviated at both ends. Normal QQ-plots of the other four years were not shown for limitation of paper length. While the analysis results found that in 2010 and 2011, soil nutrients data except AK, followed a nearly straight diagonal line just like that in 2009. And in 2012 and 2013, only pH values remained the same distribution, the others became not very accordant with that more or less.

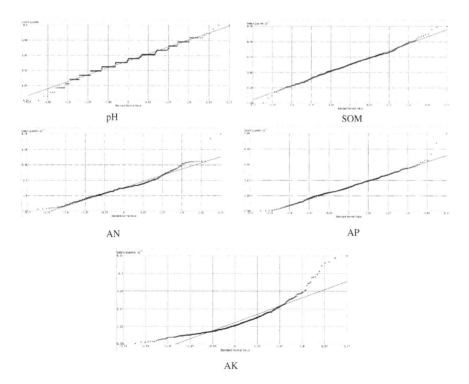

Fig. 3. Normal QQ-plots of soil nutrients in 2009

Transformation was needed for the followed analysis. With the combination of exploratory statistical analysis and normal QQ-plot analysis, different transformation method were chosen, seen from Table 3. Figure 4 shows the Normal QQ-plots of data before and after the transformation for AK values in 2011, it is obvious that after transformation, the points were more fitted with the straight line.

Table 3. Data transformation method selected

	2009	2010	2011	2012	2013
pH	a	a	a	a	a
SOM	a	a	a	b	$c(\lambda = -0.4)$
AN	$c(\lambda = 0.1)$	a	a	b	b
AP	a	a	a	a	$c(\lambda = 1.3)$
AK	b	b	$c(\lambda = -0.4)$	b	$c(\lambda = -0.5)$

a - None transformation; b - Log transformation; c - Box-Cox transformation.

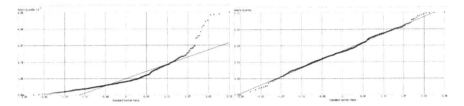

Fig. 4. Normal QQ-plots of AK raw data and transformation result Box-Cox ($\lambda = -0.4$) in 2011

4.3   Spatio-temporal Variation of Different Soil Properties

According to the theory of geostatistical, semivariance analysis was applied to soil properties from 2009 to 2013, the results indicated that spatial autocorrelation existed for the soil properties in study area, which means spatial interpolation method of kriging could be used to predict the soil nutrients in missing data area. While the step was to choose the appropriate semivariogram model for each property and each year, Fig. 5 shows the semivariogram for soil pH in 2009 (anisotropic) with different models (Spherical, Gaussian and Exponential).

In order to select the best model for following analysis, comparison of precision for different models is needed. Table 4 offered the precision analysis results of spatial data of soil nutrients in 2009. For in condition that mean standardized closer to zero, the root-mean-square was smaller, average standard error closer to root-mean-square, and root-mean-square standardized closer to one, semivariogram model may be the most appropriate one. Based on the precision errors, the best semivariogram model for soil data collected in 2009 were: Spherical for pH and SOM, Gaussian for AN, Exponential for AP and AK.

Table 5 displays the selected best models for each soil properties from 2009 to 2013 and their parameters. First of all, directional features were observed for the

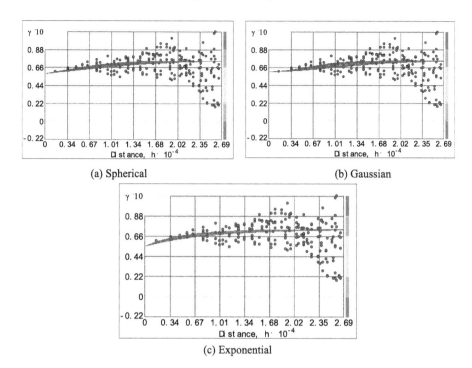

(a) Spherical   (b) Gaussian

(c) Exponential

Fig. 5. Semivariograms of different models for soil pH in 2009

Table 4. Comparison of precision analysis among different models for soil test data in 2009

	Model	Root-mean-square	Average standard error	Mean standardized	Root-mean-square standardized
pH	Spherical	0.2482	0.249	0.003599	0.9969
	Gaussian	0.2485	0.2521	0.004238	0.9863
	exponential	0.248	0.2444	0.002406	1.015
SOM	Spherical	8.133	8.304	0.000509	0.9803
	Gaussian	8.136	8.293	0.001108	0.9823
	exponential	8.115	8.27	-0.0005942	0.9823
AN	Spherical	48.07	47.88	0.001782	1.005
	Gaussian	48.3	48.54	-0.001631	0.9986
	exponential	47.8	47.67	0.002623	1.004
AP	Spherical	9.093	9.417	-0.0001489	0.9657
	Gaussian	9.098	9.525	-0.0003632	0.9555
	exponential	8.996	9.183	0.001384	0.9798
AK	Spherical	68.57	73.67	-0.009017	0.9971
	Gaussian	68.98	75.13	-0.00288	0.9827
	exponential	68.2	72.55	-0.01263	0.9979

majority soil data except for SOM values collected in 2009 and 2010, which also became the special cases of range values above 25 km. The value of Nugget/Still shows the relative size of the nugget effect among varied soil properties [15]. This value was used to define different classes of spatial dependence for the soil variables as such rules:

(a) If the ratio was smaller than 25 %, the variable was considered extremely spatially dependent;
(b) If the ratio was between 25 % and 75 %, the variable was considered to be moderately spatially dependent;
(c) If the ratio was larger than 75 %, the variable was considered to be weakly spatially dependent, which indicated that random factors were the majority factors affect the spatial variation of soil properties [16].

Table 5. Semivariogram models selected for soil nutrients and parameters of each model (2009-2013)

	Year	Model	Anisotropic	Still	Major range(m)	Nugget	Direction	Nugget/Still
pH	2009	Spherical	Yes	0.074	25577.3	0.058	277.6	0.79
	2010	Exponential	Yes	0.036	26993.5	0.019	39	0.35
	2011	Exponential	Yes	74.571	25829.5	0.021	276.8	0.45
	2012	Exponential	Yes	42.344	26322.6	0.021	336.5	0.28
	2013	Exponential	Yes	76.740	25442.6	0.011	79.6	0.31
SOM	2009	Spherical	No	0.0491	3584.73	59.571		0.80
	2010	Spherical	No	0.0045	4148.61	41.847		0.99
	2011	Exponential	Yes	0.141	26133.4	71.409	290.7	0.93
	2012	Spherical	Yes	3263.87	25427.1	0.035	312.7	0.72
	2013	Exponential	Yes	1799.66	25493.9	0.0025	56.4	0.56
AN	2009	Gaussian	Yes	0.0217	25507.6	0.126	316.3	0.90
	2010	Exponential	Yes	0.041	25902.3	3156.4	66	0.97
	2011	Exponential	Yes	99.234	25766.3	1535.3	25.8	0.85
	2012	Exponential	Yes	97.084	25321.6	0.017	307.1	0.80
	2013	Exponential	Yes	88.399	26021.5	0.027	239.6	0.66
AP	2009	Exponential	Yes	289.06	25667.5	77.23	29.4	0.78
	2010	Exponential	Yes	768.09	26219	64.194	51	0.66
	2011	Spherical	Yes	0.227	26795.5	47.283	48.8	0.53
	2012	Exponential	Yes	0.175	26322.6	250.5	59.175	0.87
	2013	Spherical	Yes	0.003	26021.5	345.93	61.6	0.45
AK	2009	Exponential	Yes	0.145	26556.1	0.169	240	0.75
	2010	Exponential	Yes	0.0009	26846.3	0.101	63.2	0.58
	2011	Spherical	Yes	0.074	26086.2	0.0018	33.2	0.64
	2012	Spherical	Yes	0.036	25407.7	0.098	24.6	0.67
	2013	Exponential	Yes	74.571	25215.6	0.0007	46.4	0.75

From this table, it is clear that all of the soil properties measured in this study were not strongly spatially dependent. From 2009 to 2013, the value of Nugget/Still for each soil property decreased in overall trend, relatively, soil pH was the considered as the most strongly spatial dependent, while soil AN was the weakest one in study area.

According to the results below, interpolation method of ordinary kriging was used to predict the spatial distribution maps for soil nutrients in study area. Figure 6 shows the prediction maps for soil pH from 2009 to 2013. For the limitation of paper length, the maps of other soil nutrients were not shown in this paper.

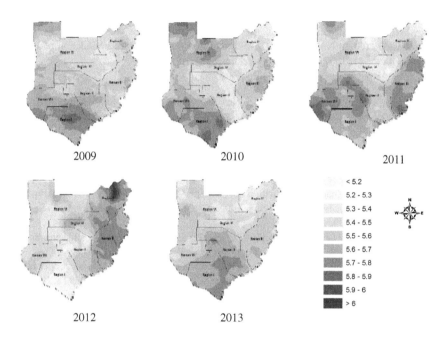

Fig. 6. Spatial distribution maps for soil pH (2009-2013)

The spatial distribution maps of soil pH shows that the soil pH in this study area were mostly acidic, or strongly acidic in some regions. From 2009 to 2011, the spatial distribution is similar in the whole area, while there was obvious changes in 2012, especially in the south area, soil pH became more acidic. Until the next year, distri-bution changed similar as that in 2009. For other soil nutrients, the spatial distribution remained not very stable in the five years, which probably because of the fertilization changed every year.

5   Conclusions

This study analyzed the spatio-temporal variation for several soil properties in a paddy rice farm from 2009 to 2013. Among these soil properties, data of soil pH collected all the five years followed a normal distribution, and had a relatively small C.V. values which decreased along the study time range. On the other hand, the spatial variation of soil pH increased, became more strongly spatial dependent. The others, such as soil AK, followed a log-normal distribution in 2009, 2010 and 2012, while in 2011 and 2013 followed neither normal nor log-normal distribution. Thus, data transformations were acquired for better analysis. Except for soil pH, raw data of other soil properties were transformed by log or box-cox method more or less. According to the analysis results, the spatial variation of soil pH, AN, SOM, AP and AK all increased from 2009 to 2013, while except soil pH, most of that were not strongly spatial dependent, which means the spatial variation was mostly based on random factors in this study area.

Acknowledgment. Financial supports from the National Science and Technology Support Project of China (Grant No. 2013BAD15B05) and the Natural Science Foundation of Beijing, China (Grant No. 4151001) are sincerely acknowledged.

References

1. Tesfahunegn, G.B., Tamene, L., Vlek, P.L.G.: Catchment-scale spatial variability of soil properties and implications on site-specific soil management in northern Ethiopia. Soil Tillage Res. **117**, 124–139 (2011)
2. Sun, B., Zhou, S., Zhao, Q.: Evaluation of spatial and temporal changes of soil quality based on geostatistical analysis in the hill region of subtropical China. Geoderma **115**, 85–99 (2003)
3. Bogunovic, I., Mesic, M., Zgorelec, Z., et al.: Spatial variation of soil nutrients on sandy-loam soil. Soil Tillage Res. **144**, 174–183 (2014)
4. Ferreira, V., Panagopoulos, T., Andrade, R., et al.: Spatial variability of soil properties and soil erodibility in the Alqueva reservoir watershed. Solid Earth **6**(2), 383–392 (2015)
5. Fu, W., Tunney, H., Zhang, C.: Spatial variation of soil nutrients in a dairy farm and its implications for site-specific fertilizer application. Soil Tillage Res. **106**, 185–193 (2010)
6. Hu, K., Li, H., Li, H., et al.: Spatial and temporal patterns of soil organic matter in the urban–rural transition zone of Beijing. Geoderma **141**, 302–310 (2007)
7. Zhang, X., Sui, Y., Zhang, X., et al.: Spatial variability of nutrient properties in black soil of northeast China. Pedosphere **17**(1), 19–29 (2007)
8. Huang, B., Sun, W., Zhao, Y., et al.: Temporal and spatial variability of soil organic matter and total nitrogen in an agricultural ecosystem as affected by farming practices. Geoderma **139**, 336–345 (2007)
9. Evans, M., Hastings, N., Peacock, B.: Statistical Distribution, 3rd edn. Wiley, New York (2000)
10. Piotrowska, A., Długosz, J.: Spatio-temporal variability of microbial biomass content and activities related to some physicochemical properties of Luvisols. Geoderma **173–174**, 199–208 (2012)
11. Burgess, T., Webster, R.: Optimal interpolation and isarithmic mapping of soil properties. J. Soil Sci. **31**(2), 333–341 (1980)
12. Liu, X., Zhang, W., Zhang, M., et al.: Spatio-temporal variations of soil nutrients influenced by an altered land tenure system in China. Geoderma **152**, 23–24 (2009)

13. Wang Z Q.: Geostatistics and Its Application in Ecology, pp. 162–192 (1999)
14. Goovaerts, P.: Geostatistics in soil science: state-of-the-art and perspectives. Geoderma **89** (1), 1–45 (1999)
15. Trangmar, B.B., Yost, R.S., Uehara, G.: Application of geostatistics to spatial studies of soil properties. Adv. Agron. **38**(1), 45–94 (1985)
16. Cambardella, C.A., Moorman, T.B., Parkin, T.B., et al.: Field-scale variability of soil properties in central Iowa soils. Soil Sci. Soc. Am. J. **58**(5), 1501–1511 (1994)

Path Planning Methods for Auto-Guided Rice-Transplanters

Fangming Zhang[1,2(✉)], Changhuai Lv[3], Jie Yang[2], Caiyu Zhang[2],
Guisen Li[2], and Licheng Fu[4]

[1] Ningbo Institute of Technology, Zhejiang University, Ningbo 315100, China
fangmingzhang@126.com
[2] Ningbo Yinzhou Maigu Technology Company, Ningbo 315100, China
{2287396064,63849331,173748207}@qq.com
[3] Ningbo Agricultural Machinery Administration, Ningbo 315016, China
779426002@qq.com
[4] Zhengda-sangtian (Ningbo) Agriculture Developing Company,
Cixi, Ningbo 315100, China
flc508@163.com

Abstract. Paddy field operation asks for robotization due to labor shortage. In this paper, a set of path planning methods were developed and tested on a guidance for transplanters. Path planning of fieldbody and headland were generated and experimented for 3 years. The path outside field consist of farm lanes connecting garage to paddy field, and a ARC-NODE structure method was proposed. Method of optimal coverage path planning should be researched in near future.

Keywords: Guidance system · Transplanter · Path planning

1 Introduction

There is an increasing demand for autonomous and robotic systems in agricultural domain, such as ridging, seeding, transplanting, and tilling. Due to constraint of field shape and machine size, some research has carried out in coverage path planning (CPP) for agriculture [1], which looks for paths from a starting point to a target point with maximum cover, minimum repetition and omissions as little as possible, while headland turning is a fact that cannot be ignored. The basic task of a CPP method is producing paths for a single machine. For example, Kise [2] set minimum turning radius and maximum rotation rate as the goal, and used a third-order spline function to create two kinds of path of end turning, namely the forward turning and back turning. Oksanen [3] divided the path planning algorithm into 2 layer: the upper one divided the whole fields into small plots, while the lower one produce path for every plot. Hou [4] proposed two path planning methods of cultivation. Huang [5] proposed turn strategies of parallel line paths of agricultural machinery with Ω shape and Π shape, and a path planning method of convex plots. Han [6] made a predefined path including C-shape headland turning for his auto-guided tractor.

A number of literatures discussed optimal algorithms. Noguchi [7] applied neural network algorithm to learn the movement of the tractor, and used genetic algorithm to find a optimal path. Jin [8] employed a genetic model to generate a optimal path. For a 3D terrain application [9], Jin modeled the terrain, and constructed a cost function based on a "seed curve" to find the optimal path. Hameed [10] also divided this problem into two stages. The first stage expressed geography relations, and the second one made optimal path. Mariano [11] developed an optimal path for reducing fuel consumption in weed and pest control. Linker [12] used the A* algorithm to determine the optimal path for vehicles operation in orchards. Jensen [13] developed an algorithmic approach for the optimisation of capacitated field operations using the case of liquid fertilising, which optimised in a post-process stage that the nonproductive travelled distance in headland turnings is further minimised. Liu [14] applied a bidirectional link table to express a complex polygon, and then it was divided into simple polygons, and made optimal path to cover subarea. Bochtis [15] generated routing plans for intra- and inter-row orchard operations, based on the adaptation of an optimal area coverage method developed for arable farming operations. However, an optimal solution cannot be guaranteed commonly due to several constraints, including soil compaction, contour of field, and agronomic constraints.

To cooperate working with other agricultural machines, or in form of multi-robot, CPP methods were also proposed. Volos [16] proposed a chaotic path generation methods to cover whole area, whose advantage is that can quickly adapt to the changes in the environment. Bochtis [17] divided the field into grids and then used optimal method to plan paths. Jensen [18] proposed a path planning method for transport units in agricultural operations involving in-field and inter-field transports with an optimization criterions of time or traveled distance.

In the case of irregular field shapes and the presence of obstacles within the field area, Zhou [19] develop a planning method with multiple obstacles.

It could be seen that CPP method for machines of paddy fields is still scarce. The objective of this paper is developing in-field and inter-field path planning methods for autonomous rice transplanters.

2   Materials and Methods

2.1   Platform

Two transplanters, one is a PZ60 (ISKI, Japan, Fig. 1.a), and another is a NSD8 (KUBATA, Japan, Fig. 1.b), were used for the verified the path planning methods. The guidance system consisted of a RTK GPS receiver (Beidou, Qiwei), an attitude sensor (Xi'an) and an electric steering wheel [20]. The RTK GPS receiver was mounted above the steering wheel with height to ground of 2.5 meters. The attitude sensor was installed in a box with shock absorption device.

As more than 90 % paddy fields are rectangle shape, it is a fundamental function to plan path for a rectangular field. In our previous research, we proposed a coordinate transformation method [20], which established a local coordinate system as shown in Fig. 2. A transplanter generally plants rice seedling row by row perfect straight, and

Fig. 1. Two test platforms, (a) the ISKI's PZ60, (b) the KUBATA's NSD8

stop planting when in headland area, which leads to two kinds of sub-fields, one is the fieldbody, another is the headland which is narrow and surround the fieldbody. Path planning in the field was then divided into these two sub-fields. Furthermore, farm lanes connecting a garage to paddy field consists of several straight roads, these paths could also be planned and the transplanter could then autonomously move between these two point.

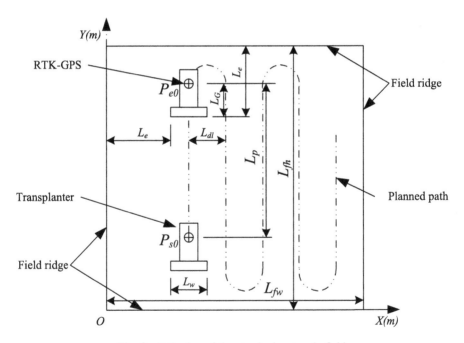

Fig. 2. Path plan of the standard rectangle field

2.2   Path Planning in Fieldbody

If the operating width is L_w, and distance from the GPS point to the seedling point is L_G, then the distance between two neighbor rows is:

$$L_{dl} = L_w \tag{1}$$

As the minimum turning radius of the transplanter is only 0.8 meters, the guidance system could rotates the steeling wheel with the maximum rotate angle that the transplanter could be in the correct position of the next row, named a pi-turn. When the transplanter is ready to plant in a new row, coordinate of the starting point should be:

$$\begin{cases} P_{snx} = L_e + L_w/2 + n \cdot L_{dl} \\ P_{sny} = L_e + L_G \end{cases} \tag{2}$$

Where, P_{snx} and P_{sny} are the coordinates of the starting point in the nth rows.
L_e is the length of track in this row.
Coordinate of the ending point of every row should be:

$$\begin{cases} P_{enx} = L_e + L_w/2 + n \cdot L_{dl} \\ P_{eny} = L_p + L_{sny} \end{cases} \tag{3}$$

Where, P_{enx} and P_{eny} are the coordinates of the ending point in the nth rows.
L_p is the length of every rows in this field, which should be:

$$L_p = L_f - 2L_e \tag{4}$$

Where, L_f is length of the field. If L_{fw} is width of the field, the total rows number is

$$N \geq \frac{L_{fw} - 2L_e}{L_{dl}} \tag{5}$$

2.3   Path Planning in Headland

After finishing planting the last row in the fieldbody, the machine should start a round planting, which might be clockwise or anticlockwise. As shown in Fig. 3, coordinate of the starting point and ending points of the new first row should be:

$$\begin{cases} P_{rs1x} = L_{fw} - 3L_{dl} \\ P_{rs1y} = 1.5L_{dl} \end{cases} \tag{6}$$

$$\begin{cases} P_{re1x} = 2L_{dl} \\ P_{re1y} = P_{rs1y} \end{cases} \tag{7}$$

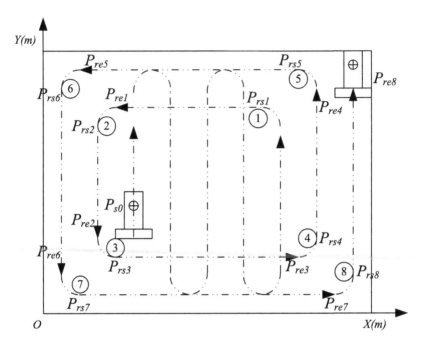

Fig. 3. Path plan in headland

Where, (P_{rs1x}, P_{rs1y}) and P_{re1x}, P_{re1y} are the coordinates of the starting point and ending point in the 1st row of SR working respectively.

Other points could be produced like similar method as Eqs. 6 and 7.

2.4   Path Plan for Triangular Shape Field

Some fields neighbor to field lanes might be triangular shape or diamond shape due to topography restraint. If the machine moves parallel to upper base line of the field, such as OA or BC shown in Fig. 4, it would cover field as much as possible. Two slopes of the bevel edges of triangular are:

$$\tan \theta_1 = \frac{y_C}{x_C} \tag{8}$$

$$\tan \theta_2 = \frac{y_B - y_A}{x_B - x_A} \tag{9}$$

If the working path parallel to line OC is $y = x \cdot \tan \theta_1 + b_1$, and $b_1 = L_w/2 \cos \theta_1$, then the equation of this road would be $y = x \cdot \tan \theta_1 + L_w/2 \cos \theta_1$ For any working path, coordinate of the starting point is:

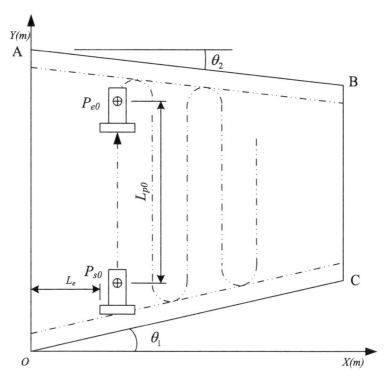

Fig. 4. Path plan of triangular shape field, where θ_1 and θ_2 are intersection angle between OC and x axis, AB and x axis respectively

$$\begin{cases} P_{snx} = L_e + L_w/2 + n \cdot L_{dl} \\ P_{sny} = P_{snx} \tan \theta_1 + L_w/2 \cos \theta_1 + L_G \end{cases} \tag{10}$$

With the similar way, we could get the working path parallel to line AB, which is: $y = x \cdot \tan \theta_2 + L_{OA} - L_w/2 \cos \theta_2$. For any working path, coordinate of the ending point is:

$$\begin{cases} P_{enx} = L_e + L_w/2 + n \cdot L_{dl} \\ P_{eny} = P_{enx} \tan \theta_2 + L_{OA} - L_w/2 \cos \theta_2 \end{cases} \tag{11}$$

2.5   Path Plan for Farm Lane

Generally, path between garage of transplanter and paddy field consists of straight line lanes, which might be two-way roads or one-way farm lanes, whose feature is that they are parallel or vertical to each other as shown in Fig. 5. If latitude and longitude of two points in every farm lane were measured and an original point was set, the road model of any farm lane could be expressed as

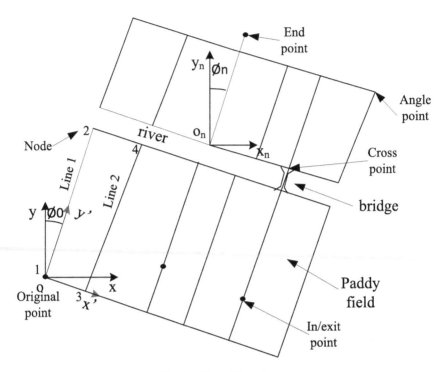

Fig. 5. Map of farm lanes

$$y = x \cdot \tan(90 - \Phi) + b \tag{12}$$

Where, Ø is angle between the road and north direction
b is intercept of this line.

ARC-NODE structure [21], consisted of 3 tables, which were node table, topology table, and road section table, is a normal method to express the spacial relationship of different lanes, in other words that these 3 tables construct a navigation map. Structure of node table was defined as

Point_ID	x	y	Point_Type	Number_Info

Where Point_Type were divided into 4 categories, which are Cross (intersection point), Angle (quarter turn crossing), Number (In/Exit station), and End (terminal of a lane). Number_Info is a serial number if the point belongs to Number type.

Structure of road section was defined as

Point_ID	Count	Point1	Point2	Point3	Point4

Where Count is the number that connected with this point.
Structure of road section was defined as

Line_ID	S_Point	E_Point	Line_Type	Heading angle/curvature radius

Where S_Point and E_Point are starting point and ending point of this Line_ID. Line_Type are divided into 3 categories, which are two-way straight road, one-way straight lane, and arc lane. Heading angle is corresponding to straight line type, while curvature radius is for arc lane.

Depth-first search method was used after a starting point and an ending point were set, which traverses from a starting point to a new point that connects with the starting one, and from the new point it would be another starting point if there was any line be not traversed yet. Two shortest paths would be constructed in the map for decision.

3   Results and Discussion

Path planning for rectangular field, including the one in fieldbody and headland, were successfully tested on in summer of 2013 with the PZ60 transplanter at Yinzhou, Zhejiang Province. The operation width is 1.8 meters, and the width of headland is 3.6 meters. Two kinds of path planning were experimented, which called three-points guidance and four-points guidance. In the experiment of three-points guidance, a driver steered the transplanter and planted the first row, which formed a reference line for the guidance system. He pressed a button on the guidance controller to get the first point 'A' when starting to plant. He pressed the button again to get the second point 'B' when arrived at end of this row. The 3rd point 'C' was got after a headland turn to the 2nd row. After that, the path planning module generated paths for fieldbody. Figure 6 is one set of testing track diagram, that the autonomous transplanter moved on north or south direction in diagram (a), and on east or west direction in diagram (b). Among them, the red is the planned path, blue is a trace of RTK - GPS. It shows the blue points do not overlap with the red line in straight moving stage, which caused by position of the RTK-GPS, who was not fixed in the middle of the transplanter precisely, and calculation error when transformed from latitude-longitude angles to Cartesian Coordinates.

In the experiment of four-points guidance, the driver got latitude and longitude of 4 corners of the field before planting operation. Point 'A', 'B', 'C', and 'D' were got with

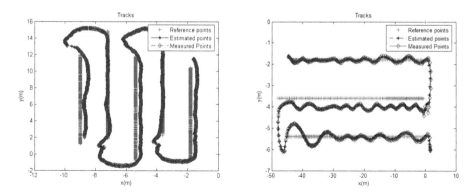

Fig. 6. Planned path and measured track in north-south and west-east direction (Color figure online)

sequence of position at southwest corner, northwest corner, northeast corner, and southeast corner. When the transplanter moved into any corner of this field, the path planning module generated paths in fieldbody and headland. The guidance system guided the machine in fieldbody firstly, and turned to headland after finishing operation in fieldbody that cover whole area automatically. In summer of 2013 and 2014 those experiments were carried out successfully too.

Path planning for farm lane have been testing on the NSD8 machine in summer of 2015 at Cixi, Zhejiang Province. A series of experiments will be done in recent days.

4   Conclusions

A path planning method for rice transplanter has been developed and validated partly. Paths for fieldbody and headland in rectangular field were generated and had been used in practice for recent three years. Offset of row following was less than 5 cm, and row-to-tow error was less than 5 cm too. From the experiments, it shows the guidance system could replace human driver for steering controlling.

The path for triangular shape field and farm lanes were proposed, while experiments will be done in autumn of 2015. However, optimal method for path planning, and method for collision avoidance in condition of cooperated multi-robots should be research in future that human can control transplanters, tractors, and combines at off-vehicle place.

Acknowledgments. The author wishes to express his sincere thanks for the financial support from Developing Foundation of NingBo (2014C10016, 2015C10012).

References

1. Galceran, E., Carreras, M.: A survey on coverage path planning for robotics. Rob. Auton. Syst. **61**, 1258–1276 (2013)
2. Kise, M., Noguchi, N., Ishii, K., et al.: Field Mobile Robot navigated by RTK-GPS and FOG (Part 3): enhancement of turning accuracy by creating path applied with motion constrains. J. Jpn. Soc. Agric. Mach. **64**(2), 102–110 (2002)
3. Oksanen, T., Kosonen, S., Visala, A. Path planning algorithm for field traffic. In: 2005 ASAE Annual Meeting, Paper number 053087 (2005)
4. Hou, J., Song, Z., Mao, E.: GIS-based path planning method for automated tractor. In: Proceeding of 2005 CSAE Annual Meeting, Beijing, pp. 141–144 (2005)
5. Huang, X., Ding, Y., Zong, W., Liao, Q.: Turning method and algorithm of agriculture vehicles. In: Proceeding of 2005 CSAE Annual Meeting, Chongqing (2005)
6. Han, X., Kim, H., Kim, J., et al.: Path-tracking simulation and field tests for an auto-guidance tillage tractor for a paddy field. Comput. Electron. Agric. **112**, 161–171 (2015)
7. Noguchi, N., Terao, H.: Path planning of an agricultural mobile robot by neural network and genetic algorithm. Comput. Electron. Agric. **18**(2–3), 187–204 (1997)
8. Jin, J., Tang, L.: Optimal coverage path planning for arable farming on 2D surfaces. Trans. ASABE **53**(1), 283–295 (2010)

9. Jin, J., Tang, L.: Optimal coverage path planning on 3D terrain. In: ASABE Annual Meeting, Paper number 1009220 (2010)
10. Hameed, I.A., Bochtis, D.D., Sorensen, C.G.: Driving angle and track sequence optimization for operational path planning using genetic algorithm. Appl. Eng. Agric. **27**(6), 1077–1086 (2011)
11. Mariano, G., Luis, E., Isaias, G., Pablo, : Reducing fuel consumption in weed and pest control using robotic tractors. Comput. Electron. Agric. **114**, 96–113 (2015)
12. Linker, R., Blass, T.: Path-planning algorithm for vehicles operating in orchards. Biosyst. Eng. **101**, 152–160 (2008)
13. Jensen, M., Bochtis, D., Sørensen, C.: Coverage planning for capacitated field operations, part II: optimisation. Biosystems Engineering, 1–16 (2015). (article in press)
14. Liu, X.: Optimal Path Planning Method for GPS Guided Tractors. Liaoning Technical University, Fuxin (2011)
15. Bochtis, D., Griepentrog, H.W., Vougioukas, S., Busato, P., Berruto, R., Zhou, K.: Route planning for orchard operations. Comput. Electron. Agric. **113**, 51–60 (2015)
16. Volos, Ch.K., Kyprianidis, I.M., Stouboulos, I.N.: A chaotic path planning generator for autonomous mobile robots. Rob. Auton. Syst. **60**, 651–656 (2012)
17. Bochtis, D.D., Sørensen, C.G., Vougioukas, S.G.: Path planning for in-field navigation-aiding of service units. Comput. Electron. Agric. **74**(1), 80–90 (2010)
18. Jensen, M., Bochtis, D., Sørensen, C., Blas, M., Lykkegaard, K.: In-field and inter-field path planning for agricultural transport units. Comput. Ind. Eng. **63**, 1054–1061 (2012)
19. Zhou, K., Jensen, A.L., Sørensen, C.G., Busato, P., Bothtis, D.D.: Agricultural operations planning in fields with multiple obstacle areas. Comput. Electron. Agric. **109**, 12–22 (2014)
20. Zhang, F., Shin, B., Feng, X.: Development of a prototype of guidance system for rice transplanter. J. Biosyst. Eng. **38**(4), 255–263 (2013)
21. Wu, X.: Study on Path Planning of Logistic System Base on Multiple AGV Systems. Jinling University, Changchun (2003)

Research of the Early Warning Model of Grape Disease and Insect Based on Rough Neural Network

Dengwei Wang[1], Tian'en Chen[1(✉)], Chi Zhang[1], Li Gao[2], and Li Jiang[1]

[1] National Engineering Research Center for Information Technology in Agriculture (NERCITA), Beijing 100097, China
{wangdw, chente, zhangc}@nercita.org.cn
[2] Fruit Service Center of Yanqing County, Beijing 102100, China

Abstract. The grape is one of the four fruits in the world and its cultivation area and production has been ranked first in the world. The area is growing in our country after the reform and opening up which is significant for the rural economic development and farmers' income. However, the growing of grape diseases and insect pests has become one of the important problems in the development of grape planting industry. In the paper, intensive and overall surveys and studies of the research progress of the early warning model of the disease and insect are made firstly and then the comparison and analysis of rough set and artificial neural network are presented. Finally, we collected a large amount of data from grape planting base of the grape and wine engineering technology research and development center in Beijing. Based on the real-time sensing data technologies of Internet of things and the intelligent grape early warning model based on rough neural network established in the paper, we did a number of experiments and its validity was verified. The model can provide beneficial references for the research of other crops diseases and insect pests.

Keywords: Rough set · Artificial neural network · Grape disease and insect · Early warning model

1 Introduction

The farming population is numerous and agriculture holds dominant position in the national economy. However, the level of agricultural science and technology information is relative backwardness in agriculture [1]. How to increase the technological content and ensure the ability to resist the disaster in the course of agricultural production have become a development trend and research focus. The greenhouse fruit and vegetable disease and especially the explosive diseases must be conducted before the disease symptom displays. Therefore, the research pays attention to emphasize on giving out the alarm before the disease symptom is shown, such as we can send out alarm through collecting and analyzing information as well as the weather forecast before the symptom is shown [2–5].

© IFIP International Federation for Information Processing 2016
Published by Springer International Publishing AG 2016. All Rights Reserved
D. Li and Z. Li (Eds.): CCTA 2015, Part II, IFIP AICT 479, pp. 310–319, 2016.
DOI: 10.1007/978-3-319-48354-2_32

At present, several countries such as Israel, America, Japan, and Spain put more research on greenhouse crop disease early warning system. In China, the introduction of agricultural early warning has elaborated the basic theoretical knowledge, as well as the natural disaster warning [6]. In recent years, the early-warning analysis theory of the macroeconomic and regional forest resource field has been applied to agriculture field [7]. For example, LiXin Bai [8] has built bollworm disaster early warning index system based on the evaluation method. According to the local climatic environment and the actual situation of crop planting in Beijing, Lili [9] set up pear plant diseases early warning system. Ming Li and ChunJiang Zhao [10] studied cucumber downy mildew early warning system in the sunlight greenhouse environment. Shuwen Liu and Qingwei Wang [11] studied the grapes diagnosis based on neural network and Ying-feng Cui and Shiping Wang [12] constructed grape intelligent diagnosis network model. However, taking into account the application cost and other factors, there is less intelligent and practical early warning system in agriculture and still less for large area promotion. In recent years, with the rapid development of networking and cloud computing technology and combined with network real time sensing data, there is a huge advantage for us to do facility agriculture disease early warning based on agricultural cloud services platform. It is still in the primary stage of the research for early warning analysis and decision based on a large number of real-time sensing data in agriculture. So it is necessary to further improve the algorithm model of early warning analysis and the precision.

The grape downy mildew is one of the most serious diseases for the grape [13]. Its incidence is closely related to the light, temperature, humidity in the micro-environment. So we took the grape in the area of yanqing country in Beijing as an example and collected real-time sensing data in grape micro environment through various sensing nodes and wireless communication network and then researched and constructed the grape diseases and insect pest warning model based on rough neural network which provides service for the intelligent analysis of grape diseases and insect pests, the agricultural precision production such as prediction and warning decision, the visual management and Intelligent decision analysis [14].

2 The Construction of Rough Neural Network-Based Early Warning Model

2.1 Comparison of Rough Set and BP Artificial Neural Network

BP neural network has strong anti-interference ability and processes large amounts of information in parallel and arbitrarily approximates non-linear function. But it can't simplify information and reduce input data. In addition, the built network is more complicated if there is large amounts of input data. Rough set theory doesn't need prior knowledge to remove redundant information without additional data in the process of data analysis and processing, but also execute algorithm in parallel. However, its anti-interference ability is insufficient [15–18].

There exist complementary relationships between rough set theory and BP neural network in many aspects. For instances, we combine them and obtain a process that concentrates the knowledge reduction and no prior knowledge and strong anti-interference performance which can improve accuracy and noise resistance of the system. The result of the comparing analysis is as shown in Table 1:

Table 1. The comparison of the rough set and BP neural network

	Rough set theory	BP neural network
Self-learning ability	Weaker	Stronger, with increasing learning network
The reasoning method	Serial reasoning way	Implement parallel computing
Data processing	Qualitative and quantitatively	Quantitatively
Prior knowledge	Without prior knowledge	Need experience and multiple trial before the network structure designed
Knowledge representation	By neatly	Connotative, through parameters
Noise proof	Affected by noise	Good ability of resisting noise
Redundant data processing	Determine the attribute and value of the decision table	Almost cannot be handled
Rule discovery method	Determine the relationships between data in the reduction process	Through a nonlinear mapping
Interpretability	Easier	The transparency is poor in knowledge acquisition and knowledge reasoning process
Knowledge maintenance	Maintenance is difficult when data size is large	The learning time is long and it is easy to be trapped in local minimum when data size is large
Integration capability	The integration with various software calculation methods is easily and closely	The integration with the fuzzy set, the rough set and genetic algorithm is easily and closely
Generalization ability	Weaker	Strong generalization ability

2.2   The Constructing Idea of RS-BPANNs Model

According to their features we build the early warning model based on rough set and BP artificial neural network (hereinafter referred to as RS-BPANNs). We collect the grape diseases information case, experimental data and expert experience over the years to build an attribute decision-based matrix which is used to reduce the attribute and then design suitable BP Neural Network to train the model. The model will become knowledge rules to judge the crop disease when it achieves certain accuracy. The knowledge model library is perfected through feedback information of diagnosis results and finally the accuracy of the early warning is improved. The decision knowledge acquisition model based on RS-BPANNs is shown in Fig. 1:

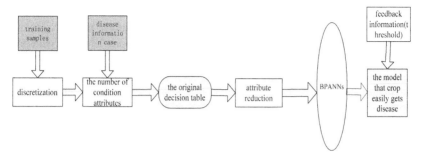

Fig. 1. Decision knowledge acquisition model based on RS-BPANNs

2.3   The Constructing Steps of RS-BPANNs Model

The paper takes grape downy mildew as an example to build RS–BPANNs-based early warning model. The rough set theory is used to reduce the attributes of the training sample data and then determine the relationship among attributes. The module is divided into two parts: the preprocessing of the rough set and the artificial neural network learning. The whole process includes the following four steps:

(1) Build the grape downy mildew case database: the way to obtain comprehensive grape downy mildew cases is various and the cases in the paper are mostly from document literature, expert discussion and experimental data at the grape planting base of the grape and wine engineering technology research and development center in Beijing that are saved in the form of vector. The data which is described using natural language can be converted into the corresponding data and is saved into the database in the form of matrix.

(2) Express the case data through the decision table. A lot of original judgment information is represented by the decision table. Among them, C represents condition attribute value. D represents the decision attribute value. All grape downy mildew cases are represented as U. V is the range of attribute values. The representation of the original case decision table is T = (U,C,D,V). The value is equal to 0 or 1.

(3) Reduce the decision table: take out the redundant attributes without changing the classification of the decision table. The reduction is realized through calculating the positive region of the decision attribute of the condition attributes in the paper. The input condition attribute set is represented as C = {C1,C2,...,Cn} and the decision attribute set is D = {D1,D2,...,Dn}. The steps of reduction algorithm are as follows:

1. Calculate the positive region of D, $pos_c(D)$
2. Remove attributes C_j from C, $C_w = C\text{-}\{C_j\}$
3. Calculate the positive region of C_w, $pos_{c_w}(D)$
4. If $pos_c(D) = pos_{c_w}(D)$. Remove C_j. Otherwise, reserved
5. Ergodic all attributes
6. Output a reduction that C is relative to D

The main idea of the reduction algorithm is that the condition can be removed if a condition in the set C is removed and it will not affect the decision result. Otherwise, the condition is reserved. The operation is complete until generating a reduction about C.

(4) The obtaining rules of neural network: take the condition attributes and decision attributes gotten after attribute reduction as the input variables and output variables of the neural network. We use the neural network to train the sample and to learn and finally get a decision information model.

On the basis of the decision information model, feedback information is added when the plant diseases and insect pests of early warning system is realized in order to record the validity of the decision information and count the accuracy of the trained decision information model, which make the decision information easy managed and optimized. The whole process of the system is shown as the Fig. 2.

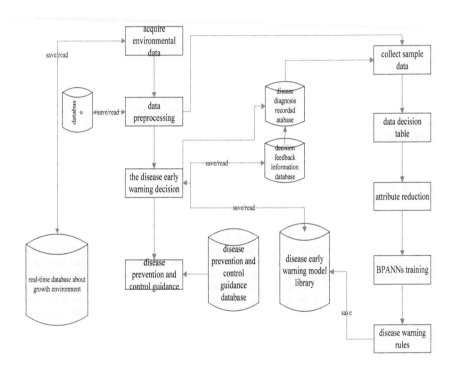

Fig. 2. The flow of the plant diseases early warning system

3 Experimental Analysis

3.1 Data Acquisition

The paper takes grape downy mildew as example to analyze rough neural network-based early warning model. The original data is gathered from air temperature and humidity, soil temperature and moisture (the depth of 20 cm), light intensity and CO_2 concentration collected every 15 min time. The following data shown in Table 2 is acquired from grape berries period in greenhouse and the acquisition time is from 7 am to 5 pm and the data collected every time is done average processing.

Table 2. The data gotten in grape berries period

Time	Air temperature/°C	Air humidity/%	Soil temperature/°C	Air moisture/%	Light intensity/lx	CO_2/ μL/L
7	15.9	92.1	10.2	35.8	0	445
8	17.3	91.6	12.5	36.4	24625	379
9	21.8	68.7	15.6	36.4	38652	324
10	26.5	51.7	21.7	36.9	45689	254
11	29.5	45.2	25.5	35.7	47895	278
12	32.8	48.9	28.8	35.7	50127	201
13	28.3	57.4	23.7	35.9	51257	156
14	28.2	55.8	23.8	35.7	38784	158
15	27.6	64.9	22.1	35.7	21784	121
16	24.3	67.5	19.0	35.7	7523	122
17	20.3	83.7	16.4	35.7	0	154

It shows that the change process of various greenhouse environmental factors in grape berries period during one day. Organizing the sensors' data and feedback information from sensor, we obtain the original decision table shown in Table 3.

The above table shows that whether grape downy mildew appears or not in the various greenhouse environmental factors. There are 100 items data which constitute the case table. Among them, U is a finite case set. $A = C \cup D$, C is the condition attribute set. Through the study of grape downy mildew material, there is the stipulation that it is low when air temperature is below 18 °C and high when it is above 25 °C. It is low when the humidity is below 50 % and high when it is above 90 %. It is low when the soil temperature is below 10 °C and high when it is above 20 °C. (We concluded that the surface air temperature is generally 5 °C higher than the soil temperature when air temperature is relatively stable from the data gathered by sensor without the case of air temperature shock.). It is low if soil humidity is below 50 % and high for above 80 %. It is low if light intensity is under 30000 lx and high for above 40000 lx. It is low if CO_2 the concentration is under $100\mu L \bullet L^{-1}$ but high for above $1000\mu L \bullet L^{-1}$.

Table 3. Grape downy mildew original decision table

Air temperature/°C	Air humidity/ %	Soil temperature/°C	Air humidity/ %	Light intensity/lx	CO_2 concentration/µL/L	Downy mildew
19.2	94.7	10.7	66.5	0	388	Y
23.5	97.3	16.4	80.1	0	421	Y
15.7	80.5	16.4	34.9	0	433	N
24.8	95.7	19.8	58.8	7468	367	Y
28.4	86.2	23.5	38.7	34954	284	N
31.2	53.5	26.1	31.8	46541	158	N
12.8	86.5	7.5	29.3	0	309	N
18	93.7	10.2	71.3	0	516	Y
21.0	90.9	16.2	75.6	7962	251	Y
...	...	...	...	...	...	...
23.2	95.3	18.6	65.4	2762	241	Y

The range of the condition attribute value is 0 represents that do not belong to the property and 1 is on behalf of belonging to the attribute. C_1 represents the berries period, C_2 represents that the air temperature is low, C_3 represents that air temperature is just right, C_4 represents the air temperature is high. C_5, C_6 and C_7 respectively represents air humidity is low, moderate and high. C_8, C_9 and C_{10} respectively represents soil temperature is low, moderate and high. C_{11}, C_{12} and C_{13} respectively represents soil moisture is low, moderate and high. C_{14}, C_{15} and C_{16} respectively C_{17}, C_{18} and C_{19} respectively represents the concentration of CO_2 is low, moderate and high. The case data table of the samples is shown in the Table 4.

It can be seen from the above table that the object $x3, x5, x10$ are the same instance because their attribute value domains are same. According to the attribute reduction method only one case is retained and the other needs removing. And classify each equivalence relation relative to the domain U after dealing with the duplicate lines.

Table 4. Grape Downy mildew case data

U	C1	C2	C3	C4	C5	C6	C7	C8	C9	C10	C11	C12	C13	C14	C15	C16	C17	C18	C19	D
x1	1	0	1	0	0	0	1	0	1	0	0	0	1	1	0	0	0	1	0	1
x2	1	0	1	0	0	0	1	0	0	1	0	0	1	0	1	0	0	0	1	1
x3	1	0	1	0	0	0	1	0	1	0	0	0	1	1	0	0	1	0	0	1
x4	1	1	0	0	0	1	0	1	0	0	0	1	0	0	1	0	0	1	0	0
x5	1	0	1	0	0	0	1	0	1	0	0	0	1	1	0	0	1	0	0	1
x6	1	0	1	0	0	0	1	0	1	0	0	0	1	1	0	0	1	0	0	1
x7	1	0	0	1	0	1	0	1	0	0	0	1	0	0	0	1	1	0	0	0
x8	1	1	0	0	0	0	1	1	0	0	0	0	1	1	0	0	0	1	0	1
x9	1	0	1	0	0	0	1	0	1	0	0	0	1	1	0	0	0	0	1	1
x10	1	0	1	0	0	0	1	0	1	0	0	0	1	1	0	0	1	0	0	1
x11	1	0	1	0	0	0	1	1	0	0	0	1	0	1	0	0	0	1	0	1
...	...	...	...	...	...	...	...	...	...	...	...	...	...	...	...	...	...	...	...	...
X100	1	1	0	0	0	1	0	1	0	0	1	0	0	0	1	0	0	1	0	0

Count the dependence degree of D relative to C and get $pos_c(D) = \{x_1, x_2, x_3, x_4, x_6, x_7, x_8, x_9, x_{11}, \ldots\}$

$$k = |pos_c(D)|/|U| = 81/81 = 1 \tag{1}$$

So we get the conclusion that decision attribute depends entirely on condition attributes and then calculate the relationship between $pos_c(D)$ and $pos_{c-c_j}(D)$ to get (2).

$$pos_{c-c_1}(D) = \{x1, x2, x3, x4, x6, x7, x8, x9, x11, \ldots\} = pos_c(D) \tag{2}$$

Finally, we get the final reduced table which includes the key factors that affect the grape downy mildew which is shown in Table 5.

Table 5. The data gotten after the attributes reduction

U	C1	C2	C3	C4	C5	C6	C7	C11	C12	C13	C14	C15	C16	D
x1	1	0	1	0	0	0	1	0	0	1	1	0	0	1
x2	1	0	1	0	0	0	1	0	0	1	0	1	0	1
x3	1	0	1	0	0	0	1	0	0	1	1	0	0	1
x4	1	1	0	0	0	1	0	0	1	0	0	1	0	0
x6	1	0	1	0	0	0	1	0	0	1	1	0	0	1
x7	1	0	0	1	0	1	0	0	1	0	0	0	1	0
x8	1	1	0	0	0	0	1	0	0	1	1	0	0	1
x9	1	0	1	0	0	0	1	0	0	1	1	0	0	1
x11	1	0	1	0	0	0	1	0	1	0	1	0	0	1
...	...	...	...	...	...	...	...	...	...	...	...	...	...	...
X81	1	1	0	0	0	1	0	1	0	0	0	1	0	0

The design of BP neural network is mainly divided into the following steps:

(1) The design of the input layer and output layer: It should be based on application need. The number of the output node is 1 in the experiment and there are 19 properties for the objects in input layer. But the number of the core attributes is taken as the number of input layer after attribute reduction. The number of the input node N is set as 12.

(2) The design of the hidden layer: It includes the design of the layer number and the neuron number each layer which mainly relies on experience. It can increase the processing capacity of neurons if we add the amount of the hidden layer. The paper uses three layers BP neural network structure. The experimental results show that single layer can meet the demand in the determination of the boundary problem of the small network.

Through the above process we can extract the knowledge rules in grape berries period:

If C_3 is between 18 °C and 25 °C/1 and

$C_7 > 90 \%/1$ and
$C_{13} > 80 \%/1$ and
$C_{14} < 30000$ lx
Then D(Downy mildew)

We can get the knowledge rules that grape downy mildew is easily appear in every period at grape growth stage which can guide the judgment of grape downy mildew. At the same time, according to the feedback information knowledge rules are optimized.

3.2   Comparison and Analysis

The paper makes comparison and analysis of BP artificial neural network model and rough BP artificial neural network model through training and testing the same sample and then analyzing of the test results of each model and the length of training time.

It is concluded that the combination of the rough set and BP neural network can decrease the redundancy attributes and optimize the structure of neural network and shorten the training time and also improve the effect of training. Therefore, the model based on the RS-BPANNs has certain advantage. Table 6 is the results of the two models in the aspect of the training time and accuracy.

Table 6. Result comparison of different methods

	The training time(s)	Accuracy %
BP artificial neural network	0.7	90
Rough BP artificial neural network	0.5	94

4   Conclusions

The facility agriculture is very important in the process of agricultural modernization in our country. Taking advantage of Internet of things, we can obtain real-time information of the infrastructure environment and build early warning model to make decision analysis of facility crop diseases whose purpose is to realize the timeliness, accuracy, convenience and practicality of the facility agriculture disease early warning.

The paper is mainly for facility agriculture and constructs a grape diseases early warning model based on rough neural network. The paper gets the effectiveness of the early warning model and provides a reference for the construction of plant diseases early warning model through analyzing the real-time sensor data from the grape downy mildew.

Acknowledgments. The research was supported by Beijing Natural Science Foundation of China (key project) - Research of adaptive the key technology and application model in agricultural cloud service (4151001), and the Technology innovation ability construction Projects of Beijing Academy of agriculture and forestry science - the research and application of the network service technology for the vegetable seeds Display and promotion (KJCX20140416), all the support is gratefully acknowledged.

References

1. Li, D.L., Fu, Z.T., Wen, J.W.: Agricultural plant diseases and insect pests of remote diagnosis and early warning technology. Tsinghua University Press (2010)
2. Yang, X., Li, M., Zhao, C., et al.: Early warning model for cucumber downy mildew in unheated greenhouses. N. Z. J. Agric. Res. **50**(5), 1261–1268 (2007)
3. Shtienberg, D., Elad, Y.: Incorporation of weather forecasting in integrated, biological-chemical management of Botrytis cinerea. Phytopathology **87**(3), 332–340 (1997)
4. Shtienberg, D., Elad, Y., Bornstein, M., et al.: Polyethylene mulch modifies greenhouse microclimate and reduces infection of Phytophthora infestans in tomato and Pseudoperonospora cubensis in cucumber. Phytopathology **100**(1), 97–104 (2010)
5. Vincelli, P.C., Lorbeer, J.W.: Relationship of participation probability to infection potential of Botrytis squamosa on onion. Phytopathology **78**(8), 1078–1082 (1988)
6. Tao, J., Chen, K.: An Introduction to Agricultural Warning (1994)
7. Haibing, G.: The Design of early warning index system for the economic and social benefits in China. Econ. Theor. Bus. Manage. **6**, 47–50 (1989)
8. Bo, L., Sun, Y.: Research on the index system of the cotton bollworm disaster monitoring and early warning and the risk warning level. Cotton Sci. **14**(2), 99–103 (2002)
9. Li, L.: Research on the Early Warning and Forecasting System of Pear Diseases and Insect Pests Based on Radial Basis Network and Support Vector Machine. China Agricultural University, Beijing (2007)
10. Li, M., Zhao, C., Li, D., et al.: Towards developing an early warning system for cucumber diseases for greenhouse in China. In: Computer And Computing Technologies in Agriculture, Volume II. IFIP 259, pp. 1375–1378. Springer, Heidelberg (2008)
11. Liu, S., Wang, Q., He, D., et al.: Research on the grape disease diagnosis systems based on Fuzzy Neural Network. Trans. CSAE **22**(9), 144–147 (2006)
12. Cui, Y., Wang, S.: Comparison of the intelligent diagnosis network models for grape diseases. J. Shanghai Jiaotong Univ. (Agric. Sci.) **29**(4), 79–86 (2011)
13. Li, H., Guo, H.: Research progress of the Grape Downy Mildew prediction models and early warning technologies. Chin. Agric. Sci. Bull. **21**(10), 313–316 (2005)
14. Wang, D., Chen, T., Dong, J.: Research of the early warning analysis of crop diseases and insect pests. In: Li, D., Chen, Y. (eds.) Computer and Computing Technologies in Agriculture VII. IFIP AICT, vol. 420, pp. 177–187. Springer, Heidelberg (2014)
15. Zhang, Y.: Research on Forest Pest Forecasting Model and Algorithm Based on Rough Set Theory. Northeast Forestry University, Harbin (2012)
16. Xing, H.: Research on the Data Fusion Method Based on Rough Set and Neural Network. Northwestern Polytechnical University, Xi'an (2007)
17. Hu, Y.: Rough Set Theory and Its Application in the Neural Network. Zhejiang University, HangZhou (2007)
18. Mao, D., Li, D.: Rough Set Theory, Algorithm and Application. Tsinghua University Press (2008)

Evaluation Model of Tea Industry Information Service Quality

Xiaohui Shi[1,2,3,4] and Tian'en Chen[1,2,3,4(✉)]

[1] Beijing Research Center for Information Technology in Agriculture,
Beijing 100097, China
[2] National Engineering Research Center of Information
Technology in Agriculture, Beijing 100097, China
{shixh,chente}@nercita.org.cn
[3] Key Laboratory of Agri-Informatics, Ministry of Agriculture,
Beijing 100097, China
[4] Beijing Engineering Research Center of Agricultural Internet of Things,
Beijing 100097, China

Abstract. According to characteristics of tea industry information service, this paper have built service quality evaluation index system for tea industry information service quality, R-cluster analysis and multiple regression have been comprehensively used to contribute evaluation model with a high practice and credibility. Proved by the experiment, the evaluation model of information service quality has a good precision, which has guidance significance to a certain extent to enhance the information service quality of tea industry information website.

Keywords: Tea industry · Information service quality evaluation · Multiple regression · R-cluster

1 Introduction

According to characteristics of tea industry information service, and based on analysis of the evaluation goal for tea industry information service quality, the typicality of index system as the main direction, also takes the information requirements characteristic of users into account, the key factors influencing the quality of information service have been analyzed from platform function, platform information, platform design, platform framework and platform efficiency, then select 5 first grades and 30 s grades. 30 s grades are clustered by R-cluster analysis, finally forming an evaluation index system consists of 5 first grades and 17 s grades. Based on the influence factor of information service quality analyzed by information service quality evaluation model built with multiple regression, which will give a scientific guidance for tea industry information service.

D. Li and Z. Li (Eds.): CCTA 2015, Part II, IFIP AICT 479, pp. 320–329, 2016.
DOI: 10.1007/978-3-319-48354-2_33

2 Analysis Methods

2.1 R-cluster

R-cluster [1] is one kinds of the cluster analysis method, is adequate for lots of variable, and variable with a high correlation, its purpose is to cluster the variables in the same group, and find the representative variables to achieve the purpose of dimension reduction. In R-cluster theory, N objects are treated as N groups, the correlation coefficient around the groups is R, two groups having the biggest related coefficient are clustered in one groups, until all groups are clustered in one big group.

2.2 Multiple Linear Regression

2.2.1 Brief Introduction for Multiple Linear Regression Model

Multiple linear regression [2] model is one kinds of math model to consider the relationship around variable x_1, x_2,..., x_p and variable y. Assuming y is change along with the changes of x_1, x_2,..., x_p.

The mathematical expression of multiple linear regression is $Y = W^{(3)} * X + C$. $W^{(3)}$ is weight matrix, C is constant term.

2.2.2 Selection Criteria of Independent Variable for Multiple Linear Regression Model

Multiple linear regression model has advantages, such as simple method and high precision, when it simulate the data owning pure linear relationship, but having bad results in this situation of data is nonlinear relationship or data missing. In order to guarantee the excellent predictive results, the selection criteria of independent variable are summarized as follows: (1) independent variable has a outstanding influence with dependent variable, and there is a close linear relationship; (2) linear relationship between independent variable and dependent variable is true, not in form; (3) a certain mutual exclusion between the independent variables is necessary; (4) independent variable must have integrated data [3].

3 Index System

3.1 Preliminary Index System

Selecting platform function, platform information, platform design, platform frame-work and platform efficiency as the first grade indexes [4], and every index is further segmented, finally forming 30 s grade indexes, which forms the service quality index system for tea industry information platform, the specific indexes are listed as follows:

3.2 Index System Selection

Normally, the method of principal component analysis is used to solve a problem which contains lots of evaluation indexes, but the results from principal component analysis depend on measured value of primitive index, so principal component analysis only solve simplification of quantitative index, not the descriptive index before quantitative [5].

In this paper, R-cluster is adopted. According to the correlation coefficient in indexes, all indexes are clustered by R-cluster, the demand of accuracy finally determines how many groups will be segmented, select the typical index in one group represent multiple indexes to achieve the purpose of simplified index [6].

Assuming there are P indexes u_1, u_2, u_3, ..., u_p; N experts score for the relative coefficient among the indexes, the relative coefficient matrix after scoring is expressed in Table 2, and the cluster results are expressed in.

To combine the results of cluster, comprehensively considering the simplicity, representation and feasibility of indexes, analysis results are carried out as follows: when the minimum distance of correlation coefficients is under 0.03, the jumping points are obvious, indexes are clustered 17 groups. So, the results represented as follows: R_1: 4, 11, 15, 7, 13, 20, 8, 18, 21; R_2: 3, 23; R_3: 5, 17; R_4: 6, 10, 14, 30; R_5: 1; R_6: 2: R $_7$: 9; R_8: 12; R_9: 24; R_{10}: 28; R_{11}: 19; R_{12}: 27; R_{13}: 16; R_{14}: 21; R_{15}: 22; R_{16}: 26; R_{17}: 29 (Fig. 1).

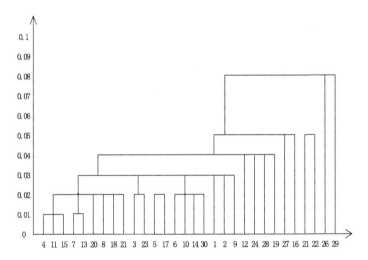

Fig. 1. Result of R-mode clustering

4   Data Acquisition

4.1   Experts Estimation

After the establishment of indexes system, expert estimation has two parts:

(1) General part indexes existing in traditional information service website or system. After the establishment of indexes system, select the typical tea information website, then let the experts who is good at agricultural informationization

estimate the quality of information service websites and the indexes that influencing the quality of information service.

(2) The indexes part reflecting the characteristics of cloud service platform. Because of tea industry information cloud service platform in domestic is still blank, there is no quantitative index to be verified, only the experts' view as the reference, that is means, when tea industry information cloud service platform is built, the 7 indexes, such as X23: Platform stabilization, X24: Platform compatibility, X25: Platform safety, X26: Platform adaptability, X27: Platform expansibility, X28: Software deployment simplicity and X29: Software operation shortcut will be fully considered.

4.2   Data Processing

First step: website selection. Tea information websites are divided into two groups: one group is comprehensive agriculture websites but covering content of tea industry information, another group is the professional tea information websites. The objects of this experiment are 40 websites in two groups, 18 comprehensive websites, such as agriculture information websites belong to province and municipal district, and 22 professional tea information websites, specific situation is listed in Table 3.

Second step: data acquisition. Experts estimate the 10 s indexes belong to Platform function, Platform information and Platform design in Table 1, information service quality and indexes estimated by quantitative evaluation, the scores are during 0–1, adopt the method of integrating multiple scores, the data is listed as Table 4 after data processing.

Table 1. Service quality evaluation index system of information platform for tea industry

	First index	Second index	After R-cluster
X_1 Service quality index system for tea industry information platform	Platform function	Function comprehensiveness	X_1
		Service interactivity	X_2
		Operation convenience	
		Service individuation	X_4
		Service professionalism	X_5
		Platform popularity	
	Platform information	Information accuracy	

(continued)

Table 1. (*continued*)

	First index	Second index	After R-cluster
		Information authority	
		Information vividness	X_9
		Information comprehensiveness	X_{10}
		Content uniqueness	
		Organization orderliness	
		Information timeliness	
		Retrieval convenience	
		Retrieval effectiveness	
	Platform design	Interface friendly	X_{16}
		Design standardization	
		Creativity uniqueness	
		Column novelty	X_{19}
		Structure clarity	
		Style unity	
		Color coordination	X_{22}
	Platform framework	Platform stabilization	X_{23}
		Platform compatibility	X_{24}
		Platform safety	X_{25}
		Platform adaptability	X_{26}
		Platform expansibility	X_{27}
	Platform efficiency	Software deployment simplicity	X_{28}
		Software operation shortcut	X_{29}
		Demand response timeliness	

Table 2. Related coefficient matrix

	X1	X2	X3	X4	X5	...	X30
X1: Function comprehensiveness	1						
X2: Service interactivity	0.15	1					
X3: Operation convenience	0.12	0.14	1				
X4: Service individuation	0.11	0.32	0.18	1			
X5: Service professionalism	0.47	0.41	0.56	0.24	1		
X6: Platform popularity	0.12	0.13	0.68	0.05	0.87	...	
X7: Information accuracy	0.21	0.89	0.12	0.98	0.94	...	
X8: Information authority	0.45	0.86	0.11	0.84	0.92	...	
X9: Information vividness	0.13	0.97	0.13	0.86	0.94	...	
X10:Information comprehensiveness	0.97	0.75	0.08	0.88	0.96	...	
X11: Content uniqueness	0.12	0.85	0.14	0.99	0.92	...	
X12: Organization orderliness	0.65	0.82	0.84	0.15	0.87	...	
X13: Information timeliness	0.78	0.94	0.89	0.98	0.82	...	
X14: Retrieval convenience	0.56	0.93	0.97	0.97	0.84	...	
X15: Retrieval effectiveness	0.42	0.84	0.85	0.99	0.86	...	
X16: Interface friendly	0.01	0.95	0.89	0.85	0.12	...	
X17: Design standardization	0.75	0.12	0.84	0.12	0.98	...	
X18: Creativity uniqueness	0.15	0.78	0.12	0.86	0.12	...	
X19: Column novelty	0.12	0.86	0.11	0.87	0.15	...	
X20: Structure clarity	0.32	0.81	0.84	0.98	0.56	...	
X21: Style unity	0.15	0.14	0.91	0.15	0.86	...	
X22: Color coordination	0.11	0.23	0.56	0.12	0.14	...	
X23: Platform stabilization	0.12	0.45	0.98	0.11	0.86	...	
X24: Platform compatibility	0.11	0.21	0.94	0.15	0.14	...	
X25: Platform safety	0.11	0.12	0.12	0.12	0.86	...	
X26: Platform adaptability	0.92	0.15	0.84	0.87	0.84	...	
X27: Platform expansibility	0.95	0.34	0.23	0.56	0.85	...	
X28: Software deployment simplicity	0.87	0.87	0.86	0.87	0.85	...	
X29: Software operation shortcut	0.12	0.15	0.92	0.86	0.92	...	
X30: Demand response timeliness	0.23	0.97	0.76	0.82	0.85	...	1

5   Evaluation Model

5.1   Experiment Data Processing

In experiment, 30 data before in Table 4 as the training sample, 11 data after in Table 4 as the test sample.

Table 3. Objects of experimental evaluation for tea information sites

Groups	Website	Website
Professional(22)	http://www.ymt360.com	http://tea.fjsen.com
	http://www.teauo.com	http://www.zgchawang.com
	http://www.i-tea.cn	http://tea.ahnw.gov.cn
	http://cy.zgny.com.cn	http://www.t0001.com
	http://www.teatea.co	http://www.fjteaw.cn
	http://www.tea160.com	http://www.chawh.net
	http://www.b-tea.com	http://www.puercn.com
	http://www.hbteainfo.com	http://www.qspg.cn
	http://www.cyppw.com	http://www.bjhjcha.com
	http://www.cs12396.cn	http://www.ctma.com.cn
	http://www.cluyu.cn	http://www.paicw.com
Comprehensive(18)	http://bb.ahnw.gov.cn	http://www.xinnong.com
	http://www.ynagri.gov.cn	http://www.sbny.cn
	http://www.cnsp.org.cn	http://www.tech-food.com
	http://www.cnluye.com	http://www.farmers.org.cn
	http://www.yuanlin365.com	http://www.8658.cn
	http://nc.mofcom.gov.cn	http://video.1kejian.com
	http://www.scnjw.gov.cn	http://www.xxsagri.gov.cn
	http://www.agronet.com.cn	http://www.qzny.gov.cn
	http://www.eagric.com	http://www.foodmate.net

Indexes' corresponding relation are: X_1: Function comprehensiveness, X_2: Service interactivity, X_4: Service individuation, X_5: Service professionalism, X_9: Information vividness, X_{10}: Information comprehensiveness; X_{12}: Organization orderliness, X_{16}: Interface friendly; X_{19}: Column novelty, X_{22}: Color coordination, Y: information service quality.

5.2 Selection of Influencing Factors

Before the influencing factors are finally selected, the correlation between information service quality and influencing factors needed to be analyzed respectively. The purpose of correlation analysis is to represent the correlation situation and its change rule between variables, and find the correlation model with each other, which can provide references for making the next decision.

Null hypothesis of test is correlation coefficient of two variables expressing 0 in totality. The process of correlation analysis in SPSS shows the probability of hypothesis formation [7].

On the SPSS 18 platform, data correlation analysis respectively is produced between Y and X_1, X_2, X_4, X_5, X_9, X_{10}, X_{12}, X_{16}, X_{19}, X_{22}. Correlation between Y and X_2, X_4, X_5, X_9, X_{10}, X_{16}, X_{19}, respectively is obvious, and shows linear relationship, and these indexes can be selected as the influencing factors for Y, so multiple regression model is established.

Table 4. Experimental data

Y	X_1	X_2	X_4	X_5	X_9	X_{10}	X_{12}	X_{16}	X_{19}	X_{22}
0.84	0.77	0.81	0.80	0.81	0.68	0.74	0.88	0.44	0.55	0.88
0.39	0.77	0.36	0.35	0.37	0.25	0.29	0.88	0.04	0.16	0.88
0.38	0.75	0.35	0.34	0.35	0.25	0.28	0.86	0.04	0.14	0.86
0.60	0.74	0.61	0.56	0.56	0.52	0.50	0.88	0.21	0.32	0.88
0.73	0.72	0.69	0.69	0.69	0.61	0.63	0.86	0.32	0.43	0.86
0.35	0.71	0.31	0.31	0.31	0.22	0.25	0.82	0.03	0.13	0.82
0.31	0.71	0.27	0.26	0.27	0.15	0.22	0.80	0.01	0.09	0.80
0.74	0.71	0.71	0.70	0.71	0.62	0.65	0.84	0.34	0.44	0.84
0.52	0.71	0.50	0.47	0.49	0.40	0.42	0.84	0.12	0.23	0.84
0.85	0.70	0.82	0.82	0.82	0.68	0.75	0.80	0.45	0.56	0.80
0.54	0.70	0.52	0.50	0.51	0.43	0.44	0.80	0.14	0.26	0.80
0.48	0.70	0.47	0.44	0.45	0.35	0.37	0.80	0.09	0.19	0.80
0.35	0.70	0.34	0.32	0.32	0.23	0.25	0.80	0.03	0.13	0.80
0.31	0.69	0.27	0.27	0.27	0.14	0.21	0.78	0.01	0.08	0.78
0.50	0.69	0.47	0.43	0.46	0.38	0.39	0.82	0.10	0.20	0.82
0.63	0.68	0.62	0.59	0.60	0.52	0.53	0.78	0.24	0.34	0.78
0.41	0.68	0.37	0.37	0.38	0.28	0.31	0.78	0.04	0.17	0.78
0.30	0.68	0.26	0.25	0.26	0.14	0.20	0.76	0.01	0.07	0.76
0.50	0.67	0.48	0.44	0.46	0.39	0.40	0.80	0.11	0.22	0.80
0.74	0.66	0.70	0.70	0.69	0.61	0.64	0.78	0.33	0.43	0.78
0.51	0.66	0.49	0.46	0.47	0.40	0.41	0.78	0.11	0.22	0.78
0.79	0.64	0.76	0.75	0.76	0.63	0.69	0.74	0.40	0.50	0.74
0.62	0.64	0.62	0.58	0.58	0.52	0.51	0.74	0.22	0.32	0.74
0.53	0.64	0.51	0.49	0.50	0.41	0.43	0.76	0.13	0.23	0.76
0.82	0.63	0.78	0.78	0.78	0.66	0.72	0.72	0.43	0.53	0.72
0.31	0.62	0.27	0.28	0.28	0.18	0.22	0.70	0.02	0.09	0.70
0.88	0.61	0.84	0.84	0.85	0.70	0.79	0.70	0.49	0.58	0.70
0.33	0.52	0.29	0.30	0.29	0.20	0.25	0.58	0.02	0.11	0.58
0.33	0.51	0.29	0.29	0.29	0.19	0.24	0.58	0.02	0.11	0.58
0.32	0.51	0.29	0.27	0.28	0.19	0.23	0.58	0.02	0.10	0.58
0.69	0.50	0.67	0.65	0.65	0.57	0.60	0.58	0.30	0.39	0.58
0.47	0.50	0.46	0.43	0.44	0.34	0.36	0.58	0.07	0.19	0.58
0.64	0.50	0.62	0.61	0.60	0.55	0.54	0.58	0.25	0.35	0.58
0.34	0.49	0.30	0.30	0.30	0.21	0.24	0.56	0.02	0.12	0.56
0.64	0.49	0.62	0.60	0.60	0.54	0.54	0.56	0.24	0.34	0.56
0.55	0.49	0.58	0.51	0.51	0.42	0.45	0.58	0.19	0.29	0.58
0.34	0.49	0.30	0.31	0.30	0.22	0.24	0.56	0.02	0.12	0.56
0.83	0.48	0.79	0.79	0.80	0.67	0.73	0.55	0.43	0.54	0.55
0.58	0.47	0.60	0.54	0.55	0.46	0.48	0.56	0.19	0.30	0.56
0.56	0.47	0.59	0.53	0.53	0.44	0.46	0.56	0.16	0.29	0.56

5.3 Experiment Analysis

On the platform SPSS 18, multiple regression model has been established, X_2, X_4, X_5, X_9, X_{10}, X_{16} and X_{19} as independent variables, Y as dependent variable. Results of multiple linear regression show that X_2, X_4, X_5, X_9, X_{10}, X_{16}, with Y shows the positive correlation respectively, but the estimated value of regression coefficient for X_{19} shows negative correlation, which is not fact, so estimated value of X_{19} is not believable [8], then eliminate variable X_{19}. Second regression is carried out.

R^2 of model is 0.999, adjusted R^2 also is 0.999. Value of F tends to rationality. To a certain extent, model reflect the relationship for information service quality and influencing factors, finally the influencing factors model of information service quality is carried out as follows:

$$Y = 0.075 + 0.100 * X_2 + 0.005 * X_4 + 0.368 * X_5 + 0.025 * X_9 + 0.481 * X_{10} + 0.020 * X_{16}.$$

5.4 Model Test

The established multiple regression model is tested by test data.

Table 5 shows that results of 10 test data present ideal results, relative errors between predictive value and real value all are less than 3 %, and average error is only 1.01 %. Experiment demonstrates that the established model is multiple fitted equation having a good reliability [9].

Table 5. Relative errors between predicted and actual values

Sequence number	True value	Predictive number	Relative error
1	0.69	0.694392	0.64
2	0.47	0.466296	1.63
3	0.64	0.640351	0.05
4	0.34	0.336517	1.60
5	0.64	0.639849	0.02
6	0.55	0.555037	0.44
7	0.34	0.336832	2.31
8	0.83	0.829506	0.54
9	0.58	0.586938	1.90
10	0.56	0.567998	0.96

6 Conclusions

All study above shows that two aspects will be paid more attention to during the building of cloud service platform:

(1) In the process of construction for information service system, these indexes such as service interactivity, service individuation, service professionalism, information vividness, information comprehensiveness and interface friendly, are important to influence information service quality.

(2) During the process of building for cloud service platform, these 7 indexes such as platform stabilization, platform compatibility, platform safety, platform adaptability, platform expansibility, software deployment simplicity and software operation shortcut should be fully considered.

Acknowledgment. Financial supports from the National Science and Technology Support Project of China (Grant No. 2013BAD15B05) and the Natural Science Foundation of Beijing, China (Grant No. 4151001) are sincerely acknowledged.

References

1. Chen, M., Deng, J.: Website quality evaluation optimization based R-cluster. Inf. Sci. **9**, 40–42 (2006)
2. Sun, H., Fang, J., Li, J.: Temperature errors modeling for micro inertial measurement unit using multiple regression method. In: Proceedings of the International Symposium on Intelligent Information Systems and Applications, pp. 411–415 (2009)
3. Sadhuram, Y., Ramana Murthy, T.V.: Simple multiple regression model for long range forecasting of Indian summer monsoon rainfall. Meteorol. Atmos. Phys. **99**, 17–24 (2009)
4. Wang, W.: Analysis and comment on study and application of EC websites evaluation. Inf. Sci. **21**(6), 640–642 (2006)
5. Zhao, Y.: Information service quality evaluation for industry information center website. Inf. Resour. Study **6**, 32–37 (2009)
6. Pei, L., Wang, J.: Research on user-oriented appraisal framework of website information service quality. Inf. Sci. **28**(5), 60–64 (2009)
7. Zhi, F., Zhou, J.: Multidimensional Statistic Analysis, pp. 188–230. Science Publisher, Beijing (2002)
8. Chen, C., Wu, H.: Agricultural information service quality evaluation model based on factor analysis and multiple regression. J. Agric. Mech. Res. **09**, 11–17 (2014)
9. Zhou, W.: Research on the quality evaluation model of mobile information service based on the consumer satisfaction. J. Acad. Libr. Inf. Sci. **02**, 74–78 (2015)

Recognition and Localization Method of Overlapping Apples for Apple Harvesting Robot

Tian Shen[1,2(✉)], Dean Zhao[1,2], Weikuan Jia[1,3], and Yu Chen[1]

[1] School of Electrical and Information Engineering,
Jiangsu University, Zhenjiang 212013, China
catcatAndrea@126.com, {dazhao,chenyu}@ujs.edu.cn,
jwk_1982@163.com
[2] Key Laboratory of Facility Agriculture Measurement and Control Technology
and Equipment of Machinery Industry, Jiangsu University,
Zhenjiang 212013, China
[3] School of Computer Science and Technology,
China University of Mining and Technology, Xuzhou 221008, China

Abstract. In order to meet the speed requirements of harvesting robot, a method of tracking and recognition of overlapping apples is proposed in this paper. First of all, the first image should be segmented and denoised, the center and the radius is determined, the template which is used for matching is extracted according to the center and radius. Then, determine the center of ten images which are taken continuously, fit motion path of the robot according to the center of each image and predict subsequent motion path. The next processing area is determined according to the radius and predicted path. Finally, overlapping apples are identified by fast normalized cross correlation match method. Experiments prove that the new method can locate the center and radius of overlapping apples correctly. Besides, matching time is reduced by 48.1 % compared with the original one.

Keywords: Overlapping apples · Fast dynamic identification · Curve fitting · Image matching

1 Introduction

It is well known that China is a big agricultural country with a long history and fruit and vegetable production plays an important part. While in fruit and vegetable production, harvesting accounts for about 40 % of the whole work, which means using robots to harvest fruit and vegetable automatically can not only reduce labor cost and damage rate of fruit picking but also improve harvesting efficiency. American scholars Schertz and Brown proposed to have robots harvest fruit and vegetable in 1960s, since then, a variety of fruit and vegetable picking robot have been studied widely. However, harvesting robots still can not be put into practice because of low recognition and picking rate and recognition problems like occlusion and overlapping fruits [1–6].

© IFIP International Federation for Information Processing 2016
Published by Springer International Publishing AG 2016. All Rights Reserved
D. Li and Z. Li (Eds.): CCTA 2015, Part II, IFIP AICT 479, pp. 330–345, 2016.
DOI: 10.1007/978-3-319-48354-2_34

As apple growing attitude is varied, the images captured by apple harvesting robot are also various, such as single apple, overlapping apples, foliage occlusion and so on. Research is relatively mature on the identification and location of single fruit [7–9]. However, the identification and location of overlapping apples is still a problem which restricts the development of harvesting robot. To solve this problem, numerous scholars have carried out a lot of researches on the identification and location of overlapping images and some progress has been made [10–17]. Among them, Petros [10] used watershed transform and gradient paths to segment touching and overlapping chromosomes. After testing on 183 Multiplex Fluorescence In Situ Hybridization images, the success rate for touching chromosomes is 90.6 % and for overlapping chromosomes is 80.4 %. Julio [11] proposed to use ellipse to approximate the leaf shape so that the complexity of it is simplified. After that, active shape models were used to recognize the clustering of the leaves. Shape model of experimental plants with 2, 3 and 4 leaves were tested and the results indicated that this method was able to identify overlapping leaves if less than 32 % of the area is overlapped. Xu [12] used histogram of oriented gradients (HOG) descriptor associated with a support vector machine (SVM) classifier to detect slightly overlapping strawberries. However, these studies are mostly under static condition and use single frame processing method, the process of harvesting robot is dynamic, so these studies can not be fully applicable to the dynamic picking of robot during moving. For dynamic recognition, many scholars begin to pay attention and gratifying results have been achieved [18, 19]. For example, Lv [18] made a preliminary research on dynamic recognition of fruits. The results indicated that the correlation of images could effectively reduce processing time. But overall, it seems that there is still much room for the improvement of dynamic recognition of overlapping fruit.

This study mainly focuses on dynamic identification and location of overlapping apples. Firstly, improved Otsu threshold algorithm is used to segment the image, after that, the image is processed by morphology so that relevant features can be extracted. Then, the center and radius of overlapping fruit are determined by local maxima method and the template can be extracted according to them. Finally, robot motion path is predicted by combining software and hardware. NCC method is used to make sure that tracking target is the same one and to accurately locate the target fruit so that the overlapping fruit can be tracked dynamically. A total of 11 images which are continuously taken by harvesting robot at a constant speed in natural scene are tested and the results show that the new method can locate the center and radius of overlapping apples correctly. Additionally, processing time of the improved method is reduced by 48.1 % compared with the original one, which means real time of harvesting robot is improved and it can better meet practical requirements.

2 Thought of Fast Tracking and Recognition

Fruit tracking recognition is based on a series of image sequences. Dynamic characteristics of fruits are obtained by searching for the correlation and difference of a range of image sequences, so that the subsequent movement can be predicted and the next range of image processing can be estimated, which reduces the time of robot recognition and improves the picking efficiency. The difficulty of tracking and recognizing overlapping fruits is to determine the center and radius of each fruit. In this paper, a method based on maxima is used to overcome this problem.

The process is shown as follows:

Step 1 Collect eleven overlapping apple images continuously;
Step 2 Segment the images by improved Otsu color difference segmentation algorithm;
Step 3 Perfect the image by mathematical morphology operation;
Step 4 Determine the centers by local maximum method;
Step 5 Determine the radii by minimun distance from the center to the edge of contour;
Step 6 Extract the template according to the centers and radii;
Step 7 Predict the motion path by least square method;
Step 8 Extract subsequent processing region according to the centers and radii;
Step 9 Conduct normalized cross correlation (NCC) fast matching algorithm for image matching.

3   Image Segmentation

3.1   Otsu Segmentation Algorithm

Overlapping fruit segmentation is a major difficulty of tracking and recognizing because segmentation result has a critical influence on subsequent steps. Although fixed threshold segmentation has the advantage of real-time. Considering that the color of fruits and background varies a lot, Otsu method based on the color characteristics is used here.

Otsu method [20] was proposed by Japanese scholar Otsu, the basic idea is dividing the image into two categories by a threshold, the optimal threshold is determined according to the variance between class of these two categories and the optimal threshold is maximal variance between class. For a specific image, set the gray-scale range of the image f(x,y) to be [0, L-1], the probability of each gray-scale is Pi. The image is divided into C0 and C1 (namely background part and foreground part). After setting a threshold t, the probability of the foreground is $\omega_0 = \sum_{i=0}^{t} P_i$, and background is $\omega_1 = 1 - \omega_0$, the average gray value is μ_0 and μ_1 respectively, while total average gray value is $\mu = \omega_0 \mu_0 + \omega_1 \mu_1$, the variance between class of these two parts is

$$\sigma^2(t) = \omega_0(\mu_0 - \mu)^2 + \omega_1(\mu_1 - \mu)^2 = \omega_0 \omega_1 (\mu_0 - \mu_1)^2 \qquad (1)$$

The t which makes $\sigma^2(t)$ the maximal value is the optimum threshold T which separates the target and background.

Although Otsu method can separate the most of the target from the background, it will easily cause over-segmentation phenomenon.

3.2   Improved Otsu Segmentation Algorithm

Although Otsu color difference segmentation algorithm can segment the apples from background to a certain extent, but it will easily cause over-segmentation or under-segmentation phenomenon. For example, Fig. 1(b) appears over-segmentation phenomenon. To overcome this problem, improved Otsu color difference segmentation algorithm is used in this paper. The main content of this algorithm is to stretch or shrink R component in the image by applying gamma conversion to it, so that the difference between R and G component in the image can be increased.

a. Original image   b. Otsu segmentation

Fig. 1. Otsu segmentation

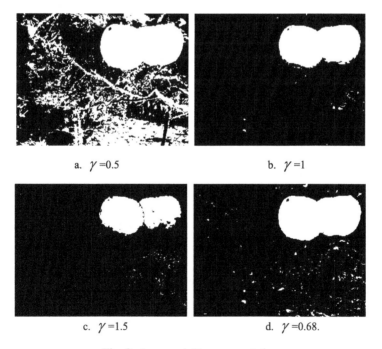

a. γ =0.5   b. γ =1

c. γ =1.5   d. γ =0.68.

Fig. 2. Improved Otsu segmentation

Gamma conversion is a nonlinear gray-scale transformation, its formula is shown as follows:

$$y = (x + K_{esp})^{\gamma} \qquad (2)$$

Where $x \in [0, 1]$; $y \in [0, 1]$; $K_{esp}-$ compensation coefficient; $\gamma-$ gamma coefficient.

When $\gamma > 1$, the contrast in high gray-scale region is enhanced and the contrast in low gray-scale region is reduced;
When $\gamma = 1$, the image is the same;
When $\gamma < 1$, the contrast in low gray-scale region is enhanced and the contrast in high gray-scale region is reduced;

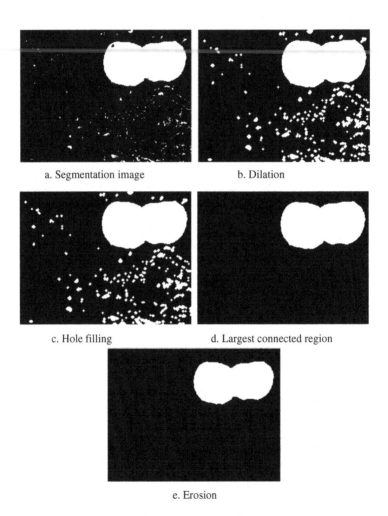

a. Segmentation image   b. Dilation

c. Hole filling   d. Largest connected region

e. Erosion

Fig. 3. Apple image perfection

Figure 2 shows the results after Otsu color difference segmentation. In Fig. 2(a), $\gamma = 0.5$; in Fig. 2(b), $\gamma = 1$; in Fig. 2(c), $\gamma = 1.5$. After several experiments, segmentation works best when $\gamma = 0.68$.

3.3 Image Improvement

There may be some holes in the calyx of apple after segmentation because the color of calyx and body of apple varies greatly. Additionally, there may also be some other holes, burrs, noise, etc. Therefore, mathematical morphology operation [21, 22] should be conducted after segmentation to perfect the image and de-noising.

The basic idea of mathematical morphology is using structural elements with a certain form to measure and extract the corresponding shape in the image, only image features which are similar to the structural elements are kept. Specific steps are:

Step 1 Firstly, the image should be dilated with a 1 radius disk shaped structural elements, so boundary points are expanded and some holes are filled;
Step 2 Then floodfill operation is used to fill the holes remained;
Step 3 After that, the maximum connected region should be obtained so that isolated burrs are removed;
Step 4 Finally, image erosion is operated to eliminate the noise around the boundary.

The results of perfection are shown in Fig. 3.

Figure 3 shows that apples are almost split from the background after dilation-hole filling-largest connection region getting-erosion and the result is satisfying.

4 Template Extraction

Apples are almost separated from the background after automatic threshold segmentation, each center and radius of overlapping apples should be obtained after that so that matching template of apples can be extracted from the image.

4.1 Determine the Center

The center is determined according to the maximum of minimum distance between the point within a circle and the edge of contour. However, calculating all the points within the circle will certainly take up a lot of memory which causes poor real-time performance, so an improved method is used to scan the points within the circle.

Define four scanning direction: $A(x+,y+)$, $B(x-,y+)$, $C(x-,y-)$, $D(x+,y-)$:

For point $E(m,n)$ which is inside the contour of a image, comparing the distance of its left $(m - 1, n)$ and lower $(m, n - 1)$ points in direction A; comparing the distance of its right $(m + 1, n)$ and lower $(m, n - 1)$ points in direction B; comparing the distance of its right $(m + 1, n)$ and upper $(m, n + 1)$ points in direction C; comparing the distance of its left $(m - 1, n)$ and upper $(m, n + 1)$ points in direction D.

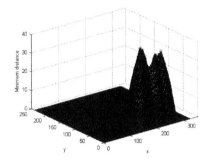

Fig. 4. Three-dimensional map of minimum distance function

Minimum distances are obtained by the examination of the points within the outline in comparison to its four-neighborhood and the minimum distance function is composed by these distances. The three-dimensional surface chart is shown as Fig. 4.

In Fig. 4, the two maxima of the minimum distance are marked with red circles, which are where the two centers of the overlapping apples are.

4.2   Determine the Radius

The radius can be determined by the center of circle, however, it can not be determined only according to the maximum distance from the center to the edge of the outline Firstly, work out the distance from the center to the edge of the contour; then the minimum distance is used as the radius. The steps involved are shown in Fig. 5.

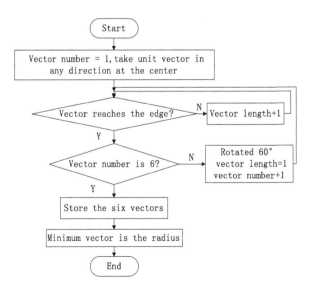

Fig. 5. Flow chart of radius determination

4.3   Template Extraction

The templates which are intercepted on the basis of the center, radius and a certain reserve value are shown in Fig. 6.

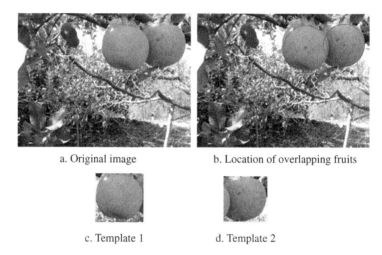

a. Original image   b. Location of overlapping fruits

c. Template 1   d. Template 2

Fig. 6. Template extraction

It can be seen from Fig. 6(b) that the centers and radii can be found accurately by above method. After that, the templates which are used in the subsequent experiments are extracted according to the center, radius and a certain reserve value.

5   Matching Recognition

5.1   Robot Motion Path Anticipation

In order to reduce the time of image processing and accelerate the speed of robot picking, motion path is predicted according to the center of apples in collected images and the location of centers in a series of images is used to narrow the processing scope of subsequent image.

Due to the complication of overlapping fruits, the case of two overlapping apples is studied in this paper. The processing steps of two overlapping apples are as follows:

Step 1 Determine the centers of two overlapping fruits in the images respectively by the method mentioned in Sect. 4.1. Then robot motion path is fitted by polynomial fitting based on the two centers respectively. The two centers in the next frame are estimated after prediction which combines robot speed and sampling time.

Step 2 The radius is determined by the method mentioned in Sect. 4.2. Find the maximum radius of the two fruits and name it rmax. A(ax, ay), B(bx, by) are the two estimated centers of the next frame. The subsequent processing area

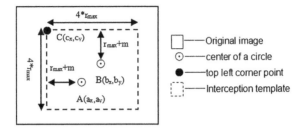

Fig. 7. Subsequent processing region extraction

is intercepted with starting point C(cx,cy). The size of it is a square with 4*rmax side length. Figure 7 shows the schematic diagram of cutting the subsequent processing area. Where, cx = min{ax,bx} − rmax − m, cy = max{ay,by} + rmax + m, m is a certain reserve value.

Least square method is used in Step 1 to fit the motion path of the robot. Specific processing steps are: set the square sum of error of the fitting curve is less than 5, then find out the best order of the fitting curve. After that, motion curve is fitted by least square method according to the best order and list the polynomial of the curve. Finally, the coordinates of the centers in the next frame are determined by combing sampling interval of robots.

5.2   Normalized Cross Correlation Matching

Subsequent processing region of the image has been extracted after the above steps. The position of apples can be located after the images conduct normalized cross correlation matching with the templates.

Normalized cross correlation (NCC) matching [23, 24] is the most classical algorithm among matching methods based on gray feature. Its basic principle is comparing the gray matrix of template with image to be searched to get the location of the most relevant match. The algorithm is simple and it eliminates the light sensitive issues which means it has good applicability for apple images under different light intensities. Besides, NCC match applies to the situation when the image is slightly displaced and rotated compared with the original one, which means it is suitable for dynamic image tracking and matching.

I is used to represent the image to be matched whose pixels are $M \times N$ and T is used to represent the template whose pixels are $m \times n$. Normalized correlation coefficient is defined as follows:

$$R(x,y) = \frac{\sum\limits_{\mu=0}^{m-1}\sum\limits_{\gamma=0}^{n-1}[I(x+u,y+\gamma) - \bar{I}_{x,y}][T(u,\gamma) - \bar{T}]}{\sqrt{\sum\limits_{\mu=0}^{m-1}\sum\limits_{\gamma=0}^{n-1}[I(x+u,y+\gamma) - \bar{I}_{x,y}]^2 \sum\limits_{\mu=0}^{m-1}\sum\limits_{\gamma=0}^{n-1}[T(u,\gamma) - \bar{T}]^2}} \tag{3}$$

Where, (x, y) is the coordinate of the top left corner of the sub-graph in the image; (u, γ) is the coordinate of the pixel in the template.

$$\bar{I}_{x,y} = \frac{1}{mn} \sum_{\mu=0}^{m-1} \sum_{\gamma=0}^{n-1} [I(x+u, y+\gamma)] \tag{4}$$

is the average value of pixel of the sub-graph $I_{x,y}$.

$$\bar{T} = \frac{1}{mn} \sum_{i=0}^{m-1} \sum_{j=0}^{n-1} T(u, \gamma) \tag{5}$$

is the average value of pixel of template T.

$R(x, y)$ is between $(0,1)$, the greater the coefficient is, the higher similarity between two matching templates will be.

However, NCC fast matching algorithm is used in this paper because NCC algorithm needs too much calculation and bad real time. Specific steps are as follows:

Step 1 Set $T'(u, \gamma) = T(u, \gamma) - \bar{T}$, then the numerator of formula (2) can be simplified as

$$\sum_{\mu=0}^{m-1} \sum_{\gamma=0}^{n-1} I(x+u, y+\gamma) T'(u, \gamma) - \bar{I}_{x,y} \sum_{\mu=0}^{m-1} \sum_{\gamma=0}^{n-1} T'(u, \gamma) \tag{6}$$

Where, $\sum_{\mu=0}^{m-1} \sum_{\gamma=0}^{n-1} T'(u, \gamma) = \sum_{\mu=0}^{m-1} \sum_{\gamma=0}^{n-1} [T(u, \gamma) - \bar{T}] = 0$, so the numerator can be simplified as

$$R(x, y)_{numerator} = \sum_{\mu=0}^{m-1} \sum_{\gamma=0}^{n-1} I(x+u, y+\gamma) T'(u, \gamma) \tag{7}$$

According to the Fourier transform, the numerator can be rewritten as

$$R(x, y)_{numerator} = F^{-1} \left\{ F\{I\} \bullet F * \{T'\} \right\} \tag{8}$$

Step 2 For the denominator, since the template is known, $\sum_{\mu=0}^{m-1} \sum_{\gamma=0}^{n-1} [T(u, \gamma) - \bar{T}]$ is a given fixed value, which means it does not affect the normalized match to find the optimal solution.

Therefore it does not have to be calculated, denominator of formula (2) can be simplified as

$$R(x,y)_{denominator} = \sqrt{\sum_{\mu=0}^{m-1}\sum_{\gamma=0}^{n-1}[I(x+u,y+\gamma)-\bar{I}_{x,y}]^2} \qquad (9)$$

In summary, normalized correlation coefficient can be simplified as

$$R_1(x,y) = \frac{F^{-1}\{F\{I\} \bullet F * \{T'\}\}}{\sqrt{\sum_{\mu=0}^{m-1}\sum_{\gamma=0}^{n-1}[I(x+u,y+\gamma)-\bar{I}_{x,y}]^2}} \qquad (10)$$

The amount of calculation of NCC algorithm is reduced significantly after simplification and real-time has been improved effectively.

6   Results and Analysis

6.1   Robot Motion Path Prediction

The operating environment of the experiment is as follows, hardware environment: windows 7 operating system, Intel(R) Core(TM)2 Duo CPU E7500 2.93 GHz processor, 2 GB memory. Software environment: Matlab R2013a.

The latest 11 images which are continuously collected by robot under uniform condition in the natural scene should be processed. Fitting the trajectory of the robot according to the previous ten images and predicting subsequent path. Specific steps are as follows.

Firstly, find each center of two overlapping apples in the 10 images, conduct polynomial fitting for the motion path of the centers. Experiment results are shown in Fig. 8.

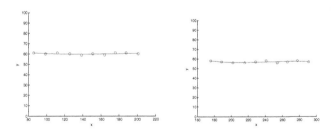

Fig. 8. Curve fitting diagram of robot motion path

Take the left apple for an example, set the square sum of error of the fitting curve is less than 5, it turns out that three order curve can fit the robot trajectory, so, it is used in this paper to fit the curve. Find the polynomial of the curve and the coordinates of the two centers of overlapping fruits in the eleventh image can be predicted according to former ten pictures combined with sampling time of robot. In this experiment, the coordinates of the predicted centers are (214,60.6) and (304,56.3) respectively.

6.2   Image Matching

According to the position of the centers predicted, a square with side length $4*r_{max}$ is intercepted as the area to be processed.

Figure 9(a) is a 320 × 240 pixel image, apples account for approximately 18.5 % of the whole image. The results show that the new method can find the apples in the eleventh image accurately.

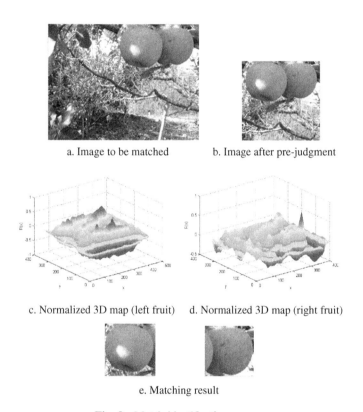

a. Image to be matched   b. Image after pre-judgment

c. Normalized 3D map (left fruit)   d. Normalized 3D map (right fruit)

e. Matching result

Fig. 9. Match identification process

Comparison experiments show that average matching time of two apples is 0.181 s before pre-judgment, while after pre-judgment, average matching time is 0.094 s, which means the speed of identification is accelerated by 48.1 %. These results indicate that it significantly accelerates image matching after pre-judgment and matching accuracy is high.

6.3   Different Proportion of Apple Area Discussion

In this paper, tracking and recognition problem caused by mutual occlusion of fruits is studied. It is a simple approach, but it is effective. The results indicate that this approach can successfully recognize the overlapping apples and track them in dynamic condition. Also it has good real-time. However, the proportion of apple area has a big influence on the matching time.

Applying this method to Fig. 10(a), (b), (c) which have the same size and the relationship between the proportion of apple area and matching time is shown in Table 1.

a. Comparison image1   b. Comparison image2   c. Comparison image 3

Fig. 10. Images of experiments on influence of proportion of apple area

Table 1. Relationship between the proportion of apple area and matching time

Image serial number	1	2	3
fruit area/image area ×100 %	18.5	55	66
Matching time without pre-judgment/S	0.181	0.185	0.188
Matching time after pre-judgment/S	0.094	0.133	0.148
Matching time acceleration ×100 %	48.1	28.1	21.3

It can be seen from Table 1 that the smaller proportion of apple fruit is, the better accelerating optimization will be.

6.4   Different Proportion of Apple Overlapping Area Discussion

Besides the proportion of apple area, the proportion of apple overlapping area also has great impact on the matching time.

The comparison experiments are done with the same two apples in different overlapping conditions (Fig. 11).

Table 2 shows that the bigger proportion of overlapping area is, the less matching time will be needed.

Table 2. Relationship between the proportion of apple overlapping area and matching time

Image serial number	1	2	3	4
Overlapping area/fruit area ×100 %	4.75	5.03	7.47	7.69
Matching time after pre-judgment/S	0.105	0.095	0.078	0.071

a. Comparison image1   b. Comparison image2

c. Comparison image 3   d. Comparison image 4

Fig. 11. Images of experiments on influence of proportion of apple overlapping area

7 Conclusion

Tracking and recognition of overlapping fruits is studied in this paper. The process of dynamic tracking and recognition is introduced, firstly, improved Otsu method based on color difference is used to segment the image, then morphology method is processed on segmented image. Secondly, the centers of overlapping fruits are determined by local maxima distance, the radii are determined by the centers and the templates are extracted according to the centers and radii. Finally, rapid normalization match is conducted to track and recognize overlapping fruits. It is proved by several tests that matching time is reduced significantly after prediction and real-time has been improved effectively. Additional hardware devices are not required in this method, lower cost is needed and it has generality for spherical fruits and vegetables picking robot. Further study is required for the following problems.

Recognize overlapping fruits in relatively more occlusion situation. The method which is based on local maxima of distance only applies to the situation when the profile of apples is a relatively complete circle. The maxima method can not be used if the profile of apple is not complete.

Improve real-time performance of template matching performance. NCC algorithm has good effect on different light and slight rotation and translation, but it needs much calculation. Although image processing time is reduced by pre-judgment, the computation is still heavy. Researches on real-time improvement can be studied in subsequent studies.

Acknowledgments. Funds for this research was provided by the National Nature Science Foundation of China (No. 31571571), Priority Academic Program Development of Jiangsu Higher Education Institutions (PAPD), The Specialized Research Fund for the Doctoral Program of Higher Education of China (No. 20133227110024).

References

1. Wei, X., Jia, K., Lan, J., et al.: Automatic method of fruit object extraction under complex agricultural background for vision system of fruit picking robot. Optik-Int. J. Light Electr. Opt. **125**(19), 5684–5689 (2014)
2. Ji, W., Zhao, D., Cheng, F., et al.: Automatic recognition vision system guided for apple harvesting robot. Comput. Electr. Eng. **38**(5), 1186–1195 (2012)
3. Zhang, L., Yang, Q., Bao, G., et al.: Overview of research on agricultural robot in China. Int. J. Agric. Biol. Eng. **1**(1), 12–21 (2008)
4. Chen, X., Yang, S.X.: A practical solution for ripe tomato recognition and localisation. J. Real-Time Image Proc. **8**(1), 35–51 (2013)
5. Ogawa, Y., Kondo, N., Monta, M., et al.: Spraying robot for grape production. In: Yuta, S., Asama, H., Prassler, E., Tsubouchi, T., Thrun, S. (eds.) Field and Service Robotics, pp. 539–548. Springer, Heidelberg (2006)
6. van Henten, E.J., Hemming, J., Van Tuijl, B.A.J., et al.: An autonomous robot for harvesting cucumbers in greenhouses. Auton. Robots **13**(3), 241–258 (2002)
7. Ji, W., Zhao, D., Cheng, F., et al.: Automatic recognition vision system guided for apple harvesting robot. Comput. Electr. Eng. **38**(5), 1186–1195 (2012)
8. Hayashi, S., Shigematsu, K., Yamamoto, S., et al.: Evaluation of a strawberry-harvesting robot in a field test. Biosyst. Eng. **105**(2), 160–171 (2010)
9. Rajendra, P., Kondo, N., Ninomiya, K., et al.: Machine vision algorithm for robots to harvest strawberries in tabletop culture greenhouses. Eng. Agric. Environ. Food **2**(1), 24–30 (2009)
10. Karvelis, P., Likas, A., Fotiadis, D.I.: Identifying touching and overlapping chromosomes using the watershed transform and gradient paths. Pattern Recogn. Lett. **31**(16), 2474–2488 (2010)
11. Pastrana, J.C., Rath, T.: Novel image processing approach for solving the overlapping problem in agriculture. Biosyst. Eng. **115**(1), 106–115 (2013)
12. Xu, Y., Imou, K., Kaizu, Y., et al.: Two-stage approach for detecting slightly overlapping strawberries using HOG descriptor. Biosyst. Eng. **115**(2), 144–153 (2013)
13. Qi, X., Xing, F., Foran, D.J., et al.: Robust segmentation of overlapping cells in histopathology specimens using parallel seed detection and repulsive level set. IEEE Trans. Biomed. Eng. **59**(3), 754–765 (2012)
14. Cloppet, F., Boucher, A.: Segmentation of overlapping/aggregating nuclei cells in biological images. In: 19th International Conference on Pattern Recognition, pp. 1–4. IEEE (2008)
15. Priya, E., Srinivasan, S.: Separation of overlapping bacilli in microscopic digital TB images. Biocybern. Biomed. Eng. **35**(2), 87–99 (2015)
16. Wang, D., Song, H., Tie, Z., et al.: Recognition and localization of occluded apples using K-means clustering algorithm and convex hull theory: a comparison. Multimedia Tools Appl. **75**, 3177–3198 (2016)
17. Lü, J., Zhao, D., Ji, W., et al.: Fast positioning method of apple harvesting robot for oscillating fruit. Trans. Chin. Soc. Agric. Eng. **28**(13), 48–53 (2012)
18. Chtcheglova, L.A., Waschke, J., Wildling, L., et al.: Nano-scale dynamic recognition imaging on vascular endothelial cells. Biophys. J. **93**(2), L11–L13 (2007)

19. Gonzalez, R.C., Woods, R.E., Eddins, S.L.: Digital Image Processing Using MATLAB. Pearson Education India (2004)
20. Serra, J., Soille, P.: Mathematical Morphology and Its Applications to Image Processing. Springer Science & Business Media, Heidelberg (2012)
21. Kaur, B., Garg, A.: Mathematical morphological edge detection for remote sensing images. In: 3rd International Conference on Electronics Computer Technology, vol. 5, pp. 324–327. IEEE (2011)
22. Heo, Y.S., Lee, K.M., Lee, S.U.: Robust stereo matching using adaptive normalized cross-correlation. IEEE Trans. Pattern Anal. Mach. Intell. 33(4), 807–822 (2011)
23. Saravanan, C., Surender, M.: Algorithm for face matching using normalized cross-correlation. Int. J. Eng. Adv. Technol. 2 (2013). ISSN: 2249-8958

Retrieval Methods of Natural Language Based on Automatic Indexing

Dan Wang[1,2(✉)], Xiaorong Yang[1,2], Jian Ma[1,2], and Liping Zhang[1,2]

[1] Institute of Agricultural Information, Chinese Academy of Agricultural Sciences, Beijing, 100081, China
{wangdan01,yangxiaorong,majian,zhangliping}@caas.cn
[2] Key Laboratory of Agricultural Information Service Technology (2006–2010), Ministry of Agriculture, Beijing 100081, People's Republic of China

Abstract. Since natural language enter the computer retrieval system, due to the natural language retrieval is not restricted by professional experience, knowledge background, retrieval experience by users, and above reasons favored by the users. As the title of the Chinese literature is the concentrated reflection of Chinese literature content, it reflects the central idea of the literature. Retrieval methods of natural language described in this article is limited to literature title in subject indexing. The basic idea of this method is, with automatic indexing methods respectively the literature title in the database of retrieval system used in natural language retrieval for automatic word indexing. To control the concept of a given keyword, namely meaning transformation, form the final indexing words. Then, using the vector space model for the index data in the database will be "or" operation to retrieve, forming a document set B. For each document title in set B for automatic indexing, the title of each article for automatic indexing, indexing terms for the formation and retrieval of natural language indexing terms similarity calculation, sorted according to similarity of each document in set B. The first best match the requirements presented to the user documentation. This method is a simple and practical method of natural language retrieval.

Keywords: Automatic indexing · Natural language retrieval methods

1 Introduction

When the computer retrieval system has just entered the practical stage, people soon find its defects in retrieval time lag, retrieval feedback results, and to develop more convenient and efficient online retrieval system in the terminal. However, in the online era, the adverse effects of a full-time retrieval personnel and user questions needs have become a new reality questions. To this end, people have developed a variety of user-friendly man-machine interface. Today, access to the network retrieval, the user of retrieval system has undergone fundamental changes, the end-users who have different ages, different occupations, different knowledge backgrounds, different experiences have become increasingly demanding for retrieval system of convenience, immediacy

D. Li and Z. Li (Eds.): CCTA 2015, Part II, IFIP AICT 479, pp. 346–356, 2016.
DOI: 10.1007/978-3-319-48354-2_35

and transparency. Thus, retrieval system (user interface) with the ability to understand natural language are increasingly welcomed by the majority of users, become an important part of the network retrieval system.

Natural language also known as "everyday language", it is a tool for expression and exchange of ideas in everyday life for the long-term social practice, it is very wide application in information retrieval. From the user's perspective, natural language search is users use the words, phrases or natural statement of daily life for questions. From the technical side, natural language retrieval is the natural language processing technology applied in information retrieval system of information organization and indexing, and output [1]. In information retrieval, the so-called natural language is relative to the case of controlled language, natural language is essentially a raw and standardized treatment of uncontrolled language. The whole process from the point of view of information retrieval, natural language search is including two aspects of natural language indexing and natural language query question. Natural language search is a direct order from the source document as the index identifies the content, users can directly use the natural language questions and complete a form to retrieve information retrieval.

2   Status of Natural Language Retrieval

Natural language retrieval was born in computer search, arising from the date it would have equal shares and information retrieval language. It is because natural retrieving language has its own advantages, it has long attracted people's attention, so that domestic and foreign scholars and experts to study it. Study abroad can be traced back in the 1960s, research focused on the automatic indexing achieve human indexing effect, the main representative of the study from the initial American scholar Salton and Bely, and later the American scholar Sparck, John, Tait, Fagan, Croft, Turtle, Lewis and so on; To the 1990s, TRC (Text Retrieval Conference Text Retrieval Conference) natural language search system began to participate in trials and competitions [2]. From TREC-1 meetings to TREC-6 meeting, the study of natural language search continues to move forward, its research focus on from the original statistical methods to the study of query expansion mode and flow index merge algorithm. The late 1990s, many foreign well-known databases such as Dialog, BIOSIS, ProQuest online, also started to provide natural language search interfaces in their own retrieval system and try natural language search. Many network-oriented information resource retrieval of test systems and the search engine uses a certain amount of natural language search technology, to a certain extent, to achieve a natural language search function, these test systems and search engines are: START, IRENA, FERRET, Ask Jeeves (http://www.ask.com), Geoquery (http://www.cs.utexas.edu/users/ml/geo.html), ixquick (http://www.ixquick.com), Northern Light (http://www.northernlight.com), Ask Northern Light a question, Electric library (http://www.elibrary.com) and so on.

Before the 1990s, in the domestic field of information retrieval research on natural language search in natural language indexing, other studies have concentrated on the theoretical discussion on the text indexing by natural language. After the middle of 1990, there have been some studies on the user interface. Professor Zhang Qiyu was the earlier focus on natural language search scholars, he made a more in-depth study on a variety of factors natural language indexing information retrieval efficiency. He proposed text type, the search range, the search terms of the degree of specificity, the wording of the text is not standardized, different indexing methods and the degree of control of natural language search system would have an impact on [3]. National Taiwan University Department of Library Chen Guanghua used LOB Corpus as the training corpus, used SUSANNE Corpus as test Corpus, studied the natural language retrieval on the syntactic level [4]. In recent years, there are some practical web search engine to provide natural language search in China, there are TRS retrieval systems, Eureka search engine and Naxun Chinese news search engine. It is worth mentioning that the "Baidu knows" is the most influential of the Q & A platform - natural language search system. As of September 15, 2012, "baidu knows" the use of natural language retrieval methods has solved the problem of 200 million [5].

3   The Key Issues of Natural Language Search

Natural language search includes two aspects, namely natural language indexing and users retrieve by using a form of natural language questions. These two aspects can work independently, in technology implementation respectively, at the same time, they have close connection each other. The former is to standardize the indexing of natural language retrieval. The latter is to provide a natural language interface for users to ask questions, make information retrieval system to retrieve user needs to understand natural language in the form of expression, and processes the user's natural language questions. To solve these two areas, we depend on the following key technology research and development [6].

3.1   Subject Indexing

One of the key issues of natural language search is how to extract most accurately fuller expression of documentation related to the topic words from the document. As well as the relationship between the words in a document expressing the theme concept, and this relationship is stored in the index, to support subsequent retrieval.

3.2   Question Treatment

Another key issue of natural language search is user's natural language questions by expressed understanding of computers. Ideal retrieval system should be able to "understand" the real search request which users use natural language expression. Not only retrieval system understands the significance of the user clear statement, but also understands the hidden meaning in natural language questions to be expressed. Thus, end users do not need to bother to go more express retrieval needs, learning tedious search command format.

3.3   Questions and Index Matching

Ask and index of effective matching is another difficulty in natural language retrieval. Specific matching algorithm depends on the structure of the index and quiz process technology. Meanwhile, adopted retrieval model will largely affect the matching algorithm and effect.

3.4   Control Concept

In essence, natural language search is a concept search, it requires a certain conceptual system or knowledge database to support. Knowledge of knowledge database can help solve the problem of natural language questions differentially expressed, that is a solution to the information source text and user questions related to the use of different words to express the concept of problem. The synonyms and near synonyms in the knowledge base can achieve control of the concept of words, to eliminate the difference of word brings the retrieval accuracy.

4   Natural Language Indexing

4.1   Factors Affecting the Quality of Indexing

In the information retrieval system, indexing methods and results have a greater impact on the retrieval results. Currently there is no indexing full-text indexing and word extraction indexing form keyword index of automatic indexing method. Quality of indexing directly affects the natural language retrieval results, influencing factors quality of indexing.

4.1.1   Indexing Depth
Indexing depth is used for indexing a document identification (keyword) number. It reflects the indexing on the degree of comprehensive and specifically to the theme of the document analysis. In general, the depth of indexing (indexing terms more) is the greater, the recall is the higher.

4.1.2   The Relationship Between Words of Indexing Words
The theme of the document is composed of multiple indexing words together common expression. There are certain grammar and restrictions relations between indexing words. For example, the position and the order of indexing words and relationship of index terms and synonyms and so on. In the process of automatic indexing, the more accurate analysis of the relationship between the word of the word, the more in the index expression, the complete retrieval result is better.

4.1.3   Indexing Size

During indexing, text block which indexing object points can be called indexing unit. Indexing unit refers to how large blocks of text to generate a set of index terms, it reflects the index object size. On the premise of indexical meaning, the smaller indexing particle size, the more sophisticated, the better retrieval refers specifically. Several factors described above is directly related to this article.

4.2   Automatic Indexing Method

Automatic indexing is to use the computer to give corresponding to deal with literature searching, the process of indexing is divided into classification indexing and subject indexing. Subject indexing is divided into title lexical, lexical unit, syria lexical and keyword method. The first three indexing methods belong assignment indexing method index, the latter belongs to the extraction indexing words. Assignment indexing requires a thesaurus for support. This article relates to the automatic indexing is assigned indexing words [7]. This article use the keywords table is by the agricultural information institute, Chinese academy of agricultural sciences "computer automatic indexing research" compiled by the "computer automatic indexing multifunctional agriculture word", there are "use, generation, genera, divide and reference" and other relations between each item, can cut out from the document keywords conversion, automatic indexing given keywords, synonym, synonyms, related words, and the quest for word to complete the indexing concept of control. There are three methods in the general automatic indexing. The first one is segmentation method based on string matching, the second one is the segmentation method based on understanding, the third one is segmentation method based on statistical word [8]. Segmentation method based on string matching is Chinese character string and entry words in dictionary match according to a certain strategy, given an indexing word after a successful match. According to different scanning direction, it can be divided into forward match and reverse match. According to the different length matching of different priority, it can be divided into the biggest (longest) and the minimum (minimum) match.

4.2.1   Forward Longest Matching Method (MM Methods)

The strings obtained by coarse segmentation have been verbatim scanned from left to right and match with Thesaurus, and the keywords of thesauri maximum matching as the primary keywords. For example, in thesaurus in the "cadres tenure" in Chinese, and also included "cadres", "office", "age". Longest matching method is that "A short length is not taken" the word extraction rules, only extracting "cadres tenure" is used.

4.2.2   Reverse Longest Matching Method (RMM Methods)

Principle of RMM with MM method is the same. Difference is that the word of the scanning direction, it is taken from right-to-left matching substring. Statistics show that simply using the forward maximum matching error rate is 1/169, simple to use reverse maximum matching error rate is 1/245. Obviously, RMM method in the segmentation accuracy than MM method has been greatly improved.

4.2.3   Minimum Segmentation

Both forward maximum matching and reverse maximum matching, guaranteed indexing words maximum benefit of indexing specificity, is advantageous to the indexing specificity. But the biggest index terms are possible split in order to extract the smaller index terms. This is likely to increase the indexing words, that is to improve the indexing depth. At the time of retrieval, which is beneficial to avoid leak phenomenon. For example, in thesauri "cadres working age" term is maximum matching and indexing, if we the further use of minimum-cut method, there are likely to increase "cadres", "working", "age" indexing terms, indexing terms has increased, can improve recall.

5   Natural Language Search Implementation

Natural language search is including natural language indexing and natural language query question two aspects. Automatic indexing of natural language are discussed earlier, the remaining problem is the problem of natural language questions and queries. Automatic indexing of natural language are discussed earlier, next we discuss natural language questions and queries and so on.

5.1   Natural Language Query Processing

One of the characteristics of natural language retrieval is to allow users to use natural language retrieval requirements directly to the system. Due to the use of natural language, in the form of natural sentences express questions and system index in the index entry form is different, the expression of document both cannot match directly, So we requires a necessary processing of natural language questions. Discussion of the relevant issues.

5.1.1   Indexing Item Level

When you retrieve using natural language questions, retrieval system requirements index entries in the language level is completely consistent with the natural language questions. In this paper the method to realize natural language retrieval is limited to the title.

5.1.2   The Control of Indexing Words

Natural language retrieval has the characteristics of concept retrieval, semantic retrieval. Therefore, in order to achieve better search results, we want to deal with natural language questions in concept. Namely, the use of a number of related linguistic dictionaries, such as thesaurus dictionary or conceptual relationships dictionary or related knowledge, we need control natural language words in document title, at the same time we also need control the language (words) on the users of natural language questions. The key words in everyday-language are converted into keywords, and give the corresponding synonyms, hypernym, synonyms, etc. In this paper, the natural language search implementation used a "Computer automatic indexing vocabulary multifunctional agriculture".

5.2 Natural Language Retrieval Match

In information retrieval systems, indexing process is the back-end processing, it participles and matches for the source document, etc. It has established the index data for retrieval. Natural language question processing is the front-end work of system. It also participles and matches for user's natural language question, etc. It provides the interaction between user and the system interface. In the process of natural language retrieval, indexing words between index data and retrieval requirements matched under certain matching control mechanism. It completed to match both generated indexing word. There are three common retrieval models that are Boolean model retrieval, vector space model retrieval and probabilistic retrieval model retrieval. Vector space model has a simple structure, formal, and easy to implement features. The vector space model retrieval was used in this article. It would document retrieval item (such as titles) and user natural language questions were compared. That is, both the indexing terms would be compared, and use weight to calculate the similarity, and to judge search results by similarity.

5.3 The Concept Control of Natural Language

In dealing with natural language question retrieval, Pattern matching calculation is commonly used, namely by keywords (index terms) in comparison to complete. In the process, if it is mechanically simple pattern matching processing, instead of using the concept of the corresponding control, pattern matching will have the following questions.

Choice of words there is no strict limit of natural language questions, words more vocabulary more cluttered, it will affect the search results. In order to solve these problems, we must further process natural language question in the matching process, namely, we must control at the conceptual level. Natural language query express retrieval concepts through natural language. The concept of a keyword in natural language is all kinds of relations with other key words. For example synonyms relations, hypernym relations, etc. For example, "microbial fertilizer" concept, based on the relationship between the upper and lower classes can be subdivided into "antibiotic fertilizer", "Rhizobium fertilizer" "nitrogen-fixing bacteria fertilizer," and so on. A simple literal match will cause missed. For example the "computer" and "electron brain" is different expression of the same thing, "potato" and "toodou" is also synonymous. If the user's natural language is "potatoes" (or computer), the word "toodou" (or electron brain) is a term describing a document will be missed. This approach can apply these same concepts are words matching retrieval, thus expanding the retrieval surface, thereby increasing the recall.

5.4 Natural Language Retrieval Method

Following implementation of natural language search will be described, in addition, natural language search to achieve this article is limited to the title of the document (title) levels.

5.4.1 Natural Language Search Process

Using natural language questions directly match and retrieval with the body or abstracts of the document, and in any case no one article is hit. Using natural language questions directly match and retrieval with the title of the document. Perhaps occasionally one article is hit.

Such as natural language search method is certainly not desirable. Therefore, we need automatic indexing process for the natural language questions and databases' document. Natural language retrieval process diagram is as follows (Fig. 1):

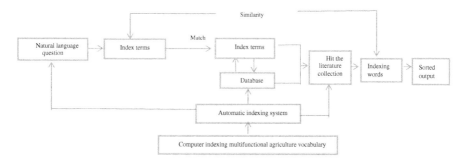

Fig. 1. Natural language retrieval process diagram

First of all, the automatic indexing does automatic word segmentation for the title of the document and matches the words of thesauri. We will do meaning conversion process to index terms matched. We give index terms. While, we give the word hypernym, hyponym, synonym, synonyms or snare words as index terms. Then we do indexing process for indexing terms. This process is the systematic background does real-time processing to the database and the new information documents. To ensure that the title of the document should be indexing and index update for every new document in a database. In natural language retrieval, the system at the front desk also use the same automatic indexing system and indexing method to natural language question statement (or phrase) for indexing, and indexing words and the information source of indexing words "or" operation matching, matching literature that is hit literature, hit literature form a collection of hit documents, known as B collection.

5.4.2 Hit Literature Output Sorting

Hit documents by the above methods are a number (n) articles in most cases. For example, indexing terms of natural language question in the statement are A, B, C three words, through A, B, C three words "or" operation to retrieve, any index terms of literature title containing A or B or C word, containing A and B or A and C or B and C words, or containing A and B and C words are retrieved. Faced with the hits literature (B collection), which was first presented to the user, but it has the sort of problem. Sort is based on the similarity of the title of hit document and the user's natural language question statement. Similarity calculation is as follows: Set the A string is indexing terms which natural language query statement provided, for a particular request statement, index terms of A string are fixed. In natural language retrieval, Set hit document is B collection, the title of each article in B collection is automatic indexed, and indexing words of every document form a C string, and indexing words of every document form a C string, so there are a number of C string in B collection, we compared A string with the C string, both indexing words overlap degree, we called similarity. The maximum similarity is 1, indexing terms of A string and indexing terms of C strings are one hundred percent the same. For each C string of the B collection, indexing words of C string and indexing words of A string one by one compares, calculate the ratio coincides both the index word (always less than 1). Thus, a characteristic natural language search results form a plurality of different sizes overlap ratio, Greater overlap percent, the higher the similarity. The coincidence percentage of high and low will be as the basis of the output sequence of the hits literature.

6   Test

Natural language retrieval was born in computer search, arising from the date it would have equal shares and information retrieval language. It is favored by the majority of users. Natural language search technology is widely used in utility database, news databases, etc. For example, in the field of agriculture practical technology database of crop management and cultivation techniques, plant protection technology, vegetable gardening management and technology and other aspects are more suitable for users to use natural language questions to ask questions for retrieval. "Baidu knows" is a very typical natural language question sentence retrieval practical database.

The author did a test by natural language retrieval methods in this paper. Test data is extracted from the CNKI database. Data mainly includes crop (wheat, corn, soybeans, cotton, potato, sesame seeds)cultivation techniques, pest control technology, gardening vegetable (cabbage, celery, tomato, carrot, rapeseed)cultivation techniques, livestock and poultry breeding, marine and freshwater aquaculture, and agro-processing, total capacity of data are more than 5,000. Automatic indexing thesaurus is Chinese Academy of Agricultural Sciences Institute of Agricultural Information compiled by the "Computer automatic indexing vocabulary multifunctional agriculture", a total of more than 40,000 keywords. There is a relation of the subject words, such as "use, generation, genera, division and reference", in addition to the category code and the corresponding net word. Using this method has been tested on the above data, the test results are satisfactory, which is a higher precision. For example, the search statement as "potato cultivation technology", from the test sample data (there are 420 potato literature) and find relevant literature. According to the provided similarity method, followed by the top ranked "potato cultivation technology", "black potato cultivation technology", "precocious potato planting technology", "Heilongjiang province potato mechanized technology", "potato cultivation comprehensive disease prevention and control measures." and so on. Because the automatic indexing words above topics and natural language search statement are "potato, cultivation techniques, planting techniques, precocious, detoxification, Heilongjiang Province, mechanization, disease, comprehensive prevention and control measures" and "potato, cultivation technology, planting, technology.

"According to the degree of overlap between the two indexing terms will have above order. Another example: Search statement was "practical agricultural technology database" only was retrieved two documents in CNKI database. That is, "the WEB retrieval of agricultural practical technology database", "the regional agricultural practical technology database under the network environment". If we use the natural language retrieval method provided by this paper, we can find out the 50 documents. Because it is not a complete word with the search statement to find, but with its indexing word "agriculture, practical technology, database," a "or" operation to retrieve. Of course, so find out more literature, this is why this method will not result in a leak (Fig. 2).

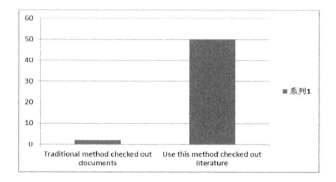

Fig. 2. Document detection number comparison

However, sometimes, check out the document number is too much, in order to save time for the user to browse the documents, the maximum number of documents and the maximum number of views in the literature will be set up. Literature amount exceeds this value no longer that shows the document content. The documents are sorted according to the similarity. In order to meet the same degree of similarity in the literature, the order is also random. In addition, using this method to retrieve must use the same indexing system and thesaurus database indexing and index, in order to ensure the accuracy of the document similarity. In the requirements of retrieval statement, both keywords (keywords), a phrase, or spontaneous statements is the same, the method is applicable.

7 Conclusion

This article provides a natural language search method based on automatic indexing is a method of natural language retrieval application method easier to achieve. This method has some advantages: (1) Simple user operation, the search condition (statement) requiring a simple and different level of users can operate; (2) System implementation easier. Systems involved in the automatic indexing algorithm module, indexing data module and so on, most of them have more sophisticated software, it is easier to integrate them together into a useful natural language search system. (3) Made some technical standardization process in natural language processing, such as between words of "use, on behalf of, divided, the Senate" and other relations, given the appropriate indexing terms, to improve the documentation of precision and recall. (4) The introduction of the concept of literature similarity. The high degree of similarity literature in the front, the user feels natural language search improved precision. In summary, this method is a simple, practical and ideal natural language search methods.

Acknowledgment. Funds for this research was provided by Technology Innovation Project of Chinese Academy of Agricultural Science.

References

1. Tang, Y., Lai, M.: Study of the application of ontology in natural language information retrieval. New Technol. Libr. Inf. Serv. **120**(2), 33–36 (2005)
2. Geng, Q., Tang, Y.: Web information oriented natural language retrieval. Inf. Sci. **22**(7), 845–849 (2004)
3. Zhang, Q.: Impact of various elements on retrieval effectiveness in natural language retrieval. Inf. Stud. Theory Appl. **5**, 257–259 (1997)
4. Chen, G.: Information retrieval queries of natural language processing [DB/OL], 20 April 2006. www.cmgt.ntut.edu.tw/cimet/92active/ntut/EC/pages/pdf/200200023_MainFile.pdf
5. Jia, J., Song, E., Su, H.: Research on assessment of answer quality in social Q&A platform. J. Inf. Resour. Manage. **2**, 19–27 (2013)
6. Geng, Q., Lai, M.-S.: The pivotal issues and implementation of natural language information retrieval. Inf. Sci. **25**(5), 733–741 (2007)
7. Wang, D., Yang, X.: Study on elimination method of ambiguous words in Chinese automatic indexing. Libr. Inf. Serv. **58**(5), 93–97 (2014)
8. He, X., Wang, W.-W.: Research and application of Chinese word segmentation technical based on natural language information retrieval. Inf. Sci. **26**(5), 787–797 (2008)

Improving Agricultural Information and Knowledge Transfer in Cambodia - Adopting Chinese Experience in Using Mobile Internet Technologies

Yanan Hu[1], Yun Zhang[1], and Yanqing Duan[2(✉)]

[1] Foreign Economic Cooperation Centre, Ministry of Agriculture, Beijing, China
huyanan05@hotmail.com, zhangyun@agri.gov.cn
[2] Business and Information Systems Research Centre, University of Bedfordshire, Luton, UK
yanqing.duan@beds.ac.uk

Abstract. Agriculture is a knowledge intensive sector. Information and knowledge plays an essential role in helping farmers to improve productivity and sustainability through promoting and adopting the most effective and relevant innovations and technologies. With rapid development and advances in agricultural science and technology, traditional agriculture practices have been transformed to knowledge based and digital agriculture production. This paper reports a project that aims to improve the agricultural information and knowledge flow in Cambodia by adopting China's success in using Mobile Internet Technologies. The paper provides a brief review of China's current achievements in using ICTs to accelerate information and knowledge flow in Agriculture and Cambodia's current status in using ICT for agriculture information disseminations. An empirical investigation was carried out in Cambodia to gain more insights into the farmers information needs and their intention to adopt Mobile Internet based information dissemination services. Based on the empirical study, Chinese experience in using Mobile App was introduced and a mobile App called AgriApp was designed and tailored to meet the Cambodia's needs and conditions. This AgriApp was initially tested and valuable feedback was collected for improvement and better deployment in Cambodia.

1 Introduction

The agriculture sector in Cambodia is the most important source of income. Cambodia's economy is still highly dependent on agriculture, which contributes close to one-third of national GDP and employs more than half of the total labour force (Yu and Diao 2011). However, as a result of the rapidly changing socio-economic conditions since 1990, Cambodia agriculture sector faces many new challenges such as high population growth, embracing a market economy, nationwide food security and decreasing agricultural production conditions. These challenges are exacerbated by the limited access to agricultural knowledge and information in the rural areas. There are limited knowledge bases and appropriate approaches for information and knowledge transfer. As a

D. Li and Z. Li (Eds.): CCTA 2015, Part II, IFIP AICT 479, pp. 357–368, 2016.
DOI: 10.1007/978-3-319-48354-2_36

result, the latest agricultural technology and market information cannot reach farmers in a timely manner. At the same time, farmers' problems and inquires cannot be addressed directly and timely. Therefore, it is imperative to improve Cambodia's agricultural productivity and economic situation of farmers through better information dissemination service.

Over the last three decades, China has been very successful in transforming its Agriculture sector through the effective deployment of Information and Communication Technologies (ICTs). Information processing and dissemination have played a critical role in this transformation process (Zhang et al. 2015). It is believed that sharing Chinese experience and best practice would help Cambodian's agriculture sector in utilising the latest ICTs to facilitate knowledge and information dissemination. This paper reports a research project that seeks to extract useful and practical Chinese experience, especially in the area of the applications of the latest mobile Internet and GIS technologies, and adapt it to the Cambodian context. The research focuses on analysing information needs, selecting information sources, developing knowledge and information repository, and designing and demonstrating technical tools and platforms for effective information dissemination. There search project combines Chinese technologies for information sharing using smart phones with Cambodian data to promote effective and efficient agricultural information dissemination, so as to improve the knowledge and information dissemination flow in the rural area of Cambodia in the long run.

2   Review of Agricultural Information Dissemination in China

This research reviews the current ICT-based information service models in China to develop a better understanding on how different ICT-based information service models are designed and adopted in China; and to share the knowledge and experience in applying emerging ICTs in disseminating agriculture information to farmers to improve productivity and economic, social and environmental sustainability.

Information services for farmers at the national and regional level are a promising new field of research and application in the emerging field of e-agriculture (Gakuru *et al.* 2009). Information services play a critical role in modern agriculture and rural development and Information management and accessibility are at the core for information dissemination (Li 2011). To improve agriculture productivity, farmers have an ever increasing demand for information because accessing information and knowledge is essential for improving their productivity (Zhang et al. 2015). Agricultural information can be effectively disseminated by television, radio, newspapers, Internet, mobile phone, Short Message Service (SMS), Mobile Internet APPs (3G, 4G smart phones and handhelds) orthecombinations of different channels.

With the development of information technology, the agriculture information service model is constantly evolved and improved. Based on extensive review of reports and literature (e.g. Li 2009, 2011; Liu 2010; MOA Information Centre 2014), the agricultural information service models in China are classified into the following types:

1. Web Portal – a collection of relevant web sites to form an one stop center for users, e.g. MOA Web Portal, etc.

2. Voice-Based Service – information dissemination through telephone, i.e. call centers, e.g. Liaoning 12316 Golden Farming Hotline.
3. Text (SMS)-Based Service – information dissemination through text message of mobile phones. This service is normally jointly operated by Agriculture sector and Telecom service providers, e.g. Hunan Agri-Telecom Platform.
4. Self-Support Online Community – information services provided by a community to its members. This is a membership based system involving all stakeholders, members share experience and exchange information through interactive service platforms, e.g. farmers Mailbox in Zhejiang Province.
5. Interactive Video Conferencing Service – using online multimedia technology to facilitate information service, e.g. Shanghai Farmers "One Click and Go" service, or Intelligent Farmers service.
6. Mobile Internet Based Service – information dissemination through smart phone service, e.g. Agribusiness price information, E-news, etc.
7. Unified Multi-Channel Service Model – utilizing multiple methods to effectively disseminate information through telephones, computers, and mobile phones., e.g. "3 in 1" service in Fujian

To select the most appropriate models, the information infrastructure, the operating costs, farmers' capabilities, farmers' information consumption behaviour and, most importantly, the local context should be taken into consideration (Zhang 2012).

The key impact of agriculture information services in China can be highlighted in the following areas (Zhang et al. 2015):

- Improved the efficiency of Agriculture Services, e.g. "12316 hotline", "Unified 3 in 1 service", etc.
- Increased farmers' income
- Improved agriculture productivity
- Reduced the digital gap between rural areas and modern cities

3   Agricultural Information Dissemination in Cambodia

Agricultural sector plays a very important role in economic development of Cambodia. Hence, the Royal Government of Cambodia (RGC) has made a strong commitment to improve the country's agriculture sector after successfully leading the Cambodian economy out of the most difficult time of recent global financial crisis and economic downturn (CARDI 2014). However, Cambodia is currently in urgent needs for developing adequate agricultural information dissemination services. For example, although effective market information systems in Cambodia can reduce information asymmetries, increase competitiveness, and improve marketing system efficiencies (EC-FAO 2015), majority of farmers are still have no access the market information in a timely way. There is, therefore, strong demand for higher efficiency and effectiveness in agricultural information dissemination within Cambodian farming communities.

The Cambodian government has realised the importance of collecting, processing, managing and disseminating agricultural information and developed a series of initiatives to improve the situation. For example, Council of Agriculture and Rural Development (CARD) has established a single entry-point called the 'Cambodian Agricultural and Rural Development information Gateway' (CARDiG). The aim of CARDiG is to provide a portal for information sharing on agriculture and rural development and boost information and knowledge management among stakeholders through better access to web-based information.

Under the supervision of Ministry of Agriculture, Forestry and Fisheries (MAFF), ICT is widely used in government organisations at all levels to enhance the information flow to farmers and other stakeholders. MAFF has been implementing some projects working on improving productivity and market information of agricultural, such as providing all related information through radio broadcasting program, mobile SMS and websites. Besides that, MAFF has also formed many agricultural cooperatives at the rural community level and equip those communities with a set of desktop computer and other agricultural lessons CD.

Cambodian Agricultural Research and Development Institute (CARDI) has set up an agriculture-related electronic library, which includes various agricultural publications and research bulletins for online reading and download. Cambodia has also developed databases on agricultural subjects, such as soil, phytopathology and rice planting. Agricultural information is usually disseminated via radio, TV, mobile SMS and websites, providing effective guidance for agricultural production.

In summary, the main methods used in disseminating the agricultural information in Cambodia are shown in Fig. 1 (CARDI 2014) which include:

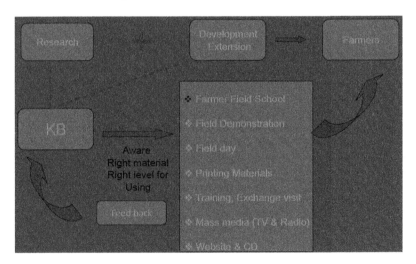

Fig. 1. Cambodia agricultural information dissemination methods (CARDI 2014)

Traditional face to face approaches – This approach disseminates information to farmers through field visit and demonstration, farmers field school and training, Agriculture forum, annual conference, monthly meetings, etc. This has been the most popular approach by farmers, but it is not the most effective and efficient method due to the financial and human resources involved.

Printed materials – Scientific and technical information was delivered to stakeholders through the periodical publication of research note, research bulletin, farmers' notes, project report, Journal of agriculture and other reading materials.

Mass media (TV and radio) – Information is broadcasted to a large number of farmers at fix times. The survey and focus group feedback indicate that this was seen as a not very popular methods because of the inflexibility of fix programmes.

Internet and CD – CARDI and other relevant bodies also publish regular information and demonstration of new development in their website for key stakeholder, such as: agricultural extension agents and key farmers to obtain the latest news and development in agriculture sector. CDs are also produced and distributed to farmers and agents for information dissemination purposes.

Though Network technology is relatively backward with low coverage, it is a fast growing area. Mobile phone penetration is significantly higher than landline. Currently, a total of 21 companies have the business license from Ministry of Posts and Telecommunications. There are eight mobile phone companies, two un-mobile phone companies, twenty eight network service companies, seven 3G service companies, 15 VOIP service companies and 16 VSAT service companies. It is predicated that Cambodian mobile phone users will up to 20 million, and mobile phone will become a communication tool widely used in daily life and work. In Cambodia, about 2.5 million people use Internet, accounting for 17.5 % of the total population, so Cambodia has the policy environment, hardware and software to build the agricultural information service platform.

However, the review of current situation suggests that although Cambodia government have realised the importance and needs for improving agricultural information dissemination and developed a series initiative in recent year, there is still a huge gap for significant improvement. The current information dissemination patterns in Cambodia are all one-way communication and lack of interactivity. They often cannot solve the specific problems of farmers in a timely and effective way. The agricultural information technology infrastructure in Cambodia is quite dated. Information dissemination mainly uses traditional media, such as radio, television, and paper based publications, so the dissemination efficiency and coverage are greatly restricted. The latest agricultural policy from agricultural sector is mainly released top down in the form of telephone. The way to promote the agricultural technology is regularly to organize trainings for farmers, so only a few farmers can get the information, and the effectiveness and pertinence is poor. For example, majority of farmers still have little information and difficult to obtain timely and accurate market and buyer information, leading to agricultural products unmarketable or the sale price is much lower than average. Aimed at these problems above, Cambodia hopes to push the latest agricultural technology to farmers through the advanced mobile Internet technology and the GMS agriculture network platform.

In China, with the development of information technology, the agriculture information service models and operating patterns are constantly developed and enhanced. It is believed that Cambodia can benefit greatly from the Chinese experience and best practice if introduced and adapted effectively.

4   Understanding the Information Needs of Cambodia Farmers – Empirical Investigations

To introduce China's success experience in agricultural information services and meet Cambodia farmers' needs and conditions, this research conducted a number of empirical investigations in Cambodia including surveys, focus groups and interviews with different stakeholders. The findings of the investigations help to gain in-depth insights into farmers' accessibility to information resources, their information needs and seeking behaviour, their intention of adopting ICT-based dissemination models, etc. The key findings also help to select the most suitable solutions for adapting Chinese information dissemination experiences using Mobile Internet technology.

4.1   Farmer Survey on Information Dissemination and Adoption Intention of Mobile Device

A questionnaire of farmer survey was developed and translated into Khmer. It consisted of two main parts which were current situation and adoption intention questions. The part one of the survey covered the following areas:

(1) Date and location of survey
(2) Personal information
(3) Farm information
(4) What are the main problems in the current information dissemination?
(5) Whom do you want to receive the information from?
(6) How important do you believe to receive the following types of information to improve your farm productivity
(7) Your preferred information delivery methods

Part two of the survey concerned the adoption intention that collect farmers perceptions and intentions in the following areas:

(1) Performance expectancy
(2) Effort expectancy
(3) Attitude toward using technology
(4) Social influence
(5) Facilitating conditions
(6) Behavioural intention to use the system if the system is available

The farmer survey was conducted in the two target pilot locations and a total of 95 questionnaires were collected. During the surveys, group discussions and interviews with farmers, farmer managers and key farmers were also conducted to explore further their information dissemination needs and preferred dissemination channels.

At the end of farmers' surveys, some of participants expressed their appreciation saying that they were very pleased to be invited and the information and demonstration were very useful for them. It was an eye-opening experience for them.

The survey results indicate that the major problems are no necessary equipment to access the electronic information, having difficulties to find the information they need, not knowing where to find information, and no relevant information available. Farmers would like to receive all relevant information that can help them improve productivity and sell their products. The adoption intention is also high although most of farmers still don't have smart phones.

4.2 Workshop on Upgrading Cambodian's Agricultural Information Dissemination System

A workshop on upgrading Cambodian's Agricultural Information Dissemination System was conducted in Phnom Penh to share experiences regarding to agricultural information dissemination system in Cambodia and the successful experiences of ICT technologies, especially mobile internet technology in China. The outcomes of the workshop were also contributing to guide the project implementation. A total of 30 participants from different institutions and organizations attended the workshop.

After introducing the Chinese current success in agricultural information dissemination using ICTs, the participants were divided into three groups to discuss: 1 - Problem and challenge of ICT in Cambodia, 2 - Adoption of ICT based information dissemination technology in Cambodia, and 3 - What is the most needed information for farmers?.

The outcome of group discussions are summarised below:

The problems and challenges of ICT based information dissemination in Cambodia:
1. Most farmers lack of knowledge in utilization of internet
2. Extension documents are not widely used
3. Not availability of internet in the rural areas
4. Not accessibility electricity in remote areas
5. Insufficient IT equipment for accessing the Internet, e.g. PC, smart phones or tablets
6. Financial problem: the rural people couldn't afford to purchase computers or phones
7. Internet service charges are too costly
8. Low literacy level
9. Market information is not regularly and widely disseminated.

The most useful ICT based on information dissemination technology in Cambodia is:
1. Mobile phone – e.g. SMS
2. Mass Media – TV, Radio and Video
3. Tele-center – Extension Hub
4. E-library
5. Internet-websites, social media, etc.

The most needed information for the farmers are:
1. National and international market information
2. Techniques of improvement product.
3. Quality and safety products
4. Weather forecasting
5. Potential product
6. Geographical potential
7. Agriculture service providers

Based on the empirical investigations, farmers have demonstrated strongest demand for the following five agricultural information services: Pest warning and control technology, agricultural price information, planting/farming technical guidance, new product and new agricultural policy.

The most important findings of the empirical investigations is the need to have information and knowledge brokers as an intermediary between information providers and farmers. Therefore, the main information users for piloting the mobile Internet based information dissemination system in the project are be the key farmers in villages and farm commune centres who have sufficient education qualifications and IT skills to understand and absorb the information and knowledge. They will then communicate the information and ideas to local farmers as an information and knowledge broker.

The content to be disseminated to farmers should be easy to understand. The field investigation analysis has suggested that farmers would like to see the information being presented using video, audio, image, simple and short text messages. The initial piloting materials can be based on the most popular paper based materials, such as farmers' notes and bulletins. Other potential piloting information would be based on the current available information resources and knowledge repository, such as the selection of the new rice varieties and use of fertilisers because farmers are very keen to adopt the new varieties and technologies to improve the rice productivity.

5   Adopting Chinese Experience and Development of AgriApp

To share the China's success story, this research project introduced China's experience and best practice in disseminating agricultural information to the vast number of farmers using ICTs in the project workshop. The presentation and demonstration in the workshop included Chinese agricultural information service models, latest development, hotline 12316, Hunan Nongxintong SMS, Nonghuibao App, Fujian 3-in-1 integrated service platform, etc. The Cambodian Agricultural representatives were very keen to learn from Chinese advanced agricultural information technology and development models. They strongly believed that conducting the agricultural information service pilot project under the GMS agriculture network platform can help Cambodia to develop the agricultural information services and improve the agricultural productivity.

The project team also conducted live demonstrations to farmers during the field trips to show how farmers in China can access important agricultural information and make online inquires via mobile Internet (see Fig. 2). Farmers were fascinated by the demonstration although most of them may not be able to benefit from this type of dissemination model directly in the near future due to their low lever education and computer literacy.

Fig. 2. Field demonstration

Based on the field investigations, China's experience, and the technical expertise of the project partners, the dissemination platform and prototype App called AgriApp were developed. The system consists of two parts: the agricultural information database management system in server-side and the agricultural information service application in mobile-side.

The mobile side is an Android program that currently have three main functions: 1. Agricultural information dissemination that is designed to show all the agricultural information in different categories; 2. Agricultural information search that allows the users to use the search function to query the information based on their interest and needs; 3. Agricultural information inquiry that allow the users to submit the problems that they encounter in their farming activities by using text description and images, so the system can provide relevant answer based on its knowledge base and users' location. User questions will be transferred to the server side along with the geographic location information to better help the administrators identify where these farmers are and provide more accurate and effective feedback.

The server side is developed to enable the management of messages from the mobile users (see sample screenshots in Fig. 3). The administrator also can edit and post his/her own message, such as some useful agricultural info and disaster alerts to the end users through this server application. More importantly, the server program can use a GIS map engine to display the locations of all message senders, thus to help the administrators to better address end users' inquires. This is very useful for future scientific analysis.

The agricultural information database management system provides data management functions to the administrator, such as data entry, data display, data modification, and data deletion (see the sample screenshots in Fig. 4). It has three main functions: 1. information content management, 2. user inquiry management, and 3. information

Fig. 3. Sample screenshots of AgriApp

display management. While the agricultural information service application provides farmers with three main functions, like agricultural information display, query/question consultation, etc. This type of information platform enables two-way communications of agricultural information between farmers and the information systems including expert team in the background, which forms an effective mechanism among information sharing and dissemination, and achieves sustainable development of the platform.

Fig. 4. Sample screenshots of AgriApp content management system

The prototype AgriApp was initially tested for its suitability in Cambodia. Project team visited Cambodia for on-site system deployment, training, and promotion. Over 30 Cambodian agricultural officials and farmers were invited to participate in the training. The project team introduced the AgriApp and provided the step-by-step demonstration on the use of the App to ensure that all participants were fully familiar with its application.

All stakeholders participated in the training workshop discussed the issues on how to effectively promote the system in Cambodia to fully benefit the farmers. The project team invited a number of local farmers to try the App on their phones. The farmers were impressed by the useful functions after going through them with the technical assistance. They showed their interest to continue using the AgriApp to better help their farming activities.

The senior officers in Cambodia government also showed their interest in promoting AgreiApp and pointed out that this new technology represents the future trend for the agricultural information dissemination and has great advantages over traditional dissemination approaches, e.g. call centre, SMS, with regard to the interactivity and costs. They stressed that this new dissemination model should be promoted widely in Cambodia.

6   Conclusion

The research reported in this paper aims to improve agricultural information dissemination in Cambodia by adapting China's experience in using Mobile Internet Technologies. To achieve this aim, the research first conducted a comprehensive review on agricultural information service models in China and their suitability for Cambodia. A number of empirical investigations on the current information provision and farmers' information needs and technology adoption intention are examined through surveys, interviews and focus groups. A prototype AgriApp is developed and tested in Cambodia and has received very positive feedback from key stakeholder. The key novelty of AgriAPP is the use of comprehensive knowledge base and the advanced GPS technologies to locate users and to provide tailor made advice suitable to the specific needs of the user. The improved App tools will be developed in the future for wider adoptions.

To improve the agricultural information and knowledge transfer in Cambodia, this research identifies two major issues to be addressed: one is to meet the farmers' urgent needs of information services; and the second is to improve the current ICT infrastructure and the outdated and ineffective information management and service methods that are not taking advantages of emerging ICTs. The mechanisms for developing effective ICT-based information dissemination models should be government-led, centralized services and market-oriented operations. Based on mobile Internet and GIS technology, the comprehensive agricultural information service platform, incorporating voice, network and mobile terminal, is an important direction for Cambodian agricultural information dissemination and mobile service applications. This interactive multi-channel service model can effectively transfer much needed information to farmers so as to improve their skills and productivity. In addition, this service model is also cost effective and needs low system maintenance and facilitate better communications.

It is believed that the experience and findings from this research provide a useful direction for researchers and practitioners in developing future ICT based information processing and dissemination systems and will help other developing countries to learn from China's experience and best practice in their endeavour of applying emerging ICTs in agriculture information dissemination and sharing.

Future research will be carried out to further improve the data sources and functionalities of the AgriApp. More field tests will also be conducted with various users, especially agriculture extentionists, to understand its potential effectiveness and impact.

Acknowledgement. The work reported in this paper is part of a research project on prompting agricultural knowledge and information flows. The authors would like to acknowledge the financial support provided by the Agricultural Technology Transfer (AgriTT) programme which is funded by UK Department for International Development (DFID). Project partners, Supermap in China, was responsible for design and development of AgriApp for the project, and, CARDI in Cambodia, provided valuable background information and assisted the Cambodian field investigation and training workshops. The authors are very grateful for their important contributions.

References

CARDI: Report on study of current agricultural information dissemination status and development environment in Cambodia. Cambodian Agricultural Research and Development Institute. Report of AgriTT project, Application of Mobile Internet Technology for Agricultural Information Dissemination in Cambodia (2014)

EC-FAO food security programme: Exchanging Agricultural Market Information through SMS in Cambodia, FAO Technical Brief, GCP/RAS/247/EC (2015). http://www.fao.org/fsnforum/sites/default/files/resources/Cambodia%20SMS%20Technical%20Brief.pdf. Accessed 15 July 2015

Gakuru, M., Winters, K., Stepman, F.: Innovative farmer advisory services using ICT. In: Cunningham, P., Cunningham, M. (eds.) IIMC International Information Management Corporation, IST-Africa 2009 Conference Proceedings, Uganda, 06–08 May 2009. ISBN: 978-1-905824-11-3

Li, D.: China Rural Informatization Development 2009 Report (in Chinese). Publishing House of Electronics Industry, Beijing (2009)

Li, D.: China Rural Informatization Development 2010 Report (in Chinese). Beijing Institute of Technology Press, Beijing (2011)

Liu, X.: Reflections on the development of agricultural informatization. Silicon Valley, 1671-7597, 1210005-01 (2010). In Chinese

MOA Information Centre: Summary of contemporary Chinese approaches of ICT-based agricultural information dissemination. Research Report published by Information Centre, Minister Of Agriculture (MOA), P.R. China (2014)

Yu, B., Diao, X.: Cambodia's agricultural strategy: Future development options for the rice sector. Cambodia Development Resource Institute (CDRI), Council for Agricultural and Rural Development (CARD), and IFPRI Special Report. Phnom Penh, Cambodia: CDRI, 1284(1283), 1282 (2011)

Zhang, J.: Formation conditions and development characteristics of Chinese agricultural information service model. Chin. Inf. Ind. (04) (2012)

Zhang, Y., Hu, Y., Duan, Y.: Supporting chinese farmers with ict-based information services: an analysis of service models. In: International Conference on E-Business (ICE-B 2015), Colmar, France, 20–23 July 2015

Principal Component Analysis Method-Based Research on Agricultural Science and Technology Website Evaluation

Jian Ma[1,2(✉)]

[1] Agricultural Information Institute of Chinese Academy
of Agricultural Sciences, Beijing 100081, China
majian@caas.cn
[2] Key Laboratory of Agricultural Information Service Technology (2006–2010),
Ministry of Agriculture, Beijing 100081, People's Republic of China

Abstract. Agricultural science and technology website is a very important supporter of driving agricultural information and servicing agriculture. An evaluation method is proposed on agricultural science and technology website based on objective data and artificial ratings, using principal component analysis method. Finally the author used the model to evaluate 18 agricultural science and technology websites, and proposed some suggestions on development of agricultural science and technology websites based on the evaluation result which would act as reference to agricultural science and technology website construction.

Keywords: Agricultural science and technology website · Principal component analysis method · Website evaluation

Agricultural science and technology website is an important window of agricultural science and technology departments and relevant units on the Internet to show their own image, also a concentrated expression of service ability and service level and service features, At the same time also as a important carrier of serviceing "three rural" and promoting agricultural informationization. From the current point of view, although the agricultural science and technology website is rich in resources, but the quality is uneven, how to better improve the website construction is a problem urgent need to solve. A Scientific evaluation method, can make the website administrator to complete understanding of the operation of the site, consummate the existing problems, improve the quality of the website. On the basis of the research of the peer, the principal component analysis method is used, exploring a new method of comprehensive evaluating websites.

1 Overview of Website Evaluation Method

At present, there are no unified standards and methods for the evaluation of the website. The author consulted a large number of literature data, summarized the following two kinds methods of the evaluation of the website:

© IFIP International Federation for Information Processing 2016
Published by Springer International Publishing AG 2016. All Rights Reserved
D. Li and Z. Li (Eds.): CCTA 2015, Part II, IFIP AICT 479, pp. 369–381, 2016.
DOI: 10.1007/978-3-319-48354-2_37

(1) web link analysis

At present, Link analysis and web impact factor measure related to it are widely used to evaluate the websites. Link analysis through the number of sites are linked to reflect the quality of the site. The evaluation is on the basis that a web site is linked another website is approval and use of this website, and the content of two website is related; the more number of external links of a web site, Explain Its influence is greater. The network impact factor measure is based on the link analysis, reflect the influence of the web site by the size of web impact factor.

Although this method is applied to a wide range of applications, but it also has shortcomings: it only start from the link analysis of the website, not comprehensive evaluation of a website, the authority is to be verified. At the same time, the data used to analysis mostly get from Alta Vista or Google and other search engines, which makes the results depends too much on the search engine, but due to the drawbacks of itself, search engine may not included all external links and internal links of a website.

(2) analytic hierarchy process

Analytic hierarchy process divide the decision problem according to total target, sub target, evaluation criteria until the specific input sequential scheme is decomposed into different levels of the hierarchy and layer by layer analysis, ultimately, get the importance weights that the lowest level factor for the highest level factor.

Analytic hierarchy process can be used to evaluate the site, but the relative importance of each factor in the same level must be evaluated in the construction of evaluation matrix. This will have error inevitably due to the subjective behavior.

Based on the above research, In this paper, we give the evaluation index and calculation method of agricultural science and technology website, and use the principal component analysis method to analyze the evaluation index. Finally, the model is used to obtain the comprehensive ranking of 18 agricultural science and technology websites, and the relevant suggestions are given according to the evaluation results.

2 Principal Component Analysis Method for the Comprehensive Evaluation of the Principles and Methods

Principal component analysis is also called the principal weight analysis, which is an important method to study how to transform the multi index problem into a less comprehensive index. Because there is a certain degree of correlation between multiple variables, people naturally want to extract information from these indicators as quickly as possible through linear combination. Principal component analysis can change the problem of high dimensional space into a low dimensional space to deal with, make the problem become more simple, intuitive, and the comprehensive index of these less interaction and provide most of the information of the original index.

In practical application, the specific steps of principal component analysis are:

(1) standardization of raw data
(2) set up the correlation coefficient matrix of variable
(3) obtain the eigenvalues and eigenvectors of the correlation matrix.
(4) the number of principal components is determined by the cumulative variance contribution rate, and the principal components are extracted.
(5) weighted synthesis of principal components, get comprehensive evaluation

.

3   Evaluation Index and Method of Agricultural Science and Technology Website

The evaluation of agricultural science and technology website mainly investigation from the content of the website, website design and user operation three major aspects. Each of the major aspects of the index selection consider of agricultural science and technology website features. The evaluation model is shown in Fig. 1:

The evaluation method of the index is too dependent on the subjective score, so the evaluation of the importance of the indicators in the evaluation of the importance of a

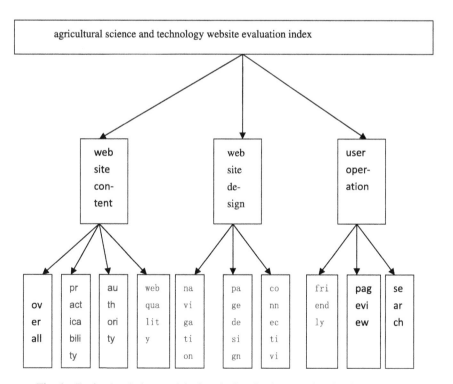

Fig. 1. Evaluation index model of agricultural science and technology website

strong subjectivity. At the same time, simply rely on the objective data can not be a comprehensive reflection of agricultural science and technology website. So we make full use of the advantages of existing systems and professional personnel, the index can be described with the objective data as an objective data, can only rely on artificial score index scoring more than re synthesis method is implemented. The index data is more scientific. Specific practices are as follows:

(1) website content

Web content is good or bad is a valuable key of a agricultural science and technology website. For the content of the site we were from the following four aspects to Investigate:

Comprehensiveness. Comprehensiveness is the breadth of agricultural science and technology website contains content. Web site can be used to quantitatively expressed by the number of web pages. The number of sites webpages can be estimated approximately using the search engine included page numbers.

Practical application. Whether the choice of the content of the site is in line with the "three rural" needs, whether it is suitable for the specific user base, which is suitable for the "three rural" information. This indicator cannot be described by objective data, and it is determined by expert scoring method.

Authority. Authority is the impact of agricultural science and technology sites and the extent of the popularity of the site. This index can be described by using the website of agricultural science and technology anti link number, which the number of links from other sites to this site. The more the number of a website's anti link, the greater the influence, the greater the authority.

The quality of Webpage. Web quality evaluation can be from the user point of view. Bounce Bounce RateBounce RateBounce RaterBounce Rateate is an important index to measure the quality of web pages. The Bounce rate is the percentage of the total number of visits from a particular portal to visit a site, which only access to one page on the number of visits to the total number of visits. When the site's bounce rate is high, the quality of the page is very poor, do not attract users.

(2) website design

Good website design should have reasonable structure, the page is simple and beautiful, easy to use. For the website design we investigate from the following three aspects:

Navigation function. Navigation function for the user to use the entire site is essential. Design good navigation function can make the user more convenient, more quick browse information. This indicator is determined by expert scoring method.

Page design. Mainly to examine the page of friendship, include that, the structure is clear, the layout is reasonable, Logo is beautiful, etc. This indicator is determined by expert scoring method.

Connectivity. Effective connectivity mainly investigate the link to the page, if the webpage have broken link or dead link. This indicator can be used to described with broken link rate, broken link rate is the number of all the broken links of the website divided by the number of all links of the website.

(3) user operation

User operation is to measure the website good or bad from the users. For the user operation we investigate from the following three aspects:

① Friendly degree. Here is not to examine the degree of friendship from web design, but examine the entire web server response time, and page download time. Obviously, the response time of the server and the web page download times shorter, the higher the site the user-friendliness. Therefore, there are two indicators measure friendliness: server response time, page download time.

② User visits. User visits Is the number of user visits the site. We can use the average daily IP traffic number to describe user visits, visitors every page view can be described by Daily average page views.

③ Search function. Search function for users to find resources is very important, well-designed site should have a comprehensive in-station search function, which should have advanced search and fuzzy queries. The index score is determined by experts. Detailed indicators and evaluation of the site's content as shown in Table 1:

Table 1. Evaluation System of agricultural science and technology website

Level indicators	Secondary indicators	Evaluation content
Website content	Comprehensiveness	Include the number of pages
	Practical application	Experts rate, standard: meet the needs of agricultural information
	Authority	Number of anti-link
	The quality of Webpage	Bounce rate
Website design	Navigation function	Experts rate, standard: ease of navigation
	Page design	Experts rate, standard:page friendly degree
	Connectivity	Broken link rate
User operation	Friendly degree	Speed (download time, response time)
	User visits	average daily IP traffic, average daily Page view
	search function	Experts rate, standard: the station search function

4 Examples of Application and Analysis

4.1 Evaluation Object

The evaluation object of this paper is 18 agricultural science and technology information website, which has 7 national agricultural science and technology website, 11 provincial agricultural science and technology website, as shown in Table 2:

Table 2. 18 evaluation objects

Number	SiteName	Site url
1	China Agricultural Science and Technology Information Network	http://www.cast.net.cn/
2	Hunan Agricultural Science and Technology Information Port	http://www.hnagri.com/
3	Guizhou Agricultural Science and Technology Information Network	http://www.gzaas.com.cn/
4	Chinese Agricultural Extension Network	http://www.farmers.org.cn/
5	Shandong Agricultural Science and Technology Information Network	http://www.saas.ac.cn/ saas.asp
6	Shanxi Agricultural Science and Technology Information Network	http://www.sxagri.ac.cn/
7	National Agricultural Science Data Sharing Center	http://www.agridata.cn/
8	Golden Agriculture Network	http://www.agri.com.cn/
9	Hangzhou Agricultural Science and Technology Information Network	http://www.hznky.com/
10	Liaoning Luyuan Agricultural Science and Technology Information Network	http://www.last.gov.cn/
11	Chinese Academy of Agricultural Sciences Network	http://www.caas.cn/
12	Anhui Academy of Agricultural Sciences Network	http://www.ahas.org.cn/
13	Zhongshan, Guangdong Agricultural Science and Technology Park	http://www.zsnk.com/
14	Hainan Agricultural Science and Technology 110 Network	http://www.hnnj110.com/
15	Nong Bo Technology	http://www.aweb.com.cn/
16	Beijing Agricultural Information Network	http://www.agri.ac.cn/
17	Tianjin Agricultural Information Network	http://www.tjagri.gov.cn/
18	Nine Hundred Million Network	http://www.new9e.com/

4.2 Data Sources

Data sources of this paper are the following four:

(1) network technology resource monitoring, analysis, evaluation system.

The system long-term monitoring 18 agricultural science and technology website in Table 2, the system provided data included the number of pages, the number of anti-link, download time and the response time four indicators.

(2) artificial: This paper invited 5 of the author's colleagues, 5 ordinary users to form a group of experts. 5 of the author's colleagues engage in agricultural science and Technology Information Research for many years, while the other 5 are ordinary users that often used agricultural science and technology website.

Respectively, usability, navigation, page layout and search functions independently scoring four indicators, score interval [0,1], 1 on behalf of full marks, 0 on behalf of 0. For each item of data to remove one of the highest and lowest scores, then averaging the remaining data is the index score.

(3) Alexa: Alexa is a site that specialized publishing website ranking. Alexa every day collect more than 1000 GB of the information, not only give billions of web site links, but also rank for each of the sites. It can be said, Alexa is currently has the largest number of URL, released ranking information the most detailed site.

Alexa provides the data for: average daily IP access, the average daily amount of page view browsing and bounce rate.

(4) Chinaz (Chinese webmaster station): is a specializes in providing information for Chinese site, technology, resources and services of the website, website existing millions of users.

Chinaz provides data for the number of broken links of the site and the total number of links to the site.

4.3 Data Extraction

Extracting data from the four data sources mentioned in Sect. 4.2, after finishing get the relevant index data of 18 agricultural science and technology site as shown in Table 3:

Table 3. Index data of 18 agricultural science and technology website

Number	Include page number	Practicability	Anti link	Bounce rate	Navigation function	Page design	Broken link rate	Download time (s)	Response time (s)	Daily IP access	Dailypage view browse	Search function
1	937	0.85	2540	33.30 %	0.9	0.92	0.146	1.18	0.02	360	2160	0.83
2	2140	0.73	1470	81 %	0.7	0.51	0.026	0.619	0.44	780	1170	0.82
3	2520	0.9	1420	78 %	0.88	0.9	0.102	0.494	0.27	300	300	0.65
4	20700	0.73	2480	60 %	0.9	0.85	0.037	0.445	0.22	300	300	0.87
5	4990	0.8	2680	37.50 %	0.85	0.8	0.4	0.864	0.01	780	4680	0.76
6	2470	0.85	1830	77.80 %	0.78	0.85	0.048	4.909	0.23	420	420	0.25
7	1060	0.87	1190	12.50 %	0.79	0.74	0.021	0.623	0.06	1500	180000	0.98
8	16400	0.92	723000	50.90 %	0.81	0.92	0.09	1.034	0.04	9600	37440	0.84
9	1040	0.78	349	50 %	0.85	0.74	0.225	0.784	2.77	180	230	0.83
10	4360	0.72	691	50 %	0.86	0.8	0.032	6.599	0.07	250	270	0.2
11	6800	0.88	20100	34.10 %	0.97	0.9	0.076	0.541	0.01	6000	23400	0.8
12	477	0.72	353	67 %	0.85	0.85	0.017	0.848	2.87	240	280	0.7
13	116	0.68	47	84 %	0.6	0.64	0.07	1.338	3.89	180	200	0.82
14	728	0.54	503	60 %	0.78	0.82	0.164	3.611	0.1	120	120	0.75
15	7580	0.71	79500	35.30 %	0.85	0.9	0.1	2.127	0.04	64200	46680	0.8
16	7080	0.68	18500	57.90 %	0.88	0.83	0.04	2.539	0.05	660	990	0.75
17	873	0.57	2360	54 %	0.95	0.8	0.069	12.244	0.08	160	220	0.85
18	2760	0.95	1560	18 %	0.8	0.75	0.09	0.829	0.17	180	210	0.8

Note: the number column is one corresponding to the number in Table 2.

4.4   Comprehensive Ranking of Agricultural Science and Technology Website Based on Principal Component Analysis

In this paper, using SPSS13.0 execute the principal component analysis, and the specific steps are:

(1) standardization of raw data

In principle component analysis method, the standard method is Normal standardization, and for the practical application, according to the difference of the index, divided the bigger the better and the smaller the better. So we take the following standard method:

With m evaluation object, n evaluation index. All data constitute a m*n order matrix X = (X1, X2,... Xn). Where Xi is a m*1 dimensional vector. Let max (Xi) is the maximum value of Xi, min (Xi) is the minimum value of Xi. Max (Xi) = (max (Xi), max (Xi),... Max (Xi)) T is a m*1 dimension vector, Min (Xi) = (min (Xi), min (Xi), ... Min (Xi)) T is the m*1 dimension vector, For the bigger the better type index, the formula is:

$$X_i^* = \frac{X_i - Min(X_i)}{max(X_i) - max(X_i)} \tag{4.1}$$

For the smaller the better type index, the formula is:

$$X_i^* = \frac{Max(X_i) - X_i}{max(X_i) - max(X_i)} \tag{4.2}$$

In this paper, n = 12, m = 18. The bigger the better type included web number, practical, the number of anti link, navigation, page design, average daily IP access, the

Table 4. Characteristic root and variance contribution rate table

Total Variance Explained

Component	Initial Eigenvalues			Extraction Sums of Squared Loadings		
	Total	% of Variance	Cumulative %	Total	% of Variance	Cumulative %
1	3.008	25.069	25.069	3.008	25.069	25.069
2	2.076	17.300	42.370	2.076	17.300	42.370
3	1.525	12.710	55.080	1.525	12.710	55.080
4	1.287	10.726	65.806	1.287	10.726	65.806
5	1.125	9.371	75.177	1.125	9.371	75.177
6	.836	6.966	82.143			
7	.725	6.042	88.185			
8	.586	4.882	93.067			
9	.306	2.549	95.616			
10	.272	2.268	97.883			
11	.159	1.326	99.210			
12	.095	.790	100.000			

Extraction Method: Principal Component Analysis.

average daily amount of page view browsing and search functions eight indexs. The smaller the better type included bounce rate, broken link rate, download time and response time four.

(2) Use SPSS 13.0 for factor analysis

Get the standardized data input SPSS data editing window, the 12 indicators were named X1 ~ X12. Select Analyze->Data Reduction->Factor menu item in the SPSS windows, Tune out factor analysis main interface and move the variable X1 ~ X12 into the variables box, click the OK button, execute factor analysis process. Get the characteristic roots and variance contribution rate table shown in Fig. 4 and factor loading matrix as shown in Table 5:

Table 5. Factor load matrix

Component Matrix[a]

	Component				
	1	2	3	4	5
X1	.582	-.074	.604	.083	-.175
X2	.526	.383	.146	-.363	.564
X3	.484	.140	.637	.130	-.079
X4	.690	.202	-.581	.001	.047
X5	.587	-.585	-.250	-.185	-.044
X6	.684	-.471	.128	-.174	-.037
X7	-.119	-.054	.262	.704	.451
X8	.223	.735	.237	-.376	.019
X9	.634	-.380	-.173	.129	.204
X10	.400	.011	.024	.411	-.471
X11	.424	.511	-.349	.489	.227
X12	.252	.592	-.197	.078	-.495

Extraction Method: Principal Component Analysis.

a. 5 components extracted.

In Table 4 in total column is each factor corresponding characteristic root, In this case extract five common factors; % of Variance column is the variance contribution of each factor; Cumulative % column is the cumulative variance contribution rate of each factor, As can be seen, the first five factors can explain 75.177 % of the variance.

(3) using factor analysis result execute principal component analysis

Get Table 5's data input SPSS Data Editor window, Column named a1 ~ a5. Then click on the menu item Transform->Compute, Recall Compute variable dialog, Enter $z1 = a1 / SQRT (3.008)$ in the dialog box, Click the OK button, you can get a

Table 6. Feature vector matrix

	z1	z2	z3	z4	z5
x1	0.336	−0.051	0.489	0.073	−0.165
x2	0.303	0.266	0.118	−0.32	0.532
x3	0.279	0.097	0.516	0.115	−0.074
x4	0.398	0.14	−0.47	0.001	0.044
x5	0.338	−0.406	−0.202	−0.163	−0.041
x6	0.394	−0.327	0.104	−0.153	−0.035
x7	−0.069	−0.037	0.212	0.621	0.425
x8	0.129	0.51	0.192	−0.331	0.018
x9	0.366	−0.264	−0.14	0.114	0.192
x10	0.231	0.008	0.019	0.362	−0.444
x11	0.244	0.355	−0.283	0.431	0.214
x12	0.145	0.411	−0.16	0.069	−0.467

first feature vector. Similarly, it can be calculated for all the feature vectors, which results are shown in Table 6:

By multiplying the matrix of the original data with the eigenvector matrix, can obtained the 5 principal components Y1–Y5. Then the 5 main components weighted comprehensive, you can get the comprehensive score of the agricultural science and technology website, Specific data shown in Table 7.

The formula for calculating the comprehensive score is

$$Y = \frac{\lambda_1}{\sum\limits_{i=1}^{5} \lambda_i} Y_1 + \frac{\lambda_1}{\sum\limits_{i=1}^{5} \lambda_i} Y_2 + \cdots + \frac{\lambda_1}{\sum\limits_{i=1}^{5} \lambda_i} Y_5 \qquad (4.3)$$

Y is a composite score, Y_i is the main component i, λ_i is the characteristic root of the main components i.

4.5   Analysis and Suggestion

We compare the ranking of this paper with the Alexa ranking, and the results are shown in Table 8:

From Table 8 we can get the following results:

(1) We can see, in this paper the first top seven sites is consistent with Alexa ranking, but the first and the second ranking is not consistent. This is because in Alexa ranking, its information flow rank is predominant factor, the impact of other parameters is very small. At the same time also can be seen, when the site information flow reach a certain degree, the greater the information flow, the better the site of the index to maintain, the higher the overall ranking.

Table 7. Agricultural Science and Technology site principal component scores and order

	Y1	Y2	Y3	Y4	Y5	Score	Order
China Agricultural Science and Technology Information Network	1.958	0.16	−0.088	−0.093	0.539	0.729	4
Hunan Agricultural Science and Technology Information Port	1.441	−0.02	0.08	0.154	0.353	0.556	9
Guizhou Agricultural Science and Technology Information Network	1.221	−0.487	−0.374	0.424	0.154	0.312	18
Chinese Agricultural Extension Network	1.659	0.504	−0.2	−0.141	0.694	0.702	6
Shandong Agricultural Science and Technology Information Network	1.47	0.091	0.201	−0.239	0.644	0.592	8
Shanxi Agricultural Science and Technology Information Network	1.641	0.274	−0.252	−0.543	0.149	0.509	11
National Agricultural Science Data Sharing Center	1.746	0.097	0.432	0.092	0.234	0.721	5
Golden Agriculture Network	2.229	0.415	0.811	0.131	0.397	1.045	1
Hangzhou Agricultural Science and Technology Information Network	1.117	−0.016	−0.14	0.108	0.096	0.373	17
Liaoning Luyuan Agricultural Science and Technology Information Network	0.786	0.575	0.172	0.257	0.449	0.517	10
Chinese Academy of Agricultural Sciences Network	1.977	0.174	−0.06	0.493	−0.087	0.749	3
Anhui Academy of Agricultural Sciences Network	1.264	−0.394	0.018	0.204	0.772	0.459	14
Zhongshan, Guangdong Agricultural Science and Technology Park	0.403	0.753	0.306	0.126	0.181	0.401	16
Hainan Agricultural Science and Technology 110 Network	1.148	0.496	−0.096	−0.327	0.167	0.455	15
Nong Bo Technology	1.866	0.931	−0.579	0.486	0.789	0.901	2
Beijing Agricultural Information Network	1.754	0.168	−0.233	−0.245	0.472	0.608	7
Tianjin Agricultural Information Network	1.166	−0.2	0.265	0.039	0.889	0.504	12
Nine Hundred Million Network	1.034	0.253	0.156	0.0006	0.377	0.477	13

(2) The websites ranked in 8–18, this article ranking and Alexa have a larger discrepancy, so we can see, Alexa for the websites comprehensive ranking is not high not has a very high reference value.

(3) Overall, the ranking of the national agricultural science and technology websites is higher than the local agricultural science and technology websites. But there are exceptions, such as nine hundred million network, so the nine hundred million network need to be further improved. Single from the national agricultural science

Table 8. Comparison of the ranking of this paper with Alexa

Website name	Article rank	Alexa rank
Golden Agriculture Network	1	2
Nong Bo Technology	2	1
Chinese Academy of Agricultural Sciences Network	3	3
China Agricultural Science and Technology Information Network	4	4
National Agricultural Science Data Sharing Center	5	5
Chinese Agricultural Extension Network	6	6
Beijing Agricultural Information Network	7	7
Shandong Agricultural Science and Technology Information Network	8	9
Hunan Agricultural Science and Technology Information Port	9	8
Liaoning Luyuan Agricultural Science and Technology Information Network	10	11
Shanxi Agricultural Science and Technology Information Network	11	10
Tianjin Agricultural Information Network	12	14
Nine Hundred Million Network	13	15
Anhui Academy of Agricultural Sciences Network	14	13
Hainan Agricultural Science and Technology 110 Network	15	12
Zhongshan, Guangdong Agricultural Science and Technology Park	16	16
Hangzhou Agricultural Science and Technology Information Network	17	18
Guizhou Agricultural Science and Technology Information Network	18	17

and technology websites: Golden agriculture network, nong bo network and Chinese Academy of Agricultural Sciences network, the comprehensive ranking is higher, the utilization of its website is also higher. From the local agricultural science and technology websites: Beijing, Shandong, Hunan stay ahead, It explained that the Ranking of Agricultural Science and Technology websites are relevant with the local information level and agriculture level. Guangdong, Zhejiang although the level of economic development is higher, the level of agricultural information needs to be further improved.

In this paper, through construction agricultural science and technology websites evaluation model, Using principal component analysis to analysis, For 18 agricultural science and technology site ranking, And based on ranking results comparison with Alexa ranking, giving agricultural science and technology websites analysis and evaluation. Through this type of evaluation, can play an important promote role to agricultural science and technology websites healthy development and agricultural Information Development.

References

1. Sha, Y.Z., Ou, Y.X.: Evaluation of China Provincial websites influence_website link analysis and web impact factor measurement. Information and Documentation Work, Beijing (2004)
2. Li, C., Yang, Y., Ge, Y.: Compare of principal component analysis and analytic hierarchy process in quantitative evaluation of comprehensive index. J. Nanjing Univ. Finances Econ., Nanjing (2005)
3. Du, J., Li, D., Li, H.: Evaluation System of Agricultural Information Websites in China. Jiangxi Agricultural Sciences, Jiangxi (2010)
4. Liu, Y., Li, H.: Evaluation Method and Evaluation Index System of Agricultural Website. Agricultural Network Information, Beijing (2010)

The Countermeasures of Carrying on Web of the Research Institutions in the Era of Big Data — Consider the Web of Chinese Academy of Agricultural Sciences

Liping Zhang[1,2(✉)]

[1] Institute of Agricultural Information, Chinese Academy
of Agricultural Sciences, Beijing 100081, China
zhangliping@caas.cn
[2] China Key Laboratory of Agricultural Information Service Technology
(2006–2010), Ministry of Agriculture, Beijing, People's Republic of China

Abstract. In recent years, large data caused great concern in industry, academia and government. As an important department for scientific research and innovation, the academy portal show the level of science research innovation ability and it is a main platform of transformation of scientific research achievement. In this paper, the construction portal of Chinese Academy of Agricultural Sciences Situation as the background, points out the shortcomings of the current construction site, and make a few suggestions for website development in the big data environment.

Keywords: Big data · Website construction · Data resources

1 Background

Big data is hot word in today's society of network. People from ordinary to the whole country enjoy the convenience of big data in varying degrees. In medical services, retail, finance, manufacturing, logistics, telecommunications and other industries, the research and application of big data has started, and has created a huge social value. Government departments also attach great importance to large data technology. In March 2012, Obama announced the U.S. Government to invest 2 billion dollar for the big data research and development program. Our government has also realized the huge potential of big data and is to make policy to promote the development and application of big data.

With application of information acquisition, mass data storage, high-speed data transmission and intelligent data analysis, it has brought a far-reaching impact on the contemporary agricultural scientific research. Chinese Academy of Agricultural Sciences (hereinafter referred to as CAAS), as the most authoritative scientific research institution in the country, have already felt the great influence on big data. In 2014,

© IFIP International Federation for Information Processing 2016
Published by Springer International Publishing AG 2016. All Rights Reserved
D. Li and Z. Li (Eds.): CCTA 2015, Part II, IFIP AICT 479, pp. 382–391, 2016.
DOI: 10.1007/978-3-319-48354-2_38

CAAS draw up "China Academy of Agricultural Sciences information development planning (2015–2019)" (hereinafter referred to as the planning). The planning put forward that the main task of is to build e-Science platform, promote[1] information of scientific research conditions, management and scientific research output and enhance capacity for the scientific data and resources acquisition, protection and utilization. The acquisition and utilization of scientific research data impartment. At the same time, the planning points out the problems in the construction of data.

2   Current Situation and Problems

CAAS Web Portal included 10 functional departments websites and 41 research institutes websites, have two major functions of department office and service research, and it is an importance display window for showing scientific research strength and service agriculture. Since it was on the line in 1997, website has played an importment role for science research and management. Now, CAAS portal has included 51 institutes secondary network. But according to statistics, total websites in CAAS has arrived 236. The specific site condition survey results are shown in Table 1.

Table 1. Number of existing sites and views

Unit	The number of sites	IP average daily traffic	Average daily page views
Agricultural Information Institute of CAAS	43	1310	7647
China National Rice Research Institute	19	1009	4211
Food Science and Technology Institute of CAAS	13	1010	3123
Farmland Irrigation Research Institute	13	977	2785
Management of CAAS	12	18901	98142
Biotechnology Research Institute of CAAS	12	812	4205
Institute of Environment and Sustainable Development in Agriculture, CAAS	11	657	1223
Biogas Institute of Ministry of Agriculture	11	612	1129
……	……	……	……

Construction of Web database has begun in the 1980's and it has arrived 185 as of 2014. Here are several examples in early phase. In the 1980s, the database of feed nutrition had been jointly build by Institute of Animal Sciences and the Computing Center head by Academician Ziyi Zhang. In the 1990s, seed resources database built by professor Xian Zhen Zhang has reached 42 million types by 2014. The agricultural science and technology data platform information presided over of a few years ago by

---

[1] Foundation item: Chinese Academy of Agricultural Sciences website content resources, research organizations and service models.

professor Xianxue Meng had been put into application in which involves 12 individual sciences, 72 subject database groups, 672 databases and 1987 TB data quantity browsed by 50–60 million people every day. Seed resources database in China has accumulated hundreds of millions of seed resource data. With development of genetics and breeding subject the gene sequence of important crops and animals measured precisely have formed a huge gene database. There are many All kinds of scale national construction scientific data and resource data center.

But, it has a weak ability in data collection and integration. There has only 14 data websites in the CAAS portal. According to statistics in 2014, the other 185 data resources websites are independent with the CAAS portal. All of these data resources cann't share and analysis. Web site data storage format and location are shown in Table 2.

Table 2. Web site data storage format and location

Number of sites	Data storage form		Storage location		
	Database	Spreadsheet	Network center in CAAS	Network in institutes	Network in the third party
236	227	9	21	22	193

2.1 Network Is Weak

2.1.1 Lack of Overall Website Construction Planning and Financial Support

Since the CAAS portal has been setup, website establishment, development and daily maintenance run by each institutes. Because there is no unified planning and special funds to support, website construction and service level is uneven and database construction and maintenance status are poor. Nothing of unified management and data sharing platform it is difficult to data sharing and analysis in the future.

According to statistics, nearly 2/5 of the institutes do not have the operation and maintenance expenses. Compared with the Chinese academy of sciences, CAAS funds in development of information is far smaller. CAS information construction began in the "fifth" period. By the end of "Thirteen Five", it has accumulated invested several million dollars.

2.1.2 Network Infrastructure Environment Is Weak

According to statistics, Chinese Academy of Agricultural Sciences 41 research institutes, 10 departments in 2014 generated approximately 10 TB of data per day, and 10 % annual growth rate, while the storage capacity of the existing network center room of the CAAS can not meet demand. All of the institutes in CAAS distributed 24 provinces in the country. Since there is no interconnection of data transmission lines, scientific research data between the institute, the base, the test station and field station can not be shared and transmitted.

2.2   Lack of Unified Planning of Data Resource Construction

Data resource construction is a long-term work and need a reasonable system to ensure data accuracy and availability. Lacking of unified standards data need to repeat collection, repeat input and maintained by multiple departments. At it caused information update asynchronously and non unified. The specific problems in following several aspects:

First, there is not a unified data format and compatible standards. According to the survey, all of 42 institutes have built their own financial and personnel platform, but because CAAS did not establish a unified system compatibility standards, data in departments with another department. As the amount of data is growing rapidly, it will increase the workload of management and work harder.

Secondly, CAAS lack research management platform, such as platform for research software and scientific instruments sharing. Instrument and software is an important part of the use of scientific research funds. Because CAAS have not scientific research software and instrument unified purchase plan and sharing mechanism it caused software and instrument purchased repeatedly and utilized lowly.

Third, there is a lack of scientific research data storage mechanism. Most of the scientific research data edit and save by each group, lacking of system and professional. According to statistics in 2014, 115 research institutes or units has built its own web site that one institute have about 3.3 of web sites on the average. But it is lack of unified management and effective utilization, scientific research data is low usage rate.

2.3   Lowly Data Processing Capacity

With the requirements of computing resources and storage resources continue to improve, high performance data processing platform is necessary condition to deal with data resources. The development and use of high performance computing in 80 years of the last century has already begun. National "973" and "863" programs had put on a lot of funds in research of high performance computing and it had applied many fields such as defense and security, oil exploration, weather forecasting, bioinformatics, gene and nanotechnology and other aspects. In "fifteen" period, CAS has made remarkable achievements in high-performance computing environment and had developed domestic-made supercomputer Shuguang 2000-II and Lenovo Pentium 6800 supercomputer. But CAAS has not its own high-performance computing center and all high-performance computing demands are dependent on other societies. Since the high-performance computing research in different directions have different software and hardware requirements and highly specialized, the results computed by outside computing center cannot reach the preset requirements, and data security can't be protected. The survey in 2014 showed that more than 2/3 institutes particularly animal husbandry, biology, vegetables and flowers, have used high performance data processing, while other institutes also expressed requirement of high performance data processing.

2.4 Website Services Model Is Single

In the current era of rapid development of the Internet, the ways people access information and communication are also diverse, new media such as Twitter and micro-channel has become an important way of communicating information. And smart phones, tablet PCs and other mobile Internet terminals grow explosively. As of mid-June 2013, China's mobile phone users reached 464 million. Mobile office has become a more relaxed and effective way of working in the future. But CAAS portal website only has a traditional Internet platform and is lack of function to diverse and release data resources.

3 Suggestions of Web Data Resources Construction in the Future

The planning in 2014 proposed that information technology development goals in the future will be around the overall goal of "build a world-class agricultural research institutes," and build domestic first-class agricultural research institutions in information service platform and to provide first-class information technology services for agricultural research through the integration and sharing of agricultural research data.

3.1 Improve System of Data Resources Construction

Construction of the database is a long-term project. Ensuring the smooth development of construction of the database, we need to establish clear rules and regulations in the fund, personnel and data collection, storage and so on.

3.1.1 Have a Realistic Implementation Plan and Funding

The planning in 2014 has listed a timetable implementation for specific information construction which including planning storage and computing platforms and data resources construction funds needed. Program is divided into two phases, the first five years will focus on the construction of network infrastructure, and the second year will deepen the service and data mining. The first task of a five-year plan have been identified, the specific implementation plan Table 3.

3.1.2 Construction of the Data Resources Needs Full-Time Institutions and Technical Personnel

We should establish academy-level information technology allied agencies and improve the information technology systems of each institute. Currently, information construction in CAAS is responsible of Agricultural Information Institute. Only two related offices including 25 employees are charge of network maintenance and site updates. Other information technology personnel is even more lacking. In the planning, CAAS will established a-hundred-person team of network operation and maintenance and website editors, and set policies to improve staffing and operational system of information technology, and regular train and assessment technical personnel. Hence we need to establish comprehensive information centralized management and to format

Table 3. The planning timetable from 2015 to 2020

No	Program	Aim	Execution
1	Rules and regulations	Formulate technology standards framework	2015.1–2015.6
		Formulate Information construction standards	2016.7–2017.12
		Formulate operation and management of information	2015.7–2016.12
2	Administrative Information Construction	Institutes information database based on completed and e-government applications.	2015.1–2016.12
3	Network-based environmental improvement	Construction of high-speed and stable network transmission channel and network bandwidth access levels and school district Wireless network coverage	2015.1–2016.12
4	High-performance computing platform	Construction of high-performance computing platform to achieve effective mining resources, integration, sharing.	2018.1–2018.12
5	Construction site of new media platforms.	Construction of multi-media platforms and carrier information dissemination, diversify information services.	2015.1–2019.12
6	Security conditions building	Through third-party security cloud platform, to all sites within the Academy belongs range safety monitoring service.	2016.1–2019.12
7	Data Sharing and Analysis Platform	Scientific Data virtual shared environment construction	2016.1–2017.12
		Data analysis platform (tools) Construction	2017.1–2019.12

a clear division working mechanism for each institute. And on this basis we would explore to establish the academy co-ordination, each institute the participation, cooperation and win-win management and service models.

3.2   To Improve Network Infrastructure Construction Environment

Network infrastructure is a prerequisite to protect the data storage and computing preconditions. In order to meet the data resource construction of network infrastructure environment we should know the needs of each network. First, it is necessary to clearly know network requirement and to overall layout design network function, structure and layout. Meanwhile we should improve power system, monitoring systems, fire systems, refrigeration systems and other facilities. The Planning put forward that CAAS will be completed in the network infrastructure to meet the next decade, the development of information technology within five years (Table 4); Secondly, we should construct high-speed data transmission channels, improve network transmission environment and establish a without barriers data sharing and transmission line to link 42 institutes distributed in the 20 provinces.

Table 4. 2015–2020 the central office-based environment construction plan

Program	Aim
Center room area	By 2020, the total area of 5000 m^2
Power Systems	Build double circuit power supply systems and power protection UPS8 h online
Security System	The establishment of advanced technology and personnel management of access control and fingerprint input facilities
Surveillance system	Establishment of an international advanced automatic monitoring system, electricity, air conditioning, UPS automatic monitoring
Fire Fighting System	The establishment of smoke, temperature sense of alarm systems and automatic fire extinguishing equipment
Cooling System	Using professional precision equipment to ensure the room temperature, humidity, fresh air status

3.3   To Build Network Platform of the Development for Data Resources

Data resources are one of the main research achievements of scientific research. According to research firm IDC predicts that the world's raw data storage capacity will be an annual increase of more than 50 % by weight, and all of data is not only growing in volume but the growing complexity of data resources. Therefore, the establishment of massive data collection, storage platforms and computing center will be a necessary condition for the era of big data research institutes and development.

3.3.1   Construction Big Data Cloud Storage Center
Big data analysis and research has been carried out and quickly deployed in a variety of different research areas, such as genomics, proteomics in particular, its data growth rate will exceed the legendary speed IT design development, because data storage capacity and data processing capacity of the existing data center can't meet the future needs of scientific computing and analysis. Survey about information technology showed in Fig. 1: This indicates that in the case of rapid growth of data the pressure storage capacity is the biggest problem of the network construction. Therefore, in order to meet the data storage demands of big data era, building cloud storage center is to meet the necessary conditions for the future of store large data.

The planning proposed to establish a data cloud storage platform for information integration and islands of information and proposed to provide long-term data preservation service, remote backup service and online storage service associated monitoring and surveillance data.

3.3.2   Construction High Performance Computing Data Center
In contemporary agricultural science research, since scientific data surge that the possibility of obtaining more and more science depends on acquiring, processing a sufficient amount of data capacity. Establishment of high-performance computing data center can provide genomics, proteomics, bioinformatics, new materials and other high-performance computing services.

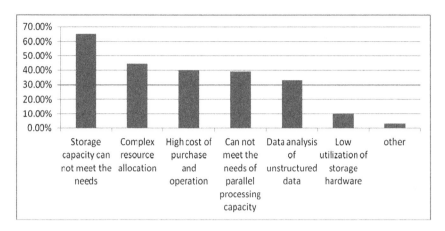

Fig. 1. Chinese Academy of Agricultural Sciences Research Network Storage Problems table

CAAS have a strong demand for high-performance computing. According to 2014 survey results, more than 2/3 of the institutes is applying high performance computing, 1/3 of the institutes also has the needs of high performance computing. Along with the agricultural development of information technology, agricultural research is becoming comprehensive and interactive. It is urgent to the needs of multi-field, regional, cooperation in collaborative research teams. All types information collection and data mining and analysis has become the main direction of agricultural research and development. Because the database has accumulated a massive data of flora and fauna resources, monitoring and sensing data, network data, traditional data mining massive data mining model can't meet the demand for computing power, thus it is the need to establish a new data mining models with high-performance computing capabilities.

Building high-performance computing centers should combine with the actual needs of research work. By the research institutes requirement of the high-performance computing application, we can build a high-performance computing platform including a scientific-basis parallel computing platform software platform, system application software platform and tools.

Build high-performance computing centers can also provide remote sensing data, sensor data, network data and other large data analysis and processing business for domestic and international users. Building cloud computing platform will achieve effective dispersion mining resources, integration, sharing.

3.3.3 Construction of the Data Collection and Processing Center Resources

Website statistics show that in 2015, all of the 236 sites, 227 sites generated data used the database management. Data collection and processing center construction can rely on the network resource data platform to meet the decentralized or centralized storage of agricultural natural resources and science and data resources to achieve centralized remote integration and data sharing agricultural scientific data resources. Construction of Agricultural Sciences data integration center, we can improve the efficiency of data query services, protect the security of the data.

For experimental data (experimental) chamber of agriculture science, agricultural science field station and observation station generated, we can deploy data collection layer, field control layer, data storage layer and business application layer architecture system, establish a wireless sensor network system, the completion of agricultural production monitoring network laying demand, eventually things will transmit information to the data collection center.

Ultimately, by constructing a data center networking platform we can provide experimental (test) data network platform for intelligence gathering and storage, and improve the processing efficiency of field observation data.

3.3.4 Established National Agricultural Resources Things Monitoring Platform

Establishment of national agricultural resources monitoring platform can integrate multi agricultural condition monitoring resources, expand and upgrade the existing infrastructure integration and software integration systems. According to the principle of cloud computing and cloud service management, we can design and build the platform architecture, deployment platform equipment, development software system, than gradually establish a national agricultural condition monitoring platform. The platform can achieve coverage area of the county's main crop types, to provide real-time dynamic for agricultural research and agricultural production and long-term accumulation of data resources, to provide support for agricultural production and scientific guidance for disaster emergency management.

3.4 Broaden the Web Services Model, Increased Resource Sharing and Communication Platform

The planning in 2014, has put forword to use the latest technology such as responsive web design to optimize the design and development of adaptive Chinese portals, and develop the network carriers to support for smartphones, tablet computers, TV, PC monitor, IOS and Android mobile phone carrier access.

New media platformsIis another task, such as building weibo WeChat which can broad the personalized information service mode and active push, diversified information services.

4 Conclusion

A large-scale production, sharing and application data era is just beginning, as Professor Victor said, the real value of big data is like an iceberg floating in the ocean, we can only see the tip of the iceberg at first glance, most hide beneath the surface. Carry out research work, will also be with the arrival of the era of big data, a huge change, and construction site data resources, but also will become all the researchers involved in scientific research is an important content.

References

1. Mayer-Schonberger, V., Cukier, K.: Big Data, pp. 127–156. Zhejiang people's Publishing House, Zhejiang (2013)
2. Huang, R., Wang, B., Zhou, Z.: A study on countermeasures for promoting scientiifc data sharing in China. Library **3**, 7–13 (2014). (in Chinese)
3. Xu, Z., Fu, Y.: Exploration of scientific research management informationization in context of big data. Chnology Innov. Manage. **2**, 112–115 (2014)
4. Zou, S.: Application of Personalized RecommendationTechnology in Agricultural Extension System-Based on the United State department of agriculture website case analysis. Central China Normal University, pp. 10–14 (2014)
5. Wen, Y.: Research on Internet Text Mining and Personalized Recommendation. Beijing Jiaotong University, pp. 18–24 (2014)
6. Feng, X., Zhu, X.: The design and implementation of information resources extractor of agricultural websites based on Spring. Inf. Res. **3**, 19–22 (2011)
7. An, Y., Li, B., Yang, R., Hu, L.: Content-based personalized recommendation on popular micro-topic. J. Intell. **2**, 155–160 (2014)
8. Xia, Q., Li, W., et al.: New social networking platform for the academic fields: research networking systems. J. Intell. **9**, 167–172 (2014)
9. Zhao, Y., Zhu, P., Chi, X., et al.: A brief view on requirements and development of high performance computing application. J. Comput. Res. Dev. **10**, 1640–1646 (2007)

The Knowledge Structure and Core Journals Analysis of Crop Science Based on Mapping Knowledge Domains

Minjuan Liu, Lu Chen, Xue Yuan, Ting Wang, Yun Yan[(✉)],
and Yuefei Wang

Agriculture Information Institution of Chinese Academy
of Agricultural Sciences, Beijing 100081, China
{liuminjuan,yuanxue,wangting,yanyun,
wangyuefei}@caas.cn, 903618592@qq.com

Abstract. This paper aims at revealing the potential structure of crop science and core journals distribution to provide reference and help for crop research and journal work. On the basis of journal co-citation analysis this paper draws the co-citation map of Crop Science Journal by means of mapping knowledge domains and information visualization technology. 86 crop science journals can be roughly divided into two groups and each group of journals can be subdivided into two regions. Plant science (including plant physiology, plant ecology, plant cell etc.) and biology and chemistry (including biochemistry and molecular biology, gene and genetics etc.) are the mainstream research field of crop science. Crop production and soil science is another important branch of crop science. In addition, there is the phenomenon of cross integration between crop science and environment, horticultural science, plant protection, food processing, animal husbandry etc.

Keywords: International · Journals · Crop science · Journal co-citation analysis · Mapping knowledge domains

1 Introduction

Crop Science is one of the core disciplines of agricultural science, the theory and technology of which plays an important role in agriculture and rural economic development as an important support. In the 21st century, the situation of international Crop Science and technology had a huge change, which is penetrated continually by the bio-technology and information technology. The combination of high-tech and traditional technology promote the rapid development of crop science and technology [1]. The rapidly change of technological development brings crop science with opportunities and challenges. If grasp the subject development dynamics, we can get opportunities in the fierce competition. Scientific researchers always want to publish the new results and findings as soon as possible to achieve the occupation of frontier research in this field, attention of academia and peer recognition and so on. Therefore, knowledge of paper, to some extent, represents the latest research level when scientific papers published in the literature. Especially in those influential journals, many

© IFIP International Federation for Information Processing 2016
Published by Springer International Publishing AG 2016. All Rights Reserved
D. Li and Z. Li (Eds.): CCTA 2015, Part II, IFIP AICT 479, pp. 392–403, 2016.
DOI: 10.1007/978-3-319-48354-2_39

outstanding scientists of the world publish high-level scientific papers. Therefore, an core journal is not only an important media to disseminate research results, but also through the periodical analysis, it can help researchers to quickly understand the current research and grasp the latest trends.

Journals without external links can be organically linked by Journal co-citation analysis, which can reveal the relationship of interdependence and cross between journals, determine the professional scope, and help to identify core journals. The combination of Journal co-citation analysis and mapping knowledge domains can show up potential subject structure and power distribution of a field in a visually intuitive graph. The journals with a closer relationship concentrate together to form a different cluster result of different research directions and fields, which vividly depict the subject structure and core journals of the field [18]. Journal co-citation analysis started to develop on the basis of literature co-citation analysis. The concept of "literature Co-citation" was put forward in 1973 by the American intelligence scientist Henry Small [2], after their studies were the first to carry out in 1974, extensive co-citation analysis followed up. In the later co-citation analysis, most of the research is literature co-citation analysis and author co-citation analysis represented by Small and White [3–9]. There were a lot of co-citation analysis and empirical research at home and abroad, but journal co-citation analysis and empirical research is not too much. For example, in 1991, McCain used journal co-citation analysis for practice to analyze the economics scholarly journals [10]; in 2000, Ding use journal co-citation analysis and visualization to study the development process of Intelligence retrieves during 1987–1997 [11]; in 2003, Tsay visualized of the semiconductor areas with journal co-citation analysis [12]; in 2004, Liu studied the literature of urban planning and visualized document structure with journal co-citation analysis[13]; in 2005, Marshakova Shaikevich used journal co-citation analysis and visualization in the subject of women's studies and library and information science, and pointed out the distribution of its subject areas [14]; in China, in 2006, Hou used co-citation analysis and draw the science map of international metrology core journals [15]; in 2008, Qiu and Zhao used co-citation cluster analysis and core-periphery model to mainly analyze the 21 editing and publishing journals and determine the core journals of the discipline [16]; in 2009, Zhao respectively used journal co-citation analysis and draw knowledge maps of Library and Information Science and biological hydrogen production [17–19]; in 2009, Qin draw the knowledge map of the relationship between agricultural history of China and neighboring discipline based on journal co-citation analysis [20], in 2010, Liang et al. used journal co-citation analysis method to learn the status of citation analysis discipline [21].

Although there is relatively little research on journal co-citation analysis, but it is not only an effective way to research subjects and the structure and characteristics of literature, but also has its unique in the study of the overall discipline structure and the nature and characteristics of professional journals [22]. However, in the previous studies, there were more empirical research on the library and information science, and lacked application and validation in other fields, especially in the agricultural area; Also, the selecting target journals of co-citation analysis more dependent on existing database journals category or used the way of the keywords retrieval, so it can not be used in the field whose journals classification are not covered in database or can not use keywords retrieval. In this paper, it delineates the core journals gradually spreading from a single female parent journal based on citation analysis, in order to try to expand the applied disciplines of journal co-citation analysis, use visualization techniques to draw crop science map on the basis of the journal co-citation analysis of crop science journals. On the one hand, it can help researchers to understand Crop science knowledge structure and research focus. On the other hand, it can help researchers to know the characteristics of the journals and give reference for selecting the appropriate journal to submit the article.

2 Data Source and Methods

The research data are all from the Science Citation Index Expanded database of Thomson Scientific, The last update time is May 2013.

This study uses the journal co-citation analysis methods to reveal the interdependence cross relationship between journals, what's more, with the emerging international method of mapping knowledge domain and information visualization technology, drawing the journal co-citation map of crop science to vividly reveal the structure of the crop science core journals groups. The mainly methods include factor analysis, cluster analysis and multidimensional scaling, which the factor analysis by principal components analysis and varimax orthogonal rotation, cluster analysis by Hierarchical Clustering and Multidimensional Scaling by ALSCAL. The research combined two analytical approaches, which are bibexcel and SPSS for obtaining visualizing information of crop science.

Analysis steps:

First, selecting CROP SCIENCE as the female parent, which is the most important journal in the field of the crop science, by the method of single co-citation analysis, there are 2008 papers with 78121 citations in CROP SCIENCE from 2008 to May 2013 were analyzed and evaluated.

Second, selecting the journals which are higher cited by Crop Science for further co-citation analysis, there are 5819 papers with 240823 citations in these journals from 2008 to May 2013, then 98 journals which the cited frequency over 300 times among these journals are chosen to do a further analysis.

Last, cleaning the data of 98 journals, and finally choosing 86 journals to do the co-citation analysis, which are much more important journals in the field of crop science. To establish co-citation matrix with the data of 86 journals by Bibexcel, then use the matrix do some factor analysis, cluster analysis and Multidimensional Scaling by SPSS to draw the journal co-citation map of Crop Science with 86 journals, which can vividly reveal the relationship between journals and disciplinary structure of Crop Science.

3   Results and Discussion

3.1   Parental Journal and Citation Analysis

The parental journal "CROP SCIENCE" which was founded in 1997, the impact factor is 1.513, published by the Crop Science Society of America (CSSA). The journal publishes crop genetics and breeding; crop physiology and metabolism; crop ecology, crop production and management; seed physiology, seed production and technology; lawn learning; genomics, molecular genetics and biotechnology; plant genetics resources and pest control and other aspects and original research papers. The journal is indexed in SCI belongs to Q2, covers most fields of crop science and well-known in crop science. Therefore, this study chosen "CROP SCIENCE" as the parental journal, which can be a good representation of crop science.

Through citation analysis, there were 2736 journals cited by "CROP SCIENCE", the total citation frequency is 62950 times. In Table 1, the top 7 journals which were cited by CROP SCIENCE more than 1000 times are listed.

Table 1. Top 7 journals cited by CROP SCIENCE

Rank	Cited frequency	Cited journal
1	11483	CROP SCI
2	3905	THEOR APPL GENET
3	2576	AGRONOMY JOURNAL
4	1658	EUPHYTICA
5	1484	GENETICS
6	1304	PLANT PHYSIOL
7	1168	FIELD CROP RES

"CROP SCIENCE", "THE THEORETICAL AND APPLIED GENETICS", "THE AGRONOMY JOURNAL" and "EUPHYTICA" are cited by "CROP SCIENCE" more than 1500 times, the total cited frequency can reach 19622, accounting for 31.17 % of all journal, and "CROP SCIENCE", its self-cited frequency is 11483 times and accounts for 18.24 %. Therefore, this study chosen the 4 journal as the parental journal to do further co-citation analysis to identify the important journals in the field of crop science.

Between 2008 and May 2013, these 4 journals has published 5819 papers with 240823 citations, there were 7113 journals cited by these 4 journals and the total cited frequency is 198362 times. Table 2 gives 98 journals which were cited more than 300. It's necessary to clean the data of 98 journals, merge the same journals, eliminate the review journals and the journal that was not indexed in SCI, eventually retained 86 journals to do co-citation analysis.

Table 2. 98 journals cited more than 300 by 4 parental journals

Rank	Cited journal	Cited frequency	Journal ID	Rank	Cited journal	Cited frequency	Journal ID
1	CROP SCI	19947	a25	50	GENOME RES	588	a35
2	THEOR APPL GENET	18898	a94	51	J CEREAL SCI	586	a47
3	AGRON J	8868	a6	52	PLANT PATHOL	583	a83
4	GENETICS	6804	a33	53	AM J BOT	573	a7
5	EUPHYTICA	6575	a28	54	COMMUN SOIL SCI PLAN	563	a24
6	PLANT PHYSIOL	3447	a84	55	ADV AGRON	563	a2
7	P NATL ACAD SCI USA	3047	a70	56	ANNU REV PHYTOPATHOL	558	a11
8	GENOME	2927	a34	57	PLANT CELL ENVIRON	549	a76
9	FIELD CROP RES	2806	a31	58	MOL ECOL	535	a61
10	PLANT BREEDING	2443	a73	59	MOL GEN GENET	527	a62
11	PHYTOPATHOLOGY	2414	a72	60	J ANIM SCI	523	a46
12	PLANT DIS	2290	a79	61	J ECON ENTOMOL	522	a49
13	MOL BREEDING	2075	a60	62	J SCI FOOD AGR	518	a56
14	PLANT CELL	1985	a75	63	ANN APPL BIOL	506	a8
15	SCIENCE	1860	a89	64	SOIL TILL RES	504	a92
16	PLANT J	1821	a80	65	ACTA HORTIC	460	a1
17	J EXP BOT	1688	a51	66	GENOMICS	456	a36
18	SOIL SCI SOC AM J	1674	a91	67	WEED TECHNOL	454	a98
19	Crop & Pasture Science	1594	a14	68	J PLANT NUTR	434	a53
20	CAN J PLANT SCI	1568	a21	69	MOL BIOL EVOL	424	a59
21	PLANT MOL BIOL	1299	a81	70	J DAIRY SCI	419	a48
22	PLANT SOIL	1273	a86	71	BIOMETRICS	415	a18
23	J HERED	1217	a52	72	T ASAE	409	a93
24	NATURE	1210	a67	73	PLANT CELL PHYSIOL	401	a77
25	ANN BOT-LONDON	1106	a9	74	PLANT CELL REP	398	a78
26	HORTSCIENCE	1089	a40	75	ANNU REV PLANT PHYS	381	a13
27	NUCLEIC ACIDS RES	1066	a69	76	WEED RES	378	a96
28	HEREDITY	925	a39	77	AGR ECOSYST ENVIRON	378	a3
29	J ENVIRON QUAL	904	a50	78	AGR FOREST METEOROL	370	a4
30	GENET RESOUR CROP EV	900	a32	79	PLANT MOL BIOL REP	369	a82
31	NEW PHYTOL	861	a68	80	J AGRON CROP SCI	368	a43
32	TRENDS PLANT SCI	854	a95	81	NAT REV GENET	368	a66
33	NAT GENET	839	a65	82	ANN EUGENICS	362	a10

(*continued*)

Table 2. (*continued*)

Rank	Cited journal	Cited frequency	Journal ID	Rank	Cited journal	Cited frequency	Journal ID
34	J AGR FOOD CHEM	794	a41	83	J AM OIL CHEM SOC	361	a44
35	J AM SOC HORTIC SCI	794	a45	84	REMOTE SENS ENVIRON	354	a88
36	J AGR SCI	782	a42	85	SOIL BIOL BIOCHEM	353	a90
37	MOL GENET GENOMICS	727	a63	86	AUST J PLANT PHYSIOL	352	a16
38	J PROD AGRIC	700	a55	87	AUST J EXP AGR	347	a15
39	PLANT SCI	699	a85	88	AGR WATER MANAGE	334	a5
40	MAYDICA	697	a58	89	ANNU REV PLANT BIOL	326	a12
41	EUR J AGRON	680	a29	90	J PLANT PHYSIOL	326	a54
42	CURR OPIN PLANT BIOL	660	a26	91	CAN J PLANT PATHOL	316	a20
43	BIOINFORMATICS	653	a17	92	HEREDITAS	310	a38
44	WEED SCI	644	a97	93	CAN J SOIL SCI	307	a22
45	PHYSIOL PLANTARUM	643	a71	94	ECON BOT	306	a27
46	MOL PLANT MICROBE IN	643	a64	95	Plant Breeding Reviews	304	a74
47	CEREAL CHEM	629	a23	96	GRASS FORAGE SCI	304	a37
48	BREEDING SCI	615	a19	97	J SOIL WATER CONSERV	300	a57
49	PLANTA	603	a87	98	EUR J PLANT PATHOL	300	a30

3.2 Journal Co-citation Analysis and Map Knowledge Domain

(1) Journal Co-citation Matrix. By bibexcel to count the cited frequency of the 86 journals, and to establish journal co-citation matrix, that is the original matrix, which laid a foundation to further reveal the relationship and structural characteristics between journals (Fig. 1. 86 journal co-citation matrix). The matrix is a symmetric matrix, is co-diagonal journal citations, diagonal value is 0, the matrix range is between 0–2379. Meanwhile, transform the original matrix into Pearson correlation matrix as a similarity matrix, where similarity is measured by the correlation coefficient, the positive correlation is stronger, the two journals of the field of study or research is more similar, also showed that more similar academic backgrounds (Fig. 1).

(2) Factor Analysis. On the basis of the Multidimensional Scaling, this study combined factor analysis and cluster analysis to supplement and improve the multidimensional scaling. This study did a principal components analysis on the Pearson

V1	a3	a4	a5	a6	a7	a8	a9	a10	a14	a16	a17	a18
1	a3	0	38	34	220	13	41	35	0	46	15	2	4
2	a4	38	0	38	143	6	16	58	0	37	35	2	4
3	a5	34	38	0	148	4	22	45	3	56	31	3	8
4	a6	220	143	148	0	60	119	315	17	359	116	37	90
5	a7	13	6	4	60	0	30	106	21	28	14	69	13
6	a8	41	16	22	119	30	0	85	14	108	40	27	21
7	a9	35	58	45	315	106	85	0	27	178	94	68	20
8	a10	0	0	3	17	21	14	27	0	61	4	48	8
9	a14	46	37	56	359	28	108	178	61	0	103	67	66
10	a16	15	35	31	116	14	40	94	4	103	0	10	8
11	a17	2	2	3	37	69	27	68	48	67	10	0	30
12	a18	4	4	8	90	13	21	20	8	68	8	30	0
13	a19	11	0	2	40	31	15	58	43	45	4	57	14
14	a20	5	5	1	29	5	24	11	33	75	2	14	7
15	a21	53	31	34	426	37	54	126	32	181	29	24	42
16	a22	42	8	14	148	1	7	15	0	12	0	0	1
17	a23	2	4	2	60	1	1	20	11	75	13	33	9
18	a24	67	17	26	324	7	20	41	0	34	9	1	5

(a)   Original matrix

V1	a3	a4	a5	a6	a7	a8	a9	a10	a14	a16	a17	a18
1	a3	1.000	.8520	.8814	.6389	.1803	.4328	.5375	.0604	.5178	.5791	.0750	.4032
2	a4	.8520	1.000	.9409	.6382	.2321	.5204	.6042	.0603	.5658	.7333	.0955	.4125
3	a5	.8814	.9409	1.000	.6900	.2812	.5598	.6829	.1189	.6003	.7920	.1616	.4636
4	a6	.6389	.6382	.6900	1.000	.4242	.6461	.6265	.3277	.6797	.6489	.3297	.6264
5	a7	.1803	.2321	.2812	.4242	1.000	.7984	.7995	.8608	.7887	.5732	.9194	.8127
6	a8	.4328	.5204	.5598	.6461	.7984	1.000	.8591	.7689	.9310	.7503	.7353	.8734
7	a9	.5375	.6042	.6829	.6265	.7995	.8591	1.000	.6826	.8613	.8569	.7371	.8077
8	a10	.0604	.0603	.1189	.3277	.8608	.7689	.6826	1.000	.7539	.4011	.9224	.7896
9	a14	.5178	.5658	.6003	.6797	.7887	.9310	.8613	.7539	1.000	.7248	.7323	.8976
10	a16	.5791	.7333	.7920	.6489	.5732	.7503	.8569	.4011	.7248	1.000	.4527	.6028
11	a17	.0750	.0955	.1616	.3297	.9194	.7353	.7371	.9224	.7323	.4527	1.000	.7990
12	a18	.4032	.4125	.4636	.6264	.8127	.8734	.8077	.7896	.8976	.6028	.7990	1.000
13	a19	.1380	.1733	.2448	.3999	.9350	.8187	.7957	.9414	.8073	.5429	.9601	.8317
14	a20	.1922	.1773	.2014	.4488	.6340	.7845	.5497	.7881	.7373	.3609	.6608	.7273
15	a21	.6805	.6986	.7360	.7228	.6934	.8864	.8472	.6299	.9184	.7654	.6160	.8654
16	a22	.8547	.6704	.6915	.4550	-.077	.1230	.2079	-.157	.2201	.2557	-.162	.1390
17	a23	.3163	.3514	.4256	.5558	.7789	.8141	.8008	.7700	.8342	.6441	.7831	.8384
18	a24	.9012	.7839	.8090	.5305	.0630	.2824	.3884	-.050	.3747	.4407	-.044	.2754

(b)   Pearson correlation matrix

Fig. 1. 86 journals co-citation matrix (parts)

correlation matrix of 86 journals by SPSS, It extract three main components factors and cumulative variance reached 86.709%, which means three components factors have been able to explain the information contained in all variables well. Table 3 make a list of variables (journals) which factor loading is over 0.7.

(3) Cluster Analysis. Meanwhile, this study did a cluster analysis on these 86 journals to further examine the similarity between the journals to supplement multidimensional scaling. Similar to factor analysis, the study choose Hierarchical Clustering to analysis the Pearson correlation matrix of 86 journals by SPSS. Figure 2 shows the result of cluster analysis. It's obviously to see that these 86 journals is better to classified into two clusters.

(4) Multidimensional Scaling Analysis. In order to reveal the affinities between journals and further determine the discipline structure of journals in crop areas, we carried multidimensional scaling analysis by putting the journal similarity matrix into SPSS, which can display the relationship between the original high-dimensional data in low-dimensional space. As the picture 3, each dot represents a journal, and the location

Table 3. Variables (journals) which factor loading is over 0.7

	Rank	Journal ID	Factor loading	Rank	Journal ID	Factor loading	Rank	Journal ID	Factor loading
Factor1	1	a82	.978	19	a89	.927	37	a58	.813
	2	a63	.974	20	a80	.924	38	a77	.813
	3	a35	.974	21	a75	.917	39	a68	.807
	4	a17	.973	22	a7	.917	40	a72	.806
	5	a65	.970	23	a34	.916	41	a94	.772
	6	a69	.968	24	a70	.909	42	a23	.769
	7	a66	.965	25	a64	.908	43	a18	.769
	8	a59	.965	26	a32	.891	44	a51	.763
	9	a36	.964	27	a85	.890	45	a79	.758
	10	a19	.964	28	a78	.890	46	a83	.752
	11	a61	.947	29	a73	.881	47	a87	.748
	12	a39	.947	30	a33	.871	48	a25	.747
	13	a52	.944	31	a27	.864	49	a49	.746
	14	a67	.941	32	a84	.851	50	a30	.728
	15	a10	.940	33	a28	.843	51	a40	.724
	16	a38	.937	34	a45	.833	52	a9	.703
	17	a81	.936	35	a41	.815	53	a14	.700
	18	a60	.930	36	a47	.814			
Factor2	1	a3	.958	8	a88	.919	15	a29	.827
	2	a24	.951	9	a53	.906	16	a37	.816
	3	a90	.949	10	a91	.900	17	a86	.811
	4	a92	.940	11	a5	.857	18	a48	.717
	5	a50	.935	12	a4	.843	19	a96	.714
	6	a57	.922	13	a98	.839	20	a43	.705
	7	a22	.922	14	a97	.829			
Factor3	1	a16	.716						

of periodicals show similarity (common disciplines or methods, etc.) between journals. The more similar the more together, and then form a knowledge group. In the knowledge group, the journal which has the closest relation with other point is in the middle of the journal position map, which shows that it is the core of knowledge group; the other hand, more in the periphery.

On the basis of the results of the factor analysis and cluster analysis, we draw the co-citation multidimensional scaling analysis diagram of 86 journals, shown in Fig. 3. The value of stress is 0.03653 and RSQ is 0.99662, so it reflects a very good fitting degree. The 86 journals are clearly divided into two parts, group 1 includes 26 journals, and group 2 includes 60 journals.

The group 1 can be divided into two parts. A1 has the most journals, which mainly includes the journals of crop production, soil science, agricultural resources and the environment and other related areas. The AGRON J, SOIL SCI SOC AM J,

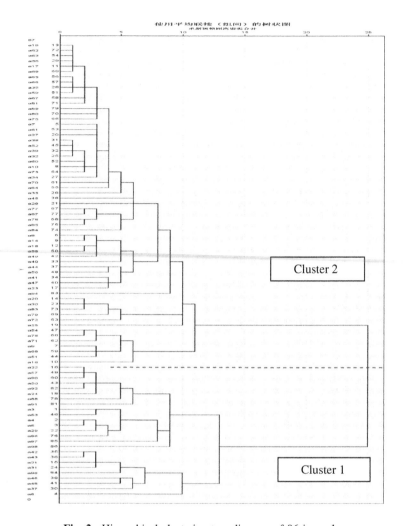

Fig. 2. Hierarchical clustering tree diagram of 86 journals

PLANT SOIL and some journals are the representations of this area, reflecting the inseparable relationship between crops and soil, water and environment. A2 area includes several agricultural comprehensive journals, and FIELD CROP RES is typical representation, which focus on crop production and cultivation, and become the connection of Journal group 1 and group 2.

The Group 2 can also be subdivided into two areas. B1 contains many journals which have tight connection and short distance from the origin, and some journals highly cited by the parental journal focused on here, which proves that this area is a core area concentrating crop science journals, covering the most important branches of Crop Science. It mainly includes the journals of crop genetics and breeding, plant science, biochemistry and molecular biology, gene and genetics, which attract more attention of

派生的激励配置

Euclidean 距离模型

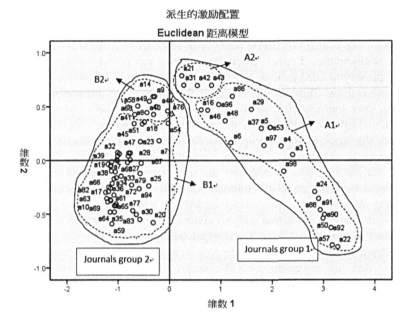

维数 2

维数 1

Fig. 3. The co-citation map of 86 important crop science journals

crop science researchers and is the most core journals. The journal of CROP SCI, THEOR APPL GENET, GENETICS, EUPHYTICA, PLANT BREEDING, GENOME, P NATL ACAD SCI USA, PLANT PHYSIOL, PHYTOPATHOLOGY, PLANT DIS and other journals are the representation. From the view of position, B2 likes the bridge of journals group 2 and group 1, which has more closely relation with A2, including the journals more closer to A2 which has Crop & Pasture Science as a representative of agriculture comprehensive journals, also including the journals of ANN BOT-LONDON and HORTSCIENCE as a representative of botany and horticulture.

4   Conclusions

According to the result of Journal Citation analysis and co-citation mapping knowledge, we can see that the international crop science research can be divided into two parts in recent years. The first part focus on plants (crops) own research (Journal Group 2), including plant physiology, plant ecology, plant cells, etc. and biochemistry and molecular biology, gene and genetics, etc., which is mainstream areas and more concerned by crop science researchers now. Another part is more concerned about the relationship between plants (crops), soil, resources and environment, including crop production, soil science and environment science, which is another important branch of crop science. In addition, some journals of resource and environment, horticulture, crop protection, food processing, animal husbandry are found to be highly cited by crop science journals, which proves that interdisciplinary integration between crop science and other subjects, and the journals from these areas have become the journals concerned by crop researchers.

Overall, the journal co-citation analysis is an effective method for revealing core journals of disciplines. Journals co-citation analysis can reveal the structure of discipline by the way of cited journals co-citation analysis, which is from the perspective of the journal analysis. Thus, through the journal position in the different disciplines, we can judge core journals of different subjects to help researchers to find more useful information [19]. Comparing with the previous research, the difference is that, it delineates the target journals group gradually spreading from a single female parent journal, namely in the way of using an important journal of crop science based on citation analysis to gradually delineate the subject core journals. From the result of International crop science related Journals co-citation analysis, it is satisfactory that it can reflect objectively the crop science underlying structure and core journals distribution by the journal co-citation analysis combined with visualization techniques. It is successful that the applied discipline is expanded.

However, when using the journal co-citation analysis, it must pay more attention on data collection and processing methods and process control, or it will have a direct impact on the accuracy of the analysis results. I believe that there are two aspects need to be taken seriously. One is periodical cleaning, which is necessary because the journals abbreviated titles may be not unified and some journal title may change, so inattention could cause distortion of data and affect the final results. The other one is the choice of the study object, factor analysis, clustering and multidimensional scaling method and the error of statistical analysis, which will affect the results objectivity, so it must carefully plan in order to ensure effective and objective analysis of the results. In the future studies, we will continue to combine journal published papers and the integration of new technologies and analysis methods to explore more accurate and reliable methods of revealing potential knowledge structure and core journal distribution in different fields.

Acknowledgment. Funds for this research was provided by the Chinese Academy of Agricultural Sciences Science and Technology Innovation Projects (CAAS-ASTIP-2015-AII).

References

1. Wan, J.M.: The development direction of crop science "11th five-year plan" in China. Crops **1**, 1–4 (2006). (in Chinese)
2. Small, H.G.: Co-citation in the scientific literature: a new measure of the relationship between two documents. J. Am. Soc. Inform. Sci. **24**, 265–269 (1973)
3. Small, H.G., et al.: Clustering the science citation index using co-citations II: mapping science. Scientometrics **8**, 321–340 (1985)
4. Small, H.G.: Multiple citation patterns in scientific literature: the circle and hill models. Inform. Storage Trisval **10**, 393–402 (1974)
5. Small, H.G.: The relationship of information science to the social science: a co-citation analysis. Inform. Prooving Manag. **17**, 39–50 (1981)

6. White, H.D., Griffith, B.C.: Author co-citation: a literature measure of intellectual structure. J. Am. Soc. Inform. Sci. **32**, 163–171 (1981)
7. White, H.D.: A co-citation map of the social indicators movement. J. Am. Soc. Inform. Sci. **34**, 307–312 (1983)
8. White, H.D., Griffith, B.C.: Authors as markers of intellectual spaces: co-citation in studies of science, technology and society. J. Doc. **38**, 255–272 (1982)
9. Miyomato, S., Nakayama, K.: A technique of two-stage clustering applied to environment and civil engineering and related methods of citation analysis. J. Am. Soc. Inform. Sci. **34**, 192–201 (1983)
10. McCain, K.W.: Mapping economics through the journal literature: An experiment in journal co-citation analysis. J. Am. Soc. Inform. Sci. **42**(2), 290–296 (1991)
11. Ding, Y., Chowdhury, G.G., Foo, S.: Journal as markers of intellectual space: journal co-citation analysis of information Retrieval area 1987–1997. Scientometrics **47**(1), 55–73 (2000)
12. Tsay, M.Y., Xu, H., Wu, C.W.: Journal co-citation analysis of semiconductor literature. Scientometrics **57**(1), 7–25 (2003)
13. Liu, Z.: Visualizing the intellectual structure in urban studies: a journal co-citation analysis (1992–2002). Scientometrics **62**(3), 385–402 (2005)
14. Marshakova-Shaikevich, I.: Bibliometric maps of field of science. Inform. Process. Manag. **41**(6), 1534–1547 (2005)
15. Hou, H.Y.: International measurement science core journals and knowledge map. Chin. J. Sci. Tech. Periodicals **17**(2), 240–243 (2006). (in Chinese)
16. Qiu, J.P., Zhao, W.H.: A metric demonstration of journal co-citation. Inform. Sci. **26**(10), 1447–1450 (2008). (in Chinese)
17. Zhao, Y.: International biohydrogen key periodical co-citation analysis and mapping knowledge. Chin. J. Sci. Tech. Period. **20**(6), 1043–1045 (2009). (in Chinese)
18. Zhao, Y., Sun, C.Q., Sha, Y.Z.: The knowledge mapping analysis on the research of biohydrogen. China Biotechnol. **29**(1), 116–121 (2009). (in Chinese)
19. Zhao, Y.: Visualizing a discipline: journal co-citation analysis of library and information science. Libr. Inform. **3**, 89–94 (2009). (in Chinese)
20. Qin, C.J.: The empirical research of the knowledge domains map between subject relationship based on journal co-citation analysis method. J. Modern Inform. **30**(5), 9–11 (2010). (in Chinese)
21. Liang, Y.X., Yang, Z.K., Liu, Z.Y.: Citation disciplinary status analytics. Inform. Stud. Theory Appl. **33**(5), 18–20 (2010)
22. Zhao, D.Z.: Journal co-citation analysis: a method of research discipline and the structure and characteristic of journals. Chin. J. Sci. Techn. Periodicals **4**(1), 55–58 (1993). (in Chinese)
23. Qiu, J.P.: Information metrology(11)–lesson information metrology in the field of library and information science-a case study on core journals and determination. Inform. Stud. Theory Appl. **24**(5), 396–400 (2001). (in Chinese)
24. Hou, H.Y.: Mapping Evolution of Scientometries. Dalian, Dalian University of Technology (2006). (in Chinese)
25. Yang, L.J., Zhang, L.Y.: An empirical research of the diagonal values of journal co-citation matrix. Libr. Inform. Serv. **54**(4), 144–148 (2010). (in Chinese)

Application of Spatial Reasoning in Predicting Rainfall Situation for Two Disjoint Areas

Jian Li[✉], Yanbo Huang, Rujing Yao, and Yuanyuan Zhang

College of Information Technology, Jilin Agricultural University,
Changchun 130118, China
liemperor@163.com, 945644996@qq.com,
562704227@qq.com, 645012393@qq.com

Abstract. In this article, we extend the 9 - intersection matrix to 27 - intersection matrix on the basis of RCC8 relations, propose 27 - intersection model between two disjoint regions and a simple region. Giving the algorithm, 31 kinds of topological relations between two disjoint regions and a simple region by specific procedure, and give schematic diagrams of the 31 kinds of topological relations, verifying 31 kinds of topological relations are achievable. In order to further research the topological relations between two disjoint regions and a simple region, in this article we have completed the reasoning for topological relations between two disjoint regions and a simple region, and given the conceptual neighborhood graph. By comparison of related work, we can draw a conclusion that the 27 - intersection model is superior to the expressive power of the 8-intersection model. Through the representation model of topological relations established in this article, we can predict the rainfall situation on two areas, and give the probability of rainfall in different regions.

Keywords: Topological relations · RCC8 · Simple region · Intersection matrix · Rainfall

1 Introduction

China is one of the most affected countries by tropical cyclones. In China, typhoon happens each year on an average of seven to eight, and China is a country that has most typhoons coming and most disasters with typhoon [1]. In the research of tropical cyclone, although some results have been made in prediction of tropical cyclone path, the predictions are still unsatisfactory for most exceptional path. At the same time, the edge of the region is not certain, all of these problems increase the difficulty in forecasting typhoon rainfall.

To establish a model that can adapt to the uncertainty of the typhoon area and improve the accuracy of typhoon rainfall warning, this article gave a new idea that forecasting typhoon rainfall is on the basis of spatial reasoning.

Spatial reasoning [2–4] is an important branch of artificial intelligence. It has become an interdisciplinary research topic involving disciplines, for example logic, algebra, topology and graph theory. Over the past three decades, the research on spatial reasoning has been extremely active, which is one of the fundamental aspects in the

© IFIP International Federation for Information Processing 2016
Published by Springer International Publishing AG 2016. All Rights Reserved
D. Li and Z. Li (Eds.): CCTA 2015, Part II, IFIP AICT 479, pp. 404–416, 2016.
DOI: 10.1007/978-3-319-48354-2_40

fields of geographic information system [5], image databases [6, 7] and CAD/CAM systems [8].

In this article, we specifically proposed a representation model between two disjoint regions and a simple region, this model is applied to predict the disaster-affected for two designated areas. This helps to improve the rainfall early warning and disaster assessment mechanism, has a guiding significance in the establishment of disaster prevention and early warning mechanisms to reduce disaster losses and casualties.

2 The Representations Model Between Two Disjoint Regions and a Simple Region

2.1 9- Intersection Model

In 1991, Egenhofer et al. [8] constructed 4-intersection model, as follows gave 4 - intersection matrix, here A^0 denotes the in-house of A and ∂A denotes the out-house of A. Then we can obtain 16 topological relations (contain the case that cannot exist in the real world):

$$\begin{pmatrix} A^0 \cap B^0 & A^0 \cap \partial B \\ \partial A \cap B^0 & \partial A \cap \partial B \end{pmatrix}$$

Based on the 4-intersection model, the complement A^- of region A is regarded as the exterior of region A, we extended the 4- intersection matrix to 9- intersection matrix [6], as follows:

$$\begin{pmatrix} A^0 \cap B^0 & A^0 \cap \partial B & A^0 \cap B^- \\ \partial A \cap B^0 & \partial A \cap \partial B & \partial A \cap B^- \\ A^- \cap B^0 & A^- \cap \partial B & A^- \cap B^- \end{pmatrix}$$

When considering region in the real world, we obtain 8 kinds of relations corresponding to RCC8 relations[9]. RCC8 relations are disjoint; meet; overlap; coveredby; inside; equal; covers; contains, as shown in Fig. 1:

2.2 The 27- Intersection Model Between Two Disjoint Regions and a Simple Region

Based on the definition of 9 - intersection model, we extend the 9 - intersection model, and get the 27- intersection model that can express the topological relations between two disjoint regions and a simple region.

2.2.1 The 27- Intersection Matrix Between Two Disjoint Regions and a Simple Region

Based on 9-intersection model, space region is divided into nine sections, as shown in Fig. 2:

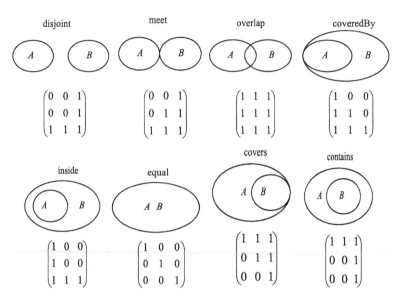

Fig. 1. 9-intersection model and 8 kinds of topological relations

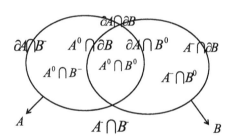

Fig. 2. 9 parts of the space is divided

Similarly, three regions of A, B, C can divide the space into 27 parts, in other word, we can get 27- intersection matrix, as follows:

$$
\begin{pmatrix}
A^0 \cap B^0 \cap C^0 & A^0 \cap B^0 \cap \partial C & A^0 \cap B^0 \cap C^- \\
A^0 \cap \partial B \cap C^0 & A^0 \cap \partial B \cap \partial C & A^0 \cap \partial B \cap C^- \\
A^0 \cap B^- \cap C^0 & A^0 \cap B^- \cap \partial C & A^0 \cap B^- \cap C^- \\
\partial A \cap B^0 \cap C^0 & \partial A \cap B^0 \cap \partial C & \partial A \cap B^0 \cap C^- \\
\partial A \cap \partial B \cap C^0 & \partial A \cap \partial B \cap \partial C & \partial A \cap \partial B \cap C^- \\
\partial A \cap B^- \cap C^0 & \partial A \cap B^- \cap \partial C & \partial A \cap B^- \cap C^- \\
A^- \cap B^0 \cap C^0 & A^- \cap B^0 \cap \partial C & A^- \cap B^0 \cap C^- \\
A^- \cap \partial B \cap C^0 & A^- \cap \partial B \cap \partial C & A^- \cap \partial B \cap C^- \\
A^- \cap B^- \cap C^0 & A^- \cap B^- \cap \partial C & A^- \cap B^- \cap C^-
\end{pmatrix}
$$

Denote $A_0 = A^0, A_1 = \partial A, A_2 = A^-$, Order $M_{ijk} = A_i \cap B_j \cap C_k, i,j,k = 0, 1$ or 2.

Then above 27- intersection matrix can be regarded as M.

$$M = \begin{pmatrix} M_{000} & M_{001} & M_{002} \\ M_{010} & M_{011} & M_{012} \\ M_{020} & M_{021} & M_{022} \\ M_{100} & M_{101} & M_{102} \\ M_{110} & M_{111} & M_{112} \\ M_{120} & M_{121} & M_{122} \\ M_{200} & M_{201} & M_{202} \\ M_{210} & M_{211} & M_{212} \\ M_{220} & M_{221} & M_{222} \end{pmatrix}$$

Definition 1. For a simple region A, define $\varepsilon(A) = \begin{cases} 1 & A \text{ is not empty} \\ 0 & A \text{ is empty} \end{cases}$. Then M is regarded as M_{ε}.

$$M_{\varepsilon} = \begin{pmatrix} \varepsilon(M_{000}) & \varepsilon(M_{001}) & \varepsilon(M_{002}) \\ \varepsilon(M_{010}) & \varepsilon(M_{011}) & \varepsilon(M_{012}) \\ \varepsilon(M_{020}) & \varepsilon(M_{021}) & \varepsilon(M_{022}) \\ \varepsilon(M_{100}) & \varepsilon(M_{101}) & \varepsilon(M_{102}) \\ \varepsilon(M_{110}) & \varepsilon(M_{111}) & \varepsilon(M_{112}) \\ \varepsilon(M_{120}) & \varepsilon(M_{121}) & \varepsilon(M_{122}) \\ \varepsilon(M_{200}) & \varepsilon(M_{201}) & \varepsilon(M_{202}) \\ \varepsilon(M_{210}) & \varepsilon(M_{211}) & \varepsilon(M_{212}) \\ \varepsilon(M_{220}) & \varepsilon(M_{221}) & \varepsilon(M_{222}) \end{pmatrix}$$

So we can represent the topological relations between two disjoint regions and a simple region with a 0-1matrix.

2.2.2 Properties of 27-Intersection Model Between Two Disjoint Regions and a Simple Region

Definition 2. We define an operation "$\vee$" on the set $\{0, 1\}$, as shown in Table 1:

Table 1. The operation on "$\vee$"

$\vee$	0	1
0	0	1
1	1	1

Definition 3. For any $m \times n$ 0-1 matrices $A = (a_{ij})_{m \times n}$ and $B = (b_{ij})_{m \times n}$, we define $A \vee B = (a_{ij} \vee b_{ij})_{m \times n}$.

According to the above definition, we can get $\varepsilon(A \cup B) = \varepsilon(A) \vee \varepsilon(B)$. Order RCC8 to Ω ,among

$$
\Omega = \left(\begin{pmatrix} 0 & 0 & 1 \\ 0 & 0 & 1 \\ 1 & 1 & 1 \end{pmatrix}, \begin{pmatrix} 0 & 0 & 1 \\ 0 & 1 & 1 \\ 1 & 1 & 1 \end{pmatrix}, \begin{pmatrix} 1 & 1 & 1 \\ 1 & 1 & 1 \\ 1 & 1 & 1 \end{pmatrix}, \begin{pmatrix} 1 & 0 & 0 \\ 1 & 1 & 0 \\ 1 & 1 & 1 \end{pmatrix}, \right.
$$
$$
\left. \begin{pmatrix} 1 & 0 & 0 \\ 1 & 0 & 0 \\ 1 & 1 & 1 \end{pmatrix}, \begin{pmatrix} 1 & 0 & 0 \\ 0 & 1 & 0 \\ 0 & 0 & 1 \end{pmatrix}, \begin{pmatrix} 1 & 1 & 1 \\ 0 & 1 & 1 \\ 0 & 0 & 1 \end{pmatrix}, \begin{pmatrix} 1 & 1 & 1 \\ 0 & 0 & 1 \\ 0 & 0 & 1 \end{pmatrix} \right)
$$

For any two ordinary regions X and Y, the corresponding matrix of their topological relations must belong to the set Ω, so we can get Theorem 1:

Theorem 1. Two disjoint regions A, B and a simple region C corresponding topological relation between any two matrixes must satisfy the following three formulas:

$$
\begin{pmatrix} \overset{2}{\underset{s=0}{\vee}} \varepsilon(M_{00s}) & \overset{2}{\underset{s=0}{\vee}} \varepsilon(M_{01s}) & \overset{2}{\underset{s=0}{\vee}} \varepsilon(M_{02s}) \\ \overset{2}{\underset{s=0}{\vee}} \varepsilon(M_{10s}) & \overset{2}{\underset{s=0}{\vee}} \varepsilon(M_{11s}) & \overset{2}{\underset{s=0}{\vee}} \varepsilon(M_{12s}) \\ \overset{2}{\underset{s=0}{\vee}} \varepsilon(M_{20s}) & \overset{2}{\underset{s=0}{\vee}} \varepsilon(M_{21s}) & \overset{2}{\underset{s=0}{\vee}} \varepsilon(M_{22s}) \end{pmatrix} = \begin{pmatrix} A^0 \cap B^0 & A^0 \cap \partial B & A^0 \cap B^- \\ \partial A \cap B^0 & \partial A \cap \partial B & \partial A \cap B^- \\ A^- \cap B^0 & A^- \cap \partial B & A^- \cap B^- \end{pmatrix} \in \Omega, \quad (1)
$$

Here $\overset{2}{\underset{s=0}{\vee}} \varepsilon(M_{00s}) = \varepsilon(M_{000}) \vee \varepsilon(M_{001}) \vee \varepsilon(M_{002})$

$$
\begin{pmatrix} \overset{2}{\underset{s=0}{\vee}} \varepsilon(M_{0s0}) & \overset{2}{\underset{s=0}{\vee}} \varepsilon(M_{0s1}) & \overset{2}{\underset{s=0}{\vee}} \varepsilon(M_{0s2}) \\ \overset{2}{\underset{s=0}{\vee}} \varepsilon(M_{1s0}) & \overset{2}{\underset{s=0}{\vee}} \varepsilon(M_{1s1}) & \overset{2}{\underset{s=0}{\vee}} \varepsilon(M_{1s2}) \\ \overset{2}{\underset{s=0}{\vee}} \varepsilon(M_{2s0}) & \overset{2}{\underset{s=0}{\vee}} \varepsilon(M_{2s1}) & \overset{2}{\underset{s=0}{\vee}} \varepsilon(M_{2s2}) \end{pmatrix} = \begin{pmatrix} A^0 \cap C^0 & A^0 \cap \partial C & A^0 \cap C^- \\ \partial A \cap C^0 & \partial A \cap \partial C & \partial A \cap C^- \\ A^- \cap C^0 & A^- \cap \partial C & A^- \cap C^- \end{pmatrix} \in \Omega \quad (2)
$$

$$
\begin{pmatrix} \overset{2}{\underset{s=0}{\vee}} \varepsilon(M_{s00}) & \overset{2}{\underset{s=0}{\vee}} \varepsilon(M_{s01}) & \overset{2}{\underset{s=0}{\vee}} \varepsilon(M_{s02}) \\ \overset{2}{\underset{s=0}{\vee}} \varepsilon(M_{s10}) & \overset{2}{\underset{s=0}{\vee}} \varepsilon(M_{s11}) & \overset{2}{\underset{s=0}{\vee}} \varepsilon(M_{s12}) \\ \overset{2}{\underset{s=0}{\vee}} \varepsilon(M_{s20}) & \overset{2}{\underset{s=0}{\vee}} \varepsilon(M_{s21}) & \overset{2}{\underset{s=0}{\vee}} \varepsilon(M_{s22}) \end{pmatrix} = \begin{pmatrix} B^0 \cap C^0 & B^0 \cap \partial C & B^0 \cap C^- \\ \partial B \cap C^0 & \partial B \cap \partial C & \partial B \cap C^- \\ B^- \cap C^0 & B^- \cap \partial C & B^- \cap C^- \end{pmatrix} \in \Omega \quad (3)
$$

Since the intersection of two sets is empty or not, we obtain the following Theorem 2.

Theorem 2. The topological relation between two disjoint regions and a simple region given by 27-intersections model are exclusive and complete.

2.2.3 Constraints of 27-Intersection Model Between Two Disjoint Regions and a Simple Region

Theorem 1 gave a indispensable condition for a 0-1 matrix corresponding to a topological relation. Therefore, not every 0-1 matrix can be represented in a topological relation, some 0-1 matrix cannot be achieved. To get achievable topological relations, we will add three constraints.

Restricted Condition 1. A 0-1 matrix corresponding to an achievable topological relation must satisfy the following three formulas.

$$
\begin{pmatrix}
\overset{2}{\underset{s=0}{\vee}} \varepsilon(M_{00s}) & \overset{2}{\underset{s=0}{\vee}} \varepsilon(M_{01s}) & \overset{2}{\underset{s=0}{\vee}} \varepsilon(M_{02s}) \\
\overset{2}{\underset{s=0}{\vee}} \varepsilon(M_{10s}) & \overset{2}{\underset{s=0}{\vee}} \varepsilon(M_{11s}) & \overset{2}{\underset{s=0}{\vee}} \varepsilon(M_{12s}) \\
\overset{2}{\underset{s=0}{\vee}} \varepsilon(M_{20s}) & \overset{2}{\underset{s=0}{\vee}} \varepsilon(M_{21s}) & \overset{2}{\underset{s=0}{\vee}} \varepsilon(M_{22s})
\end{pmatrix}
=
\begin{pmatrix}
A^0 \cap B^0 & A^0 \cap \partial B & A^0 \cap B^- \\
\partial A \cap B^0 & \partial A \cap \partial B & \partial A \cap B^- \\
A^- \cap B^0 & A^- \cap \partial B & A^- \cap B^-
\end{pmatrix}
\in \Omega \quad (4)
$$

$$
\begin{pmatrix}
\overset{2}{\underset{s=0}{\vee}} \varepsilon(M_{0s0}) & \overset{2}{\underset{s=0}{\vee}} \varepsilon(M_{0s1}) & \overset{2}{\underset{s=0}{\vee}} \varepsilon(M_{0s2}) \\
\overset{2}{\underset{s=0}{\vee}} \varepsilon(M_{1s0}) & \overset{2}{\underset{s=0}{\vee}} \varepsilon(M_{1s1}) & \overset{2}{\underset{s=0}{\vee}} \varepsilon(M_{1s2}) \\
\overset{2}{\underset{s=0}{\vee}} \varepsilon(M_{2s0}) & \overset{2}{\underset{s=0}{\vee}} \varepsilon(M_{2s1}) & \overset{2}{\underset{s=0}{\vee}} \varepsilon(M_{2s2})
\end{pmatrix}
=
\begin{pmatrix}
A^0 \cap C^0 & A^0 \cap \partial C & A^0 \cap C^- \\
\partial A \cap C^0 & \partial A \cap \partial C & \partial A \cap C^- \\
A^- \cap C^0 & A^- \cap \partial C & A^- \cap C^-
\end{pmatrix}
\in \Omega \quad (5)
$$

$$
\begin{pmatrix}
\overset{2}{\underset{s=0}{\vee}} \varepsilon(M_{s00}) & \overset{2}{\underset{s=0}{\vee}} \varepsilon(M_{s01}) & \overset{2}{\underset{s=0}{\vee}} \varepsilon(M_{s02}) \\
\overset{2}{\underset{s=0}{\vee}} \varepsilon(M_{s10}) & \overset{2}{\underset{s=0}{\vee}} \varepsilon(M_{s11}) & \overset{2}{\underset{s=0}{\vee}} \varepsilon(M_{s12}) \\
\overset{2}{\underset{s=0}{\vee}} \varepsilon(M_{s20}) & \overset{2}{\underset{s=0}{\vee}} \varepsilon(M_{s21}) & \overset{2}{\underset{s=0}{\vee}} \varepsilon(M_{s22})
\end{pmatrix}
=
\begin{pmatrix}
B^0 \cap C^0 & B^0 \cap \partial C & B^0 \cap C^- \\
\partial B \cap C^0 & \partial B \cap \partial C & \partial B \cap C^- \\
B^- \cap C^0 & B^- \cap \partial C & B^- \cap C^-
\end{pmatrix}
\in \Omega \quad (6)
$$

Restricted Condition 2. For simple bounded regions, $A^- \cap B^- \cap C^-$ must be nonempty, namely $M_{222} = 1$.

Restricted Condition 3. The regions A and B are disjoint, so we can get the following formula:

$$
\begin{pmatrix}
\overset{2}{\underset{s=0}{\vee}} \varepsilon(M_{00s}) & \overset{2}{\underset{s=0}{\vee}} \varepsilon(M_{01s}) & \overset{2}{\underset{s=0}{\vee}} \varepsilon(M_{02s}) \\
\overset{2}{\underset{s=0}{\vee}} \varepsilon(M_{10s}) & \overset{2}{\underset{s=0}{\vee}} \varepsilon(M_{11s}) & \overset{2}{\underset{s=0}{\vee}} \varepsilon(M_{12s}) \\
\overset{2}{\underset{s=0}{\vee}} \varepsilon(M_{20s}) & \overset{2}{\underset{s=0}{\vee}} \varepsilon(M_{21s}) & \overset{2}{\underset{s=0}{\vee}} \varepsilon(M_{22s})
\end{pmatrix}
=
\begin{pmatrix}
A^0 \cap B^0 & A^0 \cap \partial B & A^0 \cap B^- \\
\partial A \cap B^0 & \partial A \cap \partial B & \partial A \cap B^- \\
A^- \cap B^0 & A^- \cap \partial B & A^- \cap B^-
\end{pmatrix}
$$
$$
=
\begin{pmatrix}
0 & 0 & 1 \\
0 & 0 & 1 \\
1 & 1 & 1
\end{pmatrix}
\in \Omega
$$

$$(7)$$

According to the above three constraints, we can give specific algorithm, and then get the topology diagram.

2.2.4 Schematic Diagram of Topological Relations Between Two Disjoint Regions and a Simple Region

According to the restricted condition, we can give the main idea of algorithm, as shown in Table 2:

Table 2. The main idea of program

topologicalRelationGen(null; TR)　　　// Input: null Output: topological Relations that satisfy all the constraints $TR\ 2^{27} \leftarrow 2^{27}$ basic topological relations　　　// All basic topological relations 　$TR \leftarrow null$　　　// TR blank 　**for each** t **in** $TR\ 2^{27}$ 　　　　**if** t satisfy restricted condition 1 　　// Inspection t satisfy restricted condition 1 　　　　　**if** t satisfy restricted condition 2　　// Inspection t satisfy restricted condition 2 　　　　　　**if** t satisfy restricted condition 3// Inspection t satisfy restricted condition 3 　　　　$TR \leftarrow \{TR,\ t\}$　　　//　　Put topological Relation into TR 　　　　　　　　**end if** 　　　　　**end if** 　　　**end if** 　　**end for** 　**return** TR　　// Return results

31 kinds of 0-1 matrices were gotten through the program, and then according to the 31 kinds of topological relations matrix, we specifically draw this 31 schematic diagrams of topological relations, as shown in Fig. 3 (Specifically, we give the corresponding topological relations and the corresponding matrix), and verified that 31 kinds of 0-1 matrices can uniquely correspond to the topological relations between two disjoint regions and a simple region.

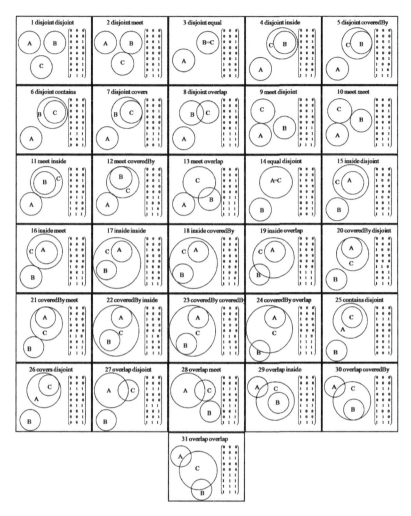

Fig. 3. Schematic diagrams of the 31 kinds of topological relations

3   The Reasoning of Topological Relations

We specially gave the schematic diagrams of the 31 kinds of topological relations between two disjoint regions and a simple region based on the content of Sect. 2.2.4, we will now complete the reasoning of topological relations. Representing the spatial topological relation between region A and region B by using $R(A, B)$, similarly representing the spatial topological relations between region A and region C by using $R(A, C)$, and representing the topological relations between region B and region C by using $R(B, C)$. By the study of $R(A, B)$; $R(A, C)$; $R(B, C)$, the reasoning table of the spatial topological relations can be gotten, as shown in Table 3. Through this table, if we know a topological relation, we will simply reasoning unknown topological relations.

Table 3. Reasoning table

R(A,B)	R(A,C)	R(B,C)
disjoint	disjoint	disjoint meet overlap coveredby inside equal covers contains
	meet	disjoint meet overlap coveredby inside
	overlap	disjoint meet overlap coveredby inside
	coveredby	disjoint meet overlap coveredby inside
	inside	disjoint meet overlap coveredby inside
	equal	disjoint
	covers	disjoint
	contains	disjoint

4   The Conceptual Neighbourhood Graph Between Two Disjoint Regions and a Simple Region

The conceptual neighbourhood graph of 31 topological relations between two disjoint regions and a simple region is given, as shown in Fig. 4. We use the straight line connected two circle to represent the corresponding two kinds of topological relations between one step to another. The digital serial number corresponding to the circle is a schematic diagram of topological relations in Fig. 4.

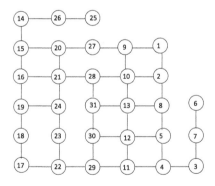

Fig. 4. the conceptual neighbourhood graph of 27-intersection model

5   Applications

5.1   Two Disjoint Areas and Rainfall Area Are Abstractly Considered as Two Disjoint Regions and a Simple Region

The studied region in this article as shown in Fig. 5, shows the possible rainfall situation in Taizhou and Ningbo. Taizhou is regarded as the region A, Ningbo is regarded as the region B, and rainfall region is regarded as region C. Therefore, the problem of predicting the possible rainfall situation in Taizhou and Ningbo is converted to the study of the topological relations between the two disjoint regions A, B and a simple region C.

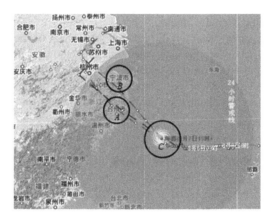

Fig. 5. the regional map

5.2   Affected Probability of the Target Region

We understand that the topological relations between rainfall region, two specified disjoint target regions and 31 kinds of topological relations are a one-to-one correspondence. It is important to note that in order to adapt to the uncertainty of abnormal typhoon landing, we stipulated that the rainfall situation associated with typhoons is completely natural and random, so we think that the probability of occurrence for 31 kinds of topological relations between two disjoint areas and rainfall area is equal, that is the probability of each situation is 1/31.

For example, in region B, we studied the situation where region B is influenced by the rainfall, region A and region B is similar. By the study of schematic diagram of the topological relations, we had found that when the topological relations of region B and region C is disjoint, meet, region B is not affected by the rainfall. There are 13 kinds of situations corresponding to both cases according to Table 2, respectively, which correspond to the number 1, 2, 9, 10, 14, 15, 16, 20, 21, 25, 26, 27, 28 of the Fig. 3, and then the probability is 41.93 %.

When the topological relation of region B and region C is contains, covers, overlap, region B is partly affected by the rainfall. There are 7 kinds of situations corresponding to three cases according to Table 2, respectively, which correspond to the number 6, 7, 8, 13, 19, 24, 31 of the Fig. 3, and then the probability is 22.58 %.

When the topological relation of region B and region C is inside, coveredby, equal, region B is perfectly affected by the rainfall. There are 11 kinds of situations corresponding to three cases according to Table 2, respectively, which correspond to the number 3, 4, 5, 11, 12, 17, 18, 22, 23, 29, 30 of the Fig. 3, and then the probability is 35.48 %.

According to the reasoning table of the topological relations, we can predict the possible rainfall situation. When we know the possible rainfall situation of an area, then we can study the possible rainfall situation of another area. For example, when we know $R(A,B)$ = disjoint and $R(A,C)$ = overlap, then $R(B,C)$ can be obtained, they are disjoint,meet,overlap,coveredby,inside.

We can further study the rainfall situation according to the conceptual neighbourhood graph, namely when we know the rainfall situation of an area, we can further study the next rainfall situation. For instance, when we know that the situation corresponding to the number 25 of Fig. 3 happens, namely contains(A,B),disjoint(A,C), then the next possible rainfall situation corresponds to the number 26 of Fig. 3, namely covers(A,B),disjoint(A,C).

6   The Comparison of Related Work

Li et al. constructed the 8-intersection model [10] that can represent the topological relations between two disjoint regions and a simple region. We can get 17 kinds of topological relations by the 8-intersection model, as shown in Fig. 6.

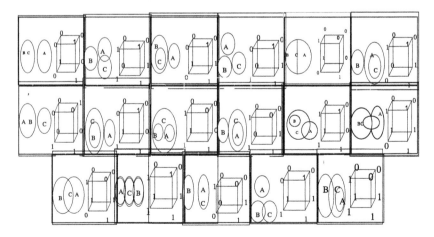

Fig. 6. Schematic diagrams of the 17 kinds of topological relations based on 8-intersection model

In this article, we can get 31 kinds of topological relations by the study of the 27 - intersection model. However, we can only get 17 kinds of topological relations by 8-intersection model. Thus, expressive power of the 27 - intersection model is superior to expressive power of the 8 - intersection model. The reason while the expressive power of the 8 - intersection model is weaker is as follows: 8 - intersection model cannot differentiate between some cases (as shown in Fig. 7: two different situations).

It is worth noting that when the specified target region is not affected by rainfall area, does not mean that the residents of the specified target region cannot take protective measures. It is different that the topological relations of region B, C are disjoint and the topological relations of region B, C are met. As shown in Fig. 7, (*a*) Figure corresponds to the schematic diagram of number 1 of Fig. 3, (*b*) Figure corresponds to the schematic diagram of number 2 of Fig. 3. Although two schematic diagrams of topological relations all represent that the specified target region is not affected by rainfall, they are very different in essence. When the situation which corresponds to (*a*) Figure happens, the residents of the designated target region cannot take protective measures. We can find that (*a*) Figure can be transformed to (*b*) Figure according to the conceptual neighbourhood graph, and target region still is not affected by rainfall. However, while the situation which corresponds to (*b*) Figure happens, the residents of the designated target region must take protective measures to reduce property damage.

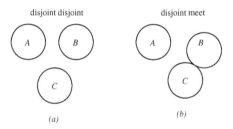

Fig. 7. Two different situations

7   Conclusions

In this article, we established a representation model between two disjoint regions and a simple region, we also got schematic diagrams of the 31 kinds of topological relations between two disjoint regions and a simple region, and then specially gave the reasoning table of the topological relations and the conceptual neighbourhood graph between two disjoint regions and a simple region. This specific theory can be applied to predict the rainfall of two disjoint areas, and we got the chance of rain for a specific area, then we can predict the rainfall situation by reasoning table and conceptual neighborhood graph of topological relations, which is of great significance for residents to take the next step of preventive measures to reduce disaster losses.

Acknowledgment. Funds for this research was provided by The research of Jilin science and technology development project (the youth fund project) (20130522110JH), Jilin province science and technology development plan project (key scientific and technological project) (20140204045NY), Agricultural science and technology achievement transformation project of national science and Technology Department (2014GB2B100021), Jilin province science and technology development plan project (key science and technology research project) (20150204058NY).

References

1. Chen, L., Meng, Z.: An overview on tropical cy-clone research progress in China during the past ten years. Chin. J. Atmos. Sci. **3**, 420–432 (2001). (in Chinese)
2. Wang, S., Liu, D.: Knowledge representation and reasoning for qualitative spatial change. Knowl. Based Syst. **30**, 161–171 (2012)
3. Wang, S., Liu, D.: An efficient method for calculating qualitative spatial relations. Chin. J. Electr. **18**(1), 42–46 (2009)
4. Liu, Y., Liu, D.: A review on spatial reasoning and geographic information system. J. Softw. **11**(12), 1598–1606 (2000)
5. Scott, J., Lee, L.H., et al.: Designing the low-power M.CORETM architecture. In: IEEE Power Driven Microarchitecture Workshop, pp. 29–33. IEEE Computer Society, Haifa (1998)
6. Chang, N.S., Fu, K.S.: Query-by-pictorial-example. IEEE Trans. Softw. Eng. **SE-6**(6), 519–524 (1980)
7. Roussopoulos, N., Faloutsos, C., Sellis, T.: An efficient pictorial database system for PSQL. IEEE Trans. Softw. Eng. **14**(5), 630–638 (1988)
8. Egenhofer, M.J., Franzosa, R.D.: Point-set topological spatial relation. Int. J. Geogr. Inform. Syst. **5**(2), 53–174 (1991)
9. Gao, Z., Wu, L., Yang, J.: Representaion of topological relations between vague objects based on rough and RCC model. Acta Sci. Nat. Univ. Pekin. **44**(4), 597–603 (2008)
10. Li, J., Ouyang, J., Wang, Z., Wang, W.: Representation model of topological relationship among three simple regions. J. Jinlin Univ. Eng. Technol. Ed. **43**(4), 117–122 (2013)

Simplifying Calculation of Graph Similarity Through Matrices

Xu Wang[1,2], Jihong Ouyang[1,2(✉)], and Guifen Chen[3]

[1] College of Computer Science and Technology,
Jilin University, Changchun 130012, China
xuwang10@mails.jlu.edu.cn, ouyj@jlu.edu.cn
[2] Key Laboratory of Symbolic Computation and Knowledge Engineering
of Ministry of Education, Jilin University, Changchun 130012, China
[3] College of Information Technology,
Jilin Agricultural University, Changchun 130118, China
guifchen@163.com

Abstract. A method to simplify the calculation in the process of measuring graph similarity is proposed, where lots of redundant operations are avoided in order to quickly obtain the initial tickets matrix. In this proposal, the element value of the initial tickets matrix is assigned to 1 when it is positive in corresponding position of the paths matrix at the first time. The proposed method calculates the initial tickets matrix value based on the positive value in the paths matrix in a forward and backward way. An example is provided to illustrate that the method is feasible and effective.

Keywords: Graph similarity measure · Paths and tickets · Adjacency matrix · Forward and backward · Simplifying calculation

1 Introduction

The complex objects with structural properties can be naturally established as graphs in chemistry, bioinformatics, data mining, social network, image processing and other fields [1–6]. For example, the science citation can be represented as a graph with the literatures as vertices and the indexes as edges. As the large amount of graph data is increasing, how to characterize the difference of the graph data becomes an important problem [7]. The graph similarity method that efficiently solves the problem can be applied to classify the graphical data.

In order to measure the graph similarity, we usually calculate the number of the common paths of the graph, such as random walk graph kernel (abbreviated as RWGK) [8,9], the shortest path graph kernel (abbreviated as SPGK) [10], and the function of the common paths or tickets [4], etc. The function of the common paths or tickets, in the spirit of the neighborhood counting and all common subsequences in measuring the sequence similarity [11–13], measures graph similarity by calculating all common paths or tickets matrices between two graphs [14–16]. Compared with RWGK and

© IFIP International Federation for Information Processing 2016
Published by Springer International Publishing AG 2016. All Rights Reserved
D. Li and Z. Li (Eds.): CCTA 2015, Part II, IFIP AICT 479, pp. 417–428, 2016.
DOI: 10.1007/978-3-319-48354-2_41

SPGK, this method considers more information of vertices and edges in graph. This method is remarkable in accuracy and running time, in addition, it is not restrictive and can be applied in larger graph dataset. Also, it avoids the tottering phenomenon [17]. In the process of calculating the tickets matrices, the initial tickets matrix is raised from all paths matrix, which is calculated by Floyd-Warshall Algorithm [18–20]. When the element value in the paths matrix is positive, by using the method in [4], the value in the initial tickets matrix is always 1 in the corresponding position, in spite of the multiple calculations of matrices multiplying and addition. However, this method still continues calculating the element value of the paths matrix, and it generates lots of redundant calculations.

For this problem, in this paper, we propose a simplified method to calculate the initial tickets matrix. When the element value in the paths matrix of graph is positive at the first time, this method initializes the value as 1 in the corresponding position in the initial tickets matrix, instead of continuing calculating in [4], and set the value of the initial tickets matrix based on the positive value of the paths matrix in a forward and backward way. This method will avoid lots of redundant calculations, and can be applied in a large-scale complex graph data. An example is provided to illustrate its feasibility and efficiency.

This paper is organized as follows: Sect. 2 describes the related concepts and notations. Section 3 proposes a simplified calculation method to obtain the initial tickets matrix. Section 4 presents the analyses process of the proposed method. The paper is concluded with a summary and an outlook for future work in Sect. 5.

2   Preliminaries

A graph is usually represented as $G = (V, E)$, where the vertices set is $V = \{v_1, \ldots, v_n\}$ and the edges set is $E = \{(v_1, v_2), \ldots, (v_i, v_j)\}$. If vertex v_i is up to the vertex v_j, then there exists a path between v_i and v_j. If the length of two paths is equal, and the vertices and edges are the same, then we call the two paths as the common paths. The more these common paths in two graphs, the more similar the two graphs are [4]. Two graphs are perfectly similar when they share all paths, while two graphs are perfectly dissimilar when they share no paths. A graph can be denoted by its adjacency matrix A as the following equation:

$$a(i,j) = \begin{cases} 1 & if\,(v_i, v_j) \in E \\ 0 & otherwise \end{cases} \tag{1}$$

$a(i, j) = 1$ denotes there exists a distinct 1-long path from v_i to v_j, $\Sigma_{ij}a(i, j)$ denotes the number of all distinct 1-long paths in graph.

For three adjacency matrices A, B, C with n rows and n columns, if $C = A \times B$, then $c(i, j)$ is the element of C in i-th row and j-th column, and it can be calculated as $c(i, j) = \Sigma_k a(i, k) \times b(k, j)$, where $1 \le i \le n, 1 \le j \le n$, $1 \le k \le n$. Obviously, it can be seen that $a^2(i, j)$ denotes a 2-long path from i to j passing the vertex k, which is calculated by the matrices multiplying of A^1 and A^1, $a^2(i, j) = \Sigma_k a^1(i, k)a^1(k, j)$. In other words, a 2-long path from i to j consists of a 1-long path from i to some k and a 1-long path pass vertex from that k to j. A 2-long path exists if and only if both $a(i, k)$ and $a(k, j)$ are positive, i.e. $a(i, k) > 0$, and $a(k, j) > 0$. The element in the matrix $A^2(A^2 = A^1 \times A^1)$ is the number of the 2-long paths in graph, so all possible 2-long paths from i to j is $\Sigma_{ij}a^2(i, j)$. In general, $a^k(i, j)$ denotes the number of distinct k-long paths from v_i to v_j, and $\Sigma_{ij}a^k(i, j)$ denotes the total number of k-long paths in graph, where $1 \le k \le n$. For example, $A^n = A^1 \times A^{n-1}$, the element in A^n is the number of the possible n-long paths in graph. If $C = A + B$, we know $c(i, j) = a(i, j) + b(i, j)$, then the number of the distinct paths from i to j can be calculated as follows:

$$P^1 = A^1$$
$$P^2 = P^1 + A^2$$
$$\vdots \tag{2}$$
$$P^n = P^{n-1} + A^n$$
$$P^n = A^1 + A^2 + \cdots + A^n, A^{n+1} = (0)$$

$P^n(i, j) \ge 0$ denotes the number of the distinct paths from i to j that are at most n-long. $\Sigma_{ij}p^n(i, j)$ denotes the total number of paths in graph.

The measurement of the paths is usually restrictive, so we need to measure the tickets instead of measuring the paths. A ticket is a contracted path by deleting any number of nodes from a path, which is obtained from graph by deleting and contracting the isolated edges [4,21]. If the length of a path between v_i and v_j is more than 1, then there exists a 1-long ticket v_iv_j. The all paths matrix is a 1-long tickets matrix that can be calculated as follows:

$$t^1(i,j) = \begin{cases} 1 & if \ p^n(i,j) \ge 1 \\ 0 & otherwise \end{cases} \tag{3}$$

By Eq. (3), there is $T^1 = P^n$. We call the matrix T^1 as the initial tickets matrix. The method of calculating all common tickets is similar to the method of calculating all the common paths. The more these common tickets, the more similar the graphs are [4].

In the paper, the graph is directed and acyclic, and the length of the path we consider is finite. If a graph is a cycle graph, the length of the path with the cycle is infinite, so it cannot be used to measure all common paths or tickets.

3   A Simplified Calculation Method to Calculate Tickets Matrix

This section proposes a simplified calculation method of measuring graph similarity in order to quickly obtain the initial tickets matrix. In the process of calculating the initial tickets matrix, when the element value of the paths matrix is positive at the first time, this method sets the value as 1 in the corresponding position of the initial tickets matrix, and does not calculate the value again. It avoids the redundant calculations of matrices multiplying and addition.

In order to obtain the initial tickets matrix, based on Eq. (2) and Floyd-Warshall Algorithm in [19], we need to calculate all paths matrix P^n. When P^n is known, we can obtain the initial tickets matrix T^1 by using Eq. (3). Obviously, we note that the element value in the initial tickets matrix is always 1 when the value is positive in the corresponding position of the paths matrix. If the value in the paths matrix is positive at the first time, in spite of lots of calculations of matrices multiplying and addition in this position, the value is always positive. So we can conclude that if the value in the paths matrix is positive at the first time, the value in the initial tickets matrix must be equal to 1. It does not need the calculation in the corresponding position of the paths matrix again. The aim of the method in the paper is to reduce the redundant calculations when the element value in the paths matrix is positive.

An adjacency matrix of the graph G is shown in the Fig. 1, where $a(i, j) = 1$ denotes that there is a distinct 1-long path from v_i to v_j. $\Sigma_{ij}a(i, j)$ denotes the number of all distinct 1-long paths in graph G.

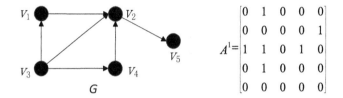

$$A^1 = \begin{bmatrix} 0 & 1 & 0 & 0 & 0 \\ 0 & 0 & 0 & 0 & 1 \\ 1 & 1 & 0 & 1 & 0 \\ 0 & 1 & 0 & 0 & 0 \\ 0 & 0 & 0 & 0 & 0 \end{bmatrix}$$

Fig. 1. The directed and acyclic graph G and its adjacency matrix A^1.

3.1   Simplified Calculation When 1-Long Paths is Fewer

When the number of 1-long paths is fewer than n^2 in the adjacency matrix of graph, where n^2 denotes the element number of the adjacency matrix, based on the adjacency matrix A^1, we calculate the initial tickets matrix T^1 as follows: $t^1(i, j) = a^1(i, j)$, where $1 \le i \le n, 1 \le j \le n$. For $a^1(i, j) = 1$, it represents that there exists a 1-long path between v_i and v_j. In other words, there exists a ticket $t^1(i, j) = 1$, and we don't calculate $t^1(i, j)$ again. It reduces the redundant calculations in i-th row and j-th column of T^1. When the element value in T^1 is 0, we can calculate the value in i backward and j forward step.

$$t^1(k, j) = \begin{cases} 1 & \text{if } a^1(k, i) = 1 \text{ and } a^1(k, j) = 0 \\ 0 & \text{otherwise} \end{cases} \tag{4}$$

If $a^1(k, i) = 1$, where $1 \le k \le n$, $k \ne i$ and $k \ne j$, it shows that $v_k v_i$ is a path in the graph, i is the backward vertex pointed by k. It implies that there exists a 1-long ticket between v_k and v_j, then $t^1(k, j) = 1$.

$$t^1(i,k) = \begin{cases} 1 & if\ a^1(j,k) = 1\ and\ a^1(i,k) = 0 \\ 0 & otherwise \end{cases} \qquad (5)$$

If $a^1(j, k) = 1$, where $1 \le k \le n$, $k \ne i$ and $k \ne j$, it shows that v_jv_k is a path in the graph, j is the forward vertex pointing to k. It implies that there exists a 1-long ticket between v_i and v_k, then $t^1(i, k) = 1$.

By using Eqs. (4) and (5), we can calculate element values of the initial tickets matrix T^1 in accordance with all paths matrix P^n. Because the number of $\{1 \le i \le n, 1 \le j \le n \mid a^1(i, j) = 1\}$ is small, the calculation of $t^1(i, j)$ is not complicated, which means this method reduces redundant calculations of matrices multiplying and addition in comparison with Floyd-Warshall Algorithm.

3.2 Simplified Calculation When 1-Long Paths is More

When the number of 1-long paths is closer to n^2 in the adjacency matrix of graph, where n^2 denotes the element number of the adjacency matrix, based on the adjacency matrix A^1, we also calculate the initial tickets matrix T^1 as follows: $t^1(i, j) = a^1(i, j)$, where $1 \le i \le n, 1 \le j \le n$. When the element value in T^1 is 1, we don't calculate the value again. It reduces the redundant calculations in the position of the value. For $a^1(i, j) = 0$, it indicates that there doesn't exist a 1-long path between v_i and v_j, we can calculate the $t^1(i, j)$ value of the initial tickets matrix T^1 in i forward and j backward step.

$$t^1(i,j) = \begin{cases} a^1(k,j) & if\ a^1(i,k) = 1 \\ 0 & otherwise \end{cases} \qquad (6)$$

If $a^1(i, k) = 1$, where $1 \le k \le n$, $k \ne i$ and $k \ne j$, it shows that v_iv_k is a path in the graph, i is the forward vertex pointing to k. The problem that whether there exists a path between v_i and v_j is changed to the problem that whether there exists a n-long path between v_k and v_j, where $n \ge 1$, then $t^1(i, j)$ recursively is equivalent to $a^1(k, j)$.

$$t^1(i,j) = \begin{cases} a^1(i,k) & if\ a^1(k,j) = 1 \\ 0 & otherwise \end{cases} \qquad (7)$$

If $a^1(k, j) = 1$, where $1 \le k \le n$, $k \ne i$ and $k \ne j$, it shows that v_kv_j is a path in the graph, j is the backward vertex pointed by k. The problem that whether there exists a path between v_i and v_j is changed to the problem that whether there exists a n-long path between v_i and v_k, where $n \ge 1$, then $t^1(i, j)$ recursively is equivalent to $a^1(i, k)$.

By using Eq. (6) and Eq. (7), we can calculate element values of the initial tickets matrix T^1 in accordance with all paths matrix P^n. Because the number of $\{1 \le i \le n, 1 \le j \le n \mid a^1(i, j) = 1\}$ in the adjacency matrix is large, then the number of $\{1 \le i \le n, 1 \le j \le n \mid a^1(i, j) = 1\}$

$n \mid a^1(i, j) = 0\}$ is small, so the calculation of $t^1(i, j)$ also is not complicated. In other words, this method also reduces redundant calculations of matrices multiplying and addition comparing to Floyd-Warshall Algorithm.

In the spirit of measuring the sequence similarity by all common subsequences, a graph similarity measure method is proposed by calculating all common tickets. By means of the measuring sequence accuracy, the accuracy is more precise by calculating all common tickets than calculating all common paths. In order to calculate all common tickets in the process of measuring the graph similarity, how to quickly obtain the initial tickets matrix is a key problem. The method mentioned above, which avoids lots of redundant calculations of matrices multiplying and addition in the process of calculating the initial tickets matrix, can effectively solve the problem.

4   Method Analyses

This section gives two algorithms to simplify calculation of the initial tickets matrix, and illustrates its feasibility by providing an example.

4.1   The Algorithm Analyses

Calculating all paths matrix P^n directly obtains the complexity is $O(n^4)$, but based on the Floyd-Warshall algorithm, an algorithm has been designed which has a complexity of $O(n^3)$ [19].

Floyd-Warshall Algorithm:

Let $A^1 = a(i, j)$ be an $n \times n$ adjacency matrix of graph.

1. $P \leftarrow A^1$
2. for $k = 1$ to n do
3.   for $i = 1$ to n do
4.   for $j = 1$ to n do
5. $p(i, j) = p(i, j) + p(i, k) \times p(k, j)$
6.   end for
7.   end for
8. end for
9. return P

If A^1 is the adjacency matrix of a graph G, then P gives the number of all paths in graph G. If A^1 is the common adjacency matrix of two or more graphs, then P gives the number of all common paths. The return matrix P is used to calculate the initial tickets matrix.

As presented in Eqs. (4) and (5), when the number of the 1-long paths is fewer than n^2, we present the idea of simplifying calculation of the initial tickets matrix as follows:

Algorithm 1: The simplified calculation of the initial tickets matrix T^1

Let $A = a(i, j)$ be an $n \times n$ adjacency matrix of graph.

1. $T^1 \leftarrow A^1$
2. for $i = 1$ to n do
3.   for $j = 1$ to n do
4.   if $t^1(i, j) > 0$
5. $t^1(i, j) = 1$;   // When $t^1(i, j) > 0$, initialize the tickets matrix T^1
6. $Q \leftarrow t^1(i, j)$; //When $t^1(i, j) = 1$, put $t^1(i, j)$ in the queue Q, i and j are known
7.   end
8.   end for
9. end for
10. while $|Q| > 0$ //Because of $|\{a^1(i, j) = 1\}|$ is small, the implement times are fewer
11. $t^1(i, j) = q \leftarrow Q$ // Take out an element from queue Q
12.   for ($k = 1$, $k \neq i$ and $k \neq j$) to n do
13.   if $t^1(j, k) = 1$ and $t^1(i, k) \notin q$ // There is a path from j to k in the matrix A
                                        and $t^1(i, k)$ is not in queue Q
14. $T^1 \leftarrow t^1(i, k) = 1$;   // $v_i v_k$ is a ticket, initialize the matrix T^1: $t^1(i, k) = 1$
15.   end
16.   if $t^1(k, i) = 1$ and $t^1(k, j) \notin q$ // There is a path from k to i in the matrix A
                                        and $t^1(k, j)$ is not in queue Q
17. $T^1 \leftarrow t^1(k, j) = 1$;   // $v_k v_j$ is a ticket, initialize the matrix T^1: $t^1(k, j) = 1$
18.   end
19.   end for
20. $Q = Q - \{q\}$ // Remove q from Q
21. return T^1

As shown in Eqs. (6) and (7), when the number of the 1-long paths is closer to n^2, we present the idea of simplifying calculation of the initial tickets matrix as follows:

Algorithm 2: The simplified calculation of the initial tickets matrix T^1

Let $A^1 = a^1(i, j)$ be an $n \times n$ adjacency matrix of graph.

1. $T^1 \leftarrow A^1$
2. for $i = 1$ to n do
3.   for $j = 1$ to n do
4.   if $t^1(i, j) > 0$
5. $t^1(i, j) = 1$;   // When $t^1(i, j) > 0$, initialize the matrix T^1
6.   else
7. $Q \leftarrow t^1(i, j)$ // When $t^1(i, j) = 0$, put $t^1(i, j)$ in the queue Q, i and j are known.
8.   end
9.   end for
10. end for
11. while $|Q| > 0$ //Because of $|\{a^1(i, j) = 0\}|$ is small, the implement times are fewer
12. $t^1(i, j) = q \leftarrow Q$ // Take out an element from queue Q
13. $m = j$;   // Assign the subscript j to m
14. $h = i$;   // Assign the subscript i to h
15.   for ($k = 1$, $k \neq i$ and $k \neq j$) to n do

16.   if $t^1(k, j) = 1$ // Find out k pointing to j in the matrix A
17. $j = k$;   // j points to the forward vertex k
18.   if $t^1(i, j) = 1$ //It shows that there is a n-long path from i to j, where $n \geq 1$,
                                so there is a ticket from i to m, in other words, $q = 1$
19. $t^1(i, m) = 1$;
20. $T^1 \leftarrow t^1(i, m)$;   // Initialize the matrix T^1: $q = t^1(i, m) = 1$
21.   end
22.   end
23.   if $t^1(i, k) = 1$ // Find out k pointed by i in the matrix A
24. $i = k$;   // i points to the backward vertex k
25.   if $t^1(i, j) = 1$ //It shows that there is a n-long path from i to j, where $n \geq 1$,
                                so there is a ticket from h to j, in other words, $q = 1$
26. $t^1(h, j) = 1$;
27. $T^1 \leftarrow t^1(h, j)$;   // Initialize the matrix T^1: $q = t^1(h, j) = 1$
28.   end
29.   end
30.   end for
31. $Q = Q - \{q\}$ // Remove q from Q
32. return T^1

The return matrix T^1 is the initial tickets matrix, in which the value of element is 1 or 0. The matrix T^1 is used to calculate the number of all common tickets in two or more graphs for measuring graphs similarity. The queue Q is used to store $t^1(i, j)$. It can be seen from line 10 in Algorithm 1 and line 11 in Algorithm 2, the smaller the length of the queue, the more quickly two algorithms implement. So, Algorithm 1 is suitable to the adjacency matrix that the number of the 1-long paths is fewer than n^2, and Algorithm 2 is suitable to the adjacency matrix that the number of the 1-long paths is closer to n^2. Compared with line 5 in Floyd-Warshall Algorithm, Algorithm 1 and Algorithm 2 can effectively solve the redundant calculations in the loop of line 12 and line 15, respectively. The two algorithms usually could be easily implemented and it can effectively handle the big graphs.

4.2   Example

For the directed and acyclic graph G as shown in Fig. 1, we calculate the initial tickets matrix of G by using Eqs. (2) and (3). Based on $A^k = A^1 \times A^{k-1}$, $A^k = (0)$ for k > 3, we can obtain the 2-long paths matrix A^2 and the 3-long paths matrix A^3 as shown in Fig. 2.

$$A^2 = A^1 \times A^1 = \begin{bmatrix} 0 & 0 & 0 & 0 & 1 \\ 0 & 0 & 0 & 0 & 0 \\ 0 & 2 & 0 & 0 & 1 \\ 0 & 0 & 0 & 0 & 1 \\ 0 & 0 & 0 & 0 & 0 \end{bmatrix} \qquad A^3 = A^1 \times A^2 = \begin{bmatrix} 0 & 0 & 0 & 0 & 0 \\ 0 & 0 & 0 & 0 & 0 \\ 0 & 0 & 0 & 0 & 2 \\ 0 & 0 & 0 & 0 & 0 \\ 0 & 0 & 0 & 0 & 0 \end{bmatrix}$$

Fig. 2. The 2-long paths matrix A^2 and 3-long paths matrix A^3.

By using Eqs. (2) and (3), we can obtain all paths matrix P^n and the initial tickets matrix T^1 as shown in Fig. 3, and its computational complexity is $O(n^4)$.

$$P^n = A^1 + A^2 + A^3 = \begin{bmatrix} 0 & 1 & 0 & 0 & 1 \\ 0 & 0 & 0 & 0 & 1 \\ 1 & 3 & 0 & 1 & 3 \\ 0 & 1 & 0 & 0 & 1 \\ 0 & 0 & 0 & 0 & 0 \end{bmatrix} \qquad T^1 = \begin{bmatrix} 0 & 1 & 0 & 0 & 1 \\ 0 & 0 & 0 & 0 & 1 \\ 1 & 1 & 0 & 1 & 1 \\ 0 & 1 & 0 & 0 & 1 \\ 0 & 0 & 0 & 0 & 0 \end{bmatrix}$$

Fig. 3. All paths matrix P^n and the initial tickets matrix T^1.

We also calculate the initial tickets matrix T^1 by using Floyd-Warshall Algorithm. By using line 2 to line 8 in Floyd-Warshall Algorithm, we can calculate the number of all paths $p(i, j)$, where $1 \le i \le n, 1 \le j \le n$. After calculating all paths matrix P, we obtain the initial tickets matrix T^1 by using Eq. (3), and its computational complexity is $O(n^3)$.

For the adjacency matrix A^1 as shown in Fig. 1, the number of 1-long paths is 6, it is fewer than $25 (n = 5, n^2 = 25)$, so we apply Algorithm 1 to calculate the initial tickets matrix T^1 as follows:

Step 1: In accordance with line 2 to line 9 in Algorithm 1, we set the initial tickets matrix $T^1 = A^1$ as shown in Fig. 4(a) and the queue $Q:\{t^1(v_1, v_2) = 1, t^1(v_2, v_5) = 1, t^1(v_3, v_1) = 1, t^1(v_3, v_2) = 1, t^1(v_3, v_4,) = 1, t^1(v_4, v_2) = 1\}$.

Step 2: For the $q \in Q$, when $q = t^1(v_1, v_2) = 1$, there exists $t^1(v_2, v_5) = 1$ and $t^1(v_3, v_1) = 1$, it shows that v_1, v_2, v_5 and v_3, v_1, v_5 are 2-long paths. In other words, $v_1 v_5$ and $v_3 v_5$ are 1-long tickets. By using the line 12 to line 19 in Algorithm 1, we can calculate new tickets $t^1(v_1, v_5) = 1$ and $t^1(v_3, v_5) = 1$ as shown in Fig. 4(b).

Similarly, we can calculate the final T^1 as presented in Fig. 4(c), the values identified by the box are the new tickets. When $t^1(v_i, v_j) = 1$, where $1 \le i \le n, 1 \le j \le n$, Algorithm 1 doesn't need to repeatedly calculate $t^1(v_i, v_j)$. It reduces redundant calculations of matrices multiplying and addition comparing to Floyd-Warshall Algorithm.

$$T^1 = A^1 = \begin{bmatrix} 0 & 1 & 0 & 0 & 0 \\ 0 & 0 & 0 & 0 & 1 \\ 1 & 1 & 0 & 1 & 0 \\ 0 & 1 & 0 & 0 & 0 \\ 0 & 0 & 0 & 0 & 0 \end{bmatrix} \quad T^1 = \begin{bmatrix} 0 & 1 & 0 & 0 & \boxed{1} \\ 0 & 0 & 0 & 0 & 1 \\ 1 & 1 & 0 & 1 & \boxed{1} \\ 0 & 1 & 0 & 0 & 0 \\ 0 & 0 & 0 & 0 & 0 \end{bmatrix} \; when \; q = t^1(v_1, v_2). \quad T^1 = \begin{bmatrix} 0 & 1 & 0 & 0 & \boxed{1} \\ 0 & 0 & 0 & 0 & 1 \\ 1 & 1 & 0 & 1 & \boxed{1} \\ 0 & 1 & 0 & 0 & \boxed{1} \\ 0 & 0 & 0 & 0 & 0 \end{bmatrix}$$

$$\quad\quad\quad\quad (a) \quad\quad\quad\quad\quad\quad\quad (b) \quad\quad\quad\quad\quad\quad\quad\quad\quad\quad (c)$$

Fig. 4. (a) Assigned T^1 using the adjacency matrix A^1; (b) when $q = t^1(v_1, v_2)$, we calculate T^1 using Algorithm 1; (c) we obtain the final T^1 based on Algorithm 1. The values identified by the box are the new tickets.

For a directed and acyclic graph G' and its adjacency matrix A'^1 as shown in Fig. 5, the number of the 1-long paths is 18, it is closer to $25(n = 5, n^2 = 25)$, so we apply Algorithm 2 to calculate the initial tickets matrix T'^1 as follows:

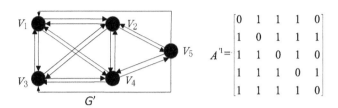

$$A'^1 = \begin{bmatrix} 0 & 1 & 1 & 1 & 0 \\ 1 & 0 & 1 & 1 & 1 \\ 1 & 1 & 0 & 1 & 0 \\ 1 & 1 & 1 & 0 & 1 \\ 1 & 1 & 1 & 1 & 0 \end{bmatrix}$$

Fig. 5. The directed and acyclic graph G' and its adjacency matrix A'^1.

Step 1: In accordance with line 2 to line 10 in Algorithm 2, we set the initial tickets matrix $T'^1 = A'^1$ as shown in Fig. 6(a) and the queue $Q : \{t'^1(v_1, v_5) = 0, t'^1 (v_3, v_5) = 0\}$. In this paper, the path doesn't contain a cycle. In other words, we don't consider the cycle $v_i, \ldots, v_k, \ldots, v_i$, where $1 \le i \le n, 1 \le k \le n$. We set $t'^1(v_i, v_i)$ as 0 in the initial tickets matrix T'^1. So, in line 7 of Algorithm 2, we don't need to put $t'^1(v_i, v_i)$ in the queue Q.

Step 2: For the $q \in Q$, when $q = t'^1(v_1, v_5) = 0$, there exists $t'^1(v_1, v_2) = 1$ and $t'^1(v_2, v_5) = 1$, it shows that there is a 2-long path from v_1 to v_5. In other words, $v_1 v_5$ is a 1-long ticket. By using the line 15 to line 30 in Algorithm 2, we can calculate new ticket $t'^1(v_1, v_5) = 1$ as shown in Fig. 6(b).

Similarly, we can calculate the final T'^1 as shown in Fig. 6(c), the values identified by the box are the new tickets. When $t'^1(v_i, v_j) = 1$, where $1 \le i \le n, 1 \le j \le n$, Algorithm 2 doesn't need to repeatedly calculate $t'^1(v_i, v_j)$. It reduces redundant calculations of matrices multiplying and addition comparing to Floyd-Warshall Algorithm.

$$T'^1 = A'^1 = \begin{bmatrix} 0 & 1 & 1 & 1 & 0 \\ 1 & 0 & 1 & 1 & 1 \\ 1 & 1 & 0 & 1 & 0 \\ 1 & 1 & 1 & 0 & 1 \\ 1 & 1 & 1 & 1 & 0 \end{bmatrix} \quad T'^1 = \begin{bmatrix} 0 & 1 & 1 & 1 & \boxed{1} \\ 1 & 0 & 1 & 1 & 1 \\ 1 & 1 & 0 & 1 & 0 \\ 1 & 1 & 1 & 0 & 1 \\ 1 & 1 & 1 & 1 & 0 \end{bmatrix} \text{ when } q = t'^1(v_1, v_5). \quad T'^1 = \begin{bmatrix} 0 & 1 & 1 & 1 & \boxed{1} \\ 1 & 0 & 1 & 1 & 1 \\ 1 & 1 & 0 & 1 & \boxed{1} \\ 1 & 1 & 1 & 0 & 1 \\ 1 & 1 & 1 & 1 & 0 \end{bmatrix}$$

(a)   (b)   (c)

Fig. 6. (a) Assigned T'^1 using the adjacency matrix A'^1; (b) when $q = t'^1(v_1, v_5)$, we calculate T'^1 using Algorithm 2; (c) we obtain the final T'^1 based on Algorithm 2. The values identified by the box are the new tickets.

Table 1. The computational complexity of calculating the initial tickets matrix algorithms

Calculating algorithms	Computational complexity
Equations (2) and (3)	$O(n^4)$
Floyd-Warshall Algorithm	$O(n^3)$
Algorithm 1	$O(n^2 + lk))$
Algorithm 2	$O(n^2 + lk))$

Let l denote the length of the queue Q, l is usually a constant in the best case. In Algorithm 1, the computational complexity is $O(n^2)$ in the loop of line 2 and line 3 and $O(lk)$ in the loop line 10 and line 12, where $1 \leq k \leq n$, $k \neq i$ and $k \neq j$. Similarly, in Algorithm 2, the computational complexity is $O(n^2)$ in the loop of line 2 and line 3 and $O(lk)$ in the loop of line 11 and line 15, where $1 \leq k \leq n$, $k \neq i$ and $k \neq j$. So the computational complexity of Algorithm 1 and Algorithm 2 is $O(n^2 + lk)$ and less than $O(n^3)$, approximately $O(n^2)$ in the best case. However, the computational complexity of using Eqs. (2) and (3) is $O(n^4)$, and Floyd-Warshall Algorithm is $O(n^3)$. In Table 1, we list the complexity of calculating the initial tickets matrix algorithms. Obviously, Algorithms 1 and 2 are more effective.

5   Conclusions

This paper proposes a simplified calculation method in the process of measuring graph similarity. This method sets the element value of the initial tickets matrix as 1 when the element value of the paths matrix is positive at the first time, and reduces lots of redundant calculations of matrices multiplying and addition comparing to Floyd-Warshall Algorithm. Depending on the number of 1-long paths in the adjacency matrix, this paper presents two algorithms to calculate the initial tickets matrix. The two algorithms calculate the element value of the initial tickets matrix based on the positive value of the paths matrix in a forward and backward way. By the algorithm analyses and the given example, these algorithms are proved to be feasible and effective. When we calculate the initial tickets matrix of the complex graph with large number of 1-long paths in the adjacency matrix, Algorithm 2 is more applicable, because the length of the stored queue is small.

In the near future, we will compare the method in this paper with graph kernel, random walk graph kernel, and the kernel function of all common paths or tickets in classification accuracy and running time through experiments. By means of easy calculation of this method, we will apply it to the undirected graph or the complex graph datasets.

Acknowledgment. This work was supported by the National Natural Science Foundation of China (Nos. 61133011, 61170092, 60973088, 60873149).

References

1. Bapat, R.B.: Graphs and Matrices. Springer Press, New York (2011)
2. Masahiro, H., Yasushi, O., Susumu, G., et al.: Development of a chemical structure comparison method for integrated analysis of chemical and genomic information in the metabolic pathways. J. Am. Chem. Soc. **125**(39), 11853–11865 (2003)
3. Volker, S.: Bioinformatics, Problem Solving Paradigms. Springer Press, Berlin (2008)
4. Cees, H.E., Hui, W.: Kernels for acyclic digraphs. Pattern Recogn. Lett. **33**(16), 2239–2244 (2012)
5. Kang, U., Hanghang, T., Jimeng, S.: Fast random walk graph kernel. In: Proceedings of the Twelfth SIAM International Conference on Data Mining, pp. 828–838 (2012)
6. Amin, K., Peyman, M.: A general framework for regularized, similarity-based image restoration. IEEE Trans. Image Process. **23**(12), 5136–5151 (2014)
7. Ye, Y., Guoren, W., Lei, C., et al.: Graph similarity search on large uncertain graph databases. Int. J. Very Large Data Bases **24**(2), 271–296 (2015)
8. Nino, S., Karsten, M.B.: Fast subtree kernels on graphs. In: The annual Conference on Neural Information Processing Systems, pp. 1660–1668 (2009)
9. Vishwanathan, S.V.N., Schraudolph Nicol, N., Risi, K., et al.: Graph Kernels. J. Mach. Learn. Res. **11**, 1201–1242 (2010)
10. Karsten, M.B., Hans-Peter, K.: Shortest-path kernels on graphs. In: IEEE International Conference on Data Mining, pp. 74–81 (2005)
11. Hui, W.: Nearest neighbors by neighborhood counting. IEEE Trans. Pattern Anal. Mach. Intell. (TPAMI) **28**(6), 942–953 (2006)
12. Hui, W.: All common subsequences. In: Proceeding of 20th International Joint Conference on Artificial Intelligence, pp. 635–640 (2007)
13. Hui, W., Fionn, M.: A study of the neighborhood counting similarity. IEEE Trans. Knowl. Data Eng. **20**(4), 449–461 (2008)
14. Jun, H., Liu Hongyan, Yu., Jeffrey, X., et al.: Assessing single-pair similarity over graphs by aggregating first-meeting probabilities [J]. Inf. Syst. **42**, 107–122 (2014)
15. Michael, K., Janusz, Dutkowski, Michael, Yu., et al.: Inferring gene ontologies from pairwise similarity data. Bioinformatics **30**(12), 34–42 (2014)
16. Andrew, R., Byunghoon, K., Jaemin, L., et al.: Graph kernel based measure for evaluating the influence of patents in a patent citation network. Expert Syst. Appl. **42**(3), 1479–1486 (2015)
17. Pierre, M., Nobuhisa, U., Tatsuya, A., et al.: Extensions of marginalized graph kernels. In: Proceedings of the 21st International Conference on Machine Learning, pp. 552–559 (2004)
18. Robert, W.: Floyd. algorithm 97: shortest path. Commun. ACM **5**(6), 345 (1962)
19. Zoltán, K.: On scattered subword complexity. Acta Univ. Sapientiae Informatica **3**(1), 127–136 (2011)
20. Wang, I.: An algebraic decomposed algorithm for all pairs shortest paths. Pac. J. Optim. **10**(3), 561–576 (2014)
21. Reinhard, D.: Graph Theory, 3rd edn. Springer Press, New York (2005)

A Systematic Method for Quantitative Diameter Analysis of Sprayed Pesticide Droplets

Wei Ma[1,2(✉)], Xiu Wang[2], Lijun Qi[1], and Yanbo Huang[3]

[1] College of Engineering, China Agricultural University, Beijing 100083, China
maw516@163.com, qilijun@cau.edu.cn
[2] National Engineering Research Center for Information Technology in Agriculture,
Beijing 100097, China
xiuwang@263.net
[3] United States Department of Agriculture, Crop Production Systems Research Unit,
Agriculture Research Service, Stoneville, MS, USA
yanbo.huang@ars.usda.gov

Abstract. In this study, a new systematic method for quantitative diameter analysis of sprayed pesticide droplets was developed. This method adopts the screw motion with precision motor and uses the vibration frequency of the precision control technology. With this method, the bubbles contained in the pesticide droplets were eliminated and the generated droplets are no longer adhere to the tip. The accuracy of this method is significantly improved compared with the previously manually operated device. The results indicated that with the calibration coefficient, the error between of the actual droplet size and the preset droplet size value was 97.2 %. This newly developed method is very valuable for future studies of droplet distribution over crop leaves for reducing the amount of pesticides sprayed on the crops.

Keywords: Quantitative diameter analysis · Pesticide droplet · Droplet spreading · Droplet evaporation · Microsyringe

1 Introduction

Pesticides are widely used in crop fields as an effective means of pest control for crop growth [1]. However, with extensive use of the chemicals, there are serious consequences. For example, soil compaction due to overuse of pesticides could reduce the fertility of soil to a certain extent, and bring yield loses of the grains. Also, atmospheric and water pollutions from pesticides could have serious impact on the health of humans and animals [2]. After years of studies worldwide, it can be determined that the off-target drift of pesticides has become one of major sources of the issues mentioned above [3]. The studies have been conducting to characterize the pesticide droplet micro behavior on plant leaf surface to reduce the stress caused by the spray drift. How to precisely determine droplet size is one of the research focuses by scientists and engineers of application technology. How to remove bubbles in the spray liquid drops is an issue to

© IFIP International Federation for Information Processing 2016
Published by Springer International Publishing AG 2016. All Rights Reserved
D. Li and Z. Li (Eds.): CCTA 2015, Part II, IFIP AICT 479, pp. 429–436, 2016.
DOI: 10.1007/978-3-319-48354-2_42

impede the study. At present, there is no instrument capable of generating pesticide droplets with specified particle size. There is no report that the instrument can remove the dissolved air bubbles in the liquid drop either [4]. The droplets are mostly still generated manually [4–8]. Because of the limited experimental means, the research accuracyabout the diffusion of the pesticide droplets and the evaporation of the single droplet in practice for crop production management is limited. [9–13]. This paper is mainly to explore a kind of systematic method for quantitative diameter analysis of sprayed pesticide droplets. Specifically, this study developed a device for generating sprayed pesticide droplet with specified diameters with the method for quantitative diameter analysis of sprayed pesticide droplets to address the issues mentioned above.

2   System and Method

2.1   Experimental Device

Figure 1 is the structural diagrammatic sketch of the device for generating the sprayed pesticide droplets with specified diameters. Because there is no instrument capable of generating pesticide droplets with specific particle size and the studies about the diffusion of the pesticide droplets and the evaporation characteristics of the single droplet have not been developed in practice for crop production management, this study developed a device for generating sprayed pesticide droplets with specified diameter with a sliding component positioned on a track and capable of sliding along the track. One end of the slider is connected with one end of a piston. The other end of the slider is connected with a driving motor through a lead screw. The other end of the piston can move in a droplet generator along with the sliding of the slider and the droplets in the droplet generator are released by a guided pipe. The droplet generator is clamped on a supporting structure. A vibration unit for removing bubbles in the droplet generator is further arranged on the structure. The vibration unit includes an eccentric counterweight module and a buffer module connected with the eccentric counterweight module. The eccentric counterweight module rotates to generate high-frequency vibration, and the vibration is slowed down to the overflow frequency of the bubbles through the buffer module. The buffer module is made of rubber. The minimal propulsion precision of the piston is 0.2 mm.

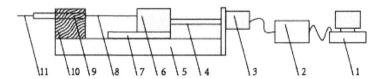

1: control center; 2: driver; 3: driving motor; 4: lead screw; 5: base; 6: sliding device; 7: track; 8: piston; 9: droplet generator; 10: support; 11: guided pipe.

Fig. 1. Illustrative diagram of the device for generating the sprayed pesticide droplets with specified diameters.

The device further contains a control center and a driver. The driver is used for receiving and obtaining instructions of the control center and controlling the driving motor according to control instructions.

The mathematical equation for specifying droplet diameters is:

$$0.2NS_n = \alpha \Pi D_g^3 \tag{1}$$

where N is the moving distance of the piston; S_n is the cross-sectional area of the extrusion droplet piston and D_g is the diameter of the droplets.

2.2 Experimental Procedure

The method developed in this study can be summarized as the procedure in the following steps:

1. Inject chemicals into the droplet generator;
2. Switch on the driving motor for the lead screw to rotate and release one droplet from the droplet generator through a guided pipe;
3. Determine the diameter of the released droplet and calculate the liquid correction coefficient through the mathematical Eq. (1) with the corresponding diameter of the droplet; and
4. Generating the next droplet with the specified diameter.

Compared with the method of generating droplet by the precision metering pump, this method is based on the spiral movement. The developed device and method in this study have the following advantages:

1. Step motor receive a certain number of driving pulse signalfrom the control center, and drives the lead screw to rotate;
2. The lead screw drives the slider to a designated position on the guiding track;
3. The exact sizedroplets are generated using adjustable vibration to eliminate bubbles. So research accuracy is improvedfor diffusion law and evaporationproperty of plant leavesdroplets.

Figure 2 is the schematic diagram of the developed vibration unit with the enlarged view of the part of the developed droplet generator. Figure 3 is the work flow of the system.

In the experiments the testing liquid used was rhodamine-B self modulated (concentration is 0.1 %). The solution reported in the literature was used to replace the pesticide. Sodium chloride salt solution (2 ml packaging, 10 %, Guangzhou Pharmaceutical Factory, Guangzhou, China) was used as a calibration.

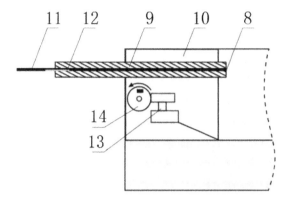

8: piston; 9: droplet generator; 10: support; 11: guide pipe; 12: calibrated scale; 13: buffer module; 14: eccentric counterweight module.

Fig. 2. A schematic diagram of the vibration unit with droplet generator.

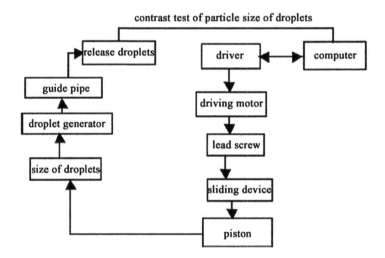

Fig. 3. System work flow.

3   Results and Discussion

3.1   Droplet Generation

The driving motor is connected with the control center through the driver. The driver is used for receiving instructions of the control center and controlling the driving motor to operate according to the control instructions. The control center is a control computer. The computer is connected with the driver through a 232 serial port and used for converting a numerical value of the diameter of the droplets, which is input from the computer, to machine instructions of the device. The instructions are sent by the computer, and the minimal propulsion precision of the piston is 0.2 mm.

As shown in Fig. 4, single droplet generation can be precisely controlled by the computer, and the precision error is less than 1.5 %. The droplets of different particle sizes are continuously generated quickly in the sequence with the rate up to three droplets per second.

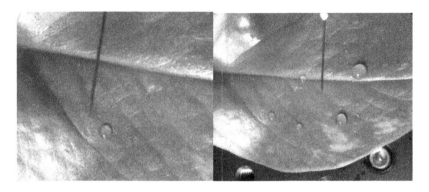

(a) Single droplet generation   (b) Continuous droplet generation

Fig. 4. Droplet generation

3.2   Accuracy Assessment

The droplets measuring instrument is Oxford Laser Imaging Division VisiSizer N60 (Oxford Lasers, Inc., Shirley, MA, USA). The vibration unit can ensure that one droplet released by the device does not contain bubbles, and the droplets can be released one by one with ensured experimental precision. If bubbles are found in chemicals in the operating process, the vibration unit can be used with an exhaust function, and the bubbles in the droplet generator can be automatically removed by extrusion and vibration. In this way the bubbles contained in the droplets can be prevented from affecting

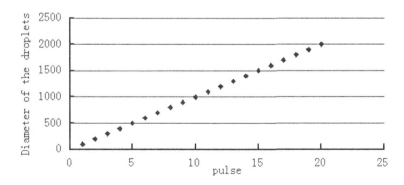

Fig. 5. Relationship between diameter of the droplets and the pulse.

the experimental precision. As shown in Fig. 5, the relationship is linearbetween diameter of the droplets and the pulsesof controller driver motor rotation.

After a user sets the size and the number of the droplets, the computer will send these parameters to the driver through the serial port, and the driver further completes the next two steps:

1. Automatically calculate the volume of the droplets, which corresponds to the diameter of the droplets;
2. Convert the volume into the extrusion distance required for producing the droplets from the droplet generator.

The control instructions are generated after calculation, and the instructions are sent to the driving motor. The driving motor rotates to drive the slider to move on the track. According to the instructions of the driver, the slider drives the piston to be slowly inserted into the droplet generator, and the insertion distance is precisely controlled.

The piston of a probe needs to advance 1 mm to generate the droplets of 200he. Generally, the minimal advancing precision of the piston is 0.2 mm.

Preferably, a calibrated scale is further arranged on the droplet generator. In the volume movement of the chemicals, whether the bubbles exist or not, other problems can be observed very conveniently through scale marks on the calibrated scale. If the bubbles exist, the bubbles can be automatically removed through the operation of the vibration unit, and then the bubbles contained in the droplets can be prevented from affecting the experimental precision.

3.3 Method Calibration

The system interferences and the error due to bubbles would affect the experimental precision. The droplet is calibrated by continuous generation from large to small droplets, every 3 s to generate a different size of droplet. In the different sizees of droplet dripping down the moment, droplet size has been measured by the droplets measuring instrument. And then, in accordance with the droplet size from big to small order by reverse and regression method, the same number and size of droplets were generated. The results of the test can be drawn in Fig. 6. Our study indicated that if the control algorithm is calibrated, the system can be with an improved control accuracy up to 97.2 %.

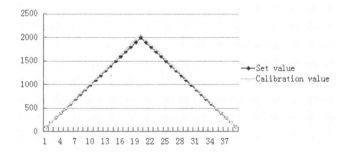

Fig. 6. Comparison of the calibration measurement value and set value

4   Conclusions

In this study, a computer-controlled mechanical compression device is used to generate pesticide droplets with specified sizes. The system performance of eliminating air bubbles in extrusionaccuratediameter droplets of pesticide was studied with a large amount of calibration data. Through this study the main conclusions are drawn as follows:

(1) The best value of the driving motor screw feeder controlled by a center motor rotation is 0.2 mm.

(2) The movement of the piston is affected by the vibration frequency of eccentric counterweight module. The slower the piston moves along with the sliding device, the smaller the error is.

(3) The experimental results indicated that high frequency vibrations were helpful to generator extrudes a non-stick droplet. Mainly because of the droplet adhesion to the tip, the module can make it fall.

So the method could be better used for studies on droplet evaporation in the tests using water-sensitive paper and other tests of the properties of the pesticides with broad application prospects.

Acknowledgment.   This research was financially supported by the Special Fund for Agro-scientific Research in the Public Interest, Ministry of Agriculture, China (Grant No. 201303031) and Expert Project Program of Beijing Innovation Team for Fruitsand Vegetables (Grant No. GC-WX2015). The authors would like to thank to Cai Jichen, Research Engineer, and Liu Chuan, Research Assistant, for their effort in acquiring experimental data.

References

1. Tsay, J., Fox, R.D., Ozkan, H.E., et al.: Evaluation of a pneumatic-shield spraying system by CFD simulation. Trans. ASAE **45**(1), 47–54 (2002)
2. Bykov, V., Glodfarb, I., Goldshtein, V., et al.: System decomposition technique for spray modeling in CFD codes. Comput. Fluids **36**(3), 601–610 (2007)
3. Sidahmed, M.M., Brown, R.B.: Simulation of spray dispersal and deposition from a forestry airblast sprayer part II: droplet trajectory model. Trans. ASAE, 44(1), 5–17 (2001)
4. Dorr, G., Hanan, J., Adkins, S., Hewitt, A., O'Donnell, C., Noller, B.: Spray deposition on plant surfaces: a modelling approach. Funct. Plant Biol. **35**(10), 988–996 (2008)
5. Dorr, G.: Minimising environmental and public health risk of pesticide application through understanding the droplet-canopy interface. Ph.D. thesis, The University of Queensland (2009)
6. Forster, W.A., Kimberley, M.O., Zabkiewicz, J.A.: A universal spray droplet adhesion model. Trans. ASABE **48**, 1321–1330 (2005)
7. Forster, W.A., Kimberley, M.O., Steele, K.D., Haslett, M.R., Zabkiewicz, J.A.: Spray retention models for arable crops. J. ASTM Int. **3**(6), 1–10 (2006)
8. Forster, W.A., Mercer, G.N., Schou, W.C.: Process-driven models for spray droplet shatter, adhesion or bounce. In: Baur, P., Bonnet, M. (eds.) 9th International Symposium on Adjuvants for Agrochemicals, p. 277. ISAA Society, The Netherlands (2010)

9. Mao, T., Kuhn, D.C.S., Tran, H.: Spread and rebound of liquid droplets upon impact on flat surfaces. AIChE J. **43**, 2169–2179 (1997)
10. Martin, J.Y., Vovelle, L.: Interaction of droplets with a surface: impact and adhesion. Agro Food Ind. Hi-Tech **10**(5), 21–23 (1999)
11. Mercer, G.N., Sweatman, W., Elvin, A., Harper, S., Fulford, G., Caunce, J.F., Pennifold, R.: Process driven models for spray retention by plants. In: Wake, G.C. (ed.) Proceedings of the 2006 Mathematics in Industry Study Group. Massey University (2007)
12. Mercer, G.N., Sweatman, W., Forster, W.A.: A model for spray droplet adhesion, bounce or shatter at a crop leaf surface. In: Fitt, A.D., Norbury, J., Ockendon, H., Wilson, E. (eds.) Progress in Industrial Mathematics at ECMI 2008. Mathematics in Industry, pp. 945–951. Springer, Berlin (2010)
13. Mundo, C., Tropea, C., Sommerfeld, M.: Numerical and experimental investigation of spray characteristics in the vicinity of a rigid wall. Exp. Thermal Fluid Sci. **15**(3), 228–237 (1997)
14. Oqielat, M.N., Turner, I.W., Belward, J.A., McCue, S.W.: Modelling water droplet movement on a leaf surface. Math. Comput. Simul. **81**(8), 1553–1571 (2011)
15. Pathan, A.K., Kimberley, M.O., Forster, W.A., Haslett, M.R., Steele, K.D.: Fractal characterisation of plant canopies and application in spray retention modeling for arable crops and weeds. Weed Res. **49**(4), 346–353 (2009)

Development of Variable Rate System for Disinfection Based on Injection Technique

Wei Ma[1,2(✉)], Xiu Wang[2], Lijun Qi[1], and Wei Zou[2]

[1] College of Engineering, China Agricultural University, Beijing 100083, China
maw516@163.com.cn, qilijun@cau.edu.cn,
[2] National Engineering Research Center of Intelligent Equipment for Agriculture,
Beijing 100097, China
zouw@nercita.org.cn, xiuwang@263.net

Abstract. A variable soil pesticide injection system was developed for control of soil pesticide amount by PWM. The paper analyzed a algorithmic model of control system, and designed hardware, algorithm and control of soil pesticide, mainly software flow and a feedback control way. In the paper, the variable-rate control system was consisted of infrared sensor, speed sensor, PWM valve, and pump motor. According to the amount of soil pesticide information, controller can automatically control flow amount by adjusting solid solenoid valve and PWM valve based on working speed, which changes the pulse duty cycle to achieve the variable work. Injection experiments of soil pesticide was pre-set different dosage, the results shown that pesticide amount was precise in fact, and the errors was less than 3.2 %. The system could achieve variable rate injection of liquid pesticide into deep soil based on infrared sensor. Fitting equation of flow amount by adjusting PWM valve based on working speed could draw the R2 value of 0.935. The chip can calculate the output PWM duty cycle according to the pre-set injection of soil pesticide amount after collected the speed of tractor. The feedback control is to regulate the PWM signal duty cycle according the real liquid flow obtained by the microcontroller chip which collected the output signal of liquid sensor which fixed on pesticide pipeline.

Keywords: Crop protection · Soil-borne disease · Variable rate injection · Control system

1 Introduction

In the suburbs of Beijing in China, greenhouses for agricultural production has developed very rapidly, mainly because of the rapid urban development and the population explosion in Beijing (Wang *et al.* 2008). Many farmers have planted with a variety of tomatoes for many years in the same piece of ground. And there are a lot of similar things. Such soil diseases cause huge losses to farmers, while an enormous impact on produce planting structure of the entire region. Due to a large area of crops reduction caused by continuous soil diseases, farmers abandoned cultivation of eggplant.

© IFIP International Federation for Information Processing 2016
Published by Springer International Publishing AG 2016. All Rights Reserved
D. Li and Z. Li (Eds.): CCTA 2015, Part II, IFIP AICT 479, pp. 437–444, 2016.
DOI: 10.1007/978-3-319-48354-2_43

Farmers try to use pesticide on soil disinfection, taking examples as bellow: a. By the method of solution to fill the root with pesticide, but it will reduce the quality of vegetables; b. With the method of changing new soil, but it can lead to cost doubled.

Precision agriculture, is the development trend of future agriculture, and a implementation of a set of modern farming operation technology and management mode, which is supported by the information technology according to the spatial variation, positioning, timing, quantitative (Yao et al. 2001). Precision agriculture technology has been used to solve farmers' problems in the greenhouse bases on the outskirts of Beijing. The technology, injecting pesticide directly, makes the storage of pesticide liquid be separated from water, the pesticide liquid, which is controlled by a peristaltic pump, flow into the main water pipe quantitatively to mix with water. It solved the issue of pesticide liquid treatment and equipment cleaning (Landers 1992). Underground spay machinery need to solve the problem of soil environment. Pesticide air jet injector cultivated shovel is a kind of device to inject pesticide into the soil by high-speed air jet. No target spray causes pesticide deposition outside spraying target, it is one of the important reasons of low utilization rate of pesticide. While target spraying is a technology, which identify weeds by the use of near infrared light reflection, to spray selectively with targets through the control circuit and spray system (Cooke et al. 1996). It can save 60 %–80 % of pesticide by using target spraying technology (Hanks et al. 2009). It is feasible to separate plants and soil by using camera sensitive to near infrared wave band when they studied the reflection characteristics of plant canopy and soil in visible light and near infrared band (Guyer 1996). Fehon designed a kind of photoelectric sensor which identify soil and weeds by sensing reflectance on the weeds and background in the visible and infrared light section, its recognition rate of the soil and weeds is as high as 95 % (Fehon et al. 1991). Deng Wei designed an infrared detection system targets based on infrared sensor technologies, different probe group with different coding modulation pulse infrared signal, can eliminate the interference of light paths between probe groups and other light signal (Wei et al. 2008). Chen Zhigang measured and analyzed the different influence factors of infrared target detection effect in spray pesticide technology by actual plant, and indicated that plant appearance, light intensity, walking speed of detectors and plant spacing is relatively significant effect on the detection performance, while particularly the effects of plant shape and light intensity (Zhigang et al. 2009).

This paper studies the test of soil disinfection by injection technology to develop a variable control system to realize the precise control and variable regulation of soil disinfectant.

2   Experiments and Methods

2.1   Experimental Samples

Experimental crop: eggplant, Bliss tower (Rijk Zwaan seed company, Netherlands), engraftment time is August 20, 2013, planting base is Beisishang village of Da xing district of Beijing (E:116.6295498320, N:39.69441545747).

The test land was being used for eggplant planting from 2011. Using Wang Binglin method to measure the height, fresh weight (FW) and root size of eggplant, the root activity was determined by TTC method. 20 samples were collected on March 31th, 2014, and of which four are infested plant samples by root-knot nematode. The roots of normal plants and diseased plants were measured after excavation. The photograph of real object as shown in Fig. 1. There is visible difference in height and root between the normal and diseased eggplant plant. The left side in picture a is the healthy plants excavated, with large distribution root and high plant. The right side of in picture a is the plants infested by Meloidogyne, with small roots and low plant. It is easy to distinguish due to obvious difference respectively. Picture b is a comparison of roots. It is need to increase the dose for roots will stop growing after illness. Figure c is disease roots measuring. Figure d is normal roots measuring.

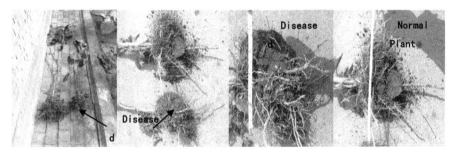

a. plant height compare   b. root compare   c. diseased roots measuring   d. normal roots measuring

Fig. 1. The photograph of real object

2.2 Experimental Methods

The experiment was done in two parts according to the plant detection and precision dosage control. For the first experiment, six infrared sensors, 0.5 m, E3F3-DS50N1 (Yueqing city Gaode electric co., LTD, China), was used to get a infrared diffuse reflectance light feed with difference sensor fixed at different heights. Delayed response time is less than 2.5 ms. The part used a group spectral sensors which is to detect target crops. Six sensors was fixed to both front sides of the tractor. The mounting height of the bottom sensor was set to be 220 mm from the ground because the weeds height was not higher than 200 mm.

The second part used PC software processing signals of infrared sensor for target detection and the speed sensor, and calculating the interval of acquisition time, obtaining the data of pesticide volume for different target root and sending it to the lower control computer. PWM was applied by the controller to regulate the dose. The controller calculated out the injection start and end time to control the injection volume.

As shown in Fig. 2. The controller of fixed-point pesticide injection acquires data of target plant and adjusts different volume of pesticide for different target root injected into soil.

Fig. 2. Infrared sensor for target detection

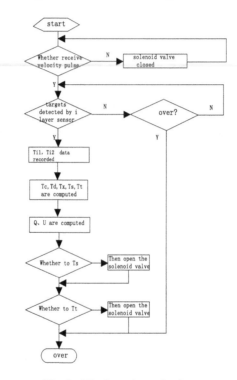

Fig. 3. The flow chart of software

A control system for the PC control software was developed by using SamDraw3.3 development platform (Shenzhen display control automation technology co., LTD). PC software issue commands to control the duty cycle of spraying pump voltage after its calculation base on the data of distance from infrared sensor to the nozzle, the correlation coefficient between the output pulse of wheel speed sensor and the vehicle speed, the transfer function between the biomass detected by infrared sensors and the duty cycle of the solenoid valve on output pipeline, the base reference dosing per plant, the acquisition of biomass signals of the infrared sensor, speed pulse signal of the vehicle speed sensor and quantitative flow rate signal of the flow sensor. The flow chart of fixed-point pesticide injection expressed the logic of control program intuitively. The dose and the injection position is computed out quickly according to this process (Fig. 3).

3   Results and Discussion

The test results were similar to Wang BingLin's results, after 30 days of inoculation, compared to normal eggplant plants, and plant height by infection of root knot nematodes decreased by 9.3 %, stem and leaf fresh weight decreased 18.21 %, and root vigor decreased 26.67 %. A solution for injection into soil of root zone is demonstrated, which is used for crop infected by Meloidogyne detected by six sets of infrared sensor. The six sets of sensors detected targets independently and obtained corresponding scanning signal to each layer separately and sent the signal to the controller. As per circuit theory to analysis, Due to the response time is less than 2.5 ms for sensor to detect the target and the computing time is less than 1 ms for single-chip, the reaction time was very fast. So only one sensor could be used to detect at the same time, interfere with each other could be avoided completely in the actual test with multiple sensors used. Soil variable injection system included disc, plow and injector, and traveled by traction in the field. The outline of the soil variable injection system could be figured out (Fig. 4).

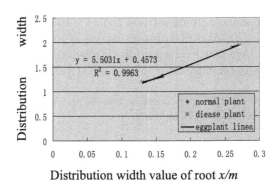

Fig. 4. The relationship between eggplant height and root

In the test, walking speed was 1 m/s, the degree of fit of the actual target signal was not less than 95 % after the signal be treated by controller model, and the relative error was not more than 3.2 % for dosing adjustment by the controller. It is linear relationship between the duty cycle for flow control and per unit of time, the linear regression equation was $y = 12.163x + 7.6$, where R2 = 0.935. The R^2 gives how much variance is explained by the model in term of overall variance in data. The R^2 is relative measure and is very intuitive. Fitting curve equation of duty ratio and flow linear relationship is $y = 12.163x + 7.6$, of which $R^2 = 0.935$, R^2 for the quadratic fitting curve equation is 0.9714 (Fig. 5).

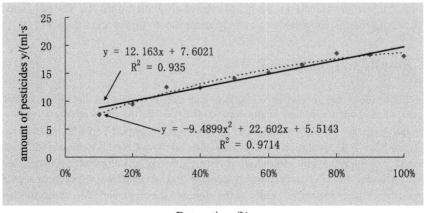

Fig. 5. Linear relationship of Duty ratio and flow

In order to test the precision of the dosing quantitative adjustment module quantitative, the pc controller software sent commands to the lower computer, the lower computer control dose automatically after receiving commands. The test data about the liquid quantitative adjustment are shown as in Table 1 with 3.2 % of maximum error of quantitative adjustment, 0.4 MPa of the maximum system pressure, 3.07 % of error after correction. Duty ratio and flow have a good linear relationship. Fitting curve equation of duty ratio and flow linear relationship is $y = 12.163x + 7.6$.

Table 1. At the pressure of 0.1 MPa, the controller controlled the injection quantity automatically according to the calculated value by PC software.

Set press/MPa	Volume of liquid/(ml)					Relative error/%
	Set volume/ml	Test value 1/ml	Test value 2/ml	Test value 3/ml	Average value/ml	
0.1	50	52.3	50.6	51.9	51.6	3.20 %
0.1	100	107.2	104.5	102.7	104.8	2.47 %
0.1	150	151.3	152.4	153.5	152.4	1.60 %
0.1	200	202	200.3	205.2	202.5	1.25 %
0.1	250	255	250.9	253.6	253.2	1.27 %
0.4	250	239.3	245.8	241.9	242.3	-3.07 %

In order to test the target crop model, time tree was set up. The detect time of start T_{i1} and end T_{i2} for each six layers, the root zone injection time of start T_s and end $T_{t,}$ and intuitive analysis of location and width of the injection zone. If the crop is tilted, the exact injection area of the root zone should be obtained relying on the model. Prediction of relative position of injection range for soil root zone is a very difficult thing. It is a certain uncertainty that the disease severity, of the underground root invaded by root knot nematode, was predicted according to the growth vigor of the target eggplant.

The problem was to be solved in two steps, the first step was to detect target and obtain the height information and position of the branches and leafs. Wherein the height of which was used to determine the prevalence of root crops and to calculate the pesticide amount need to be injected, and the position of the branches and leafs was used to build the computation model of time tree and calculate the injection width of the root and the relative benchmark time point. The second step was dynamic online adjustment of the flow and dose. Wherein the flow was calculated by controller according to the linear relationship to duty cycle, and sent commands to lower computer to realize. And the dose was obtained from the corresponding number of pulse to certain flow volume fed back by flow pulse sensor. Soil variable injection system was traction to travel in the field.

4   Conclusion

From measuring data, a linear relationship could be seen between the diseased plant height and root growth range. The relationship between eggplant height and root could be obtained by measuring the root excavated. And it is linear for eggplant with the same species and planted at the same time.

The relationship between duty and amount of pesticide is similar to linear approximation in experiment. The proposed approach is meant to be used in mathematical model for computing the relative time according to benchmark time point.

Acknowledgment. This research was financially supported by the National High Technology Research and Development Program of China (Project Number: 2013AA102406) and Youth Fund Project (QNJJ201525) and Science and technology innovation team Project (JNKST201619) of Beijing Academy of Agriculture and Forestry Sciences. The authors would like to thank to engineer Liang Xiaofei and my assistant Liu Chuan for spending time acquiring experimental target data.

References

Brown, L., Giles, K., Oliver, N.: Targeted spray technology to reduce pesticide in runoff from dormat orchards. Crop Prot. **27**(3), 545–552 (2008)

Chueca, P., Garcera, C., Molto, E., Gutierrez, A.: Development of a sensor-controlled sprayer for applying low-volume bait treatments. Crop Prot. **27**(10), 1373–1379 (2006)

Fang, X., Qiu, R.: Behavior of pesticide in soil environment. Soil Environ. **11**(1), 94–97 (2002)

Hanks, J.E.: "Seeing-eye" sprayer for weeds. Agric. Res. **46**(7), 22–25 (1998)

Bora, G.C., Schrock, M.D., Oard, D.L., et al.: Reliability tests of pulse width modulation (PWM) valves for flow rate control of anhydrous ammonia. Appl. Eng. Agric. **21**, 955–960 (2005)

Landers, A.J.: Direct injection system of crops. Agric. Eng. **47**(1), 9–12 (1992)

Li, L., Tian, Z., Xu, F.: Effect of different soil disinfectant on soil temperature and soil nutrient in sunlight greenhouse. Acta Agric. Boreali Occidentalis Sin. **12**(6), 328–331 (2009)

Li, Y., Cao, H., Xu, F.: Effects of different forms of soil disinfection on soil physical properties and cucumber growth. Chin. J. Eco Agric. **18**(6), 1189–1193 (2010)

Llorens, J., Gil, E.: Variable rate dosing in precision viticulture: use of electronic devices to improve application efficiency. Crop Prot. **29**(3), 239–248 (2013)

Linch, C.: Smart sprayer select weeds for elimination. Agric. Res. **44**(4), 15–18 (1996)

Wang, L., Qiu, L., Guo, S.: The present situation and countermeasure development of greenhouse in China. J. Agric. Mech. Res. **12**(10), 207–209 (2008)

Yao, M., Zhang, B.: Development of precision agriculture and agricultural ecological environment protection in China. Chin. Popul. Resour. Environ. **12**(S1), 114–115 (2001)

Zhou, C., Yao, F.: Organic cultivation techniques of more crop in a year for greenhouse vegetable. JiLin Vegetable **12**(12), 38–39 (2012)

Zhai, C., Zhao, C., Wang, X.: Design and experiment of young tree target detector. Trans. CSAE **28**(2), 18–22 (2012)

Zhong, G.: Development status and trend of domestic and overseas greenhouse. Agric. Sci. Technol. Equipment **12**(9), 68–69 (2013)

Zuo, Y.: Soil disinfection and improved technology for facility **12**(1), 11–12 (2014)

Establishment and Optimization of Model for Detecting Epidermal Thickness in Newhall Navel Orange

Yande Liu$^{(\boxtimes)}$, Yifan Li, and Zhiyuan Gong

Institute of Electrical Machinery, East China Jiaotong University,
Nanchang 330013, China
jxliuyd@163.com

Abstract. Diffuse transmittance spectra in the near-infrared scope as a prevalent sensitivity method carried out to test epidermal thickness of 'Gannan' navel oranges. In order to lay a good foundation for accurate and rapid online classification, variable selection methods was intervened for navel orange model optimization. In spectral range of 900–1650 nm, navel orange in thick skin depth chosen arbitrarily were set up the qualitative models for both calibration and prognostication sets in this experiment. Firstly, different pretreatment methods such as the Savitzky-Golay, the first derivative and so on were compared by PLS modeling results. Then GA and SPA were brought in to improve predictive models. Compared with results, light scattering can be effectively eliminated by the standard normal variate transformation (SNV). Moreover, fewer variables and model optimization were carried out by GA. The supreme calibration model procured with GA-PLS approach had the *Rp* of 0.864, RMSEP of 0.290, R_C of 0.882 and RMSEC of 0.264. The experiment showed the detection of epidermal thickness of navel orange is completely feasible.

Keywords: Newhall · Navel orange · Near-infrared diffuse transmittance spectra · Epidermal thickness · Variable selection

1 Introduction

Navel orange is a kind of comprehensive nutritional food, containing various essential nutrients in human body. A great source of vitamin C and carotenoids can be taken in through eating navel orange. The ripe navel orange is a popular fruit for its benefits, such as: seedless, juicy flesh, tasting good et al. When placed indoor, they can radiate enticing aroma and the rind's color and lustre is charming [1]. Near infrared spectrum detection technology has been used for a wide range of applications, especially in the quality test of thin skin fruit such as apple, peach, pear et al. [2]. In general, the average consumers take cortical thickness as one of the important considerations when they buy orange, water melon and other thick skin fruits. In the process of fruit grading, if the skin is too

Fund Project: "863" high-tech Research and Development Program (2012AA101906).

Published by Springer International Publishing AG 2016. All Rights Reserved
D. Li and Z. Li (Eds.): CCTA 2015, Part II, IFIP AICT 479, pp. 445–454, 2016.
DOI: 10.1007/978-3-319-48354-2_44

thick, the fruit can't be treated as optimal fruit. Even if the internal quality of fruit is excellent, the fruit also be degraded, as peel thickness directly influences its edible rate. Currently, the application of near infrared spectral detecting the thickness of the fruit skin also has not been reported. As a consequence, in this paper the navel orange peel thick testing research was discussed. Through a variety of spectral data processing method to establish the optima navel orange skin thickness detection model of Newhall Navel Orange. The researchers at home and abroad processed a series of researches in near infrared spectrum technology area detecting internal quality of both thin skin fruit and thick skin fruit [3–6]. Taking the application of 3 different wavelength of LED lights to estimate SSC and size of Shuijing pears, PLS and LS - SVM models were established. The model obtained by PLS was preferable than that of LS – SVM. The R of forecast set was respectively 0.86 and 0.90 for soluble solids content and size [7]. In order to implement modern management for pear orchards, the technology of NIR was applicated in the quality detection for Dangshan pear. PLS, LS - SVM and GRNN model was established to portend the SSC of Dangshan pear. Then uninformative variables elimination (UVE) was introduced to simplify the former models. UVE-LSSVM had a great advantage in terms of regression model [8]. In the wavelength of 1200–2200 nm, transmission spectrum of the mangos were collected for measurement during SSC and potential of hydrogen (PH) experiments. Infinite variety of pretreatment techniques was employed to process all wave spectral for the processing efforts. MLR based on PLS was applied to build calibration models [9]. In the wavelength of 400–1000 nm, visible/near-infrared spectroscopy of valencia oranges were applied to conduct the SSC and TA experiments. The prediction models especially the model of the fruits taste best characteristic value stand (BrimA) on PLS and PCR are developed. Based on these results, visible/near-infrared spectroscopy technology is a promising and feasible method for detecting the BrimA of valencia oranges [10]. Analyzing the above scholars' researches comprehensively, relevant testing models are established using fruits in regular shape, smooth surface and high internal quality [11–18].

The experiment take Newhall navel oranges which have thick wooden peel and poor uniformity of internal quality as objects of study. After different spectral preprocessing methods and the variable selection method, the raw navel orange spectrum data was to establish the optimal PLS prediction model. Compared with the origin model, it promotes the discriminative and forecasting ability.

2   Materials and Methods

2.1   Materials

One hundred twenty samples of Newhall were procured from a fruit wholesale marketplace in NanChang (NanChang, China) in January, 2015. The samples were wiped up with distilled water and then naturally dried. All samples were be numbered consecutively and stored in the experimental environment at room temperature and 60 % relative humidity (RH) for 24 h. Spectral collection were carried out on the next day and SSC measurement performed soon thereafter. Newhall navel oranges were divided by proportion of 3:1 in the light of calibration and prediction [19].

2.2 Spectral Collection

Portable fruit quality detection device based on Android system consisted mainly of optical module, table computer with spectrum acquisition unit and power supply unit was set up. The optical module (Micro-NIR 1700, JDSU) kept an account of wavelength from 950 nm to 2150 nm. The table computer with Atom Z2580 processor displayed the measurement results on a 9.7 in. capacitive touch screen. The JDSU Micro-NIR and table computer are workable under 5 V model via 10000 mA portable power source.

Reference spectral of a white Teflon tile and dark current were gathered before spectral acquisition of test samples. Diffuse reflection spectra of Newhall were gathered at about 20°C. Acquisition spectrum evenly picked around the equator equidistantly (approximately 120°). Optical gain had been set to 'low gain' (Figs. 1 and 2).

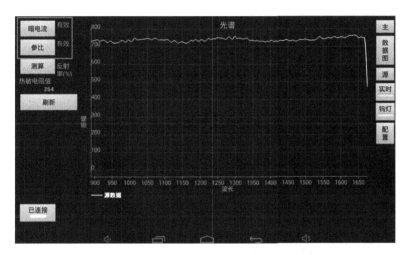

Fig. 1. Collecting the effective reference and dark current

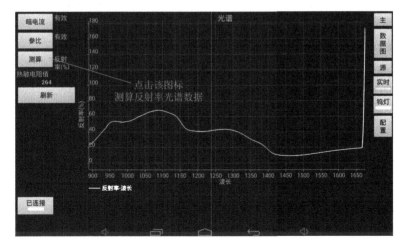

Fig. 2. Gathering fruit spectrum data

2.3   Measurement of Soluble Solid Content and Skin Depth

In the spectral acquisition area, several drops of filter juice extracted by manual compression were used for soluble solid content measurement using the PR-101αCat refractometer. Observed value of skin depth was conducted after soluble solid content measurement via vernier caliper immediately. When the flesh of the orange was removed from the peel clearly, the peel was to be hold pressure level off to measure the skin depth. The thickness value at three equidistant positions were recorded exactly during the experiment. The average value of the three marked points was taken as the thickness values of navel orange skin.

2.4   Data Processing and Model Evaluation

All of the R, RMSEC and RMSEP together assess the performance of the model built in the experiment. The number of the best principle component achieved when coupled with the RMSEP which had reached the minimum value [20–22].

In this paper, all the RMSEC and RMSEP were processed by the formula (1) and (2) respectively.

$$\text{RMSEC} = \sqrt{\frac{\sum_{i=1}^{I_c}(\hat{y}_i - y_i)^2}{I_c - f - 1}} \tag{1}$$

$$\text{RMSEP} = \sqrt{\frac{\sum_{i=1}^{I_P}(\hat{y}_i - y_i)^2}{I_P - 1}} \tag{2}$$

Where y_i was the model's actual measured value; $\hat{y}_i$ was model's predictive value; f was used as dependent variables

3   Results and Discussion

3.1   Analysis of Newhall Navel Orange

Figure 3 is the original absorbance spectra chart of 120 ripe Newhall navel orange. Figure 4 is the original absorbance spectra chart of 120 ripe Washington navel orange. It is obviously that two kinds of samples have similar spectrum shape and location of wave crest and trough. Because of the O-H, C-H or N-H stretching vibration [10, 23], there is a obvious absorption peak near the location of 970, 1090, 1220, 1285, 1470 nm.

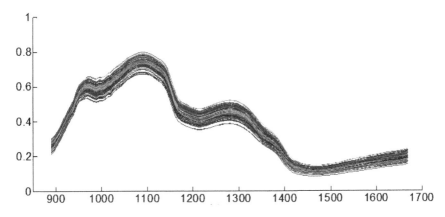

Fig. 3. The original spectra of ripe Newhall navel orange

3.2   Measurement Results of Soluble Solid Content and Skin Depth

The samples of Newhall navel orange had certain representativeness as the sizes of them contain the dimension from small size to large size. Form Table 1 available, the skin thickness range of calibration set is 1.98–5.57 mm meanwhile the prediction set is 2.63–5.43 mm. The data structure make better fit for prediction set to predict using the model calibration set made (Table 2).

Table 1. Sugar measurement results of Newhall navel orange

Sample sets	Min (°Brix)	Max (°Brix)	Mea (°Brix)	SD (°Brix)	CV (%)
Calibration set	9.70	18.83	14.37	1.97	13.68
Prediction set	10.50	18.63	14.43	1.99	13.81

Table 2. Skin thickness measurement results of Newhall navel orange

Sample sets	Min (°Brix)	Max (°Brix)	Mea (°Brix)	SD (°Brix)	CV (%)
Calibration set	1.98	5.57	3.72	0.57	15.42
Prediction set	2.63	5.43	3.76	0.58	15.58

3.3   Comparison of the Results of Spectral Preprocessing

Table 3 illustrates that the model PLS established through base line, 1st derivatives, SNV and MSC improve the model prediction compared with the model built by raw spectral data. The prediction abilities of SNV also with MSC after pretreatment, while SNV had the better result. Take Newhall Navel Orange as research objects to discuss the outcome. The result showed that the R_C reached 0.883 and of prediction set reached 0.815. It is better to see that RMSEC was 0.268 and of PMSEP was 0.333. Therefore, the process of spectra band screening in the further adopted the spectra data preprocessed by SNV.

Table 3. The effect of different pretreatment methods on PLS modeling results for Newhall

Preprocessing method	Calibration set		Prediction set	
	R_C	RMSEC	R_P	RMSEP
Raw	0.881	0.270	0.804	0.343
S-G	0.880	0.271	0.803	0.344
1st derivatives	0.880	0.271	0.808	0.339
Base line	0.880	0.271	0.812	0.337
SNV	**0.883**	**0.268**	**0.815**	**0.333**
MSC	0.882	0.269	0.815	0.333

3.4   Method of Spectral Band Selection

To test and attest the practicability of near infrared diffuse reflection technology to inspect Newhall Navel Orange. In the hope of providing the theoretical basis for online detection through setting up a corresponding optimal prediction model simultaneously. As a consequence, testing requires not only high accuracy, but also higher detection efficiency. Spectral band selection methods are needed so as to acquire fewer variables and good prediction ability. Genetic algorithm is an algorithm proposed on basis of imitating the preferred choice of biology and genetic principle. With the advantages of high efficiency of global searching, keeping the optimal variables, eliminating worse variables, establishing model more conveniently and predicting more accurately. According to the reference [24–26], in the process of GA variable screening of 127 wavelength points, the main parameters are set as follows. Initial value of the group was 30; Adaptive crossover probability and mutation probabilities was set to 0.5 and 0.01 separately; The RMSECV was regarded as fitness function of genetic algorithm. After the iteration reached 100 times, Fig. 4 illustrated that the minimum value corresponding to the RMSECV was 0.292 when the variable number was 13. For this reason, the 13

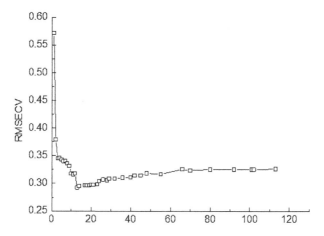

Fig. 4. Relation between RMSECV and wavelength variables for epidermal thickness of Newhall navel orange

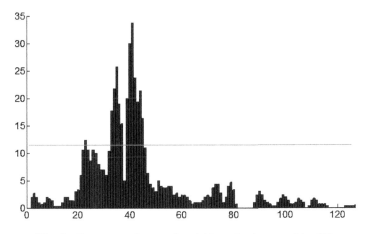

Fig. 5. Frequency of spectral variables selection runed by GA

variables was the characteristics screened out by GA method. The processed result can be seen from Fig. 3. The variable frequency was more than 11 times which were chosen as characteristic variable of the model in all 127 variables (Fig. 5).

Successive projection algorithm randomly select a variable from spectrum matrix and calculate projection of the variable on the other variables afterwards. This method is based on the minimum RMSECV generally choose the characteristic variables within 20. Calculation procedure and selection of numbers of variables of SPA consult to relevant literature [27–29]. The leading eigenvalue is set to 20 and the least eigenvalue is 1. Figure 6 was the diagram of characteristic variables of screened out by applying SPA for epidermal thickness of Newhall Navel Orange. There were 6 variables filtered out from the total variables.

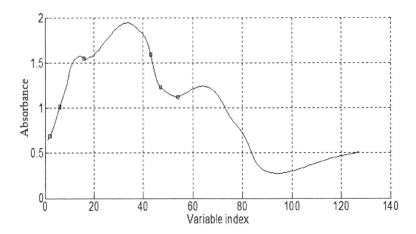

Fig. 6. Variables selected by SPA for epidermal thickness of Newhall navel orange

3.5   Comparison of Models

Table 4 provides a comparison of forecasting results of different forecasting models made by spectral data processed through method of GA and SPA which were the method of artificial screening in the multi-variable analysis and full spectral of samples. For the PLS model of Newhall navel orange skin thick in mature stage, research findings shows that GA-PLS models outperformed the SPA-PLS models and full wave band models. SPA-PLS models had the least spectral variables but the ability to predict relatively weakened. The reason may be that some useful variables are eliminated, so that the model precision is lower. It can be observed that GA-PLS models made the variables from 127 down to 13 and improved Rp from 0.815 to 0.815 at the same time. It's very gratifying to see that RMSEP reduced from 0.333 to 0.290. So from the experimental results, GA-PLS models increased the optimization efficiency of spectral models and produced the best Rp and RMSEP for calibration models. Performance of 30 navel oranges peel thickness in calibration set had been presented in Fig. 7. Finally practical data are compared to predicted ones to assess the quality of calibration models [29, 30].

Table 4. Modeling results and comparison of different PLS

Model	Variables	R_C	RMSEC	R_P	RMSEP
PLS	127	0.883	0.268	0.815	0.333
GA-PLS	**13**	**0.882**	**0.269**	**0.864**	**0.290**
SPA-PLS	6	0.859	0.293	0.811	0.339

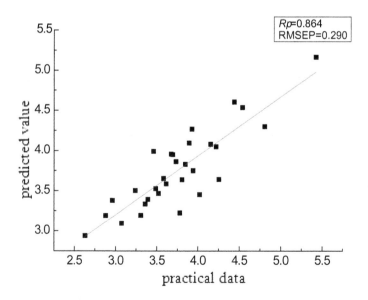

Fig. 7. Comparison of predicted values and measured values of 'Gannan' navel orange in prediction set by GA-PLS mode

4 Conclusion

This study aimed to establish prediction model to detect peel thickness of Newhall navel orange accurately. To make a thorough acquire for the means of simplifying and improving model, comparison of spectral preprocessing methods and the band selection method of GA and SPA were introduced. The predictions suggested that the best model with Rc, Rp, RMSEP and RMSEC of 0.882, 0.864, 0.290 and 0.296 were established through spectral data processed by SNV and spectral variables screened out by GA. Based on these results, SNV was a effective way to greatly reduce the effects of light scattering. Instead of using full spectral data, GA models simplified models and improved the ability to predict with decreasing the extent of the spectral variables.

Acknowledgment. The authors gratefully acknowledge the financial support provided by National High Technology Research and Development Program of China (863 Program).

References

1. Sun, X.D., Zhang, H.L., Liu, Y.D.: Int. J. Agric. Biol. Eng. **2**(1), 65 (2009)
2. Jie, D.F., Xie, L.J., Rao, X.Q., et al.: Trans. Chin. Soc. Agric. Eng. **29**(12), 264 (2013)
3. Sun, T., Lin, J.L., Xu, W.L.: J. Jiangsu Univ. Nat. Sci. Edn. **34**(6), 663 (2013)
4. Ma, G., Sun, T.: Trans. Chin. Soc. Agric. Mach. **44**(7), 170 (2013)
5. Liu, Y.D., Shi, Y., Cai, L.J., et al.: Trans. Chin. Soc. Agric. Mach. **44**(9), 138 (2013)
6. Cai, L.J., Liu, Y.D., Wan, C.L.: J. Northwest A&F Univ. Nat. Sci. Edn. **40**(1), 215 (2012)
7. Liu, Y.D., Peng, Y.Y., Gao, R.J., et al.: Trans. Chin. Soc. Agric. Eng. **26**(11), 338 (2010)
8. Wang, M.H., Guo, W.C., Liu, H., et al.: J. Northwest A&F Univ. Nat. Sci. Edn. **41**(12), 113–119 (2013)
9. Shyam, N.J., Pranita, J.: Sci. Honiculturae **138**, 171–175 (2012)
10. Bahareh, J., Saeid, M., Ezzedin, M., et al.: Comput. Electron. Agric. **85**, 64–69 (2012)
11. Greensill, C.V., Walsh, K.B.: J. Near Infrared Spectrosc. **10**, 27–35 (2002)
12. McGlone, V.A., Fraser, D.G., et al.: J. Near Infrared Spectrosc. **11**, 323–332 (2003)
13. Saranwong, S., et al.: J. Near Infrared Spectrosc. **11**, 175–181 (2003)
14. Walsh, K.B., Golic, M., et al.: J. Near Infrared Spectrosc. **12**, 141–148 (2003)
15. Gomez, A.H., He, Y., Pereira, A.G.: J. Near Infrared Spectrosc. **77**, 313–319 (2006)
16. Nicolaï, B.M., Beullens, K., et al.: Postharvest Biol. Technol. **46**, 99–118 (2007)
17. Hashimoto, T., et al.: US: 6754600B2, 22 June 2004
18. Liu, Y.D., Sun, X.D., et al.: LWT Food Sci. Technol. **43**, 602–607 (2010)
19. Li, H., Wang, J.X., Xing, Z.N., et al.: Spectrosc. Spectral Anal. **31**(2), 362–365 (2011)
20. Han, D.H., Chang, D., Song, S.H., et al.: Trans. Chin. Soc. Agric. Mach. **44**(7), 174 (2013)
21. Barbin, D.F., ElMasry, G.: Food Chem. **138**(2–3), 1162–1171 (2013)
22. He, H.J., Wu, D., et al.: Food Chem. **138**(2–3), 1162–1171 (2013)
23. Sun, X.D., Dong, X.L.: Spectrosc. Spectr. Anal. **29**(14), 262 (2013)
24. Goicoechea, H.C., Olivieri, A.C.: Chemometr. Intell. Lab. Syst. **56**(2), 73–81 (2001)
25. Zhu, W.X., Jiang, H., et al.: Trans. Chin. Soc. Agric. Mach. **41**(10), 129 (2010)
26. Zhang, S.J., Zhang, H.H., et al.: Trans. Chin. Soc. Agric.Mach. **43**(3), 108 (2012)
27. Kamruzzaman, M., ElMasry, G., et al.: Food Chem. **141**(1), 389–396 (2013)
28. Wu, D., Sun, D.W., et al.: Innovative Food Sci. Emerg. Technol. **16**, 361–372 (2012)

29. Paz, P., Sánchez, M.T., et al.: Comput. Electron. Agric. **69**, 24–32 (2009)
30. Yun, Y.H., Wang, W.T., et al.: Analytica Chimica Acta 10–17 (2014)
31. Liu, Y.D., Gao, R.J., et al.: Food Bioprocess Technol. **5**, 1106–1112 (2012)
32. Piyamart, J., Yoshinori, K., Sumio, K.: J. Near Infrared Spectrosc. **22**, 367–373 (2014)
33. Jiang, H., Zhu, W.X.: Food Anal. Methods **6**, 567–577 (2013)
34. Chen, Q.S., Zhao, J.W., et al.: Anal. Chim. Acta **572**, 77–84 (2006)

Design and Implementation of Greenhouse Remote Monitoring System Based on 4G and Virtual Network

Guogang Zhao[1,2], Yu Lianjun[4], Haiye Yu[1,2(✉)], Guowei Wang[1,2,3], Yuanyuan Sui[1,2], and Lei Zhang[1,2]

[1] College of Biological and Agricultural Engineering, Jilin University, Changchun 130022, China
zhaoguogang2000@qq.com, haiye@jlu.edu.cn, 41422306@qq.com,
suiyuan0115@126.com, z_lei@jlu.edu.cn
[2] Key Laboratory of Bionic Engineering, Ministry of Education, Changchun 130022, China
[3] School of Information Technology, Jilin Agricultural University, Changchun 130118, China
[4] Changchun City Academy of Agricultural Sciences, Changchun 130111, China
120142901@qq.com

Abstract. In modern agriculture, the temperature of the greenhouse is one of the main factors that affect the growth of crops, which plays an important role in the growth of crops. Based on 4G and virtual network technology, this paper designed greenhouse remote monitoring system, which can automatically collect, remotely transfer, automatically store, analyze and process temperature data of greenhouse.

Keywords: Agricultural modernization · 4G · Virtual network · Automatic collection

1 Introduction

China is a populous country, the stable development of agriculture, and the stability of society. In the 2015 government work report, Premier Li Keqiang made it clear that: "to accelerate the agricultural modernization". Agricultural modernization is the main way to realize the output of agricultural products, increase the quality and increase the income of the farmers. Greenhouse, also called a glasshouse. In not suitable for crop growth season, crops provide a suitable environment for the growth of plants and ensure the crop normal growth. The physiological activities of crops must be carried out at a certain temperature, the temperature is too high, the physiological activity of the crops is accelerated, the temperature is too low, the physiological activities of the crops become slow. So the change of crop growth temperature has an obvious effect on the growth, yield and quality of crops [1–3].

With the rapid development of information technology, it provides a strong support for the modernization of agriculture. The modern greenhouse temperature collection has not need manual collection, can realize the temperature of the automatic collection through the sensor [4, 5].

The transmission of data acquisition is from the original wired network to wireless network, such as ZigBee, WiFi, Bluetooth, 3G and so on, more advanced technology to accelerate the development of agricultural modernization [6–8]. In recent years, with

D. Li and Z. Li (Eds.): CCTA 2015, Part II, IFIP AICT 479, pp. 455–462, 2016.
DOI: 10.1007/978-3-319-48354-2_45

the popularization of Internet, the greenhouse temperature collection system also from the past service in a small range of greenhouse, and gradually developed into the same service multi region, a large range of greenhouse. The existing greenhouse temperature monitoring system, the data receiving server access to the Internet, to open a monitoring program, the data packets sent to receive, analyze and store. Such a design can be convenient and quick to achieve data collection and storage, but because the Internet has interoperability, so in addition to the normal data communication Internet, there will be some non normal communication, these non normal communication most are malicious communication. Malicious communication will affect the normal operation of the system, so that the temperature monitoring system can not get normal information, resulting in the normal operation of the greenhouse, the growth and development, yield and quality of crops have a very significant impact.

Using virtual network technology, it can effectively solve this problem, the traditional temperature monitoring system data receiving server is not directly connected to the Internet, instead of using the virtual network to verify the server access Internet, the virtual network authentication server provides authentication function in Internet. The use of virtual network technology to improve the security of the monitoring system, the system's stable operation, to ensure the normal growth of crops in the greenhouse, laid a solid foundation for agricultural modernization.

2   System Structure

The system structure is shown in Fig. 1, the system is based on 4G and virtual network technology, the end of the data acquisition is achieved, and the traditional data acquisition method can effectively ensure the security of data.

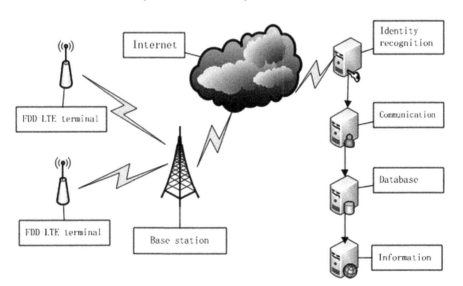

Fig. 1. The overall architecture of the system

3   Temperature Collection

3.1   DS18B20

DS18B20 is a commonly used temperature sensor with small size, anti disturbance and high accuracy. Measuring temperature range old-55 DEG C — +125 DEG C, work in the voltage 3.0 v–5.5 v, using the sensor, can effectively guarantee the accuracy of the temperature data [9].

3.2   Core

In the core chip, often used in hardware and arm. SCM has the advantages of low price, low power consumption, in many ways, SCM is a good choice, but because of its own architecture, the network is not very good, but also the need to update each application. Therefore, the microcontroller does not apply to the design of this paper. ARM processor is a microprocessor which is designed by Acorn Computer Co., Ltd., the ARM processor is designed in 32 place. It can run the operating system Linux, and the complex hardware is developed. Since Linux supports the completion of the TCP/IP protocol, the choice of ARM is more in line with the design of this paper.

S3c2440A, which is produced by the Samsung Corp, based on the ARM920T kernel, has a dominant frequency of 533 MHz, supports WinCE system and embedded Linux system, RJ45 interface network controller, USB interface [10].

3.3   System

S3c2440A can run WinCE system and embedded Linux system. The embedded Linux system is based on Linux, and can be run on the S3c2440 Linux operating system. Embedded Linux system is also open source, free of charge, with excellent performance. Therefore, the choice of embedded operating system is more in line with the needs of this paper.

4   Network

4.1   Wireless Networks

In the greenhouse, if the use of the traditional wired network, due to the environment is relatively complex, the installation of the cable network to increase the difficulty, but also in the greenhouse layout too many lines, the watering and fertilization also increased the difficulty. So in the greenhouse, is not suitable for the installation of cable network. Mainstream wireless network is divided into two kinds of wireless LAN and public mobile communication network. In wireless information systems, the common wireless local area network has ZigBee and Wifi, they are able to provide wireless network services, but can not provide the service to Internet alone. So the use of the public mobile communication network to achieve the wireless network, can provide wireless network services, and wireless network can be directly connected to the Internet.

4.2   4G

LTE, WIMAX and UMB technology are often referred to as 4G technology, in the past 3G technology, while providing voice and data communications, and to the 4G, no voice communications, only data communication. In our country, the 4G technology is only one kind of LTE. LTE is a global standard, including FDD and 3GPP two models, in this paper, the use of FDD-LTE. Because FDD-LTE developed earlier than TD-LTE, the technology is more mature, access terminal more, faster, and more suitable for FDD-LTE in wide area coverage. In the access equipment, is the use of HUAWEI's B310, the device supports FDD-LTE and VPN services, can provide a stable wireless network services.

4.3   Virtual Network

In the greenhouse, if the use of the traditional wired network, due to the environment is relatively complex, the installation of the cable network to increase the difficulty, but also in the greenhouse layout too many lines, the watering and fertilization also increased the difficulty. So in the greenhouse, is not suitable for the installation of cable network. Mainstream wireless network is divided into two kinds of wireless.

5   Server

5.1   Virtual Implementation

Linux when the operating system used in this system, so virtual network authentication, is also based on Linux platform implementation, implementation, use the following software: dkms-2.0.17.5-1.noarch, kernel_ppp_mppe-1.0.2-3dkms.noarch, pptpd-1.4.0-1.el6.x86_64, ppp-2.4.5-5.AXS4.x86_64. Will they package uploaded to the server's/TMP directory.

```
#installation package
rpm -ivh *.rpm,

#Add in the last
vi /etc/pptpd.conf
#server ip
localip 172.16.26.3
#Assign IP

remoteip 192.168.26.200-230

# start server
service pptpd start
Starting pptpd: [OK]    # start server
```

#Configure the user information
vi /etc/ppp/chap-secrets

```
# client   server    secret        IP addresses
wendu      pptpd     wendu         192.168.26.200

client   #   user
server   # server
server   #password
server   # get ip of virtual
```

5.2   Virtual Test

Client using the Windows operating system test, and can easily show as a result, the input connection of the user name and password, as shown in Fig. 2.

Fig. 2.   Input user name and password

After the success of the virtual network link status, as shown in Fig. 3.

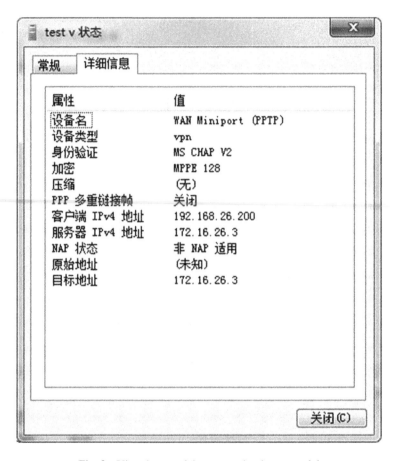

Fig. 3. Virtual state of the connection is successful

Can execute commands under Windows, ipconfig, as shown in Fig. 4.

In the virtual network authentication server, you can perform commands, ps - ef |
grep 172.16.26.11, check the server connection user state, as shown in Fig. 5.

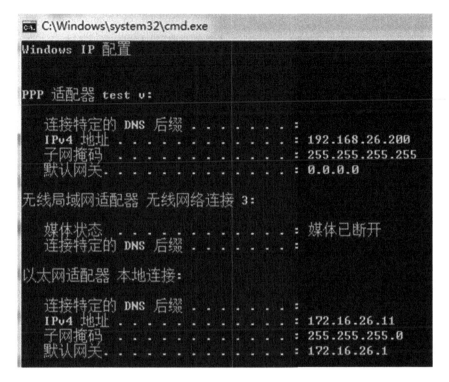

Fig. 4. ipconfig command execution results

```
[root@vendu 20150721]# ps -ef|grep 172.16.26.11
root      19328 19292  0 18:54 ?        00:00:00 pptpd [172.16.26.11:BD3A - 0080]

root      19329 19328  0 18:54 ?        00:00:00 /usr/sbin/pppd local file /etc/ppp/options.pptpd 115200 172
.16.26.3:192.168.26.200 ipparam 172.16.26.11 plugin /usr/lib64/pptpd/pptpd-logwtmp.so pptpd-original-ip 172
.16.26.11 remotenumber 172.16.26.11
root      19358 17842  0 18:56 pts/3    00:00:00 grep 172.16.26.11
[root@vendu 20150721]#
```

Fig. 5. User to connect this virtual server

6   Conclusions

Through the embedded terminal system, the real-time collection of greenhouse temperature data is realized through the 4G wireless network access Internet, and a special virtual network is established. The embedded terminal is connected with the greenhouse temperature monitoring system through the virtual network, which can ensure the effective transmission of the data. Compared with the traditional greenhouse temperature collection system, the security of greenhouse temperature system based on virtual network is obviously higher than that of the traditional network, which makes up the deficiency of the existing system, and provides a more secure and effective guarantee for the development of precision agriculture.

Acknowledgment. Funds for this research was provided by National 863 subjects (2012AA10A506-4, 2013AA103005-04), Jilin province science and technology development projects (20110217), China Postdoctoral Science Foundation the 54th surface funded (2013M541308), Jilin University Young Teachers Innovation Project (450060491471).

References

1. Yu-jun, W., Ben-hua, Z.: The current situation and developing trend of greenhouse technology. J. Agric. Mechanization Res. **1**, 249–251 (2008)
2. Xinkun, W., Hong, L.: Current research status and development trend of greenhouse in China. J. Drainage Irrig. Machn. Eng. **28**(3), 179–184 (2010)
3. Fei, Q., Xinqun, Z., Yuefeng, Z., et al.: Development of world greenhouse equipment and technology and some implications to China. Trans. CSAE **24**(10), 279–285 (2008). (in Chinese with English abstract)
4. Juan, Z., Jie, C., Zhenjiang, C.: Greenhouse temperature collection based on multi-sensor date fusion technology. Microcomput. Inf. **23**(1), 153–154 (2007)
5. Xinyu, W., Weiqing, Y.: A new temperature collection & monitor system base on AT89S51 for greenhouse. J. Agric. Mechanization Res. **9**, 107–110 (2010)
6. Changchun, B., Ruizhen, S., Yuquan, M., et al.: Design and realization of measuring and controlling system based on ZigBee technology in agricultural facilities. Trans. CSAE **23**(8), 160–164 (2007). (in Chinese with English abstract)
7. Wei-bo, Z., Zhong-mei, L., Jie, S., et al.: Design and implementation of ZigBee-WIFI gateway for facility agriculture. Comput. Sci. **41**(6A), 484–486 (2014)
8. Ping, S., Yangyang, G., Pingping, L.: Intelligent measurement and control system of facility agriculture based on ZigBee and 3G. Trans. Chin. Soc. Agric. Machn. **43**(12), 229–233 (2012)
9. Jun, Z.: Smart temperature sensor DS18B20 and its application. Instrum. Technol. **4**, 68–70 (2010)
10. Hao, Z., Chun-yan, Y., Xiao-yang, W.: Introduction and application of CMOS chip S3C2440A. Electron. Des. Eng. **19**(24), 26–29 (2011)

The Study of Farmers' Information Perceived Risk in China

Jingjing Zhang[✉]

College of Computer and Information Engineering,
Tianjin Agricultural University, No. 22 Jinjin Road,
Xiqing District, Tianjin 300384, China
zhangjingjing1982@126.com

Abstract. The study provides insights into the perceived risk in the course of farmers' agriculture information adoption in China. The information perceived risk model is gotten with the approach of the factor analysis and six aspects constitute the model. The regression equation of farmers' information perceived risk is obtained with the multiple regression analysis in this research. The results show that there are many aspects that farmers consider about when they are applied with new agricultural information and technologies, and information service departments may pay more attention to those especially.

Keywords: Perceived risk · Information adoption · Factor analysis

1 Introduction

Information plays an important role in the agricultural development. Farmers are provided with a variety of information. Meanwhile, there are many obstacles in the process of the information adoption due to some elements, such as information asymmetry, different education and economic levels and so on. The existence of the obstacle factors, leads to the risk considerations. However, there are differences between the perceived risk and the actual risk, and the real respect that affects the decisions is the perceived risk normally. Therefore, it is necessary to resolve and eliminate farmers' perceived risk aiming to take full advantage of the information and technologies.

Perceived risk is an important content of consumer behaviors, which was originally developed by Bauer Raymond from Harvard University. Perceived risk is proposed as the unpredictable result of the product's quality and the feeling of uncertainty which is resulted from the unsuccessful purchase (Li 2007). Previous researchers show that the perception of risks and types of risks all influence the intention significantly (Wu and Ke 2015).

This paper aims to carry on the theoretical and empirical studies in order to examine and predict farmers' perceived risk of agricultural information.

D. Li and Z. Li (Eds.): CCTA 2015, Part II, IFIP AICT 479, pp. 463–468, 2016.
DOI: 10.1007/978-3-319-48354-2_46

2 Experiments and Methods

A small investigation is designed and interviews about agricultural information perceived risk are carried out. The items in the questionnaire are measured on a 5-point Likert-type scale (1 = strongly disagree to 5 = strongly agree). Some inappropriate questions are corrected and 16 questions are retained, which constitute the risk measurement of farmers' perceived risk. The 16 items are proposed as follows.

Item 1: I worry about that information is not timely
Item 2: I worry about that information can not meet the needs
Item 3: People around who have used the information product think the effect is not evident
Item 4: I worry about that I am not able to use the information product
Item 5: I worry about that it is difficult to communicate with the information servers
Item 6: I worry about that information is not helpful enough as expected
Item 7: I worry about that persistent and convenient services are not available
Item 8: I worry about that I am not able to understand the information
Item 9: There are not many people around that have used the information product
Item 10: I worry about that the information is fake
Item 11: I worry about that the information server's introduction is not true
Item 12: Anxiety may appear in the course of using the information product
Item 13: If the effect is disappointing, my confidence will be reduced
Item 14: I do not tend to take a risk
Item 15: I am not able to pay for the information product
Item 16: I worry about that information searching and obtaining may cost me much.

A total of 231 effective questionnaires are returned. The sample comes from 13 provinces in China. The characteristics are presented in Table 1.

Table 1. The characteristics of the sample

Categories	Number	Percentage %
Gender		
Male	161	69.7
Female	70	30.3
Age group		
18–30	44	19.0
31–40	77	33.3
41–50	69	29.9
51–60	34	14.7
Over 60	7	3.03
Education level		
None	22	9.52
Primary education	50	21.6
Secondary education	105	45.5
Higher education	42	18.2
Over	12	5.19

3 Results and Discussion

The Cranach's alpha coefficient of the scale is 0.78, which has a good credibility. Bartlett's test and KMO test are carried out in order to confirm whether the factor analysis is suitable to be performed.

The results show that the KMO value is 0.783, and the factor analysis is fit to be carried out generally when the KMO value is more than 0.7. The outcome of the Bartlett's test is 0.000, and it is less than the significance level, which is 0.01. The results of KMO test and Bartlett's test both show that the factor analysis is suitable. Results are listed in Table 2.

Table 2. The test of KMO and Bartlett's

Kaiser-Meyer-Olkin measure of sampling adequacy		.783
Bartlett's test of Sphericity	Approx. chi-square	344.344
	df	28
	Sig.	.000

Six factors are extracted from the 16 items based on the factor analysis. The total variance is 82.316 %, which shows that these 6 factors have covered all those items. The results are described in Table 3.

Table 3. The variance contribution rates

Factors	Variance contribution rates	Cumulative variance contribution rates
1	23.871	23.871
2	18.756	42.627
3	15.010	55.637
4	10.332	67.969
5	8.801	75.770
6	6.546	82.316

In this research, farmers' perceived risk is divided into six parts, which is shown in Table 4. The first part, including item 2, item 7 and item 11, is related to the information service departments, which may be summarized as the service risk. The second part, including item 10, item 1 and item 6, reflects that farmers worry about the function of the information product, which may be summarized as the function risk. The third part is summed up as the individual risk, which is related to the ability of information understanding and adoption. In the fourth part, the two factors are related to the economic condition, so it is named as the cost risk, and it is also contains the cost of the skill training and the time in the process of searching. The fifth part mainly shows that farmers are anxious for the effect of the information product, so it may be summarized

Table 4. Results of the factor analysis

	Fart 1	Fart 2	Fart 3	Fart 4	Fart 5	Fart 6
Item 2	0.802					
Item 7	0.723					
Item 11	0.652					
Item 10		0.765				
Item 1		0.638				
Item 6		0.547				
Item 4			0.754			
Item 8			0.632			
Item 5			0.552			
Item 16				0.694		
Item 15				0.488		
Item 12					0.611	
Item 13					0.493	
Item 14						0.578
Item 3						0.472
Item 9						0.337

as the psychological risk. The sixth part is mainly about the social relationship of farmers, which reflects the influence of other people, so it is named as the social risk. Farmers' information perceived risk model is shown in Fig. 1.

The causal relationship between the dependent variable and the independent variable is determined with the regression analysis, which is a statistical analysis method. The multiple regression analysis is adopted to calculate the influence of each

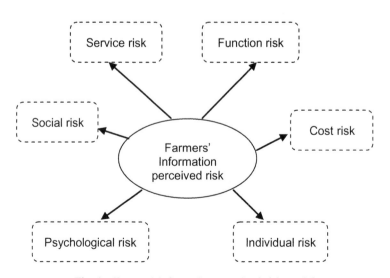

Fig. 1. Farmers' information perceived risk model

Table 5. Weights of farmers' information perceived risk

	Elements	Regression coefficient	T test	Significance level
X_1	Service risk	0.287	9.544	0.000
X_2	Function risk	0.256	5.476	0.001
X_3	Individual risk	0.214	3.246	0.003
X_4	Cost risk	0.125	2.674	0.006
X_5	Psychological risk	0.076	1.235	0.008
X_6	Social risk	0.042	0.348	0.033

component of the perceived risk on the total perceived risk. The result of the multiple regression analysis is shown in Table 5.

Therefore, the regression equation of farmers' information perceived risk is as Eq. (1). It shows that the service risk and the function risk may affect farmers' information perceived risk evidently. Farmers are likely to pay more attention to these aspects, such as whether the service is timely and reliable, whether the information and technologies are useful and so on.

$$PR = 0.287X_1 + 0.256X_2 + 0.214X_3 + 0.125X_4 + 0.076X_5 + 0.042X_6 \qquad (1)$$

4   Conclusions

This paper studies farmers' information adoption from the perspective of the perceived risk based on the investigation and the statistical analysis. Six aspects are concluded, which are the service risk, the psychological risk, the function risk, the society risk, the individual risk and the cost risk. The regression equation is obtained and the result indicates that the service risk and the function risk are the most important determinants that farmers consider about. Information servers may think more about these aspects when they provide farmers with information and technologies. The agriculture information service is a persistent project and it needs the joint effort of the government, social organizations, enterprises and farmers for a long period.

The outcomes might not be very accurate due to the small sample. Farmers may be divided into several types, and different kinds of farmers pay attention to different risks. The next research will focus on the risk preference with a larger sample to discuss about the information adoption of farmers thoroughly.

References

Matrin, J., Mortimer, G., Andrews, L.: Re-examining online customer experience to include purchase frequency and perceived risk. J. Retail. Consum. Serv. **25**(7), 81–95 (2015)

Jianjun, J., Yiwei, G., Xiaomin, W., Nam, P.K.: Farmers'risk preferences and their climate change adaptation strategies in the Yongqiao District, China. Land Use Policy **47**(9), 365–372 (2015)

Horst, M., Kuttschreuter, M., Gutteling, J.M.: Perceived usefulness, personal experinces, risk perception and trust as determinants of adoption of e-government services in The Netherlands. Comput. Hum. Behav. **23**, 1838–1852 (2007)

Lynch, N., Berry, D.: Differences in perceived risks and benefits of herbal, over-the-counter conventional, and prescribed conventional, medicines, and the implications of this for the safe and effective use of herbal products. Compl. Ther. Med. **15**, 84–91 (2007)

Li, Q.: The empirical study of risk perception of shopping online of consumers. Econ. Soc. Dev. **5**(12), 55–57 (2007)

Yang, Q., Pang, C., Liu, L., Yen, D.C., Tam, J.M.: Exploring consumer perceived risk and trust for online payments: an empirical study in China's younger generation. Comput. Hum. Behav. **50**(9), 9–24 (2015)

Wang-Yih, W.U., Ching-Ching, K.E.: An online shopping behavior model integrating personality traits, perceived risk, and technology acceptance. Soc. Behav. Person. **43**(1), 85–98 (2015)

Lu, X., Xie, X., Liu, L.: Inverted U-shaped model: how frequent repetition affects perceived risk. Judgment Decis. Making **10**(3), 219–224 (2015)

Dynamic Changes of Transverse Diameter of Cucumber Fruit in Solar Greenhouse Based on No Damage Monitoring

Ruijiang Wei[1,2(✉)], Xin Wang[1,2], and Huiqin Zhu[3]

[1] Meteorological Institute of Hebei Province, Shijiazhuang 050021, China
weirj6611@sina.com
[2] Key Laboratory of Meteorology and Ecological Environment of Hebei Province,
Shijiazhuang 050021, China
wonghsin@163.com
[3] Meteorological Bureau of Gaoyi County, Gaoyi 051330, China
zhuhuiqin510@sohu.com

Abstract. In order to master the dynamic changes of cucumber fruit growth in solar greenhouse and its relationship with meteorological elements of greenhouse, continuous no damage dynamic monitoring for transverse diameter of Cucumber fruit collected by every 10 min were carried out by the crop growth monitoring instrument, air temperature, air relative humidity, solar radiation and other factors in the greenhouse were also continuous observed by micro climate observing system. Studies have shown that cucumber fruit diameter hourly increment is most at 17:00 or 18:00 in the fair-weather, but least at about 11:00, with a sensible diurnal variation; in the cloudy weather, the growth law of cucumber fruit diameter is similar to which in the fair-weather; the hour increment has a no apparent diurnal variation in the overcast weather; in the overcast to clear weather, there was a noticeable transformation in the diurnal variation of cucumber fruit diameter hourly increment, the diurnal peak value usually appears at 17:00 or 18:00, but before and after noon rebound phenomenon could occurs obviously. The good correlation with the Cucumber fruit diameter hourly increment should be in the sequence below: the average temperature four hours before, the average relative humidity four hours before and the average solar radiation five hours before; The significantly correlation with the Cucumber fruit diameter daily increment should be in the sequence below: the daily minimum temperature that day, the daily average temperature the day before, the daily average relative humidity the day before, the daily average radiation the day before, and daily temperature range the day before.

Keywords: Solar greenhouse · Cucumber fruit diameter · No damage monitoring · Dynamic change

1 Introduction

It is essential for mastering vegetable growth and greenhouse micro climate conditions, and then doing related scientific work that real-time monitoring on vegetables growth

© IFIP International Federation for Information Processing 2016
Published by Springer International Publishing AG 2016. All Rights Reserved
D. Li and Z. Li (Eds.): CCTA 2015, Part II, IFIP AICT 479, pp. 469–478, 2016.
DOI: 10.1007/978-3-319-48354-2_47

increment and greenhouse micro climate. It is also an important basis for targeted greenhouse management. Previous studies showed that observation on the growth of greenhouse vegetables were applied by artificial methods(measure growth increment in a (configurable) time interval). Such as Cheng Zhihui et al. [1], Zhang Zhiyou et al. [2], Pan Xiuqing et al. [3], they measured transverse diameter and vertical diameter of tomato or eggplant fruit every 5 to 7 days with vernier caliper for researching the relationship between environmental factors and tomato or eggplant fruit growing; Wang Xiufeng et al. [4] measured daily growth increment of tomato to research influence of greenhouse temperature and light conditions on tomato vegetative growth and fruit enlargement, although they observed the growth of vegetables varies with time and its accuracy can meet requirements, but also no damage, they get the data in the scale of days, reflecting the diurnal variation of vegetable growth, but not embody changes in smaller scales. Along with the development of protected agriculture, its management also needs refinement, to grasp subtle changes in the growth process, but these studies are rare, only Ma Pengli et al. [5], Hu Xiaotao et al. [6] they applied the Plant Physiology monitoring system for observing the growth of tomato fruits, researching the tomato Physiological characteristics, but observation only for certain period of tomato, rather than a period from fruit setting to harvesting would been tracking. It has not been reported that a continuous non-invasive dynamic growth of the entire cucumber fruit observation from fruit setting to harvesting.

This paper intends to adopt the fruit growth observation instrument to follow-up continuous non-invasive observations of the growth of cucumber fruit transverse diameter from cucumber fruiting to picking, and then analyze the relationship between subtle changes of cucumber fruit growth under different weather conditions and the greenhouse micro climate, so as to provide a basis for the fine management of the micro climate inside the greenhouse and meticulous management of protected agriculture.

2   Materials and Methods

The greenhouse in Gaoyi County (Hebe Province) is sit in the north facing the south, It is 9 m from north to south and 32 m from east to west. Its ridge height is 3.5 m, wall height is 2.8 m. The back wall and east-west wall of the greenhouse are made of earth with thickness of 1.5 m, the front roofing of the greenhouse is a single arc-shaped structure, covered with poly vinyl plastic film. The plastic film was covered with a straw mat, which every day opened after sunrise and covered before sunset. There is a vent at the top of the greenhouse, which opening and closing depends on the temperature, humidity and ambient weather conditions. Solar greenhouse cultivation management is in accordance with local general technical requirements.

The Cucumber fruit diameter growth increment monitoring is applied by DEX20 fruit - trees grow stems measuring instrument manufactured by Dynamax company (USA). The instrument consists of two parts, one part is the data collector, the other part is the sensor. When the Cucumber fruit began to bear fruit, dynamic monitoring of cucumber fruit diameter changes would be running with the aid of tension of the spring sheet (led to the sensor) which is clasp on the fruit, the data was collected by every

10 min, and automatically stored in the collector with monitoring precision 0.01 mm. In order to grasp the quantitative relationship between cucumber fruit growth and meteorological conditions, only one cucumber fruit measured is allowed to leave in the monitoring process, the rest flower buds in the plant would be removed so as to eliminate the effect of dry matter partitioning on cucumber fruit growth. Just when the monitored Cucumber fruit grow to be picked, another fruit would be measured. This work started from the beginning of fruit setting and finished after cucumber harvest.

The net growth of cucumber fruit transverse diameter within one hour would be defined as hourly increment and the net growth within one day as daily increment.

The greenhouse micro climate data including air temperature, relative humidity, solar radiation was measured by AR5 micro climate observation system(manufactured by Beijing Yugen Technology Development Co., Ltd.), with its observation accuracy was $\pm 0.2°C$, ± 2 % and ± 3 W/m^2. Air temperature and relative humidity sensor probe were located in the 1.5 m height and solar radiation sensor probe in 2.0 m height, with every 10 min collecting a data and automatically storing in the collector.

3   Conclusion and Analysis

3.1   Cucumber Fruit Transverse Diameter Dynamic Changes

3.1.1   In the Fair-Weather

By analyzing the dynamic change data of cucumber fruit transverse diameter by every 10 min, it concluded that if the cucumber fruit was growing in fair weather from cucumber fruiting to picking all the time, the cucumber fruit diameter has no rebound phenomenon during the daytime, but at night there was rebound phenomenon, generally appeared in 20:00 ~ the next morning 08:00 (Fig. 1). Figure 1 shows Cucumber fruit transverse diameter growth increment curves change in the fair-weather.

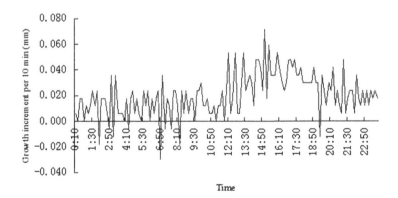

Fig. 1. Cucumber fruit transverse diameter growth increment curves change in the fair-weather

Figure 2 shows cucumber fruit transverse diameter hourly increment diurnal variation curves from cucumber fruiting to picking, which fruit-setting day is on November

27 and harvest day on December 3. Because monitoring data in these two day was in less than one day length, the data was deleted. From November 28, 2008 to December 2 2008, there were consecutive sunny days in this cucumber fruit growth period, the daily sunshine hours was 6.9 to 8.5 h, and the maximum greenhouse solar radiation received was between 372 ~ 459 W/m^2. Figure 2 shows that cucumber fruit diameter hourly increment in sunny days had significant diurnal variation, the diurnal peak value generally appeared at 17:00 or 18:00, hourly maximum growth increment would be 0.3 mm or more, It would decline in volatility from 20:00 to 11:00 the following day, and least at about 11:00.

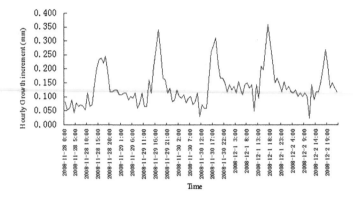

Fig. 2. Cucumber fruit transverse diameter hourly increment diurnal variation curves in the fair-weather from cucumber fruiting to picking

3.1.2   In Cloudy Weather

On both day and night, the cucumber fruit diameter occur rebound phenomenon occasionally in cloudy weather, and even occur in successive two 10 min.

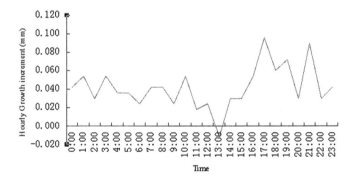

Fig. 3. Cucumber fruit transverse diameter hourly increment diurnal variation curves in cloudy weather

Figure 3 shows cucumber fruit transverse diameter hourly increment diurnal variation curves in cloudy weather. The daily sunshine hours is 4 h on January 2, 2009, the maximum greenhouse solar radiation received was 305 W/m^2. It is showed that the maximum value of cucumber fruit diameter hourly increment generally appear at 17:00, hourly maximum growth increment would be 0.1 mm or less, and least at about 13:00.

3.1.3   In Overcast Weather

Cucumber fruit diameter growth increment was related to the degree of overcast days. If the degree of overcast weather was heavy, the growth is highly irregular; if the degree was light, the growth law is similar to which in the cloudy weather.

Figure 4 shows cucumber fruit transverse diameter hourly increment diurnal variation curves on December 12, 2008. The daily sunshine hours was 0, solar radiation occur in only 11:00 to 14:00, the maximum greenhouse solar radiation received was 182 W/m2. Seen from Fig. 4, cucumber fruit diameter hourly increment was volatile, no significant changes in the trend.

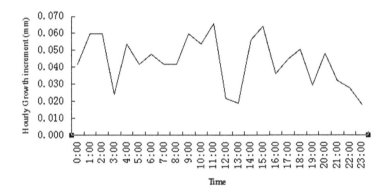

Fig. 4. Cucumber fruit transverse diameter hourly increment diurnal variation curves in overcast weather

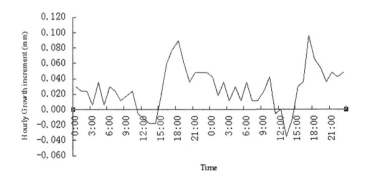

Fig. 5. Cucumber fruit transverse diameter hourly increment diurnal variation curves in overcast weather

Figure 5 shows cucumber fruit transverse diameter hourly increment diurnal variation curves. December 27 and 28 were overcast days, solar radiation occur in only 10:00 to 15:00, the maximum greenhouse solar radiation received was 245 W/m² and 190 W/m². Seen from Fig. 5, the cucumber fruit diameter hourly increment reached a maximum at about 16:00 and then drops down at a rapid speed, and least at about 23:00.

3.1.4   In the Overcast to Clear Weather

The cucumber fruit diameter has rebound phenomenon both on day and night in the overcast to clear weather with its hourly increment has a sensible diurnal variation. Figure 6 shows cucumber fruit transverse diameter hourly increment diurnal variation curves on January 9,2008 and January 10, 2008. There were overcast or foggy weather for the 6 consecutive days during January 3 ~ 8 days. The daily sunshine hours was 2.2 h on January 6 and 0 h in other days. The maximum greenhouse solar radiation received was 440 W/m² and 404 W/m². Seen from Fig. 6, the cucumber fruit diameter hourly increment reached a maximum at about 17:00 or 18:00, hourly maximum growth increment would be 0.25 mm or less. But a significant rebound phenomenon would be occurred around noon (11 to 15), when hourly increment was negative.

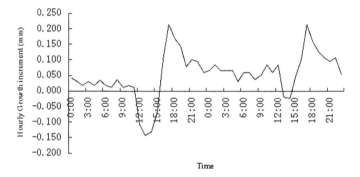

Fig. 6. Cucumber fruit transverse diameter hourly increment diurnal variation curves in the overcast to clear weather

3.2   Relationship Between Cucumber Fruit Diameter Growth Increment and Meteorological Conditions

3.2.1   Relationship Between Cucumber Fruit Diameter Hourly Increment and Meteorological Conditions

A mass of greenhouse micro climate data with every 10 min collecting were arranged into the hourly data (exactly on time), and the correlation coefficient respectively between cucumber fruit diameter hourly increment and each meteorological elements at the same time and a few hours before. The result was shown in Table 1.

Table 1. Correlation coefficient respectively between cucumber fruit diameter hourly increment and each meteorological elements at different times (n = 971).

Meteorological element	Average temperature	Average relative humidity	Average solar radiation
The same time	0.3025**	0.1118	0.1100
1 h before	0.4486***	0.3095**	0.0412
2 h before	0.5873***	0.4601***	0.2326*
3 h before	0.6033***	0.4641***	0.3574***
4 h before	0.6940***	0.5387***	0.5031***
5 h before	0.6317***	0.4584***	0.5262***
6 h before	0.5202***	0.3347***	0.4611***

Note: *, * *, * * * represents the significant test of the level of 0.02, 0.01 and 0.001 respectively.

Seen from Table 1, the correlation between Cucumber fruit diameter hourly increment in Greenhouse and meteorological elements at different times were mostly passed the reliability test of 0.001, significant correlation. The good correlation with the Cucumber fruit diameter hourly increment should be in the sequence below: the average temperature four hours before, the average relative humidity four hours before and the average solar radiation five hours before. Correlation diagram is given in Fig. 7.

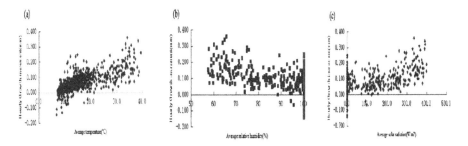

Fig. 7. Correlation diagram of respectively between cucumber fruit diameter hourly increment and each meteorological elements at different times (a) The average temperature four hours before as the abscissa, (b) The average relative humidity four hours before as the abscissa, (c) The average solar radiation five hours before as the abscissa

3.2.2 Relationship Between Cucumber Fruit Diameter Daily Increment and Meteorological Conditions

A mass of greenhouse micro climate data with every 10 min collecting were arranged into the daily data, and the correlation coefficient respectively between cucumber fruit diameter daily increment and each meteorological elements in the same period and a few days before. The result was shown in Table 2.

Table 2. Correlation coefficient respectively between cucumber fruit diameter daily increment and each meteorological elements at different times (n = 35)

Meteorologic al element	Maximum temperature	Minimum temperature	Minimum relative humidity	Maximum solar radiation	Daily average temperature	Daily average relative humidity	Daily average solar radiation	Daily temperature range
That day	0.6075***	0.7983***	0.5119**	0.4293**	0.8239***	0.5124**	0.4338**	0.3868*
1 day before	0.7705***	0.6458***	0.7026***	0.6378***	0.8770***	0.7306***	0.6802***	0.6054***
2 days before	0.7031***	0.5839***	0.6233***	0.6409***	0.7897***	0.6663***	0.6328***	0.5634***
3 days before	0.6432***	0.5483***	0.5699***	0.6108***	0.6989***	0.5975***	0.5715***	0.5134**

Note: *, * *, * * * represents the significant test of the level of 0.02, 0.01 and 0.001 respectively.

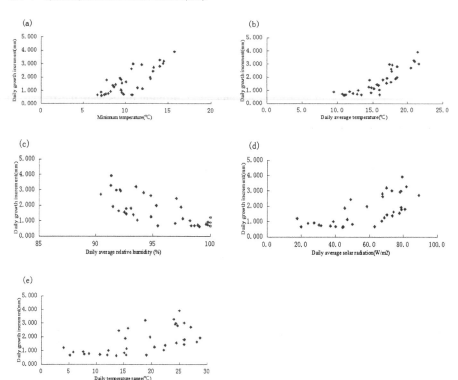

Fig. 8. Correlation diagram of respectively between cucumber fruit diameter daily increment and each meteorological elements at different times (a) The daily minimum temperature that day as the abscissa, (b) The average temperature the day before as the abscissa, (c) The average relative humidity the day before as the abscissa, (d) The average radiation the day before as the abscissa, (e) Daily temperature range the day before as the abscissa

Seen from Table 2, the correlation between Cucumber fruit diameter daily increment in Greenhouse and meteorological elements at different days(the same day, one day before, two days before and three days before) were all passed the reliability test of 0.02, and most of the elements passed the test of 0.001. The significantly correlation with the Cucumber fruit diameter daily increment should be in the sequence below: the daily minimum temperature that day, the average temperature the day before, the average relative humidity the day before, the average radiation the day before, and daily temperature range the day before. Correlation diagram is given in Fig. 8. It was showed that the growth of cucumber fruit would be greatly affected by the minimum temperature that day, the average temperature the day before, the average air relative humidity the day before, the average solar radiation the day before, and daily temperature range the day before.

4   Conclusion and Discussion

Cucumber fruit growth is a complex process, change in the fruit growth increment is a result of the accumulation of plant photosynthesis and respiration. The continuous observation on cucumber fruit diameter growth increment collected by every 10 min can record the dynamic changes of cucumber fruit. Solar greenhouse cucumber fruit diameter has no rebound phenomenon during the daytime in the continuous sunny day; on both day and night in the cloudy, overcast and the overcast to clear weather the cucumber fruit diameter would occur rebound phenomenon occasionally.

The cucumber fruit diameter hourly increment is most at 17:00 or 18:00 in the fair-weather, the cloudy weather and the overcast to clear weather, but least at about noon, with a sensible diurnal variation, especially on the noon of overcast to clear weather, cucumber fruit diameter has obvious rebound phenomenon, and the hourly increment is negative; in the overcast weather, the diurnal variation of cucumber fruit diameter hourly increment is not obvious. This is in agreement with the physiological activity of cucumber. Because in the sunny day, the straw mat cover would be lifted around 8:30 generally, the temperature inside the greenhouse would began to rise, the plants start doing photosynthesis and accumulation of organic matter, but due to the closed greenhouse only around noon leaked, carbon dioxide within a greenhouse in the morning should be relative scarcity [7–10], so the growth increment is at least around noon and from noon until this afternoon, the environmental conditions are relatively suitable for cucumber plant growth, so the maximum fruit diameter hourly increment usually appears at 17:00 or 18:00. In the overcast to clear weather, the plant physiological function is still weak, the photosynthesis accumulation is less than plant respiration, so around noon the fruit would register negative growth.

Cucumber fruit growth process is affected by many factors, both personal factors and environmental factors, under the influence of the elimination of dry matter distribution, the good correlation with the Cucumber fruit diameter hourly increment should be in the sequence below: the average temperature four hours before, the average relative humidity four hours before and the average solar radiation five hours before. The significantly correlation with the Cucumber fruit diameter daily increment should be in the sequence below: the daily minimum temperature that day, the average temperature the day before, the average relative humidity the day before, the average radiation the day before, and daily temperature range the day before.

Except by the influence of meteorological factors, cucumber growth in greenhouse would be affected by cucumber varieties, management level, the growth stage and other factors, because of less relevant research, the subtle changes of cucumber fruit growth still need further study and discussion.

References

1. Zhihui, C., Xuejin, C., Linling, L., et al.: The relationship between tomato fruit growth and environmental factors in greenhouse. Ecol. J. **31**(3), 0742–0748 (2011). (in Chinese)
2. Zhiyou, Z., Hongxin, C., Binglin, C., et al.: Simulation model of tomato fruit growth and yield in greenhouse. Jiangsu Agri. J. **28**(1), 145–151 (2012). (in Chinese)
3. Xiuqing, P., Xiurui, G., Yanrong, W., et al.: A study on the content variation of soluble saccharide and proteins in the eggplant Parthenocarpy fruit. Acta Agriculturae Boreali-Sinica **22**(2), 50–52 (2007). (in Chinese)
4. Fanyang, Z., Xiufeng, W., Xue, H., et al.: Effect of temperature and light conditions on the growth and fruit enlargement of Tomato in greenhouse. Chin. Veg. **12**, 66–70 (2013). (in Chinese)
5. Pengli, M., Dong, Y., Xudong, Z., et al.: Study on observation of greenhouse tomato physiological and ecological characteristics of observation. Agri. Res. Arid Areas **24**(1), 142–146 (2006). (in Chinese)
6. Xiaotao, H., Zhenchang, W., Hua, M.L.: The relationship between the change of stem diameter and plant water content of eggplants. Trans. Chin. Soc. Agri. Eng. **30**(12), 87–92 (2014). (in Chinese)
7. Min, W., Yuxian, X., Xiufeng, W., et al.: Research on the change of CO2 concentration in solar greenhouse. Appl. Ecol. J. **14**(3), 354–358 (in Chinese)
8. Ruijiang, W., Xiping, W., Yun, X., et al.: Research on the C02 fertilization technology system by industrial exhaust gas to greenhouse vegetable. Eco-agri. Study. **8**(1), 53–55 (2000). (in Chinese)
9. Qiwei, H., Xizhen, A., Xiaolei, S., et al.: Study on the growth and decline of CO2 concentration in the cultivation of Cucumber in greenhouse Cucumber. Chin. Veg. **1**, 7–10 (2002). (in Chinese)
10. Pengli, M., Li, Y.X., Xiaojuan, L.: Effect of greenhouse application of CO2 gas fertilizer on cucumber growth development. Agri. Meteorol. **24**(4), 48–50 (2003). (in Chinese)

Study on Laos-China Cross-Border Regional Economic Cooperation Based on Symbiosis Theory: A Case of Construction of Laos Savan Water Economic Zone

Sisavath Thiravong[1,2,3], Jingrong Xu[1,2(✉)], and Qin Jing[4]

[1] Business School of Hohai University, Nanjing 211100, China
xujr1989@163.com
[2] Jiangsu Provincial Collaborative Innovation Center of World Water Valley and Water Ecological Civilization, Nanjing 211100, China
[3] National Electrical Company of Laos, Vientiane, Laos
[4] China Institute of Water Resources and Hydropower Research, Beijing 100044, China

Abstract. The strategy of "one belt one road" has provided an opportunity of accelerate economic growth for China's surrounding countries. Laos is actively joining the initiative, but how to cooperate with China is a difficulty. Biological symbiosis shares similar characters with cooperation and mutual benefits among cross-border countries. So taking the construction of Savan-Seno Special Economic Zone of Laos as example, we introduced the symbiosis theory and analyzing symbiosis units, models and interfaces between Laos-China regional economic cooperation, it led to the generation, optimization and mechanism under which the symbiotic system was determined. Then we should improve the Laos-China regional economic cooperation symbiosis system by training and harvesting favorable environments, expanding symbiotic interface so as to accelerate the cooperative relationship.

Keywords: One belt one road · Laos-China · Symbiotic theory · Cross-border regional economic cooperation

1 Introduction

Chinese President Xi Jinping has put forward the strategic proposal of "Build Silk road economic belt and maritime silk road in the 21st century based on the innovation cooperation mode" in the year of 2013, which has attracted worldwide attention and positive response. That strategy has became the broad consensus and guide of actual action for the countries in Southeast Asia. Chinese government has presented the policy of "The vision and action of accelerating the construction of silk road economic belt and maritime silk road in the 21st century based on the innovation cooperation mode" in March of 2015 for guiding the path of this strategy.

Laos is a landlocked country with advantages of location in northern Indo-China Peninsula, whose south is Cambodia, north is China, northwest is Myanmar, and

© IFIP International Federation for Information Processing 2016
Published by Springer International Publishing AG 2016. All Rights Reserved
D. Li and Z. Li (Eds.): CCTA 2015, Part II, IFIP AICT 479, pp. 479–486, 2016.
DOI: 10.1007/978-3-319-48354-2_48

southwest is Thailand. The "great Mekong sub-regional economic cooperation" (GMS) has an important strategic position within the economic cooperation between China and ASEAN. Compared with other countries in the "neighborhood" all the way along the route of "One Belt One Road", Laos is considered as an all-weather strategic partner of China in Indo-China Peninsula because of its political and social stability, economic growth speed, high degree of political trust and highly complementary economic cooperation with China. In addition, Laos is in poverty but not in infertility, which is rich in natural resources, whose economy gained a momentum of rapid growth after joining the WTO in 2012. The 9th National Congress of Laos Revolutionary Party put forward the national development strategy to "get rid of the less developed countries state" in 2020 based on Chinese reform and opening up and development experience to Laos by "closed country" into an "open country". That also advocates accepting and agreeing the ideas and values of "regional development", all of these strategies become the solid material foundation and strategic environment for the economic cooperation and implement cross-border collaboration between Laos and China.

Laos Savan Water Economic Zone, located in south central of Laos, mainly including Khammouane and Savannakhet, is next to Vietnam in the east and Thailand in the west. Khammouane is located in the middle part (near southern part) of Loans, Thailand is located in its west, Vietnam is located in its east, Bolikhamsai is located in its north and Savannakhet is located in its south, Savannakhet is in its south, Savannakhet has 9 counties, an area of 16,315 km^2, population of 298,000 and provincial capital is Thakhek; Savannakhet is located in southern part of Laos, facing Thailand in the west across Mekong River, next to Vietnam in the east, has 15 counties, it enjoys an area of 21,774 km^2, the population of about 954,000, the capital is Kaysone Phomvihane. National territorial areas of two provinces account for 16 % of that of Laos and the population accounts for 20 %. Khammouane and Savannakhet enjoy the good geographical location with No. 9, No. 12 and No. 13 national highways passing through them, and are important geographic centers and main agricultural areas of production.

Savan-Seno Special Economic Zone is built by Laos and China together, involved two main subjects. There is a big gap between the two main economic dimensions, compared to China, Laos has lower technical level, but is richer in resources, How to coordinate the relationship between the two subjects? Therefore, this paper studies the symbiotic theory of ecology and from a new perspective to studies the relationship between the two subjects.

2   Literature Review

Cross-border economic cooperation is refers to the adjacent two countries on different levels of resource, technology, market and capital. The elements of the two countries can make use of their complementary advantages of production factors of forms of cooperation that enjoy preferential policies such as export processing zone, bonded zone, free trade area to promote economic development between the two countries. For cross-border research of regional economic cooperation, the earliest dates back to international trade theory of classical economics. Classical trade theory achieves the objective of the international economic cooperation based on the international movement of goods. Proposed the theory of "boundary effect" of the early economic geography, think that the existence of national boundaries will affect the free flow of goods, which influence the development of trade, inhibit the development of national economy, such as Losch (1944), Ciersch (1949195) and Heigl (1978) [1]. Brocker (1984) using a district-level trade gravity model, empirically the effect of the European Community boundaries, think the border trade barriers produced obvious effect [2]. Domestic scholar Li Tianzi (2015) by using gravity model, make empirical analysis of the boundary effect of the cross-border economic cooperation by using the 199 Chinese border city of data. He believe China's cross-border economic cooperation has higher boundary effect [3]. Yang Rong-hai, Li yabo (2014) in Yunnan region of Vietnam, Laos and Myanmar cross-border economic zone as the object, the empirical test the existence of border effect of economic cooperation zones between China and Vietnam, Laos and Myanmar, the boundary effect is shown by the mediation effect, while the intermediary effect is positive, can effectively promote the level of open of bambina in southwest China [4].

Regional economic theory holds that due to the differences in natural resources, social cultural and the level in factors of production, different economic regions has the characteristics of imbalance. Cross-border economic cooperation is emerging a new pattern of regional economic cooperation in recent years, hang Lijun (2014) is analyzed the inevitability of China's cross-border economic cooperation and the objective requirements on the basis of the theory of regional economic integration, and proposed the measures of China's cross-border economic cooperation [5]. From the perspective of regional economics, Jiang Yongming (2009) holds that the theoretical basis of inter-national economic cooperation included regional spatial interaction theory, regional spatial structure theory and transnational regional governance theory, he focused on the European Union, North America and east Asia international economic cooperation zone and analyzed that the cooperation mode and path selection, system is important to attacked in the international regional economic cooperation [6]. From the perspective of regional economic theory, three scholars who Fure J S (1997) [7], Blatter J (2000) [8], Hanson G H (2001) [9], analyzed the influence factors of cross-border economic coop-eration and cross-border economic cooperation were the major influencing factors of political factors, institutional factors and transportation cost. Tang Zhongjian, Zhang Bing, Chen Ying (2002) have viewed the east Asian region as the research object and putted forward the three modes of cross-border economic cooperation: channel models, trade port mode and development zone [10].

"Symbiosis Theory" thinks that symbiosis is a common phenomenon between nature and human society while the nature of symbiotic is consultation and cooperation and mutualism is the inevitable trend of human society and the natural symbiosis. Mutually beneficial relationship is the strategic basis for cross-border economic cooperation, as well as the premise of countries` cooperation. Zhiming Leng and Heping Zhang (2007) regarded the core issues of regional economic cooperation as a starting point and found that regional economic problems and symbiosis theory has a strong consistency and applicability on the objectives and mechanisms of regional economic cooperation [11]. Baojie YI and Jieyan Zhang (2015) analyzed the applicability of symbiosis theory on "One Belt and one Road" cooperation in Northeast Asia, and put forward the idea that the establishment of symbiosis mechanism and cooperation model innovation of "One Belt and one Road" in Northeast Asia should be focused on deepening the system platform, enhancing interaction and cooperation in a comprehensive upgrade regional cooperation in the Tumen River development [12].

Throughout the literature and we can find that scholars focus more on regional economic cooperation with the research literature symbiosis theory but rarely the study of cross-border regional economic cooperation. Cross-border economic cooperation involves a number of economic entities and "Laos - China" cross-border economic cooperation faces the problem that what the mode of cooperation between the various economic agents is. After the analysis of the literature we can find that regional economic cooperation can be used on symbiosis theory. Therefore, this article intends to use the symbiotic theory to research "Lao - Chinese" cross-border economic cooperation, focusing on the Laos Haven water economic district as the specific objects.

3 The Generation of "Laos - China" Regional Economic Cooperation Symbiotic System

(1) At least one group of quality parameters of the cross-border collaboration concert is compatible.
(2) Cross-border collaboration subject shall at least form one symbiosis interface, while the subject factors for the collaboration can act freely in the symbiosis interface.
(3) Subject factors of cross-border collaboration parties have flowability.
(4) Symbiosis units need energy for interaction through symbiosis interface.
(5) Information richness between entities for cross-border collaboration not less than a certain threshold.

With the analysis of the above 5 conditions, for the "Laos - China" Symbiosis system, the key to generate symbiotic system is that each participant economic agents must be compatible with each other, the degree of homogeneity should be higher than the critical value, cross-border economic cooperation body (the main parameter) information on the other side of the master degree (information abundance), their own benefits and win-win benefits (symbiotic energy) by the cooperation. Since the establishment of diplomatic relations between Laos and China, the two countries have established a very strong economic ties after decades of development. The changes of industrial structure and business structure of one country will result in the related economic benefits changes in the other country. Regarding to the Laos Haven water economic district, Lao Chamber of Commerce and the China Power Construction Group Co., Ltd. Kunming Survey and Design Institute has already launched a number of cooperation projects, which shows that Laos and China can express the related quality parameters each other and the economic symbiotic relationship has been established between Laos and China. Laos and China are bounded and have been developing the economic cooperation for a long history. During such a Laos – China cross-border regional economic cooperation system, the subject to establish "One Belt and one Road" is to choose a practical and efficient cooperation way to seek common development, enabling cooperation to get better economic utility.

4 "Lao - China" Symbiotic System Optimization of the Regional Economic Cooperation

Incentive efficiency is the important condition of symbiotic system optimization; the ideal incentive efficiency can make the symbiosis body common symmetric mutual symbiotic evolution towards integration, to attain the ideal incentive efficiency, symbiotic system works the same production of energy in different symbiotic unit [13, 14], Is $E_{si}/E_{ei} = E_{sj}/E_{ej}$, Laos - China's "economic cooperation system is a two-dimensional symbiotic system, in certain cases of the symbiotic interface. When the main quality parameters is fixed of Z_i and Z_j. Symbiosis energy is respect of $E_{si} = S_i(\rho_i, \rho_j, \eta_i, \eta_j)$, $E_{sj} = s_j(\rho_j, \rho_i, \eta_j, \eta_i)$. The loss of symbiosis energy is respect of $E_{ei} = C_i(\rho_i, \rho_j, \eta_i, \eta_j)$, $E_{ej} = C_j(\rho_j, \rho_i, \eta_j, \eta_i)$. The net energy of symbiotic system is $\Delta E = (E_{si} + E_{sj}) - (E_{ei} + E_{ej})$.

The optimization of the conditions in Laos - "China economic cooperation system is:

$$Max\Delta E = (E_{si} + E_{sj}) - (E_{ei} + E_{ej})$$
$$s.t. E_{si}/E_{ei} = E_{sj}/E_{ej}$$

Using the Lagrange multiplier method to calculate the optimized conditions:

$$\partial E_{si}/\partial \rho_i = \partial E_{ei}/\partial \rho_i \tag{1}$$

$$\partial E_{sj}/\partial \rho_j = \partial E_{ej}/\partial \rho_j \tag{2}$$

$$\partial E_{si}/\partial \eta_i = \partial E_{ei}/\partial \eta_i \tag{3}$$

$$\partial E_{sj}/\partial \eta_j = \partial E_{ej}/\partial \eta_j \tag{4}$$

$$E_{si}/E_{ei} = E_{sj}/E_{ej} \tag{5}$$

(1) and (2) are symbiosis density equilibrium conditions. (3) and (4) are the symbiotic dimension equilibrium conditions. When a symbiotic system of symbiotic energy with the depletion of the symbiotic energy satisfy this five equations of (1), (2), (3), (4) and (5) at the same time. There are symbiotic system presents the structural matching, symbiotic system maximize the interests of symbiosis on any subjects to achieve the goal of convolution. Based on the above analysis, for "Laos - China" symbiotic system optimization, regional economic cooperation is the optimization of its economic main body density and symbiosis symbiotic dimension, but the most important point is structural optimization, specific to shawan water in terms of economic zone, economic zone is wading industries and gigantic symbiotic system density, wading enterprises economic zone of China. Cooperation deeply continued and symbiotic system dimensions are more and more widely.

5   The Generation of the Symbiotic Energy of the "Lao China" Regional Economic Cooperation

One of the most important essential characteristics is produced new energy symbiotic process, called symbiosis energy. If the symbiotic system S has a quality parameter Z_S, and there are $m(m \geq 2)$ symbiotic units, then:

$$Z_S = f\left(Z_1, Z_2 \cdots Z_i \cdots Z_m\right)$$

The total factor of system δ_s is:

$$\delta_s = \frac{1}{\lambda} \sum_{i=1}^{m} \delta_{si}$$

λ is the characteristic coefficient of symbiotic interface, and δ_{si} is one of the symbiotic system.

Symbiosis energy (E_S) is the concrete embodiment of the existence and value added ability of the symbiotic system. The principle of symbiosis energy production shows that the system which does not produce symbiotic energy can not be added and developed. Under normal circumstances, the generation of symbiotic energy is mainly influenced by the total factor, when $\delta_s > 0$, will produce symbiotic energy [14]. But at the same time, the symbiotic energy is related to the density and the symbiotic dimension, then $E_S = f\left(\delta_s, \rho_s, \eta_s\right)$ or $\delta_s^m = \sum_{i=1}^{m} \delta_{si}$. ρ_s is the symbiotic density, and η_s is the symbiotic dimension. For the "Lao-China" economic cooperation, such a two-dimensional symbiotic system, the symbiotic system E_s can be expressed as followed:

$$E_s = f_s(Z_a^m, Z_b^m, \theta_{ab}, \lambda, \rho_{sa}, \rho_{sb}, \eta_{sa}, \eta_{sb}, F, \delta_s)$$

F is the symbiotic environment of symbiotic system.

Therefore, "Laos – China" Economic Cooperation symbiosis energy generation is to be maximize, not only to conforms to the symbiotic system optimization conditions, also with main quality parameters of governance, symbiotic interface, symbiosis of the quality of the environment related. Put it differently: the quality of main quality parameters, symbiotic density, symbiotic dimension, the symbiotic interface and symbiotic environment are jointly decided to symbiosis energy size. How to improve economic cooperation, the "Laos China" symbiosis energy, specific to the Savan-Seno Special Economic Zone, its intrinsic requirement is to improve the cross-border economic subject characteristics, so as to achieved the structural matching; improved the cross-border economic entities symbiotic density and dimension; the expansion of the cross-border economic entities symbiotic interface; cultivated and selected the symbiotic environment conducive to cooperation.

6   Conclusions

This article explores the symbiotic relationship between Laos and China economies under "One Belt and one Road" Project. Through analysis, we find the typical symbiotic relationship between Laos and China economic cooperation. To establish "One Belt and one Road" together is internal demand of the two countries` economic development strategy as well as the external requirement of regional economic cooperation.

The establishment of Laos Haven water economic district is the specific applications in Laos and China cross-border economic cooperation, whose symbiotic system has symbiotic units consisting of "three levels, six co-factor": government level, social organization level and the level of industry, university, research and financial; education, science and technology, entrepreneurship, finance, industry, and culture. The two governments should strengthen communication and collaboration: to optimize the allocation of the six elements, and promote the flow of the six elements. So that the two governments can work and combine with each other. The two governments should strengthen the regulatory role: to guarantee the enterprises settled in the region in line with the requirements of the economic area, to draft more policies and measures, and to promote bilateral business cooperation, in order to improve the density of the symbiosis density and dimensions of symbiotic systems of water economic area and to optimize the industrial structure in the area. In this way, provide support for the energy cogeneration system and promote the continuous optimization of the symbiotic system.

Acknowledgment.   Funds for this research was provided by the Special Program For Key Program for International S&T Cooperation Projects (2012DFA60830), the National Planning of Social Science, Funding Project (11&ZD168) and Supported by Program for Changjiang Scholars and Innovative Research Team in University (IRT13062)

References

1. Feng, G.-Q., Ding, S.-B.: Retrospect and prospect of the cross-border cooperation study. World Reg. Stud. **14**(1), 53–60 (2005)
2. Bröcker J.: How do international trade barriers affect interregional trade[J]. In: Regional and Industrial Development Theories, Models and Empirical Evidence, pp. 219–239 (1984)
3. Li, T.-Z.: Boundary effect of Chinese cross-border economic cooperation. Econ. Geogr. **10**, 5–12 (2015)
4. Ronghai, Y., Li, Y.: Will border effect restrict the construction of China's cross-border cooperation zone?: an analysis based on the date of China and Vietnam, Laos and Burma. Int. Econ. Trade Res. **30**(20703), 73–84 (2014)
5. Zhang, L., Zheng, Y.: Achievements, problems and countermeasures of border trade of Yunnan province. J. Minzu Univ. China(Philos. Soc. Sci. Ed.) **41**(21302), 43–51 (2014)
6. Yongming, J.: CROSS-National Regional Economic Cooperation and Development. Jilin University (2009)
7. Fure J.S.: The German-Polish Border Region: A Case of Regional Integration[M]. ARENA (1997)
8. Blatter, J.: Emerging cross-border regions as a step towards sustainable development. Int. J. Econ. Dev. **2**(3), 402–439 (2000)

9. Hanson, G.H.: US–Mexico integration and regional economies: evidence from border-city pairs. J. Urban Econ. **50**(2), 259–287 (2001)

10. Tang, J.-Z., Zhang, B., Chen, Y.: The boundary effect and cross-border subregional economic cooperation-a case study of East Asia. Hum. Geogr. **01**, 8–12 (2002)

11. Zhi-ming, L., Zhang, H-P.: The cooperating mechanism of regional economic development based symbiosis theory. Econ. Rev. **2**, 32–33 (2007)

12. Baozhong, Y., Zhang, J.: Study on "one belt one road" cooperation symbiotic system of North-East Asia. Northeast Asia Forum **24**(11903), 65–74 (2015)

13. Chen, S.: PPRD economic cooperation in the perspective of the symbiotic theory. J. Yunnan Nat. Univ.(Soc. Sci.) **02**, 115–123 (2012)

14. Ling D.: The Study on Supply Chain Alliance based on the Symbiosis Theory. Jilin University (2006)

Study on Mode of Laos-China Cross-Border Collaboration Strategy Facing Symbiosis Relation

Sisavath Thiravong[1,2,3], Jingrong Xu[1,2(✉)], and Qin Jing[4]

[1] Business School of Hohai University, Nanjing 211100, China
xujr1989@163.com
[2] Jiangsu Provincial Collaborative Innovation Center of World Water Valley and Water Ecological Civilization, Nanjing 211100, China
[3] National Electrical Company of Laos, Vientiane, China
[4] China Institute of Water Resources and Hydropower Research, Beijing 100044, China

Abstract. Laos is a country with abundant natural resources, low labor costs and land rents, which has natural geo-economic relationship with China, and gradually became one of the important destinations of Chinese enterprises' foreign direct investment. Meanwhile, China has become the largest overseas investment country in Laos. Based on the symbiotic relationship between Laos and China, this paper studies the dynamics of Laos-China cross-border cooperation from three aspects, including Laos's reform, China's "One Belt One Road" and "going abroad" strategy, and the multi-agent coordination of external demand respectively. This paper presents the "overseas Jiangsu" mode of Laos-China cross-border collaboration strategy and offers some elicitation to the cross-border cooperation in Laos-China. Finally, this paper proposed to establish the "overseas Jiangsu" mode of Laos-China cross-border collaboration strategy .

Keywords: Symbiosis relationship · Cross-border collaboration · Overseas Jiangsu · Synergetic net

1   Introduction

As the largest neighboring country of Laos, China has maintained close economic connections with Laos for a long time. In the 21st Century, in particular, Laos and China successively entered into agreements in economic cooperation and trade investment protection, the agreement is enhanced transparency, eliminated non-tariff barriers and simplified trade formalities, optimized trade environment and greatly promoted the bilateral trade development between the two countries. Laos is different from China in accumulation of factors such as resources, capital, technologies and resource endowment, while cross-border cooperation can sufficiently make use of resource advantages of the countries, especially under the collaboration of multi-entity factors. Enterprise cooperation between Laos and China enables the realization of collaboration and sharing of factors such as resource capital and technology and promotes the economic development of the countries, and particularly, the resource factors can reduce the economic

D. Li and Z. Li (Eds.): CCTA 2015, Part II, IFIP AICT 479, pp. 487–495, 2016.
DOI: 10.1007/978-3-319-48354-2_49

cooperation cost between Laos and China, reduce the investment and improve the efficiency of cooperation between enterprises in Laos and China. Building a symbiotic relationship oriented Laos-China cross-border collaboration strategy can drives the economy complementary development, promoted two-way investment and pushed forward economic development in Laos and transformation and update of economy in China, and promote the international political and economic status of Laos.

2   Literature Review

Collaboration is the inherent law and substantive characteristic of change and development of complex system, H. Igor Ansoff [1] thinks that collaboration is the enterprise successfully expand new causes by recognizing the matching relation of its capacities and opportunities, emphasizing the relation between acquisition of tangible and intangible economic benefits and the enterprise's capacities. Hiroyuki Itami, a Japanese strategy expert, thinks that the collaboration is "thumb a lift", namely, when the resources produced by a department in its development can be used by other departments at the same time without any cost, the additional benefit produced by this circumstance is the collaboration effect, the collaboration effect can create true competitive advantages for the whole group. Robert Buzzell and Breadley Gale [2] who studied from the perspective of enterprise group, thinks that collaboration can be shown as all enterprises jointly sharing the cost of a kind of business, economies of scale lead to cost shared by every enterprise is lower than the cost when they undertake this business independently. Restricted by natural, social, economic and historical factors, huge gaps among speed, levels, structure exist in economically underdeveloped areas compared with developed areas. There are rare researches about constructions in development zones of underdeveloped areas, merely scattering in individual papers about development zones' growth, such as Levine's [3]. He pointed out that those underdeveloped areas needed excavating comparative advantage, promoting technological innovation and creating competitiveness to achieve great-leap-forward development. Different from other economic development zones, cross-border collaborative strategy located in collaboration between the subject elements among countries, emphasizing mutual benefits, mutual learning from respective strengths and featuring innovation driving and collaborative innovation. J.A. Schumpeter and C. Freeman, representatives of innovation theory, this school studied technological innovation effect on economic growth from the view of endogenous (Iammarino) [4]. And they discussed the internal cause for strong relation between innovative activity space and specialized areas' location factors. However, Schumpeter came up with the concept of "innovation cluster" while explaining aggregation phenomenon of industrial innovation. Freeman [5] reckoned that as the result of common choices made by multiple enterprises, industrial aggregation in reverse would promote investment environment by external effects through systematic research so as to more innovative activities will be attracted. Scott [6] said that innovative activities were inclined to the concept of aggregation and took innovation as driving forces in the course of industrial aggregation. Based on that theory, cross-border collaborative strategy provided a new path for development and construction in underdeveloped areas.

Laos is the neighbor of China at the water's edge, and the bilateral trade is an important component of the economic and trade cooperation between the two countries. China is the third largest export market and the second largest exporter of Laos. The domestic and foreign scholars mainly study the bilateral economic cooperation from the perspective of the cross-district economic cooperation, for example, from the point of resource endowment, the degree of element copiousness in resource, technology, market and the capital between the two countries is different, Laos and China can make their respective advantages complementary to each other in the cooperation [7]. Relying on the China - ASEAN free trade area construction and the Mekong sub-regional cooperation platform for the research of economic cooperation content and path between Laos and China, and there is few research on bilateral economic cooperation from the perspective of cross-border collaborative strategy [8]. The paper starts from the economic cooperation symbiotic relationship between in Laos-China, through the study of Cross-border collaboration strategy, established the cross-border collaboration strategy mode of Laos and China with the orientation of symbiosis relationship.

3 Coordinating Motivation for Laos-China Cross-Border Collaboration Strategy

3.1 Reform and Opening up of Laos Providing Conditions for Laos-China Cross-Border Collaboration Strategy

In 1991, the Lao People's Revolutionary Party (LPRP) determined the "principled comprehensive reforming route" on 5th National Congress, proposed the six basic principles, including persisting to the leadership of the party, persisting to socialist direction and etc. and implemented the policy of opening up. In 2001, the LPRP formulated the struggling target of eliminating poverty by 2010 and getting rid of less developed status by 2020 on 7th National Congress. In 2011, the theme of the 9th National Congress of LPRP is to "enhance national unity and party unity, carry forward the Party's leadership and capability, realize renovation of reform route so as to lay a solid foundation for getting rid of the status of less developed country and continuing to march toward the target of socialist". The Laos Government believes that the sustainable development and reduction of poverty shall be based on trade, then try to acquired sustainable international aid and technical support. In February 2013, Laos became the 158th member state of WTO, which is an important step for promoting Laos to connect with global economics. As a less-developed country, Laos is also benefited from the unilateral trade with other 36 countries.

3.2 The "Going Out" Strategy and "One Belt One Road" Strategy of China Providing Driving Force for Laos-China Cross-Border Collaboration

The reform and opening of China in 1978 mainly focused on market opening and importing foreign technologies, funds and administration methods. Through nearly 40 years of development, China has accumulated some industry energy, and can transfer some mature technology to the demands of the international marketplace. So as to make up insufficient domestic resources and markets, take out domestic technologies, equipment, labors and products to leave space and funds to develop new industries, and form Chinese transnational companies in international market and fully participate in international division of labor. Based on this, China is actively implementing the strategy of "going out" since entering the 21st century. The "going out" strategy is also called transnational business strategy, internationalization business strategy, overseas operation strategy or globalization operation strategy, mainly referring to all external businesses of China, including foreign trade, foreign contracting engineering, foreign labor cooperation, foreign investment and etc. With the promotion of further accelerating the "going out" strategy on the 16th Party's Congress and the 3rd Plenary Session of 16th Party's Congress, the transnational business operation of enterprises in China began to participate in a wider range of international economic and technical cooperation and competition in broader fields and higher level and the enterprises are entering a new stage of forming powerful transnational companies.

3.3   The Multi-agent Coordination of External Demand Became External Motivation for Laos-China Cross-Border Collaboration Strategy

In the course of globalization, the external environment of enterprises are changing and the market demand, competitors and government policies are dynamically changing, and the impact and influence of these changes to the enterprises are external stimulation of environment. The enterprises can only survive by adapting to these changes. At present, Laos-China economic cooperation is still in its initial stage, since the economic development status and market order are not matured in Laos, while enterprises in China are also lacking of experiences in "going out", many problems has risen in initial cooperation between enterprises in Laos and China. From the experiences of economic cooperation practices between developing countries, government shall lead in the trade investment between developing countries and coordinate and deal with problems thereof, and followed by the enterprises' initiative "going out" behaviors after establishment of smooth collaboration mechanism. Besides, the enterprises' "going out" also need the support of other entities, which is beneficial to reduce the risks of cross-border operation, such as the participation of scientific research institutes, various intermediaries, so external demand coordinated by multiple entities is an important external motivation for Laos-China cross-border collaboration strategy.

4   Construction of Model for Laos-China Cross-Border Collaboration Strategy

4.1   Conception of "Overseas Jiangsu" Mode for Laos–China Cross-Border Collaboration Strategy

Aiming to the problems that Chinese enterprises have fought alone and got no results in the process of "going out". Professor Zhang Yang, chief expert of decision-making research base of Jiangsu Province (internationalization development of enterprise) proposed to implement "overseas Jiangsu" and "going out together to invest in designated place" in 2014. Essentially, "overseas Jiangsu" mode belongs to cross-border collaboration, which mainly refers that under the background of "One Belt One Road", based on multiple demand of social and economic development of countries along the "One Belt One Road" in technology, capital, talent, project and culture, as well as Jiangsu Province's endogenous development demand caused by excess production capacity and industry structure adjustment, all sectors of Jiangsu Province integrate the resources and actively went out to search for larger development space and seize overseas market for Jiangsu. However, this "Jiangsu" is not limited to "Jiangsu Province", it in generally refers to a region in China, and which proposes to take China as a regional unit and to invest to a designated place in cross-border collaboration country.

In the "overseas Jiangsu" mode of cross-border collaboration, the multi-agent cooperation including the government, industry, university, research, finance etc. there are formed a cooperation network, depending on the guidance of government, taking finance organization as the impeller, coordinating six major factors such as education, technology, entrepreneurship and industry and gather talent, research and development, capital, industry, culture to organize enterprises in China to go out and developing overseas innovation and entrepreneurship, forming Jiangsu overseas scientific parks, Jiangsu overseas industrial parks, Jiangsu overseas cultural parks and Jiangsu overseas urban areas marked with "overseas Jiangsu".

4.2 "Overseas Jiangsu" Is the Symbiosis Interface of Laos-China Cross-Border Collaboration

Since entities such as government, industry, university, research, finance of "overseas Jiangsu" are regional entities and have the characteristics of typical symbiosis unit, symbiosis entities such as government, industry, university, research, and finance have energy exchange and interaction relations in "overseas Jiangsu", and in behavior modes, it also possesses all the relations of parasitism, commensalism, symmetric mutualism and asymmetric mutualism. As a symbiosis interface, "overseas Jiangsu" has provided material, information and energy transmission media, channel or carrier for entities such as government, industry, university, research, finance in Laos-China cross-border collaboration strategy that taking Chinese region as unit, and maintained the form and development of Laos- China cross-border collaboration symbiosis relation.

Therefore, from the symbiosis system of cross-border collaboration, "overseas Jiangsu" mode can be applied in the process of Laos-China cross-border collaboration. From the view of Laos, "overseas Jiangsu" mode will enhance Chinese enterprise's market penetration depth and range in Laos as well as the success rate of cooperation between enterprises in Laos and China, and also push China's entity factors to enter Laos and greatly promote the effectiveness of Laos-China cross-border collaboration.

4.3 Characteristics of "Overseas Jiangsu" Mode for Laos-China Collaboration Strategy

(1) "Overseas Jiangsu" actively explores overseas new space by taking provincial-level division as a unit

"Overseas Jiangsu" is a new carrier that developed region in China leads in exploring overseas market, and it takes province (city, district) level administration region as a unit to establish regional cluster abroad for enterprises. It takes industrial parks as carries to gather various production factors, put emphasis on the characteristics of regional industry transfer and optimize function layout. It is a new form and new path for Chinese regions to explore new overseas space in Laos.

(2) Government construct "overseas Jiangsu" and guide Chinese enterprises to "going out together and invest to designated place"

Construction of "overseas Jiangsu" is different from construction of "overseas Japan". "Overseas Japan" is the outcome of industry transfer in Japan and growth of overseas capital, and it is the results of "going out" of Japanese enterprises. "Overseas Jiangsu" is a behavior before "going out". In accordance with the characteristics of regional industry transfer and internal demand on economic development and through cooperation with entities such as government, industry, university, research and finance in different layers of Laos, developed provincial or municipal government in China carries out scientific integration, put emphasis on industry factor collaboration, avoid "fight alone" of enterprises and give prominence to cross-border collaboration.

(3) "Overseas Jiangsu" is a method for Laos-China cross-border collaboration strategy, and it is also a path for Chinese region to join with China's "One Belt One Road" strategy

Cross-border collaboration is a new type of economic cooperation model. Overseas Jiangsu effectively collects the resources, factors and methods for cross-border collaboration and sufficiently releases energy to explore a new type of material basis for cross-border development path. "Overseas Jiangsu" pays attention to the strategic allocation of "entity-factor" in the cooperation region between countries, focuses on sufficiently driving the flow of entity factors such as talent, capital, information, technology resources and etc. on the basis of mutual benefits, and realizes advantage complementary of entity factors between countries and pushes forward the development of cross-border cooperation region through bringing into play each other's advantages and integrating resources.

(4) "Overseas Jiangsu" is not the superposition of overseas assets in Chinese region, it is the carrier for "going out" collaboration network in Chinese region.

Although Japan is the largest overseas assets country in the world, its assets are still staying at the bottom stage of "one-way street" layer. In 2014, balance of current items of national trade is JPY26266 billion, showing the lowest value since 1985. "One-way street" means that overseas assets of Japan are mainly maintained by the net inflow of income and expenditure obtained in international income and expenditure, and it is lacking of two-way interaction with international economy, essentially the outcome of developing the country with trade. Overseas assets of old capitalist countries such as the USA and England are acting as a hub for interaction of internal and external economies. Therefore, "overseas Jiangsu" is a new type of "going out", which is different from the "one-way street" model of "overseas Japan" and the hub model of "overseas USA" and "overseas England". It is the "government, industry, university, research and finance" "going out" collaboration network in Chinese region, which can avoid the defects of "one-way street" of "overseas Japan" and realize strategic alignment of the "going-out" "subject-factor" in designated place and region.

5   Policy Suggestion

All the entities coordinated in the process of Laos-China cross-border collaboration, such as government, industry, university, research and finance, shall play important roles, and these entities interact and rely on each other, each entity is depending on the collaboration of other entities to bring into play its functions and the disharmony of any entity will influence the acts of overall function., therefore, integration and interaction between multiple entities shall be strengthened to realize smooth cooperation between multi entities.

5.1   Integrate to Establish a Promoting Organization for Construction of "Overseas Jiangsu"

Construction of "overseas Jiangsu" is a "government, industry, university, research and finance" going out collaboration network. The "entity-factor" collaboration of "overseas Jiangsu" involves the collaboration between "going out" and economic transition and updating of enterprises, belonging to a complicated symbiosis system in which multiple entities and factors coordinate with each other and involving several departments of Laos and China that are responsible for internal and external economy development, including National Development and Reform Commission, Commercial Department, foreign exchange administration department and etc. Besides, system establishment and innovation driving for "overseas Jiangsu" also need the participation of departments such as regional public opinion and power organs (e.g. the National People's Congress of China and Council of Laos), technology department, education department and foreign affairs departments and etc. From the practical experience of enterprises' "going out" and economic transition, the approval and management of overseas investment and economic transition in both Laos and China are disordered, which shows that the existing administration systems cannot support the Laos-China cross-border collaboration strategy. By referring to the high-efficient experiences of "one window and division management" in the process of Japan's overseas investment and based on the actual demand of "overseas Jiangsu". Laos and China shall establish a special collaboration mechanism for relevant departments as soon as possible. So as to overall coordinate the cross-border collaboration of entity factors in local regions of Laos and China, and establish corresponding offices for "overseas Jiangsu" construction commission to take charge of concrete matters raised during local collaboration in the process of Laos-China cross-border collaboration.

5.2 Cultivate Social Organizations and Intermediary Service Institutions to Participate in Laos-China Cross-Border Collaboration Strategy

Social organization is the outcome of social development. The social organization acts as the connection and link between the labors, residents and citizen and the government and society, it balances the equity of labors and enterprises and public institutions and it is also the stabilizer and bumper for the society. The government organization can't replace the functions of social society, the higher is the level of social civilization, the more developed will the social organization be, and it's promotion ability to the society will be greater. As part of social activities, Laos-China cross-border collaboration strategy won't do without the promotion and function of social organization. Currently, there are industry associations and professional commercial chamber in various industries in both Laos and China, and these social organizations play an important role in pushing forward the economic cooperation of enterprises in Laos and China. Professional organizations providing consultants and training to enterprises also play an important role in promoting the "going out" and updating of enterprise. However, "overseas Jiangsu" is a new model for cross-border collaboration, it is a "government, industry, university, research and finance" collaboration network, and its construction requires the participation of various industry associations as well as some professional intermediary institutions, such as investment bank, international law firm and etc. as well as many factor service organizations, including intelligent property transaction, technology transfer, property transfer and etc. Therefore, we need to accelerate the cultivation of social organizations and intermediary service organizations as well as a batch of professional talents in the construction area of "overseas Jiangsu", so as to provide professional services for construction of "overseas Jiangsu".

5.3 Building the "Government, Industry, University, Research and Finance" Collaboration Network Mechanism

"Overseas Jiangsu" is different from "overseas Japan" and "overseas USA" and is also different from overseas industrial parks and Singapore Industry Park in general meaning. Construction of "overseas Jiangsu" is a Jiangsu overseas collaboration innovation complex which is the outcome of coordinated "going out" of "government, government, industry, university, research and finance", with the following functions: firstly, exploring new development space for economy development in China and promote the economy transformation and upgrading speed in Jiangsu Province; secondly, pushing forward fast development of economy in Laos and increase Laos' ability to undertake industry transfer.

The construction of "government, industry academy and resource" coordinated "going out" network can avoid the defects of single Chinese entity possessing limited resources, and by integrating resource factors such as information, technology, talents, funds and etc. for "going out" through the collaboration network and combining basic characteristics of Laos, it can realize maximum of efficiency in resource allocation. For instance, during the construction of "overseas Jiangsu", Jiangsu can transfer excessive industry to national level development zone of Laos, and in the transfer process, various entities are formed through network organization relation, government, enterprise, universities, scientific research institutes and finance organization coordinate with each other to cooperate through crisscrossed connection relation, wherein, the government leads with policies, universities and scientific research institutes provide professional knowledge and technologies support to industry transfer, the finance organization provides fund support to drive the enterprises in the up and down streams of the industrial chain to transfer excessive production capacity to Laos, and realize optimal allocation of factor resources in the process.

Acknowledgment. Funds for this research was provided by the Special Program For Key Program for International S&T Cooperation Projects (2012DFA60830), the National Planning of Social Science, Funding Project (11&ZD168) and Supported by Program for Changjiang Scholars and Innovative Research Team in University (IRT13062).

References

1. Ansoff, H.: Corporate Strategy. Revised edn., pp. 35–83. McGraw Hill Company, New York (1987)
2. Gale, B.T., Swire, D.J.: Business strategies that create wealth. Plann. Rev. **16**(2), 6–47 (1988)
3. Levine, S., White, P.E.: Exchange as a conceptual framework for the study of interorganizational relationships. Adm. Sci. Q. **5**, 583–601 (1961)
4. Iammarino, S., McCann, P.: The structure and evolution of industrial clusters: transactions, technology and knowledge spillovers. Res. Policy **35**(7), 1018–1036 (2006)
5. Freeman, C.: Networks of innovators: a synthesis of research issues. Res. Policy **20**, 499–514 (1991)
6. Scott, A.J.: New industrial spaces: flexible production organization and regional development in North America and Western Europe. Pion Ltd. (1988)
7. Zhu, D., Du, D.: Analysis of Laos Foreign open environment. Around SE Asia **5**, 3–7 (2014). (in Chinese)
8. Yang, R., Li, Y.: Will border effect restrict the construction of China's cross-border cooperation zone?: an analysis based on the data of China and Vietnam, Laos and Burma. Int. Econ. Trade Res. **30**(3), 73–84 (2014). (in Chinese)

Research of Fractal Compression Algorithm Taking Details in Consideration in Agriculture Plant Disease and Insect Pests Image

Qiao Deng[1], Chunhong Liu[1,2(✉)], and Liting Fu[1]

[1] College of Information and Electrical Engineering, China Agriculture University,
Beijing 10083, People's Republic of China
sophia_liu@cau.edu.cn
[2] Beijing Engineering and Technology Research Center for Internet of Things in Agriculture,
Beijing 10083, People's Republic of China

Abstract. Agriculture plant disease and insect pests is varied and tremendously harmful, so experts are needed to diagnose. But experts can't have the energy and time to the field for the majority of farmers to guide, they can only get the image through remote. However, agriculture plant disease and insect pests image is rich in detail, while transmission bandwidth and storage is limited, it is necessary to compress image to make sure of image quality. This paper proposes an improved method based on Jacquin theory to reduce coding time. Encoded sub block is classified into detailed block and non-detailed block, so we can reduce the encoding time. Experimental results show that, for the agriculture plant disease and insect pests image which is rich in detail, the number of encoded blocks reduces to 31.45 % and the encoding time reduces to 32.9 % of the original one.

Keywords: Plant disease and insect pests image · Detail · Fractal encoding · Image compression

1 Introduction

China is an agricultural country with the frequency of crop disasters. The great harm of crops diseases and insect pests is an important restriction factor of agricultural product quality improvement. On the other hand, the scientific and technological cultural quality of farmers is generally low, and they are urgent to be guided by experts who have knowledge of crop diseases and insect pests. But experts can't have the energy and time to the field for the majority of farmers to guide, they can only get the image through remote. However, agriculture plant disease and insect pests image is rich in detail and it contains a lot of information. So it need much transmission bandwidth and storage. At present, because the cable communication cable is unable used to cover the majority of farmland, the wireless wide area network technology can be used to transmit the field images. The wireless communication network is narrow of bandwidth, easy of jitter, low of transmission efficiency, so it is not conducive to transmit a large amount of data.

D. Li and Z. Li (Eds.): CCTA 2015, Part II, IFIP AICT 479, pp. 496–504, 2016.
DOI: 10.1007/978-3-319-48354-2_50

Therefore, it is very important to develop an algorithm of high compression ratio, good quality in transmitting agricultural images [1].

There are many classification methods for image compression, the most commonly used one is lossless compression and lossy compression. Lossless compression is a technique of image compression and decompression without any loss of data. Image obtained by decompressing the image is completely equal to the original image, but the compression ratio is about 2:1, which is a very low ratio [2]. Lossy compression can be reconstructed only by the approximation of the original one, the reconstructed image is similar to the original one, but it is not the exact copy. Lossy compression can get higher ratio compare to lossless compression [3].

In decades of image compression research, great progress has been made with image compression technique, and a series of international compression standards were formed, such as JPEG based on DCT and JPEG2000 based on wavelet transform that have been widely used [4]. Although these methods have achieved a higher compression ratio, however, compared with a rapid development of multimedia technology, a development of image compression is relatively backward. Moreover, these traditional compression algorithms still have some problems, such as low compression ratio, the block effect, etc. In recent years, in order to overcome the shortcomings of traditional image compression, experts and scholars proposed many new compression and broke the limitation of traditional entropy coding, such as fractal image compression [5].

Fractal compression method is a new method of compression in the 90's of last century. It uses the similarity of image itself to achieve compression, so we can achieve a high compression ratio. In 1990, A.E. Jacquin, one of Barnsley's student, using local iteration function theory, put forward a full automatic fractal image compression based on block and make fractal image compression from manual coding to automatic coding become a reality [6].

When small scale for sub block was used in Jacquin algorithm encoding, lost details were small and encoding time is short, but the compression rate is low. While the use of large scale for sub block encoding is the opposite. To improve the efficiency of Jacquin algorithm, quartering method, adaptive search method and other fractal generation algorithms have been proposed in recent years. But when encoding image of rich details, compression rate of these algorithms is almost same with the original one, and the encoding time is too long [7]. Therefore, this paper proposes a fractal image algorithm based on image details to shorten compression time and improve the efficiency of encoding [8].

2   Jacquin Algorithm

In corresponding decompression process, fractal compress is used as operator to calculate fixed point, which is suitable for the situation of one compression, multiple decompression. The basic compression algorithm is shown in Fig. 1. First, original image was divided into blocks of equal size, which was called range block (R) or sub block, then image region matching for each range block R was found, called domain block (D) or parent block. The length of parent block is generally 2 times of the sub block [9]. Then it shrinks to the same size to match the range blocks, then rotates and transforms symmetricly to get 8 transformed domain blocks. Comparing transformation of domain blocks with every range block to search the best match of the domain block. Compression coefficients and offset were calculated, so that the error of R and sD+oI is the smallest, I is a unit vector whose size is the same as D. After calculating the whole image, the encoding parameters are obtained [10].

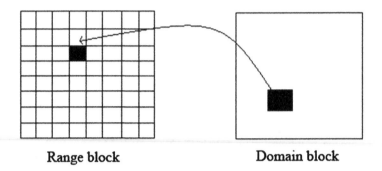

Range block   **Domain block**

Fig. 1. Regional matching image

Jacquin compression algorithm solves the image segmentation problem of Barnsley compression algorithm, by taking use of local self-similarity to construct affine transformation, which is the first practical image segmentation method. However, the searching volume of sub block is very large and searching time is very long. For example, an original image A, if there are $2^N \times 2^N$ pixels, by using Jacquin's compression algorithm, can be divided into sub block of $2^N \times 2^N$. Sub blocks were not relevant, the number of sub blocks were:

$$n = \frac{2^N \times 2^N}{2^R \times 2^R} = (2^{N-R})^2 \tag{1}$$

If A is divided into parent block of $2^D \times 2^D$, D > R, the number of parent blocks is: $(2^N - 2^D + 1) \times (2^N - 2^D + 1)$.

In Jacquin algorithm, 8 kinds of transformations, such as rotation and reflection, are performed to a parent block. So 错误！未找到引用源。 $8 \times (2^N - 2^D + 1) \times (2^N - 2^D + 1)$ times searching for computation is needed to complete encoding. The magnitude of searching is about 错误!未找到引用源。. For example, a gray image of 512×512 pixels is divided into sub blocks of 8×8 and parent blocks of 16×16. Each sub block is compared with $(512 - 16 + 1) \times (512 - 16 + 1) = 247009$ parent blocks, and each time there are 8 kinds of operations such as rotation, reflection, in the comparison with parent blocks. So each sub block is compared $8 \times 247009 = 1976072$ times. All sub blocks are compared $4096 \times 1976072 = 8093990912$ times [11]. 错误!未找到引用源。

Thus, the main reason for the slow speed of the fractal image encoding is that time searching for the best parent block consume too much. Therefore, researching for a fast and effective method of encoding is an important direction of fractal encoding [12].

The above discussion shows that, using Jacquin algorithm, for each sub block 错误!未找到引用源。, the number of parent block D_i needed to search is:

$$K = 8 \times (2^N - 2^D + 1) \times (2^N - 2^D + 1) \tag{2}$$

Fixed network algorithm is that position of the parent block is fixed on the image grid to reduce the number of K. There is no need to search for each possible parent block but to search for the parent block on the grid. So the number of searching times can be reduced not to produce much impact on the compression quality [13]. The grid interval is l, which is distributed in image A, 错误!未找到引用源。. The upper left corner of parent block is located on the grid. This number of parent block is:

$$K_1 = \frac{1}{l^2} K \tag{3}$$

When 错误!未找到引用源。, the number of parent blocks in the parent block library are 1/4 of the original number. But this method is not combined with the characteristics of different images respectively, and it is not adaptive. Therefore, there is great significant to find a kind of image compression algorithm with short encoding time, high image quality and adaptability. 错误!未找到引用源。

3 Fractal Compression Algorithm Taking Details in Consideration

A sub block segmented from an image can be divided into two categories. One is rich in detail, the other does not contain details [14]. These two kinds of sub blocks have their own features in the encoding. Sub blocks are rich in detail, which have high complexity and much time to encode. While non-detailed sub blocks have low complexity and high degree of similarity. According to the differences, different methods can be used to encode. The detailed sub blocks have much impact on the quality of the image, so the detailed sub block need to be fully encoded. Non-detailed sub blocks have a high degree of similarity, so that we can only encode a small part of detailed sub blocks to reduce encoding time and encoding space.

In this paper, fractal image algorithm based on details is proposed. Firstly, the edge of the image was detected which is the main component of the image details. Sobel algorithm was used to detect the edge of the image. The operator of Sobel algorithm is:

$$\begin{aligned} D(x,y) = &\ |f(x-1,y-1) + 2f(x-1,y) + f(x-1,y+1) - f(x+1,y-1) - 2f(x+1,y) - f(x+1,y+1)| \\ &+ |f(x-1,y+1) + 2f(x,y+1) + f(x+1,y+1) - f(x-1,y-1) - 2f(x,y-1) - f(x+1,y-1)| \end{aligned} \tag{4}$$

To filter out the details which can't be resolved by human eyes, we treat them as non-detailed part, a threshold value need to be set. The formula is:
错误!未找到引用源。

$$D'(x,y) = \begin{cases} D(x,y) & (D(x,y) \geq T) \\ 0 & (D(x,y) < T) \end{cases} \tag{5}$$

T is the threshold value, 错误!未找到引用源。is the gradient after setting the threshold value. The selection of T is based on adaptive method and the value is 1/5 of the max gradient value. After the details of the image was detected, the encoding detailed sub blocks were selected. The selection method is to calculate the sum of gradient of each pixel in the sub block and calculate whether the sub block contains the details. The mathematical expression is: 错误!未找到引用源。

$$s_{ij} = \sum_{x=0}^{M-1} \sum_{y=0}^{N-1} D'(x, y) \tag{6}$$

s_{ij} is the sum of internal pixel's gradient value of sub block whose coordinate is (i,j). If s_{ij} is not 0, then the sub-block is considered containing details. After analyzing all the sub blocks, we start encoding from the first sub block. First sub blocks containing details were encoded, then a representative of non-detailed sub block for encoding was selected. The method to select detailed sub block is to calculate the variance of each pixel value and average gray value in each sub block and select the block which has mini variance as a non-detailed block encoding.

First, calculate the average gray value $U(s_{ij})$ 错误!未找到引用源。of sub blocks whose coordinate is (i,j):

$$U(s_{ij}) = \frac{1}{M^2} \sum_{n=0}^{M-1} \sum_{m=0}^{M-1} s_{ij}(m, n) \tag{7}$$

Then calculate the variance 错误!未找到引用源。according to its gray mean value:

$$D(s_{ij}) = \sum_{n=0}^{M-1} \sum_{m=0}^{M-1} (s_{ij}(m, n) - U(s_{ij}))^2 \tag{8}$$

错误!未找到引用源。

Because there are only difference 错误!未找到引用源。on the gray level between other no-encoded sub blocks and the sub blocks of mini variance, we can get the encoding parameter p', g' of other no-encoded sub blocks according to the parameters 错误!未找到引用源。that sub block D had calculated:

$$\begin{cases} p'(i,j) = p \\ g'(i,j) = g + \Delta N(i,j) \end{cases} \tag{9}$$

4   Experiment and Analysis of Results

To verify the effect of proposed algorithm, encoding and decoding test were carried for Jacquin algorithm and the proposed algorithm, the computer is Intel(R) Core(TM) i3-2120 3.30 GHz. The size of test image is 256 × 256 pixels, which is shown in Fig. 2, the image is a leaf of northern leaf blight of corn. The sub block size is 4 × 4, and is divided into 4096 sub blocks and 1024 parent blocks.

Fig. 2. Original image

In the experiment, the number of detailed sub blocks, which is obtained by the edge detection algorithm is 1287, about 31.42 % of the total blocks. Non-detailed blocks is 2809, about 68.91 % of the total blocks. There is a sub block has to encode of the non-detailed blocks, so the actual encoding number is 1288. In this paper, the actual encoding number is 31.45 % of Jacquin algorithm. The result of Sobel operators is shown in Fig. 3, The distribution of details is shown in this image. The result of 10 times decoding image using Jacquin algorithm is shown in Fig. 4. The result of 10 times decoding image using algorithm of this paper is shown in Fig. 5.

Fig. 3. The detailed image after calculating by Sobel operator

Fig. 4. 10 times decoding image using Jacquin algorithm

Fig. 5. 10 times decoding image using proposed algorithm

The comparison of proposed algorithm with the original Jacquin algorithm is shown in Table 1, the *PSNR* (peak signal-to-noise ratio) is calculated as follows:

$$PSNR = 10 \lg \frac{f_{max}^2}{\frac{1}{MN} \sum_{x=0}^{M-1} \sum_{y=0}^{N-1} [f(x, y) - f_0(x, y)]^2} \tag{10}$$

Table 1. The comparison of proposed algorithm and Jacquin algorithm

Encoding algorithm	Encoding time/s	Decoding time/s	Encoding blocks	PSNR
Jacquin algorithm	440.862	0.880	4096	38.755
Proposed algorithm	145.239	0.683	1288	37.863

错误!未找到引用源。$f(x, y)$ is an image to be evaluated, and $f_0(x, y)$ is a referencing image. The image size is $M \times N$.

The *PSNR* got by using Jacquin algorithm is slightly greater than the proposed one, which is shown in Table 1. While the encoding blocks of the proposed algorithm is much less than Jacquin's, so time of the proposed algorithm is less compare to Jacquin algorithm. The decoding time is both short. The number of encoded blocks reduces to 31.45 % and the encoding time reduces to 32.9 % of the original one. This algorithm is better than Jacquin algorithm.

5   Conclusions

Fractal image algorithm has much potential of increasing compression ratio, and decoding time is very short. Therefore, in image and multimedia, the fractal image compression algorithm is promising in application. But the process time of the fractal image algorithm is too long to be done, which greatly limits the application of this method.

In this paper, the fractal image compression method based on details of the image was proposed. So the number of sub block encoding is significantly less than the existing method in the image which is rich in detail. Encoding time of proposed method is greatly shorter than the original method, and it improved the compression effect of encoding.

Acknowledgement.   This paper was financially supported by the National International Cooperation Special Program (Grant No. 2013DFA11320).

References

1. Salar, N.M., Nadernejad, E., Najmi, H.M.: A new modified fast fractal image compression algorithm. J. Imaging Sci. **61**(2), 219–231 (2013)
2. Fisher, Y.: Fractal Image Compression: Theory and Applications, vol. 34. Springer, New York (1995)
3. Lin, Y.L., Wu, M.S.: An edge property-based neighborhood regions search strategy for fractal image compression. Comput. Math Appl. **62**(1), 310–318 (2011)
4. Shin, C.W., Chun, H.C., Chen, Y.M., et al.: The effectiveness of image features based on fractal image coding for image annotation. Expert Syst. Appl. **39**(17), 12897–12904 (2012)
5. Seeli, D.S., Jeyakumar, M.K.: A study on fractal image compression using soft computing techniques. Int. J. Comput. Sci. Issues **9**(6), 420–430 (2012)
6. Jacquin, A.E.: Image coding based on a fractal theory of iterated contractive image transformations. IEEE Trans. Image Process. **1**(1), 18–30 (1992)
7. George, L.E., Minas, N.A.: Speeding up fractal image compression using DCT descriptors. J. Inf. Comput. Sci. **6**(4), 287–294 (2011)
8. PeiXin, N.: Image compression research based on fractal theory. Chongqing University (2012). (in Chinese)
9. Fei, S.: Fast encoding algorithm for fractal image compression. Nanjing University of Posts and Telecommunications (2013). (in Chinese)

10. Tang, W., Wu, X., Yu, Y., Luo, D.: Fractal encoding compression method based on detail block of image. Comput. Eng. **09**, 217–218+221 (2010). (in Chinese)
11. Wang, X.Y.: A fast fractal coding in application of image retrieval. Fractals **17**(4), 441–450 (2009)
12. Wee, Y.C., Shin, H.J.: A novel fast fractal super resolution technique. IEEE Trans. Consum. Electron. **56**(3), 1537–1541 (2010)
13. Huang, X.Q.,Yu, S.L., Zhang, W.: A noval method for embedding gray image watermarking into orthogonal fractal compression image. In: Proceedings of International Conference of Communications, Circuits and Systems, pp. 485–488 (2009)
14. Guangchao, F., Qiyan, J.: An improved fractal image compression algorithm. Commun. Technol. **04**, 109–111 (2013). (in Chinese)

Commentary on Application of Data Mining in Fruit Quality Evaluation

Jinjian Hou[1,2], Dong Wang[2,3], Wenshen Jia[2,3], and Ligang Pan[1,2,3(✉)]

[1] China Three Gorges University, Yichang 443002, China
[2] Beijing Academy of Agriculture and Forestry Sciences, Beijing 100097, China
[3] Risk Assessment Laboratory for Agro-Products of the Ministry of Agriculture,
Beijing 100097, China
hou_jinjian@163.com, nirphd@163.com, jiawenshen@163.com,
panlg@nercite.org.cn

Abstract. In order to provide reference for the fruit quality research and fruit selective breeding, in this paper, data mining methods of fruit quality in recent years, including fuzzy comprehensive evaluation method, analytic hierarchy process method, gray correlation degree analysis method and so on, which were compared for the characteristics of advantages and disadvantages. Furthermly, the main evaluation factors of the common fruits were summarized. Finally, the research on data mining methods of the fruit quality was summarized and prospected. This review indicated that data mining methods could evaluate multi-index of fruit quality comprehensively, which will provide reference for rapid detection of fruit quality and cultivation of the excellent species. Meanwhile, it will be a new direction in the field of fruit quality research by studying more main factors of fruit quality and simplifying the evaluation procedures in the near future.

Keywords: Fruit quality · Data mining · Fuzzy evaluation · Analytic hierarchy process · Principal component cluster analysis

1 Introduction

Fruit quality is one of the most important factors that dictates the economic value and directly affects the market competitiveness of the fruit. China firmly remains as the world's superpower in terms of fruit trees, fruit yield, and planting area [1, 2]. However, the proportion of fruit exports trail behind those of other countries [3]. A study found that, on one hand, given the country's lagging fruit quality evaluation and sorting technology [4], grading of fruit quality is disordered, and resulting quality is uneven. Single fruit indices, such as strawberry [5] and apple [6] soluble solids, sweet orange titratable acid [7], and kiwi fruit hardness [8], which were analyzed by scholars. However, single indicators can only evaluate the quality of one aspect of the fruit but fails to meet the requirements of fruit quality evaluation, because these indicators cannot be used in comprehensively evaluating fruit quality and present certain limitations. On the other hand, fruit quality evaluation contains numerous factors, including the internal and

© IFIP International Federation for Information Processing 2016
Published by Springer International Publishing AG 2016. All Rights Reserved
D. Li and Z. Li (Eds.): CCTA 2015, Part II, IFIP AICT 479, pp. 505–513, 2016.
DOI: 10.1007/978-3-319-48354-2_51

external quality factors of fruit, are involved in evaluating fruit quality, and each factor present close correlation and relative independence, resulting in difficulty in conducting fruit quality evaluation and grading work. In view of the above problems, the fruit quality data mining analysis of the indicators of hierarchical method and classification system, and relevant evaluation methods can be used to simplify quality indicators, extract the main evaluation factors, and simplify the evaluation process. The use of data mining study on comprehensive evaluation methods for fruit quality has become a research hotspot in recent years, scholars have used data mining methods on fruit quality of Nanfeng tangerine [9], apple [10], pineapple [11], pear [12] and other fruit. Results show that data mining method can be used to effectively evaluate fruit quality. At present, the literature on fruit quality data mining is rarely reported. In this paper, data mining methods applied on fruit quality in recent years were reviewed and analyzed. Finally, the main evaluation factors of common fruits were consolidated for evaluating fruit quality research and providing a reference.

2   Main Fruit Quality Indicators and Access Methods

Fruit quality includes the appearance and the intrinsic qualities. The main evaluation indices include fruit shape index, fruit weight, fruit color, fruit firmness, soluble solids, vitamin C, and others, as shown in Table 1. These indicators are representative of the different aspects of fruit characteristics. Close relationships exist among these indicators, such as total sugar, including soluble solid matter represented by sucrose and other reducible carbohydrate carbonyl components, which denote different attributes that are also related. At present, methods for obtaining quality indices mainly involve chemical and instrument measurement methods, but the difficult quantitative analysis indices of fruit flavor can be obtained only by depending on expert scoring.

Table 1. Main index and access methods of fruit quality

Main indexes of fruit quality	Access methods
Fruit shape index	Fruit vertical diameter divided by transverse diameter
Fruit flavor	Evaluation score method
Fruit weight	Electronic balance measurement
Fruit color	Colorimeter measurement
Edible rate	Peel fruit weight divided by the weight after to go nuclear
Fruit hardness	Fruit hardness tester
Soluble solid content	Refractometer determination
Titratable acid content	Potentiometric titration or NaOH titration method
VC content	Automatic potentiometric titration or 2,6-dichloroindophenol titration method
Soluble sugar content	Anthrone colorimetry
Sugar acid ratio	Soluble total sugar content divided by content of soluble acid
Solid acid ratio	Soluble solids content divided by titratable acid content
Carotene	Spectrophotometric method, thin layer chromatography, high performance liquid chromatography

3   Data Mining Overview

3.1   Simple Mathematical Method

For simple-featured and small amounts of data, existence of the unknown and potential information can be handled by simple mathematical processing method, such as mean, percentage, classification method. A simple data processing method can mine the data set of potential, valuable information.

3.2   Mathematical Statistics Method

Statistical analysis is mainly used to complete knowledge summary and relational knowledge mining. For some data, a function or relationship that cannot be expressed in a function exists. At this point, implicit data information can be excavated by using mathematical statistics method. Common methods include regression analysis, correlation analysis, and principal component analysis.

3.3   Artificial Intelligence Method

For large amounts and particularly complex data sets, a general data mining method cannot obtain the data set of implicit information. At this point, we can use artificial intelligence method of data mining, which is extremely complex. The main methods include fuzzy evaluation, association rules, and clustering analysis.

4   Fruit Quality Data Mining Method

4.1   Single Evaluation Method

(1) Fuzzy evaluation method

Fuzzy evaluation method is influenced by numerous factors so as to conduct a comprehensive evaluation of a highly effective multi-factor decision method. One characteristic of this method is that, instead of an absolutely positive or negative evaluation result, fuzzy sets are used to represent the results [13]. The advantages of fuzzy evaluation are that we can quantify several qualitative indices, overcome the disadvantages of qualitative analysis, and objectively and accurately evaluate the pros and cons of varieties [14]. Its disadvantages include information duplication problem, which is caused by the unresolved correlation between the evaluation indexes. Thus, the confirmation of membership function and fuzzy correlation matrix, among others should be studied in the present research [15]. This method is mainly used in fruit quality identification and breeding of good varieties, and is presently applied in the quality evaluation of longan [16], persimmon [17], and other fruits.

(2) Analytic hierarchy process method

The analytic hierarchy process is a multi-objective decision analysis method that combines qualitative and quantitative analysis methods [18]. The main concept of this method is to decompose the complex problem of fruit quality evaluation into several levels and factors. Comparison between two indices is essential for judgment. The judgment matrix is established by computing the largest eigenvalue in the matrix, and corresponding eigenvectors can indicate the different degrees of importance weights and provide a basis for selecting the optimal evaluation index. One advantage of the analytic hierarchy process is that not only the weight coefficient of each evaluation index is obtained but simultaneous filtering by accidental factors determines the perception of differences and the different dimension of factors in a unified evaluation system with high reliability and small error. On the other hand, one disadvantage is the limited number of fruit indicators, with the maximum generally being 9. This method has been applied in the cultivation of good varieties of jinxixiaozao [19], pear [20], and other fruits.

(3) Gray correlation degree analysis

Correlation analysis is the main tool in grey correlation analysis method using the grey system theory for the comprehensive evaluation of the research object. The correlation coefficient and correlation between the sequence of numbers and the reference sequence are compared to determine the primary and secondary factors and their correlation degree [21]. This method offers the advantages of simplicity, ease of operation, and intuitiveness. On the other hand, its disadvantages include strong subjectivity and difficulty in determining certain optimal values. This method is mainly used in the situations where in the index correlation between is too high. Grey correlation degree analysis method has performed an important function in the comprehensive evaluation on muskmelon [22], peach [23], amomum [24], and other fruits.

(4) Principal component analysis

The goal of principal component analysis is to secure the data under the principle of minimum information loss and convert the more original data and related indicators into new, fewer data at smaller orthogonal transforms to each other or comprehensive indices with slight correlation to simplify the evaluation process [25, 26]. Principal component analysis presents advantages of calculating the comparison standard, capability of being realized on the computer and using special software for analysis. Its disadvantage is that the new comprehensive index is difficult to explain, and the general method of combining clustering is used. Principal component analysis method is mainly used for more quality indicators, and the correlation among the indices of strong case and multiple correlated stochastic variables according to the main component of the contribution rate are simplified into several variables to avoid traits and related traits caused by error evaluation [29]. At present, the analysis is used for the comprehensive evaluation for selecting fruit quality evaluation factors and fruit quality [27, 28].

4.2 Hybrid Evaluation Method

(1) Principal component cluster analysis method

For the multi-index evaluation of sorting fruit quality, the variance contribution of the first principal component F1 rate is not sufficiently high. In other words, the first principal component expression of original data information is not large enough, only the first principal component scores for evaluating the sample sort are one-sided. At this point, the two methods of combining principal component analysis and clustering analysis are combined to form "principal component clustering analysis method". As an advantage, the method can extract multiple indicators simultaneously with most of the information, prevent the artificial selection evaluation factor of subjectivity, and provide a true reflection of varieties of comprehensive characteristics so as to offer an objective basis for breeding materials [29]. One disadvantage is clustering difficulty when the data is too large. Principal component cluster analysis method can effectively extract the main quality factors, simplify the fruit quality evaluation work, and provide theoretical basis for fruit speed measurement. The method has been used for tomato [30] and Lee apricot [31] quality rapid detection.

(2) Rationalization-satisfaction degree and multiple value method

The so-called "reasonable–satisfaction" refers to fruit varieties that demonstrate the characteristics of satisfaction that people need. The reasonable degree is1 if a characteristic species is in full compliance with the "rule". If not in line with "rule", then the reasonable degree is 0 [32]. The advantages of the algorithm are simplicity, ease of calculation, and the ability to distinguish between good quality and poor quality. Its disadvantage is larger algorithm error. The algorithm objectively and accurately reflects the people's needs and satisfaction degree of fruit quality. The method can not only be used as a method to identify fruit quality but also can be used as a reference value of fruit tree breeding species, especially for the breeding of commercial varieties. At present, the method has been used on pear [33] and other fruits for the cultivation of good varieties.

(3) Principal component cluster combined with rationalization-satisfaction multidimensional value analysis theory of merger rules

The algorithm presents new ideas and methods of comprehensively evaluating fruit quality in combination with principal component analysis, cluster analysis, and multidimensional value theory "reasonable–satisfaction" composite evaluation method. This method can be used to extract the main factors of common fruit so as to simplify the evaluation process. Moreover, the method can be used for fruit breeding. Its disadvantage is that the method computation is trivial, complex, and requires large amounts of calculation. This combined method has been applied in mango [34] fruit quality assessment factor selection and simplify the work of mango fruit quality evaluation.

4.3 Parts of Comprehensive Fruit Evaluation Factors

Table 2 summarizes the evaluation factors and the use of the method of data mining for fruit quality after the main evaluation factors of fruit in certain literature. Numerous fruit quality indicators, the presence of fruit quality evaluation using a single index presents certain limitations and evaluation of all indicators inevitably requires too much work. Fruit quality data mining methods can effectively reduce fruit evaluation indices and simplify the evaluation process. Jiyun Nie, et al. [35]. used principal component analysis to select five indicators of the contribution rate of more than 95.75 % of the previous four components reflecting apple quality as the main evaluation factors. Haying Zhang, et al. [36], simplified 19 peach quality indicators for five items according to principal component analysis, clustering analysis, and the national standard of GB - 10653-1989 regarding "the fresh peach" requirements indicators. Table 2 shows that the use of data mining methods can effectively reduce the evaluation index, provide good evaluation of fruit quality, and solve the problems of limited single index evaluation and hefty workload of multi-index evaluation. The approach provides new ideas and methods for the evaluation of fruit quality.

Table 2. Part of the fruit of evaluation factors and main evaluation factor

Fruit	Evaluation factors	Main evaluation factor
Apple	Soluble solid content, soluble sugar content, solid acid ratio and other seven quality index	Fruit hardness, soluble sugar content, titratable acid content, Sugar acid ratio, VC content
Pear	Fruit weight, fruit shape index, fruit hardness and other eight quality index	Fruit weight, soluble solid content, Sugar acid ratio, titratable acid content, fruit hardness
Peach	Soluble solid content, titratable acid content, Solid acid ratio and other nineteen quality index	Fruit weight, fruit hardness, moisture content, solid acid ratio and flavor
Mango	Fruit weight, fruit hardness, soluble solid content and other nineteen quality index	Sugar acid ratio, VC content, fruit hardness, Carotene and other nine quality index

5 Conclusion and Prospect

As people's living standards continue to improve, the demand for fruit quality keeps growing. The search for rapid, simple methods of evaluating fruit quality has become a hot topic in the field of fruit quality analysis. The composition of fruit quality evaluation factors is too numerous, and different degrees of correlation and relative independence exist among and between different quality factors. The use of single quality index to evaluate the quality of fruit exist certain limitations. Moreover, a single indicator can only explain the quality of fruit in a certain aspect but cannot evaluate the overall quality of fruit. Through the use of data mining methods can combine multiple quality metrics for the comprehensive evaluation of fruit quality to obtain a comprehensive and objective assessment. In actual fruit quality assessment process, the use of relevant data mining method to determine the main fruit quality evaluation factors of common fruits can substantially reduce the workload of fruit quality appraisal.

Data mining method provides a new thinking and approach to the selection and breeding of fruit. First, data mining can identify the good traits of prominent fruit varieties, which can provide hybrid parent reference for the improvement of the fruit quality. Second, for single specific varieties, data mining methods can distinguish between fruit quality and provide a basis for directional breeding, further improvement of fine varieties, thereby yielding more excellent varieties. For the comprehensive evaluation of overall poor quality or general varieties and given the highly prominent individual quality, selection of a specific function and strong varieties not only will aid in improving the level of comprehensive utilization of fruit but is also conducive to determining the different uses and maximizing the performance of fruits on the basis of quality characteristics.

In recent years, with the development of cloud platforms, computers, massive databases, networking, and other technologies, data mining will perform a more important function in fruit production, distribution, sales, and consumption sectors. Although various studies have thorough, all kinds of fruit quality assessment method of research are more mature, but the existing methods are still hard to meet the needs of actual production and consumption. Therefore, finding a rapid and easy method for evaluating fruit quality remains a hot topic in the field of fruit quality analysis. The application of data mining method to more fruits and to extract major evaluation factors and simplify the evaluation process will become a new direction in fruit quality research.

Acknowledgment. Funds for this research was provided by the Beijing Municipal Science and Technology Commission "The capital of food safety science and technology excellence special cultivation" project – "near infrared fruit quality rapid nondestructive testing equipment research and development" (Z141100002614021123).

References

1. Zhang, J., Hu, J., Zhang, Z.: Research on the fruit industry development in China. J. Shandong Agric. Univ. (Soc. Sci.) **4**(3), 31–34 (2002)
2. Chen, M.: Non-destructive Detection of Fruit Internal Quality Based on Portable Near Infrared Spectrometer. Zhejiang University, Hangzhou (2010)
3. Li, J.: Research on the Quality Grading of Fruit Based Machine Vision and NIRS. Nanjing University of Aeronautics and Astronautics, Nanjing (2011)
4. Li, G., Wei, X., Li, L., et al.: Reserching actuality and development of fruit grader. J. Agric. Mech. Res. **9**, 20–23 (2009)
5. Sánchez, M.-T., José De la Haba, M., Benítez-López, M., et al.: Non-destructive characterization and quality control of intact strawberries based on NIR spectral data. J. Food Eng. **110**(1), 102–108 (2012)
6. Aiguo, O., Xiaoqiang, X., Yande, L.: Selection of NIR variables for online detecting soluble solids content of apple. Trans. Chin. Soc. Agric. Mach. **45**(4), 220–225 (2014)
7. Sun, Q., Xie, R., Deng, L., et al.: Analysis of fruit quality and acid constituents of three bred cultivars selected from Jiangjin sweet orange. Food Sci. **36**(6), 124–129 (2015)

8. Liu, H., Guo, W., Yue, R.: Non-destructive detection of Kiwi fruit firmness based on near-infrared diffused spectroscopy. Trans. Chin. Soc. Agric. Mach. **42**(3), 145–149 (2011)

9. Ni, Z., Zhang, S., Gu, Q., et al.: Evaluation of fruit quality of Nanfeng Tangerine based on multivariate statistics. J. Fruit Sci. **28**(5), 918–923 (2011)

10. Dong, Y., Zhang, Y., Liang, M., et al.: Selection of main indexes for evaluating apple fruit quality. Acta Agric. Boreali Sin. **26**(S1), 74–79 (2011)

11. Lu, X., Sun, D.: Evaluation on fruit quality of 12 pineapple cultivars introduced from Thailand. Chin. J. Trop. Crops **32**(12), 2205–2208 (2011)

12. Tian, R., Hu, H., Yang, X., et al.: Selection of factors for evaluating pear fruit quality. J. Yangtze Univ. (Nat. Sci. Edit.) **6**(3), 8–11 (2009)

13. Li, L., Shen, L.: An improved multilevel fuzzy comprehensive evaluation algorithm for security performance. J. Chin. Univ. Posts Telecommun. **13**(4), 48–53 (2006)

14. Shen, Q., Zhu, J., Peng, H., et al.: Fuzzy comprehensive evaluation of fruit characteristics in early mature seedling Litchi resource of Southwestern Guangxi. Southwest Chin. J. Agric. Sci. **24**(4), 508–515 (2011)

15. Meng, L., Chen, Y., Li, W., et al.: Fuzzy comprehensive evaluation model for water resources carrying capacity in Tarim River Basin, Xinjiang, China. Chin. J. Geogr. Sci. **19**(1), 89–95 (2009)

16. Zhu, J., Yu, P., Huang, F., et al.: Quantifying analysis of main fruit characters of Longyan germsplasm in Guangxi. Southwest Chin. J. Agric. Sci. **19**(2), 283–286 (2006)

17. Deng, L., He, X., Xu, J., et al.: Diversity analysis of fruit traits and fuzzy comprehensive evaluation of persimmon germplasm resources in Guangxi. Guihaia **33**(4), 508–515 (2013)

18. Saaty, T.L.: Decision making – the analytic hierarchy and network processes (AHP/ANP). J. Syst. Sci. Syst. Eng. **13**(1), 1–35 (2004)

19. Liu, N., Zhang, Y., Zhao, Z., et al.: Quality evaluation of Jinsixiaozao fruits and its relationship with soil body configuration. J. Agric. Mech. Res. (12), 125–129 (2009)

20. Zhao, J., Qiao, J., Li, H., et al.: Comprehensive evaluation on the internal quality of Yali' Pear fruit from the main production area in Hebei Province. Northern Hortic. (16), 33–35 (2010)

21. Li, X., Wang, W.: Estimation of apple storage quality properties with mechanical property based on grey system theory. Trans. Chin. Soc. Agric. Eng. (2), 80–86 (2005)

22. Lin, B., Gao, S., Lin, F., et al.: Grey relational grade analysis on yield characteristics of inbred melons. Fujian J. Agric. Sci. **23**(2), 178–181 (2008)

23. Zhang, X., Sun, Y., Wang, Y., et al.: Determination of freezing point temperature of different yellow peach cultivars and correlation analysis of impact factors. J. Food Sci. Technol. **31**(4), 37–41 (2013)

24. Zhang, W., Wei, X., Long, Y., et al.: Gray correlation degree analysis on main agronomic traits of Amomum tsao-ko fruits. J. Honghe Univ. **11**(2), 52–53 (2013)

25. Sai, T.Y., Li, J.S., Wang, S.B.: Principal component analysis in construction of 3D human knee joint models using a statistical shape model method. Comput. Methods Biomech. Biomed. Eng. **18**(7), 721–729 (2015)

26. Xu, C., Gao, D.: Comprehensive evaluation on fruit quality of peach cultivars in green house based on principal component analysis. Sci. Technol. Food Ind. **23**, 84–94 (2014)

27. Qin, H., Xu, P., Ai, J., et al.: Diversity of fruit quality and phenotypic traits of Actinidia arguta Planch Germplasm resources and their principal component analysis. Chin. Agric. Sci. Bull. **31**(1), 160–165 (2015)

28. Wang, X., Bi, J., Liu, X., et al.: Different origin Fuji apple quality evaluation factors choice. J. Nucl. Agric. Sci. **27**(10), 1501–1510 (2013)

29. Ding, C.: Principal component analysis of water quality monitoring data in XiaSha region. Environ. Transp. Eng., 2321–2324 (2011)

30. Zhao, S., Yuan, D., Zhang, L., et al.: Study of the screening of sand pear (Pyrus pyrifolia) cultivars. J. Cent. South Univ. Forest. Technol. (Nat. Sci.) **27**(1), 30–34 (2007)
31. Han, Z., Jiang, B.: A study on comprehensive evaluation of the processing tomato varieties multiple traits. Sci. Agric. Sin. **47**(2), 357–365 (2014)
32. Niu, J., Liu, M., Peng, Q.: Analysis of main index components and cluster of Plumcot "Weihou" during different harvest periods. Xinjiang Agric. Sci. **52**(1), 33–36 (2015)
33. Ayiguli, T., Yusufu, A., Patiman, A., et al.: Analysis and evaluation on of fruit quality of main Pear varieties in Xinjiang. Xinjiang Agric. Sci. **51**(3), 417–422 (2014)
34. Xin, M., Zhang, E., He, Q., et al.: Selection of evaluation factors for mango fruit quality. J. South. Agric. **45**(10), 1818–1824 (2014)
35. Nie, J., Li, Z., Li, H., et al.: Evaluation indices for apple physicochemical quality. Sci. Agric. Sin. **45**(14), 2895–2903 (2012)
36. Zhang, H., Tao, H., Wang, Y., et al.: Selection of factors for evaluating peach (Prunus persica) fruit quality. Trans. CSAE **22**(8), 235–239 (2006)

Study on Identification of Bacillus cereus in Milk Based on Two-Dimensional Correlation Infrared Spectroscopy

Zizhu Zhao[1], Ruokui Chang[1(✉)], Yong Wei[1], Yuanhong Wang[2], and Haiyun Wu[1]

[1] College of Engineering and Technology, Tianjin Agricultural University,
Tianjin 300384, China
274636904@qq.com, changrk@163.com, 595183963@qq.com,
haiyunwu2013@163.com
[2] College of Horticulture and Landscape, Tianjin Agricultural University,
Tianjin 300384, China
529007475@qq.com

Abstract. Bacillus cereus is a kind of common food-borne pathogen that can cause vomiting and diarrhea shortly after ingestion. The spectroscopy properties of *Bacillus cereus* were measured using the infrared reflectance spectroscopy, which was a nondestructive technology. The influence of *Bacillus cereus* concentration on the spectroscopy was explored based on the two-dimensional (2D) correlation spectroscopy method. The results showed that some functional groups of capsule in pure Bacillus cereus culture, such as carboxyl, protein amide I and amide II, C-H methyl and methylene could induced some self-correlation peaks near 1592 cm^{-1}, 1652 cm^{-1}, 1512 cm^{-1} and 1412 cm^{-1} respectively. Some functional groups of spore, such as COO- group and C-H bond could induced some self-correlation peaks near 1348 cm^{-1}, 1616 cm^{-1} and 1592 cm^{-1} respectively. In milk sample, the functional groups of capsule and spore could induced some self-correlation peaks too. The infrared spectroscopy combined with 2D correlation spectroscopy analysis method could be a effective method for the Bacillus cereus detection.

Keywords: Bacillus cereus · Infrared spectroscopy · Two-dimensional correlation spectroscopy · Milk

1 Introduction

Bacillus cereus is a common aerobic spore-forming rod-shaped bacteria that can cause food poisoning, which was Widely distributed in dust, sewage, soybean products, flour rice, and dairy products. Traditional methods for detection of Bacillus cereus such as PCR technology, immunological techniques, and enzyme reaction, have high accuracy. While they need pre-treatment sample and were not suitable for real time detection. Hence development of methods for the real-time detection of Bacillus cereus has attracted the considerable interest [1].

© IFIP International Federation for Information Processing 2016
Published by Springer International Publishing AG 2016. All Rights Reserved
D. Li and Z. Li (Eds.): CCTA 2015, Part II, IFIP AICT 479, pp. 514–521, 2016.
DOI: 10.1007/978-3-319-48354-2_52

Bacillus cereus cell structure contains a special cell structure such as capsule, spores, and other energy storage material, analysis by IR (known as "molecular fingerprint" [2]. These peaks can be quickly used to find structural features contained in the chemical composition, which determine the presence of these special structural form. It will provide important reference information for molecular microbial, cytology and taxonomy. At the same time it does not destruct the cells and add any chemicals, which greatly improves the authenticity and reduce the cost of analysis [3–5] in the analysis of the constitution.

In this work, the infrared spectroscopy technology combined with the two-dimensional (2D) correlation spectroscopy were used to detect of Bacillus cereus. The relationship between spectroscopy and Bacillus cereus concentrations was explored.

2   Experiments and Methods

2.1   Experimental Material and Instruments

Equipment: The spectroscopy properties measurements were performed with Spectrum 100 FTIR (USA PerkinElmer Inc).

Microbial Sample: The Bacillus cereus is provided by Beijing North Carolina Chuanglian Biotechnology Research Institute. Potassium bromide, tryptone, yeast extract, sodium chloride and distilled water are provided by the plant protection laboratory of Tianjin Agriculture University. A Yili nonfat pasteurized milk sample was purchased from a local supermarket.

2.2   Experimental Methods

The activated cells were incubated at 30°C for 24 h. The stock cultures were serially diluted with distilled water. A conventional spread plating method was used for bacterial counts. The plate was showed in Fig. 1. In analysis of real-life samples, the milk samples were spiked with different concentrations of *Bacillus cereus* cells.

Fig. 1. Plate count

For spectral measurements, settings were made to provide 16 measurements by the FTIR spectrometer Spectrum (100 type, USA PerkinElmer Inc.) with a scale from 4000 to 650 cm^{-1} and a resolution of 4 cm^{-1}. All experiments were carried out at room temperature. The sterile water was used as back. The infrared spectrum analysis software OriginPro7.5 and 2Dshige two-dimensional correlation spectra processing were used in analysis.

3   Results and Discussion

3.1   One-Dimensional Infrared Spectroscopy of Bacillus cereus

The one-dimensional spectrum of different concentrations of pure *Bacillus cereus* and *Bacillus cereus* mixed with milk were shown in Figs. 2 and 3 respectively. It can be found that a direct one-dimensional spectrum was can not discriminate the concentrations of *Bacillus cereus,* since there were some spectral overlaps.

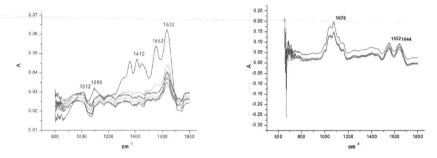

Fig. 2. The one-dimensional spectra of pure *Bacillus cereus*

Fig. 3. The one-dimensional spectra of *Bacillus cereus* mixed in milk

3.2   The Analysis of Two-Dimensional Correlation Spectroscopy

3.2.1   The 2D Correlation Spectroscopy Analysis of Bacillus cereus

Experiments have found that, according to the autocorrelation peak of the two-dimensional correlation diagram can clearly distinguish the diagonal position *Bacillus cereus* spores and capsule in the corresponding group vibration peaks.

The two-dimensional spectral signal can be expanded to the second dimension, which has a higher resolution, and can be distinguished on the one-dimensional spectrum covered with small peaks and weak peak. Experiments have proved that, the diagonal position *Bacillus cereus* spores and capsule in the corresponding group vibration peaks can clearly be distinguished according to the autocorrelation peak of the two-dimensional correlation diagram [6, 7].

Therefore the two-dimensional correlation spectroscopy of *Bacillus cereus* was analysis in this study. Synchronous and asynchronous 2D correlation spectra of pure Bacillus cereus capsule from 1450 cm^{-1} to 1700 cm^{-1} were shown in Fig. 4(a) and (b) respectively. The self-correlation spectroscopy was showed in Fig. 4(c).

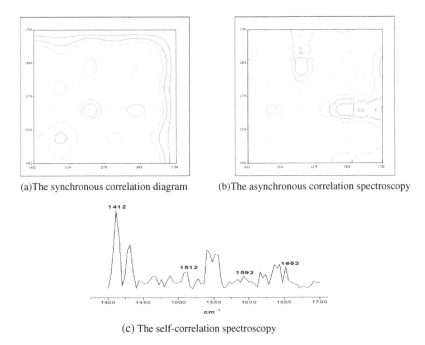

(a)The synchronous correlation diagram   (b)The asynchronous correlation spectroscopy

(c) The self-correlation spectroscopy

Fig. 4. 2D correlation diagrams of Bacillus cereus capsule after the second derivative

It can be seen clearly from Fig. 4(c) that there were two self-correlation peaks near 1652 cm^{-1} and 1512 cm^{-1}, which indicated that the 2D correlation spectroscopy was sensitive to the changes of Bacillus cereus on the concentration. That was induced by the protein amide I band and amide II band contained in capsule. The self-correlation peak near 1592 cm^{-1} was induced by the carboxyl stretching vibration in capsule. And the self-correlation peak near 1412 cm^{-1} was produced by the bending vibration of C-H methyl and methylene. These self-correlation peaks changes of characteristic functional groups can be clearly proved that the capsule existed in the pure *Bacillus cereus* culture.

The synchronous and asynchronous 2D correlation spectra of Bacillus cereus spore from 1300 cm^{-1} to 1750 cm^{-1} were shown in Fig. 5(a) and (b) respectively. The self-correlation spectroscopy was showed in Fig. 5(c).

It can be seen clearly from Fig. 5(c) that there were two self-correlation peaks near 1348 cm^{-1} and 1616 cm^{-1}, which indicated that the 2D correlation spectroscopy was sensitive to the changes of Bacillus cereus on the concentration. That was induced by the COO- group stretching vibration in Bacillus cereus spore. And the self-correlation peak near 1592 cm^{-1} was produced by the C-H bond contained in spore. These self-correlation peaks changes of characteristic functional groups can be further proved that the spore existed in the pure *Bacillus cereus* culture.

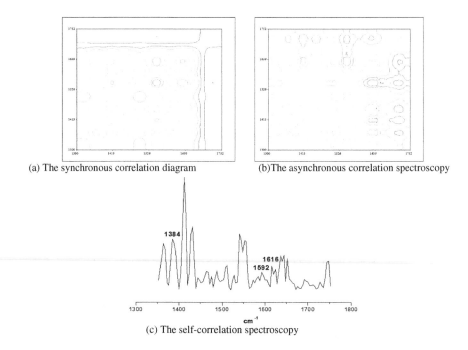

(a) The synchronous correlation diagram   (b)The asynchronous correlation spectroscopy

(c) The self-correlation spectroscopy

Fig. 5. 2D correlation diagram of Bacillus cereus bacteria's spore after the second derivative

3.2.2   The 2D Correlation Spectra Analysis of Bacillus cereus in Milk

Bacillus cereus is often foodborne, and unpasteurized milk and dairy products are common vehicles of transmission. In order to test the sensitivity of the method, different concentrations of *Bacillus cereus* cells were spiked with milk to simulate the real milk sample. The synchronous and asynchronous 2D correlation spectra of Bacillus cereus capsule mixed in milk from 1400 cm^{-1} to 1700 cm^{-1} were shown in Fig. 6(a) and (b) respectively. The self-correlation spectroscopy was showed in Fig. 6(c).

It can be seen clearly from Fig. 6(c) that there were four self-correlation peaks near 1692 cm^{-1}, 1504 cm^{-1}, 1600 cm^{-1} and 1432 cm^{-1}, which was maybe indicated that there was some capsule in milk sample. In order to improve it, the asynchronous 2D correlation spectra were explored at this wavelength range. According to Fig. 6(b), the self-correlation peaks near 1692 cm^{-1} and 1504 cm^{-1} was induced by the protein amide I band and amide II band contained in capsule. The self-correlation peaks near 1600 cm^{-1} was induced by the carboxyl stretching vibration in capsule. And the self-correlation peak near 1432 cm^{-1} was produced by the bending vibration of C-H methyl and methylene.

The synchronous and asynchronous 2D correlation spectra of Bacillus cereus spore mixed in milk from 1350 cm^{-1} to 1700 cm^{-1} were shown in Fig. 7(a) and (b) respectively. The self-correlation spectroscopy was showed in Fig. 7(c).

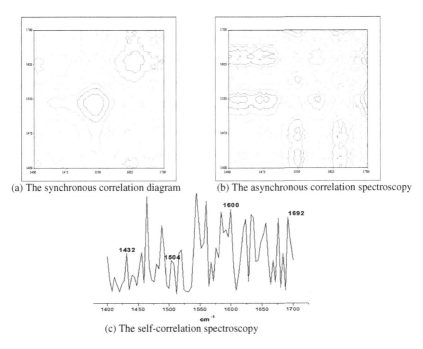

(a) The synchronous correlation diagram   (b) The asynchronous correlation spectroscopy

(c) The self-correlation spectroscopy

Fig. 6. 2D correlation diagrams of Bacillus cereus capsule in milk after the second derivative

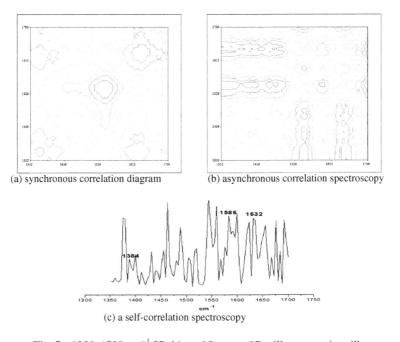

(a) synchronous correlation diagram   (b) asynchronous correlation spectroscopy

(c) a self-correlation spectroscopy

Fig. 7. 1350–1700 cm^{-1} 2Dshige of Spores of Bacillus cereus in milk

It can be seen clearly from Fig. 7(c) that there were four self-correlation peaks near 1632 cm^{-1}, 1384 cm^{-1} and 1586 cm^{-1}, which was maybe indicated that there was some spore in milk sample. In order to improve it, the asynchronous 2D correlation spectra was studied at this wavelength range. According to Fig. 7(b), the self-correlation peaks near 1632 cm^{-1} and 1384 cm^{-1} was induced by the COO- group stretching vibration in Bacillus cereus spore. The self-correlation peaks near 1586 cm^{-1} was produced by the C-H bond contained in spore. These self-correlation peaks changes of characteristic functional groups can be further proved that the spore existed in the milk sample.

4   Conclusion

The spectroscopy properties of capsule and spore in different concentrations of Bacillus cereus were studied using infrared spectroscopy and 2D correlation spectroscopy analysis method. The conclusions are as follows:

(1) In pure Bacillus cereus culture, some functional groups of capsule, such as carboxyl, protein amide I and amide II, C-H methyl and methylene could induced some self-correlation peaks. There was a carboxyl stretching vibration self-correlation peak near 1592 cm^{-1} and it was gradually weakened with the concentration of Bacillus cereus increased. The self-correlation peak of protein amide I and amide II in capsule was presented near 1652 cm^{-1} and 1512 cm^{-1} respectively. And the self-correlation peak near 1412 cm^{-1} was produced by the bending vibration of C-H methyl and methylene. Some functional groups of spore, such as COO- group and C-H bond could induced some self-correlation peaks. Two self-correlation peaks near 1348 cm^{-1} and 1616 cm^{-1} was induced by the COO- group stretching vibration in Bacillus cereus spore. And the self-correlation peak near 1592 cm^{-1} was produced by the C-H bond contained in spore.

(2) In milk sample, the functional groups of capsule could induced some self-correlation peaks too. There was a carboxyl stretching vibration self-correlation peak near 1600 cm^{-1} and it was gradually weakened with the concentration of Bacillus cereus increased. The self-correlation peak of protein amide I and amide II in capsule was presented near 1692 cm^{-1} and 1504 cm^{-1} respectively. And the self-correlation peak near 1432 cm^{-1} was produced by the bending vibration of C-H methyl and methylene. Some functional groups of spore, such as COO- group and C-H bond could induced some self-correlation peaks. Two self-correlation peaks near 1632 cm^{-1} and 1384 cm^{-1} was induced by the COO- group stretching vibration in Bacillus cereus spore. And the self-correlation peak near 1586 cm^{-1} was produced by the C-H bond contained in spore.

These self-correlation peaks changes of characteristic functional groups can be further proved that the capsule and spore existed in the pure Bacillus cereus culture and milk sample. Thus the infrared spectroscopy combined with 2D correlation spectroscopy analysis method could be use for the Bacillus cereus detection.

Acknowledgments. This research was performed with financial support from National Natural Science Foundation of China under the project No. 31171892, Tianjin Science and technology project 13JCYBJC25700, and the Innovation and Entrepreneurship Training Plan for Undergraduate in Tianjin of China through grant number No. 201410061075.

References

1. Mareike, W., Herbert, S., Siegfrids, K.: Appl. Envir. Microbiol. **68**(10), 4717 (2002)
2. Royston, G., Eadaoin, M.T., Paul, J., et al.: FEMS Microbiol. Lett. **140**, 233 (1996)
3. Goodacre, R., Timmins, E.M., Burton, R., et al.: Microbiol. **144**, 1157 (1998)
4. Ngothi, N.A., Kirschner, C., Naumann, D.: J. Mol. Struct. **662**, 371 (2003)
5. Li, H.: Improved method for plate count. Biol. Bull. **41**(1), 51 (2006)
6. Zhou, D.: Microbiology Tutorial, pp. 183–190. Higher Education Press, Beijing (1992)
7. Chang, C., Ying, T.: Spectroscopy and Spectral Analysis **25**(1), 36 (2005)

Stimulating Effect of Low-Temperature Plasma (LTP) on the Germination Rate and Vigor of Alfalfa Seed (Medicago Sativa L.)

Xin Tang[1], Fengchen Liang[1], Lijing Zhao[2], Lili Zhang[2], Jing Shu[1],
Huamei Zheng[1], Xu Qin[2], Changyong Shao[2(✉)], Jinkui Feng[3],
and Keshuang Du[3]

[1] Shandong Agricultural Administrators College,
Jinan 250100, People's Republic of China
txsd@163.com, liangfc01@sina.com
[2] Shandong Province Seeds Group, Jinan 250100, People's Republic of China
zhaolijing008@163.com, zhagnll20020201@163.com,
shaochangyong68@163.com
[3] College of Engineering, China Agricultural University,
Beijing 10083, People's Republic of China

Abstract. Low-temperature plasma (LTP) treatment was applied to stimulate the seed of alfalfa (Medicago sativa L.) for better germination rates and vigor with the use of an inert gas, namely neon, at various discharge power levels. The seed passed through a neon-plasma glowing zone between two horizontally parallel electrode plates within the glow discharge chamber of a seed processing machine with an internal $2 \sim 8$ mm atmospheric pressure. The seed were treated for 20 s. The LTP-treated seeds were germinated at 20 °C in a germination chamber, and seedling emergence was evaluated at 24-h intervals for up to 10 days. LTP stimulation significantly increased the germination rate and vigor versus an untreated sample. Among ten discharge power levels, the 20 W treatment had the most significant effect on the rate and vigor. The use of 20 W treatment increased the germination rate and vigor by 11 % and 22 %, respectively, relative to the control. LTP treated seeds that were stored for 20 days also had higher germination vigor than those stored for 4 days. These results suggest that the use of the LTP technique with a seed processing machine is effective and practical for the purpose of stimulating crop seeds for improved germination.

Keywords: LTP treatment · Stimulating effects · Alfalfa seeds · Germination rate · Germination vigor

1 Introduction

Low-temperature plasma (LTP) as a physical method has been used experimentally in agriculture since the 1970s (Hao and Qin 1998). The technique was first applied in seed treatments to improve germination rate and vigor in Russia and the Commonwealth of Independent States (CIS), and has been delved further into seed processing in China,

Canada and United States. In recent years, some small scale seed enterprises with specialized LTP systems are starting to take shape as new uses of LTPs are developed (Li 2010; Yin 2006). Although observations of the interaction of plasma with seed or living tissue have been primarily empirical, it has been reported that the LTP treatment could not only stimulate seed germination and make crops mature earlier, but also increase flower number and yield (Zhang et al. 2005; Shi et al. 2010). Based on current studies from around the world, LTP stimulates crop seeds by ionizing radiation, mass deposition and charge exchange of ion implantation to biomolecules during the glow discharge process (Dhayal et al. 2006; Volin et al. 2000; Grzegorzewski 2010; Grzegorzewski et al. 2009; RDEOFESWLTP 2008). As a result of the LTP treatment, seed coat is softened, and organic compounds and bio-reactions in embryos are activated, hastening germination and seedling emergence, and increasing crop earliness, yield and resistance to drought and certain diseases (Xu et al. 2011; Liu et al. 2007; Shao et al. 2012; Hu et al. 2007; Yin et al. 2005; Filatova et al. 2010). Various techniques of seed stimulation such as microwave heat radiation, salt stress, water stress, electric-field induction, lanthanum ion radiation, ultraviolet ray irradiation and drought shocks are currently used to treat crop seed, but there has been relatively little research on the application of LTP machines for seed processing, especially alfalfa seed treatment (Fu et al. 2006; Luo et al. 2012; Li et al. 2006; Huo et al. 2011; Qin 2004; Zheng et al. 2002). Our goal in this study was to evaluate the effect of LTP treatment on alfalfa seed germination rate and vigor by subjecting the seed to a simulated outer-space environment within the evacuated and neon-filled glow discharge chamber of a LTP seed processing machine.

2 Materials and Methods

Equipment and Seed Material. An LTP seed processing machine HL-2 N (Fig. 1), provided by Chang Zhou Zhong Ke Chang Tai Plasma Technology Co., was used to perform the treatment of alfalfa seed. The primary part of the machine was the glow discharge chamber which consisted of a pair of electrodes with horizontal plates parallel to each other, and a conveyer belt system (Fig. 1-1). The conveyer belt was positioned between the two planar electrodes and capable of horizontal movement driven by a wheel. The electrodes were connected to a high-tension AC power supply to generate a glow discharge for gas plasma. It was capable of treating a large capacity of seed materials in a continuous or batched process.

The seed material used in this study was a material of purple-flowered alfalfa (*Medicago sativa* L.) with main-season maturity, which was Zhongmu No. 6, bred by the Institute of Animal Sciences (IAS), Chinese Academy of Agricultural Sciences (CAAS). Only plump and viable seeds with a uniform size (2.3 g./1000) were selected for the test.

LTP Seed Stimulation. After the selected alfalfa seeds were loaded in the receiving hopper Chamber I (Fig. 1-6), the machine was evacuated by the vacuum air pumps (I, II, III and IV) and filled with neon (an inert gas) as plasma gas. Neon conducts

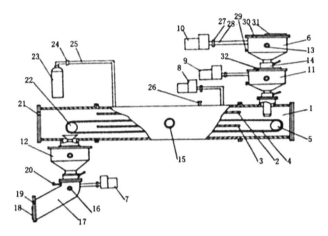

Fig. 1. Experimental configuration for low-temperature plasma treatment 1. Glow discharge chamber; 2. Top electrode plate; 3. Bottom electrode plate; 4. Conveyer belt; 5. Driven wheel; 6. Receiving hopper Chamber I; 7. Vacuum air pump I; 8. Vacuum air pump II; 9. Vacuum air pump III; 10. Vacuum air pump IV; 11. Receiving hopper Chamber II; 12 Seed-discharging hopper chamber; 13. Detector tube I; 14. Butterfly gate; 15. Detector tube II; 16. Detector tube III; 17. Discharging Pipeline; 18. Discharging gate; 19. Thermocouple gauge tube I; 20. Exhaust valve I; 21. Thermocouple gauge tube II; 22. Driving wheel; 23. Gas tank; 24. Gas valve; 25. Intake pipe; 26. Exhaust valve II; 27. Electromagnetic relief valve; 28, Vacuum tube; 29, Exhaust valve III; 30, Thermocouple gauge tube III; 31, Hopper cap; 32, Thermocouple gauge tube IV.

electricity 75 times more efficiently than air and has stronger penetrating force. To obtain uniform and stable glow discharge between the two electrode plates, an internal $2 \sim 8$ mm atmospheric pressure within the machine was maintained. Opening the butterfly valve initiated the glow discharge between the two electrodes, which in turn generated the gas plasma of neon which consisted of many plasma-created species, such as: electrons, ions, photons, and neutrals, with a variety of energies and momenta, chemical activities, and transport characteristics. After the seeds went through the first two receiving hopper chambers and landed on the conveyer belt in the glow discharge chamber, they passed through the plasma glowing zone and were affected/energized by the synergistic effects of the multiple plasma-created species driven by externally-applied electric fields. Ten electric wattages (20 W, 40 W, 60 W, 80 W, 100 W, 120 W, 140 W, 160 W, 180 W, 200 W) as discharge power were used as the stimulating levels of LTP treatment, and an untreated (0 W) sample of seeds was kept asVolin et al. 2000; Grzegorzewski 2011 the control for later seed germination. The seeds were treated with the LTP in the chamber for 20 s before they were moved to the seed-discharging hopper chamber (Fig. 1-12).

Germination Tests. After the LTP stimulation, the treated seeds were divided into two parts and arranged in a nested design for the germination tests, in which the first half of the seeds was stored on shelf for four days as the first time-nest (T1) and the second half for twenty days as the second time-nest (T2). Each nest consisted of ten LTP-treated seed samples and the control. Before the germination test, the stimulated seeds were pre-chilled to break up their dormancy. The germination test was conducted at 20 °C in a dark germination chamber. 100 seeds from each of the 10 treated levels and the control were laid in 12-cm Petri dishes lined with filter paper. The Petri dishes were irrigated with 4 ml of distilled water and constantly maintained sufficient moisture for the seeds to germinate. Germinated seeds or seedlings were counted at 24-h intervals for up to 10 days. Since germination rate and vigor are both important indications of seed quality, viability and the potential of emergence, they were adopted to determine the effects of the LTP treatment on the stimulation of alfalfa seed in this study. Germination rate (GR) is the percentage of seeds that germinate in a specified time (10 days after sowing for alfalfa seed according to GB/T 3543.4-1995, Rules for Agricultural Seed Testing-germination Test) and germination vigor (GV) is the percentage of seeds that germinate within a short time (4 days after sowing for alfalfa seed in accordance with GB/T 3543.4-1995), which are defined as:

GR (%) = (Number of seeds germinated in 10 days/total number of seeds) × 100 %
GV (%) = (Number of seeds germinated in 4 days/total number of seeds) × 100 %.

Because GR and GV are proportional data which rarely produce normal distributed residuals and stable variances, the two values were transformed by logit transformation during data analysis and mean separation.

3 Results and Discussion

Effects of Discharge Power levels and Storage Time of LTP Pre-treated Seeds on Germination Vigor. Two discharge power levels (20 W and 100 W) had the most significant effect, at $P = 0.005$, on the germination vigor of alfalfa seed (Table 1). Six discharge power levels (40 W, 60 W, 80 W, 120 W, 160 W and 200 W) resulted in a significantly higher range of germination vigor than the control (0 W), but no differences among their respective influence. The effect of 180 W was low and the seed stimulation of 140 W was not significantly different from the control.

When considering the effect of storage time on LTP pre-treated seeds, it was found that the germination vigor of the LTP pre-treated alfalfa seeds which were stored for 20 days was significantly higher than that of seeds stored for 4 days (Table 1). This suggests that LTP pre-treated seeds should be stored longer in order to improve germination vigor before planting. It can be deduced that the synergistic effects of chemical activities, energy momenta, mass deposition and ion implantation of the plasma-created species (electrons, ions, photons and neutrals) on biomolecules are complex and need a relatively long period of time to be revealed through modifying and energizing a series of biochemical reactions in the seeds. More research is necessary to determine whether there were interactions between discharge power levels and storage time of the LTP pre-treated seed influence seed vigor.

Table 1. Effects of discharge power levels and the storage time of LTP treated seeds on the germination vigor of alfalfa seed.

Power levels	GV Values[W]		
	T1	T2	Mean[Z]
0 W	43	51	47.00 c
20 W	57	58	57.50 a
40 W	53	54	53.50 abc
60 W	47	57	52.00 abc
80 W	50	56	53.00 abc
100 W	54	57	55.50 ab
120 W	52	54	53.00 abc
140 W	45	50	47.50 c
160 W	50	53	51.50 abc
180 W	47	52	49.50 bc
200 W	50	53	51.50 abc
Mean	49.82 b	54.09 a	

[Z]Tested by logit-transformed GV data. Mean separation within columns and rows (a, b, c) by least significant different ($P = 0.05$). The means bearing the same letters were not significantly different at the 5 % level.
[W]Germination rate values for T1 (LTP treated seeds were stored for 4 days) and T2 (LTP treated seeds were stored for 20 days).

Effects of Discharge Power levels and the Storage Time of LTP Pre-treated Seeds on Germination Rate. Of the ten treatments, the 20 W, 80 W, 180 W and 200 W had the highest stimulating effect ($P = 0.05$) on the germination rate of alfalfa seed in comparison to the others (Table 2). However, none of these four treatments resulted in a significantly different rate from one another. The levels of 40 W, 100 W, 120 W and 160 W brought about an intermediate influence on the rate, but were significantly better than the two treatments 60 W and 140 W, which showed no differences from the control.

As the seed germination rates were observed in the 4-day-storing lot and 20-day-storing lot, the two mean rates were 59.46 % and 58.36 % (Table 2). The results showed that the effects of all the LTP treatments on the mean rates of the two time lots were nonsignificant. The lack of influence of seed storage time levels after the LTP stimulation indicates that certain germination rates can be eventually achieved no matter how long the LTP-treated seeds are stored, so long as adequate time and appropriate conditions are met for germination. In other words, the effect of LTP seed stimulation on germination rate is not correlated with the storage time of pre-treated seeds, although it could make significant differences in germination vigor.

Table 2. Effects of discharge power levels and the storage time of LTP treated seeds on the germination rate of alfalfa seed.

Power levels	GR Values[W]		
	T1	T2	Mean[Z]
0 W	55	56	55.5 b
20 W	63	60	61.50 a
40 W	61	57	59.00 ab
60 W	55	56	55.50 b
80 W	59	62	60.50 a
100 W	62	56	59.00 ab
120 W	61	58	59.50 ab
140 W	57	55	56.00 b
160 W	59	60	59.50 ab
180 W	61	61	61.00 a
200 W	61	61	61.00 a
Mean	59.46 a	58.36 a	

[Z]Tested by logit-transformed GR data. Mean separation within columns and rows (a, b) by least significant different ($P = 0.05$). The means bearing the same letters were not significantly different at the 5 % level.
[W]Germination rate values for T1 (LTP treated seeds were stored for 4 days) and T2 (LTP treated seeds were stored for 20 days).

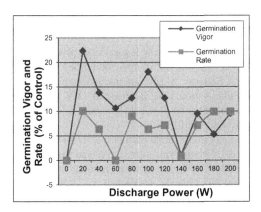

Fig. 2. Effect of discharge power (W) on the seed germination rates and vigor of alfalfa, expressed percent of the control (0 W).

Germination Rate and Vigor Relative to the Control. In comparing the effects of all ten levels (20 W, 40 W, 60 W, 80 W, 100 W, 120 W, 140 W, 160 W, 18 W and 200 W) of discharge power to the control (0 W), a trend of fluctuating effects is displayed in Fig. 2 that the LTP treatments resulted in higher (at 20 W), intermediate (between 80 W and 100 W), and lower (between 160 W and 200 W) increments of germination rates and vigor as the discharge power levels increased. Two obvious peaks (11 %, 22 %) of the increments of germination rates and vigor both occurred at the 20 W and 100 W levels. The chart also demonstrated that the LTP treatment had higher effects on the germination vigor than rates in the low range of discharge power, but no obvious differences in the high range.

4   Conclusion

The application of new agricultural technologies has resulted in impressive gains in crop productivity. LTP is a technique that will likely play an important role in future gains in the improvement of crop seed production and quality, especially in stimulating germination and seedling emergence, and possible also in increasing crop earliness, yields and resistance to drought and certain diseases. Unlike non-LTP seed stimulating methods discussed in the introduction, the LTP seed treatment for germination stimulation in this experiment could not only be conducted mechanically, but also very effectively. Our study provides that LTP seed treatment significantly increased the germination rates and vigor of alfalfa seed in comparison to an untreated control. More importantly, this experiment has established a powerful model for seed businesses and farmers to stimulate crop seeds for better germination rates, vigor and higher yields.

References

Dhayala, M., Leea, S.-Y., Parkb, S.-U.: Using low-pressure plasma for *Carthamus tinctorium* L. seed surface modification. Vacuum **80**(5), 499–506 (2006)

Filatova, I., Azharonok, V., Kadyrov. M.: Rf and microvawe plasma application for pre-sowing caryopsis treatments. Publ. Astron. Obs. Belgrade **89**, 289–292 (2010)

Fu, S., Zhang, F., Li, J., et al.: Several physical techniques in agriculture and prospects. Agric. Mech. Res. **1**, 36–38 (2006). (in Chinese)

Grzegorzewski, F.: Influence of non-thermal plasma species on the structure and functionality of isolated and plant-based1, 4-benzopyrone derivatives and phenolic acids (2010). www.0pus4. kobv.de/opus4-tuberlin/files/2839/grzegorzewski_franziska.pdf. Accessed 30 Nov 2013

Grzegorzewski, F., Schluter, O., Ehlbeck, J., Kroh, L.W., Rohn, S.: Influence of non thermal plasma-immanent reactive species on the stability and chemical behavior of bioactive compounds (Talk). In: Euro Food Chem. XV - Food for Future, Copenhagen, Denmark (2009)

Hao, X.J., Qin, J.G., et al.: Preliminary Study of low temperature plasma seed treatment. Shanxi Agric. Sci. **26**(2), 39–41 (1998)

Hu, L., Tian, L., Hu, Z., et al.: Application of physical agriculture techniques in cleaned seeds treatments. J. Anhui Agric. Sci. **13**(3), 3778–3779 (2007). (in Chinese)

Huo, P.H., Li, J.F., Shi, S.L.: Germination and growth of ultra-dried alfalfa seeds under salt stress. Grassland Turf. 01, 13–17 (2011). (in Chinese)

Li, B., Jiao, D.Z., Zhan, C.Y.: Effect of microwave treatment on the germination and seedling drought resistance of Alfalfa. Seed **12**, 28–30 (2006). (in Chinese)

Li, R.: Plasma machine seed treatment technology. North Rice **4**(4), 52–53 (2010). (in Chinese)

Liu, S., Ouyang, X., Nie, R.: Application status and development trend of the physical methods in the crop seed treatment. Crop Res. **5**(2), 520–524 (2007). (in Chinese)

Luo, H., Ran, J., Wang, X., et al.: Comparision study of dielectric barrier discharge in inert gases at atmospheric pressure. High Voltage Engineering,2012, (38)5:1070–1077

Qin, F.: Effect of microwave treatment on drought resistance of four local alfalfa cultivars of Gansu Province. Pratacultural Sci. **11**, 41–43 (2004). (in Chinese)

RDEOFESWLTP: Low Temperature Plasma Science: Not only the fourth state of matter but all of them. Report of the Department of Energy Office of Fusion Energy Sciences Workshop on Low Temperature Plasmas (2008). www.scence.energy.gov/fes/about/ ~ /media/fes/pdf/about/low_temp_plasma. Accessed 10 Nov 2013

Shao, C., Wang, D., Tang, X., et al.: Arc plasma system and its application & development treads on pre-sowing seeds treatment. China Seed Ind. **8**, 1–3 (2012). (in Chinese)

Shi, Y.H., Fang, X.Q., Xu, D.H.: Effect of plasma seed treatment with different radiation intensity on the biological traits, yields and output values of soybean. Jilin Agric. Sci. **35**, 6–7 (2010). (in Chinese)

Volin, J.C., Denes, F.S., Yong, R.A., Park, S.M.: Modification of seed germination performance through cold plasma technology. Crop Sci. **40**, 1076 (2000)

Xu, Z., Chen, B., Wei, Z.: Various seed treatments on corn yield. Agric. Sci. Technol. Equip. **4**, 15–16. (in Chinese)

Yin, M.: Research of Magnetized Arc Plasma on Seeds Biological Effects. Dalian University of Technology, Dalian (2006). (in Chinese)

Yin, M., Huang, M.: Stimulatnig effects of seed treament by magnetized plasma on tomato growth and yield. Plasma Sci. Technol. **6**(7), 3143–3147 (2005)

Zhang, Y., Zhang, J., Wang, Q.: The application of physical methods in sugar beet seed treatment. China Beet Sugar **2**, 20–22 (2005). (in Chinese)

Zheng, R.Y., Xu, Y., Yang, T.Q.: Influence of treating Lucerne seed with electric field on the seedling growth. Scientiarum Naturalium Univ. Neimongol. **03**, 359–362 (2002). (in Chinese)

Evaluation of Timber and Carbon Sequestration Income of Cunninghamia Lanceolata Timber Forest and Management Decision Support

Yan Qi[1], Baoguo Wu[2], and Shanghong Li[1(✉)]

[1] International College Beijing, China Agricultural University, Beijing, China
meganqiyan@sina.cn, shanghongli@cau.edu.cn
[2] Beijing Forestry University, Beijing 100083, China
wubg@bjfu.edu.cn

Abstract. Carbon sequestration exchange has just started in China. This study was based on the general plantation regions in Fujian China where Cunninghamia lanceolata is quite popular and has great economic and ecological significance for farmers and the society at large. This study used Hartman model to forecast the income of timber sales and carbon sequestration so as to integrate the model in the decision support system to find a practical tool for farmers to project their future monetary income to make sound operating decisions in terms of plantation density and rotation age, which is of practical significance and innovation.

Keywords: Evaluation · Timber income · Carbon sequestration income · Cunninghamia lanceolata · Management decision support

1 Introduction

Facing the huge social demand for industrial forest products, as well as the serious challenge of the environmental protection put forward by the global climate warming, the sustainable management of artificial timber forest has the important dual role of economic benefits creation and environmental protection. The foreign countries advanced in forestry have achieved the 60 %–93 % of the total industrial forest products by using the artificial timber forest which takes only 1.2 %–16 % of their total forest area [1], which poses a challenge to China and at the same time provides very good experiences too.

The Cunninghamia lanceolata (China fir) is one of the unique fast-growing commercial timber tree species. It has fast growth speed, good timber material of being light and toughening at the same time, moderate intensity, and high quality. It is fragrant, resistant to insect decay and easy to process. Its broad economic use can be found in construction, furniture, appliances, shipbuilding, etc. The Cunninghamia lanceolata is widely planted in the 16 provinces and regions in south China. It is the one of the important forest resources in south China as one of the most important afforestation and fast-growing timber tree species in southern collective forest region of China. And it is the most widespread plantation and ecological system in Fujian province.

© IFIP International Federation for Information Processing 2016
Published by Springer International Publishing AG 2016. All Rights Reserved
D. Li and Z. Li (Eds.): CCTA 2015, Part II, IFIP AICT 479, pp. 530–538, 2016.
DOI: 10.1007/978-3-319-48354-2_54

The Chinese fir plantation has great ecological, economic and social benefits. Therefore, it is of great significance to study on the sustainable management of the Chinese fir artificial timber forest.

2 Literature Review

With the background of global climate change and environmental protection, research on carbon sequestration has drawn a lot of attention in the academic field, including the carbon sequestration related research about Cunninghamia lanceolata. Bayer et al. highlighted the importance of forest biomes for maintaining and increasing biogeochemical carbon sequestration in their study, considering the biome-specific response to current climate and land use, and their projections for the future [2].

Huang et al. analyzed the carbon sequestration in living biomass and soil organic carbon pools using the carbon estimation model and reached the amount of carbon stock up to date and in both the past and the future. According to them, from 1950 to 2012, plantations in China sequestered 1.686 Pg carbon by net uptake into biomass and emissions of soil organic carbon. The carbon stock of China's present plantations was 7.894 Pg, 21.4 % as forest biomass and 78.6 % as soil organic carbon. They projected that China's carbon stock will reach a level of 3.169 Pg by 2050, and it will amount to 10.395 Pg [3].

There have been some comparison studies. Yen and Lee conducted a comparison study on the aboveground carbon sequestration between moso bamboo (Phyllostachys heterocycla) and Chinese fir (Cunninghamia lanceolata) forests based on the allometric model [4]. NIU et al. conducted a comparison study on the carbon storages in Cunninghamia lanceolata and Michelia macclurei plantations during a 22-year period in south China [5].

Chen et al. investigated changes of carbon stocks in Cunninghamia lanceolata plantations converted from a natural broadleaved forest, based on a typical chronosequence in mountain land of subtropical China, which includes six first- generation Chinese fir stands at different development stages. It was concluded that over-mature tree plantations had a limited role in continuously sequestering carbon, and that the soil organic pools of tree plantations can be hardly recovered to those of natural forests due to a large initial loss and a low late gain in soil organic carbon following tree plantation establishment [6].

The research on management of timber forest is of more significance to this study. In Wang et al's study, field data were combined with the forest ecosystem management model FORECAST to estimate the impacts of different forest management strategies, that is, the choice of different combinations of planting densities, rotation lengths, and different harvesting intensities, on carbon sequestration of Phoebe bournei plantations in south-eastern China [7].

From Noormets et al's study, it is amazing to find that managed forests generate about 50 % lower carbon stocks than unmanaged forests. It was found the same with the gross primary productivity (GPP) and total net primary productivity (NPP), but relatively more of the assimilated carbon is allocated to aboveground pools in managed than in unmanaged forests, whereas allocation to fine roots and rhizosymbionts is lower [8].

Therefore, a good management strategy should be able to balance the benefits of timber sales and carbon sequestration for the long-term sustainability.

3 Materials and Methods

The Hartman model was proposed by Richard Hartman in the middle of the 1970 s. He introduced the amenity service function of forests to the Fautsmann model so as to combine the assessment of economic income with the ecological income of forests. The Hartman Model is an objective function equation taking a rotation period as the variable rendering a rotation age at which the integrated income can be maximized [9, 10].

$$\max_{T \geq T^0} LEV(T) = \frac{1}{1 - e^{-rT}} [-C + \int_{t=1}^{T} f(t)e^{-rT}dt + p(T)V(T)e^{-rT}]$$

where,

T = rotation age
R = discount rate
C = regeneration cost
$p(T)$ = stumpage price at age T (Yuan/m^3)
$V(T)$ = growing stock of timber at age T (m^3/mu)
$p(T) V(T)$ = the economic income of timber
$f(t)$ = function of amenity or ecological income

By calculating land expected values (LEV) at different rotation ages, a maximized expected value can be reached. In the meantime, the rotation age when LEV can be maximized can be found. This is helpful for reaching the reasonable management strategy.

Including the income of carbon sequestration in the Hartman model, the model looks like [9]:

$$\max_{T} LEV(T) = \frac{-c + \sum_{t=0}^{T} p_c g(t)e^{-rt} + p(T)V(T)e^{-rT} - e^{-rT} \sum_{t=0}^{d} p_c V(T)q(t)e^{-rt}}{1 - e^{-rT}}$$

Where,

C = regeneration cost (Yuan/mu)
$p(T)$ = stumpage price at age T (Yuan/m^3)
p_c = social benefits of carbon sequestration (Yuan/m^3)
r = discount rate
$V(T)$ = growing stock of timber at age T (m^3/mu)
$g(t)$ = timber growth at age t (m^3/mu/year)
$q(t)$ = rate of decay of harvested timber at time t (%)
d = time period during which the harvested timber completely decays

Before reaching the yield of carbon sequestration, the general biomass of forests should be calculated based on the volume of timber.

Following that, the carbon sequestration reserves can be calculated according to the total biomass volume. Usually the calculation is based on the proportion of carbon storage in the dry weight of plant organics. The international practice is to adopt the rate of 0.50, thus total carbon sequestration reserves = total biomass*0.5 [11].

The monetary income of carbon sequestration then can be calculated finally. According to the first transaction practice of farmers' carbon sequestration exchange in China in 2014, 42 farmers sold 4,285 tons of carbon sequestration for the monetary income of 128,550 yuan in total. Thus in this study, the price of forest carbon sequestration was set as 30 yuan/ton [12]. And in this paper, the parameter d was set to 1.

The economic parameters have been reached by visiting farmers and markets in the general plantation regions in Fujian.

i. According to the investigations on the forest land and mountain land in general regions, the land rent of the medium level is selected, that is, the opportunity cost of land use is 122.45 yuan/mu/year. In other words, the farmer would have a rental income of 122.45 yuan/mu/year if he or she let this land out to others.

ii. Renminbi benchmark interest rates of loans by the financial institutions in China are as follows.

Annual rate %

Item	Since June 28, 2015
Interest rates of loans	
Short-term loans	
Within 1 year (including 1 year)	4.85
Mid/Long-term loans	
1 to 5 years (including 5 years)	5.25
Above 5 years	5.40

Considering the assumption of continuous operation and land management, where farmers contract the land and land use rights and may extend the contract after the expiration of the original contract, so this study adopted mid/long-term interest rates of 5.40 % as the expected rate of return.

iii. The initial plantation density is found at 3 levels based on the practice of the general regions of plantation in Fujian, 133 plants/mu at the lowest level, 267 plants/mu at the medium level and 400 plants/mu at the highest level.

iv. Afforestation cost: Seedling cost of 0.20 yuan/plant (20−50 CM of height, 0.3 CM of ground diameter); soil preparation, plantation, machinery and utility costs of 380 yuan/mu; Total afforestation costs at 3 levels of initial plantation desinty are 406.60 yuan/mu, 433.40 yuan/mu and 460 yuan/mu.

v. Cost of forest culture and management: nurture and protection cost 40 yuan/mu/year averagely during the different stages of growth of the timber forest; cost of fertilization and pest control is 100 yuan/mu/year averagely.

vi. Commercial timber price of Cunninghamia lanceolata is 900 yuan/m^3

vii. Trees growth model and the natural growth process table

This study adopted the tree growth model and table of volume given by Zongming He et al., where were based on the large amounts of data and information of 665 fixed and temporary HWWCCSC standard plantation plots of the general regions of Cunninghamia lanceolata in Fujian province. They established the natural growth model and the natural growth process table of Chinese fir timber forest in Fujian general plantation regions. This study used the natural growth process table based on the status index 16 to reach the Chinese fir artificial timber forest tree volume, which laid the foundation for further calculation of the commercial timber income and carbon sequestration income [13].

viii. To convert the tree volume of Cunninghamia lanceolata to the general biomass of forests, this study adopted the following function [14]:

$$B = 12.82556 + 0.4621618V$$

Where, B represents the general biomass,
V represents the tree volume

4 Findings and Analysis

Using the Hartman model, both the expected timber income and expected carbon sequestration of the general plantation regions of Cunninghamia lanceolata timber forests in Fujian at three different levels of plantation density were found respectively. And more importantly, the highest land expected values of the Cunninghamia lanceolata timber forests and the corresponding ages to reach the land expected values were found at the same time.

From Table 1, at the plantation density of 133 plants/mu, the highest land expected value including the expected value of both timber income and carbon sequestration income can be reached at the forest age of 20 at 3,020.13 yuan/mu. It was also found that the highest expected value of timber income would be reached in the same year as the land expected value at year 20. At year 20, the expected value of the timber income is 2,262.63 yuan/mu, and expected value of the accumulated carbon sequestration of 757.50 yuan/mu.

From Table 2, at the plantation density of 267 plants/mu, the highest land expected value including the expected value of both timber income and carbon sequestration income can be reached at the forest age of 15 at 4,373.85 yuan/mu, different from that at the density level of 133 plants/mu. The highest expected value of timber income would be reached, again, in the same year as the land expected value at this density level. And at year 15, the expected value of the timber income is 3,859.08 yuan/mu, and expected value of the accumulated carbon sequestration of 514.77 yuan/mu.

Table 1. Land Expected Value (LEV) of Cunninghamia lanceolata in general plantation regions of Fujian at the density of 133 plants/mu

Age (year)	Tree volume (m3/mu)	NPV of accumulated annual cost (yuan/mu)	NPV of accumulated carbon sequestration (yuan/mu)	NPV of timber income (yuan/mu)	EV of timber income (yuan/mu)	EV of accumulated carbon sequestration (yuan/mu)	LEV (yuan/mu)	Discounting factor
5	0.728000	1119.25	150.71	−1025.68	−4334.71	95.26	−4239.45	0.763379
10	4.516667	1973.66	278.13	−11.39	−27.29	400.88	373.59	0.582748
15	10.275333	2625.90	381.47	1081.46	1948.08	507.51	2455.59	0.444858
20	16.440000	3123.81	588.74	1494.25	**2262.63**	**757.50**	**3020.13**	0.339596
25	22.054667	3503.90	752.04	1235.21	1667.50	912.47	2579.96	0.259240
30	26.805333	3794.05	1003.89	573.61	715.14	1172.22	1887.36	0.197899
35	30.735333	4015.55	1200.38	−243.23	−286.52	1352.63	1066.12	0.151072

Table 2. Land Expected Value (LEV) of Cunninghamia lanceolata in general plantation regions of Fujian at the density of 267 plants/mu

Age (year)	Tree volume (m3/mu)	NPV of accumulated annual cost (yuan/mu)	NPV of accumulated carbon sequestration (yuan/mu)	NPV of timber income (yuan/mu)	EV of timber income (yuan/mu)	EV of accumulated carbon sequestration (yuan/mu)	LEV (yuan/mu)	Discounting factor
5	1.227333	1119.25	153.36	−709.42	−2998.13	96.93	−2901.20	0.763379
10	6.769333	1973.66	287.86	1143.28	2740.01	405.64	3145.66	0.582748
15	12.992000	2625.90	392.63	2142.34	**3859.08**	**514.77**	**4373.85**	0.444858
20	18.244667	3123.81	604.83	2019.02	3057.25	776.40	3833.65	0.339596
25	22.392667	3503.90	766.93	1287.27	1737.78	931.87	2669.65	0.259240
30	25.606667	3794.05	1021.61	333.32	415.56	1196.05	1611.61	0.197899
35	28.094667	4015.55	1215.38	−629.07	−741.02	1373.08	632.06	0.151072

From Table 3, at the plantation density of 400 plants/mu, the highest land expected value including the expected value of both timber income and carbon sequestration income can be reached at the forest age of 15 at 4,402.07 yuan/mu. Again, the highest expected value of timber income is reached in the same year as the land expected value. And at year 15, the expected value of the timber income is 3,885.21 yuan/mu, and expected value of the accumulated carbon sequestration of 516.87 yuan/mu.

Figure 1 shows the different levels of land expected values of the timber forests at different ages and different density levels. As shown in Chart 1, it is obvious that the highest land expected value is generated at the plantation density level of 400 plants/mu. Again the highest expected value of timber income and carbon sequestration income can be found with the plantation density of 400 plants/mu.

Table 3. Land Expected Value (LEV) of Cunninghamia lanceolata in general plantation regions of Fujian at the density of 400 plants/mu

Age (year)	Tree volume (m3/mu)	NPV of accumulated annual cost (yuan/mu)	NPV of accumulate carbon sequestration (yuan/mu)	NPV of timber income (yuan/mu)	EV of timber income (yuan/mu)	EV of accumulate carbon sequestration (yuan/mu)	LEV (yuan/mu)	Discounting factor
5	1.620000	1119.25	155.43	−466.24	−1970.42	98.24	−1872.17	0.763379
10	7.420667	1973.66	290.98	1458.28	3494.97	407.77	3902.74	0.582748
15	13.094667	2625.90	394.06	2156.84	**3885.21**	**516.87**	**4402.07**	0.444858
20	17.748000	3123.81	605.89	1840.62	2787.11	779.52	3566.63	0.339596
25	21.333333	3503.90	765.29	1013.51	1368.21	931.84	2300.05	0.259240
30	24.013333	3794.05	1018.87	22.93	28.59	1194.96	1223.55	0.197899
35	25.990000	4015.55	1209.41	−941.83	−1109.43	1368.24	258.81	0.151072

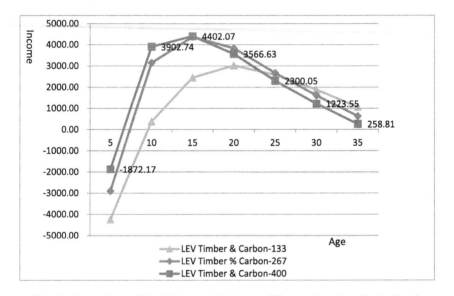

Fig. 1. Comparison of Land Expected Values at different plantation density levels

5   Conclusion

Using Hartman model to calculate and forecast the land expected value of the Cunninghamia lanceolata timber forests is effective and important in identifying the ecological income of the forest operations. By doing this, the age of forests when the different levels of timber income and carbon sequestration income can be realized will be able to be found too. All the above proves the validity and usefulness of the Hartman model as the tool of the Cunninghamia lanceolata timber forests' income prediction.

This evaluation model will be integrated in the decision support system to assist farmers' operating decision making. Based on the evaluation and forecast results, the farmers are able to have an estimate of their future income under different operation schemes, that is, at different plantation densities and the arrangement of rotation, etc. And this is exactly the purpose of this study, that is, to apply this evaluation and forecast module in the decision support system so as to find a practical tool for the farmers to assist their decision making.

To conclude, China is expecting a complete national carbon sequestration exchange system, including the market system and the carbon emission quota allocation system, etc. Both theoretically and practically, carbon sequestration income is an important component of the integrated income of farmers of Cunninghamia lanceolata timber forests. It is important to educate the farmers to know about the ecological and economic value of carbon sequestration, and more importantly, to know well how to improve this part of income with the aid of related techniques.

Acknowledgements. This work was supported by National Natural Science Funds of China: Research on the Management Scheme Optimization Decision Model of Cunninghamia Lanceolata Timber Forest under grant [No. 31170513].

References

1. Zhou, T. The sustainable operation of artificial forests in China (in Chinese). The Chinese academy of forestry science (2008)
2. Bayer, A.D., Pugh, T.A.M. Krause, A., et al.: Historical and future quantification of terrestrial carbon sequestration from a Greenhouse-Gas-Value perspective. Glob. Environ. Change **32**, 153–164 (2015)
3. Huang, L., Liu, J.Y., Shao, Q.Q., Xu, X.L.: Carbon sequestration by forestation across China: Past, present, and future. Renew. Sustain. Energy Rev. **16**(2), 1291–1299 (2012)
4. Yen, T.M., Lee, J.S.: Comparing aboveground carbon sequestration between moso bamboo (Phyllostachys heterocycla) and China fir (Cunninghamia lanceolata) forests based on the allometric model. For. Ecol. Manag. **261**(6), 995–1002 (2011)
5. Niu, D., Wang, S.L., Ouyang, Z.Y.: Comparisons of carbon storages in Cunninghamia lanceolata and Michelia macclurei plantations during a 22-year period in southern China. J. Environ. Sci. **21**(6), 801–805 (2009)
6. Chen, G.S., Yang, Z.J., Gao, R., Xie, J.S., et al.: Carbon storage in a chronosequence of Chinese fir plantations in southern China. For. Ecol. Manag. **300**, 68–76 (2013)
7. Wang, W.F., Wei, X.H., Liao, W.M., et al.: Evaluation of the effects of forest management strategies on carbon sequestration in evergreen broad-leaved (Phoebe bournei) plantation forests using FORECAST ecosystem model. For. Ecol. Manag. **300**, 21–32 (2013)
8. Noormets, A., Epron, D., Domec, J.C., et al.: Effects of forest management on productivity and carbon sequestration: a review and hypothesis. For. Ecol. Manag. (2015). In Press, Corrected Proof, Available online 26 June
9. Amacher, G.S., Ollikainen, M., Koskela, E.A.: Economics of Forest Resources. The MIT Press, Cambridge (2009)
10. Qi, Y., Wu, B.G.: The application of the Hartman Model in the assessment of agro-forestry operations' income. Sens. Lett. **10**(1–2), 660–665 (2012)
11. Xu, T.: The estimating technology of forest biomass and carbon storage based on RS information (in Chinese). For. Invent. Plann. **33**(3), 11–13 (2008)

12. Xu, J.: 4,000 tons for 120 thousand RMB: the first successful transaction of carbon sequestration of farmers' forest management (in Chinese) [EB/OL], 16 October 2014/7 March 2015. http://www.nbd.com.cn/articles/2014-10-16/869259.html
13. He, Z.M., Lin, S.Z., Yu, X.T., et al.: Study on natural growth model of Chinese fir plantation in ordinary growing area of Fujian (in Chinese). J. Fujian Coll. For. **17**(3), 231–234 (1997)
14. Wen, Y.G., Qin, W.M., Wei, S.Z.: The tentative trial of using tree volume to estimate forest biomass (in Chinese). For. Sci. Technol. **7**, 7–10 (1989)

Researches on the Variations of Greenhouse Gas Exchange Flux at Water Surface Nearby the Small Hydropower Station of Qingshui River, Guizhou

Lei Han[1], Xuyin Yuan[1(✉)], Jizhou Li[1], Yun Zhao[2], Zhijie Ma[2], and Jing Qin[2]

[1] College of Environment, Hohai University, Nanjing 210098, Jiangsu, China
netyxy@263.com
[2] China Institute of Water Resource and Hydropower Research, Beijing 100044, China

Abstract. Greenhouse gas (GHG) emission from water surface nearby the small hydropower station is a rising problem of concern. This paper studied the daytime changes of GHG flux of Fujiang hydropower station (FJHPS) and Xiasi hydropower station (XSHPS) located on Qingshui river in Guizhou by the static float chamber sampling and the gas chromatography analysis method in Autumn. Data showed, the fluxes of CO_2, CH_4 and N_2O ranged from -43 to 72, -23 to 15 and -0.016 to 0.13 mmol·$(m^2 \cdot d)^{-1}$, respectively. Overall, the GHG fluxes in the downstream of the station were slightly higher than the upstream, which manifested the downstream released more GHGs. The CO_2 exchange fluxes in FJHPS were higher than XSHPS, while CH_4 and N_2O fluxes showed a reverse situation. The fluxes of GHG had a positive correlation with DO and pH. Compared with other lakes and reservoirs, smaller releasing rates of GHG were existed in the small hydropower station.

Keywords: Greenhouse Gas · Small hydropower station · Water surface · Qingshui River

1 Introduction

The major greenhouse gas concentrations in the atmosphere have reached the highest point since data was recorded at 2011, with average 390.9 ppm, 1813 ppb and 324.2 ppb for CO_2, CH_4 and N_2O respectively and increased by 40 %, 159 % and 20 % comparing with the period of Industrial Revolution based on the World Meteorological Organization [1]. The continuous increase of atmospheric greenhouse gas concentrations and the consequent global warming raises attentions to greenhouse gases produced by the running hydropower station.

Previous studies have shown that the rapid increase of greenhouse gas concentration is closely related to human activities, of which, reservoir is considered to be an important source of greenhouse gas emission. It is estimated that the CO_2 exchange flux of world's freshwater reservoirs through accounts for 4 % of total anthropogenic CO_2 exchange flux [2]. For a long time, the hydropower has been considered a clean, carbon-free energy

© IFIP International Federation for Information Processing 2016
Published by Springer International Publishing AG 2016. All Rights Reserved
D. Li and Z. Li (Eds.): CCTA 2015, Part II, IFIP AICT 479, pp. 539–547, 2016.
DOI: 10.1007/978-3-319-48354-2_55

and gets extensive development [3]. However, some literatures have reported that the reservoir is likely to be an emission source of greenhouse gas [4, 5], so the greenhouse gases release from the reservoir has become a controversial problem.

The GHG emissions from reservoir (CO_2, CH_4) is mainly caused due to the mineralization of organic matter [4, 5] which originate from the reservoir water and sediments. The existing researches have showed that their exchange processes are closely related to reservoir age, soil properties of flooded area, vegetation coverage and regional climate conditions [6–8]. The static floating box technology was used to observe the GHG variations of Dongting Lake, Poyang Lake, Dianchi Lake and the Three Georges Reservoir by Chen et al. [9]. But the data of greenhouse gas emission in small reservoirs is lacked. In addition, as an important greenhouse gas, the warming potential of N_2O is about 310 times [10] than that of CO_2, so it needs to be paid more attention in small reservoirs.

This paper studies the GHG variations in water surface nearby two small hydropower stations in Qingshui river, Guizhou province. Then we link the climatic conditions, water chemical parameters and GHG fluxes to discuss their potential relationships. These results will promote a better understanding of GHG exchange fluxes in the water surface areas of small hydropower stations.

2   Materials and Methods

2.1   Sampling Locations

Qingshui river is located in the Guizhou province, southwest China, which is 459 km long and cover 17,145 square kilometers in the watershed. Duyun and Kaili are two major cities in this watershed. In this study, GHGs were collected at the water surfaces of upstream and downstream of Fujiang hydropower station (weak human activity) and Xiasi hydropower station (strong human activity). Locations of the study site were shown in Fig. 1.

2.2   Greenhouse Gas Collection and Detection

GHGs were collected at water surface of two small river reservoir, in a daytime of the Autumn, with the static float chamber [11]. The collection times ranged from 09:00 am to 17:00 pm with one hour interval. The gas samples were extracted by a 30 ml plastic syringe, and quickly transferred to a vacuum-sealed glass vial. Four parallel samples were collected at one time. After collection, samples were quickly returned to the laboratory, and detected within 48 h. Agilent 7890 A gas chromatograph was used for the simultaneous determination of GHGs.

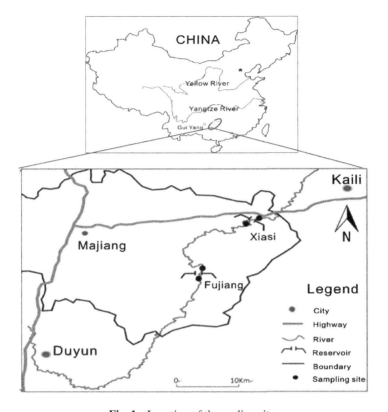

Fig. 1. Location of the sapling sites

2.3   Calculation of GHG Exchange Flux

Based on the average concentration of four samples and the background values of local atmosphere, the CO_2, CH_4 and N_2O exchange fluxes were calculated. The calculation formula was as follow:

$$F = (F1 \times F2 \times V \times \Delta c)/(F3 \times S \times \Delta t)$$

Where F represents the gas exchange flux [$mg \cdot (m^2 \cdot d)^{-1}$]. F_1 is the unit conversion factor between ppm and $\mu g \cdot m^{-3}$. F_2 is the conversion factor between min and d. $V(m^3)$ represents the volume of air in the floating container. S is the superficial area of water surface inside the floating container. F_3 is the unit conversion factor between μg and mg. $\Delta c/\Delta t (10^6 \cdot min^{-1})$ implies the slope of the greenhouse gas concentration versus time during the observation time, which were calculated in each time period. If the exchange flux is positive, it indicates the waterbody release GHGs into the atmosphere. The negative value of exchange flux represents the waterbody absorption from the atmosphere. In this study, the gas concentration calculated by the above formula was divided by the molar mass of each gas, and the unit $mg \cdot (m^2 \cdot d)^{-1}$ was transformed into unit $mmol \cdot (m^2 \cdot d)^{-1}$.

2.4   Environmental Parameters Monitoring

The on-site monitoring of relevant environmental parameters were carried out at the same time of GHG collection. A SX-751 portable multi-parameter water quality monitor was used to measure water quality parameters temperature, DO and pH.

3   Results and Discussions

3.1   Variations of GHG Exchange Flux in FJHPS and XSHPS

3.1.1 CO_2 Exchange Fluxes

Monitoring results in XSHPS showed that CO_2 exchange fluxes in the downstream water surface was higher than the upstream water surface (most values > 0), which indicated that XSHPS was a carbon source of CO_2. For FJHPS, CO_2 exchange fluxes appeared positive or negative values at different times. The CO_2 exchange fluxes ranged from -43.15 mmol·$(m^2 \cdot d)^{-1}$ to 71.17 mmol·$(m^2 \cdot d)^{-1}$, with the range of -43.15 to 37.25 mmol·$(m^2 \cdot d)^{-1}$ and -39.21 to 71.17 mmol·$(m^2 \cdot d)^{-1}$ for the upstream water surface and downstream water surface respectively. Obviously, concentrations of CO_2 released in FJHPS were larger than XSHPS. This is because that FJHPS has run to generate electric power, which stirs water by turbines and releases the dissolved CO_2 in water [12]. A part of water without disturbance directly ran into river channel in the downstream over the rubber dam. This process can release dissolved CO_2 in water again. But XSHPS, has not yet run which doesn't disturb water to release CO_2.

3.1.2 CH_4 Exchange Fluxes

In the downstream water surface of FJHPS, the concentrations of CH_4 exchange flux changed slightly, which ranged from -2.43 mmol·$(m^2 \cdot d)^{-1}$ to 0.94 mmol·$(m^2 \cdot d)^{-1}$. The maximum concentrations of exchange fluxes in the upstream water surface and the downstream water surface were 0.94 mmol·$(m^2 \cdot d)^{-1}$ and 0.72 mmol·$(m^2 \cdot d)^{-1}$, respectively. In XSHPS, the changes of exchange fluxes fluctuated significantly, which ranged from -22.63 mmol·$(m^2 \cdot d)^{-1}$ to 14.71 mmol·$(m^2 \cdot d)^{-1}$ with a maximum in the noon. Friedl et al. have found that warm water can further promote the generation of CH_4 at the water surface [13]. It is obvious that the CH_4 exchange flux in XSHPS was greater than FJHPS. XSHPS has deeper waterbody, submerging more vegetation in the shore. When all the plants decay, they produce more CH_4 [14]. Meanwhile, XSHPS is located in the center of Xiasi Town, domestic sewage can directly enter into the reservoir and led to release more methane gas in waterbody [15] (Fig. 2).

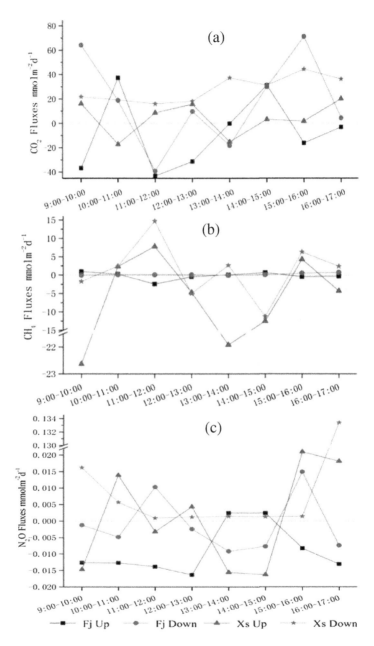

Fig. 2. Day time variations of CO_2, CH_4 and N_2O flux in FJHPS and XSHPS (a) Variations of CO_2 emission fluxes (b) Variations of CH_4 emission fluxes (c) Variations of N_2O emission fluxes.

3.1.3 N₂O Exchange Fluxes

The N_2O exchange fluxes in both hydropower stations were small, which ranged from -0.016 to 0.133 mmol·$(m^2·d)^{-1}$. In FJHPS, the N_2O exchange fluxes ranged from 0.016 to 0.015 mmol·$(m^2·d)^{-1}$, which were obviously smaller than those in XSHPS. The main cause is that XSHPS is located in Xiasi Town, where waterbody receives domestic sewage with high nitrogen. And the amount of N_2O is closely related to TN [16] in the waterbody.

3.2 Correlation Analysis Between Water Chemical Parameters and GHG Fluxes

Water temperature, dissolved oxygen and pH were selected as three water chemical parameters to study their correlations with GHG fluxes.

Water temperature can not only directly affect the gas exchange flux by changing the speed of gas molecule diffusion and solubility in water, but also indirectly affect geochemical process of GHG by changing the metabolic activity of microorganisms in water [17]. In addition, the water temperature can affect photosynthesis and respiration of aquatic plants, resulting in variations of CO_2 exchange flux [18]. The average temperature in both areas respectively was 17.75 °C and 18.5 °C. No correlations between the water temperature and greenhouse gas exchange flux were showed in the studied area (Table 1). It was indicated the water temperature was not a major affecting factor in Autumn.

Table 1. Correlation coefficients between the fluxes of CO_2, CH_4, N_2O and the environmental factor Temperature, DO and pH in waterbody of small hydropower stations in Qingshui River.

Item	Fujiang hydropower station			Xiasi hydropower station		
	CO₂	CH₄	N₂O	CO₂	CH₄	N₂O
Temperature	−0.078	0.057	0.192	−0.173	−0.076	−0.010
DO	0.428	0.414	0.557*	0.331	0.307	0.377
pH	0.506*	0.274	0.495	0.257	0.162	0.102

* Significant correlation at the 0.05 level (bilateral); ** Significant correlation at the 0.01 level (bilateral).

The concentration of DO determines the ways and products of organic matter degradation in water. In the cycle of carbon, organic matters mainly generate CO_2 in aerobic environment and CH_4 in anaerobic condition. In the nitrogen cycle, organic matters exercise the aerobic nitrification and anaerobic denitrification [19]. Correlation analyses showed the DO had a good positive correlation with three parameters, which implied high DO can promote the release of GHGs in waterbody.

The pH value is another important chemical parameter of waterbody, which can affect the release of CO_2 in the reservoir by changing the carbonate balance. It is a significant factor affecting the produce and release of CO_2 and CH_4, which has close relationships with the decomposition of organic matter, microbial activity and biological metabolic activity in waterbody [20]. Results indicated that correlations between pH and GHGs in FJHPS had better positive correlation than in XSHPS

(Table 1). It can be explained that the pH value of waterbody in XSHPS is disturbed by organic matter from sewage.

3.3   Comparisons of GHGs in the Study Area and Other Areas

The GHG flux data summarized in Table 2 were obtained from literatures of different areas. We found that GHG fluxes of tropical lakes or reservoirs of were significantly higher than the other temperature Zone. CO_2 fluxes of frigid zone and temperate were relatively lower. And the CO_2 flux ranges in our studied hydropower stations changed from -43 to 72 mmol·$(m^2 \cdot d)^{-1}$, which was similar with those of frigid and temperate reservoirs. The values of CH_4 flux ranged from -23 to 15 mmol·$(m^2 \cdot d)^{-1}$, significantly larger than the those of other lakes or reservoirs. But the N_2O fluxes ranged from -16 to 130 μmol·$(m^2 \cdot d)^{-1}$, which were significantly lower than those of other lakes or reservoirs. Therefore, we can conclude the running of small hydropower station only promote the rising of CH_4 flux, but slightly affect the CO_2 and N_2O fluxes.

Table 2. Range of greenhouse gas fluxes from the studied area and other areas. (mmol/$m^{-2}d^{-1}$)

Name	Temperature zone	CO_2	CH_4	N_2O	References
Lokka Lake	Frigid zone	11.23−73.44	0.328−7.430	−0. 500−5.797	21
Kevätön Lake	Frigid zone	−1.81−25.06	0.276−12.09 6	−1.702−0. 440	22
Arrow Lake	Temperate	13.82−25.92	0.216−0.665	−0. 023−0.079	20
Cabonga Lake	Temperate	5.01−78.62	0.190−3.370	−0. 390−6.099	6
Maotiao River	Subtropics	−9−77	nd	nd	23
Donghu Lake	Subtropics	−31.97−87.2 6	0.086−8.294	nd	16
Qingshui River	Subtropics	−43−72	−23−15	−0.016−0.13	This study
Curua-Una Lake	Tropical	7.5−227.27	0.125−42.5	nd	24
Petit Saut Lake	Tropical	13.18−238.6 4	0.313−237.5	nd	25

Note: nd is not detected.

4   Conclusions

The fluxes of greenhouse gas (CO_2, CH_4 and N_2O) fluctuate at the water surface of small hydropower stations Qingshui river of Guizhou province in Autumn, during a daytime continuous monitoring. On the whole, the GHG fluxes at the downstream water surface of hydropower station are slightly higher than the upstream water surface, manifesting the more GHG releases at downstream waterbody. The turbine rotating can significantly influence the CO_2 exchange fluxes of waterbody nearby hydropower stations. The domestic sewage and submerged plants can result in decomposition of organic matter to release more CH_4 and N_2O. Through the correlation analysis, the dissolved oxygen and pH influences apparently the GHG in Autumn. The running of small hydropower station only influences obviously the CH_4 fluxes, and therefore can be considered as an effective way to clean energy utilization.

Acknowledgment. This study was financially supported by the International Technology Cooperation and Exchange Fund from the Chinese Ministry of Science and Technology (2012DFA60830) and the National Natural Science Foundation of China (41372354).

References

1. Administration C M: Greenhouse gases status of China atmospheric. China Greenhouse Gas Bull. **12**(1), 38–41 (2012)
2. Stlouis, V.L., Kelly, C.A., Duchemin, E., et al.: Reservoir surfaces as sources of greenhouse gases to the atmosphere: a global estimate. Bioscience **50**(9), 766–775 (2000)
3. Victor, D.G.: Strategies for cutting carbon. Nature **395**(6705), 837–838 (1998)
4. Kelly, C.A., Rudd, J.W., Stlouis, V.L., et al.: Turning attention to reservoir surfaces, a neglected area in greenhouse studies. EOS Trans. Am. Geophys. Union **75**(29), 332–333 (1994)
5. Macintyre, S., Wanninkhof, R., Chanton, J.: Trace gas exchange across the air-water interface in freshwater and coastal marine environments. In: Biogenic Trace Gases: Measuring Emissions from Soil and Water, pp. 52–97 (1995)
6. Louisv, L.S., Kelly, C.A., Duchemin, É., et al.: Reservoir surfaces as sources of greenhouse gases to the atmosphere: a global estimate reservoirs. Bioscience **50**(9), 766–775 (2000)
7. Abril, G., Gurin, F., Richard, S., et al.: Carbon dioxide and methane emissions and the carbon budget of a 10-year old tropical reservoir (Petit Saut, French Guiana). Global Biogeochemical Cycles **19**(4), GB4007 (2005). doi:10.1029/2005GB002457
8. Roland, F., Vidal, L.O., Pacheco, F.S., et al.: Variability of carbon dioxide flux from tropical (Cerrado) hydroelectric reservoirs. Aquat. Sci. **72**(3), 283–293 (2010)
9. Chen, Y.G., Li, C.H., et al.: Carbon dioxide flux on the water-air interface of the eight lakes in China in winter. Ecol. Environ. **15**(4), 665–669 (2006)
10. Jain, A.K., Briegleb, B.P., Minschwaner, K., et al.: Radiative forcings and global warming potentials of 39 greenhouse gases. J. Geophys. Res. Atmos. **105**(D16), 20773–20790 (2000)
11. Lambert, M., Frchette, J.-L.: Analytical techniques for measuring fluxes of CO_2 and CH_4 from hydroelectric reservoirs and natural water bodies. In: Tremblay, A., Varfalvy, L., Roehm, C., Garneau, M. (eds.) Greenhouse Gas Emissions—Fluxes and Processes, pp. 37–60. Springer, Heidelberg (2005)
12. Kemenes, A., Forsberg, B.R., Melack, J.M.: CO_2 emissions from a tropical hydroelectric reservoir (Balbina, Brazil). J. Geophys. Res. Biogeosciences (2005–2012) **116**(G3), 206–216 (2011)
13. Friedl, G., West, A.: Disrupting biogeochemical cycles-consequences of damming. Aquat. Sci. **64**(1), 55–65 (2002)
14. Huntington, T., Aiken, G.: Export of dissolved organic carbon from the Penobscot River Basin in North-Central Maine. In: Proceedings of the AGU Spring Meeting Abstracts (2009)

15. Daelman, M.R., Van Voorthuizen, E.M., Van Dongen, U.G., et al.: Methane emission during municipal wastewater treatment. Water Res. **46**(11), 3657–3670 (2012)
16. Xing, Y.P., Xie, P., Yang, H., et al.: Methane and carbon dioxide fluxes from a shallow hypereutrophic subtropical lake in China. Atmos. Environ. **39**(30), 5532–5540 (2005)
17. Singh, S., Kulshreshtha, K., Agnihotri, S.: Seasonal dynamics of methane emission from wetlands. Chemosphere Glob. Change Sci. **2**(1), 39–46 (2000)
18. Patra, P.K., Lal, S., Venkataramani, S., et al.: Seasonal and spatial variability in N_2O distribution in the Arabian Sea. Deep Sea Res. Part I **46**(3), 529–543 (1999)
19. Huttunen, J.T., Vaisanen, T., Hellsten, S.K., et al.: Methane fluxes at the sediment-water interface in some boreal lakes and reservoirs. Boreal Environ. Res. **11**(1), 27–34 (2006)
20. Krumbein, W.E.: Photolithotropic and chemoorganotrophic activity of bacteria and algae as related to beachrock formation and degradation (Gulf of Aqaba, Sinai). Geomicrobiol. J. **1**(2), 139–203 (1979)
21. Huttunen, J., Mntynen, K., Alm, J., et al.: Pelagic methane emissions from three boreal lakes with different trophy. In: Proceedings of the 4th Finnish Conference of Environmental Science, Tampere, Finland (1999)
22. Huttunen, J.T., Alm, J., Liikanen, A., et al.: Fluxes of methane, carbon dioxide and nitrous oxide in boreal lakes and potential anthropogenic effects on the aquatic greenhouse gas emissions. Chemosphere **52**(3), 609–621 (2003)
23. Wang, F., Wang, B., Liu, C.-Q., et al.: Carbon dioxide emission from surface water in cascade reservoirs–river system on the Maotiao River, southwest of China. Atmos. Environ. **45**(23), 3827–3834 (2011)
24. Duchemin, E., Lucotte, M., Queiroz, A., et al.: Greenhouse gases emissions from a 21 years old tropical hydroelectric reservoir, representativity for large scale and long term estimation. Veranlundgen Int. Vereinigung Theor. Angew. Limnol. **27**(1391) (2000)
25. Galy-Lacaux, C., Delmas, R., Jambert, C., et al.: Gaseous emissions and oxygen consumption in hydroelectric dams: a case study in French Guyana. Global Biogeochem. Cycles **11**(4), 471–483 (1997)

The Application of Internet of Things in Pig Breeding

Minghua Shang[1], Gang Dong[2], Yuanjie Mu[1], Fujun Wang[1], and Huaijun Ruan[1(✉)]

[1] Institute of S&T Information, Shandong Academy of Agricultural Sciences,
Jinan 250100, China
seqsoft@163.com, myj2437@163.com, 191810604@qq.com,
rhj64@163.com
[2] Shandong Institute of Agricultural Engineering, Jinan 250100, China
telagram@126.com

Abstract. A pig breeding IoT system is designed, in view of the human resources, natural resources consumption, the quality and safety problems occurred frequently, the management mode is backward and so on. In this paper, the system architecture, information awareness, system application of the three aspects of pig farming system is introduced. The system can use all aspects of pig farming to sales, has some reference to the intensive farming of pigs.

Keywords: The Internet of Things · Pig-breeding · Large-scale · Application

1 Preface

China is the world's largest pork producer and consumer, the demand of pork promoted the rapid development of pig breeding industry. First, the number of pig herds are increasing. In terms of Shandong, the number of pigs for 24.0181 million in 2000, by the end of 2013 pigs grew to 29.3141 million [1]. Second, the pig breeding way has changed. The traditional way of farming is to take individual farmers as the unit of extensive farming methods [2, 3], farming scale is small, due to the benefits and risks of breeding to promote the industry to accelerate the withdrawal of the family, large-scale breeding has been rapid expansion [4–7]. With the pig population increased and the large-scale farming expansion, the problem of pig breeding is becoming more and more prominent: Pig farming consumes a lot of natural resources, such as human resources and feed, feeding costs greatly increased [8, 9]; In the process of pig breeding, the nitrogen compounds, H2S, NH3, etc., are not strictly monitored, this is not conducive to the growth of pigs, but to some extent, the environment has caused a certain impact. Pork quality and safety problems occur frequently; Traditional production management mode is relatively backward, the data is difficult to collect and deal with it in time, which leads to the lag of production statistics, which has not adapted to the needs of modern production management [10].

Today, the development of Internet of things technology is very fast, has been in traffic, medical, security, environmental protection and other industries have been very good application. Apply the Internet of things technology to all aspects of pig breeding,

D. Li and Z. Li (Eds.): CCTA 2015, Part II, IFIP AICT 479, pp. 548–556, 2016.
DOI: 10.1007/978-3-319-48354-2_56

through a variety of terminals, a variety of forms to allow producers, managers, anywhere access to the relevant information and daily management, will bring unprecedented opportunities for the pig breeding [11].

2   System Framework

Apply the Internet of things technology to the whole process of pig breeding. System is mainly composed of three parts: information perception, data transmission and system application. The system architecture is shown in Fig. 1. Information perception is a collection of information on the breeding environment and individual information, data transmission means that the environmental information is uploaded to the server through the gateway node. Pig individual information through smart phones, PAD, PDA and other handheld devices to collect and through mobile communication networks, the Internet to upload to the server. System application layer is a monitoring platform for pig breeding, it can provide a variety of terminals, a variety of methods of data viewing, equipment management, early warning parameters set, quality traceability and other functions.

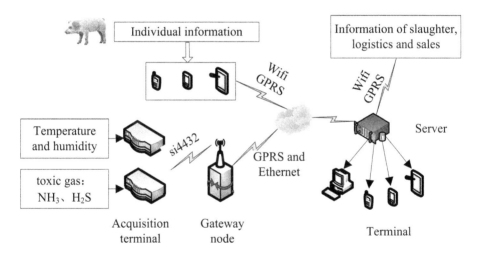

Fig. 1. System architecture

3   Information Perception

3.1   Environmental Information Perception

Environmental information has a certain effect on the growth of pig: pig is a constant temperature animal, if the temperature is too high, the pig will be adjusted to heat, but the environment temperature is too high will make the adjustment range of pigs narrow; if the temperature is too low, it will lead to the consumption of pigs is too large, which leads to the increase of feed consumption. If the humidity in the pigsty is too high, it will reduce the pig's resistance, increase the probability of illness; if humidity is too low, it will cause respiratory disease in pigsty. The harmful gas such as H2S content is too high will cause inflammation and respiratory diseases, high concentration of NH3 will stimulate the respiratory tract mucosa and tuberculosis membrane of pigs, causing conjunctivitis, bronchitis and other diseases. Environmental information has become more and more attention.

3.1.1 Design of Acquisition Node

In order to achieve better monitoring of environmental information, environmental information acquisition node is designed. The acquisition and processing of the multi parameter data of pig breeding environment and the function of wireless transmission are realized. The acquisition nodes mainly collect air temperature, air humidity, H2S, NH3 four parameters, the acquisition of real-time data processing, and is responsible for communicating with the gateway node. H2S, NH3 acquisition nodes as shown in Fig. 2(a), the air temperature and humidity acquisition nodes, as shown in Fig. 2(b). The acquisition node is made of four parts (see Fig. 3), which is composed of the core processing unit, signal processing unit, communication unit and power supply circuit.

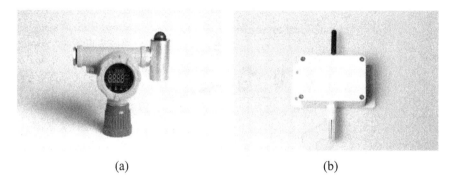

(a)   (b)

Fig. 2. Physical picture of the collection node

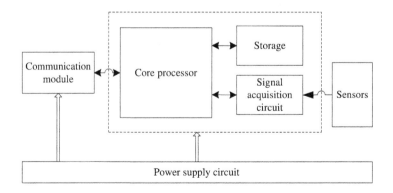

Fig. 3. Structure of acquisition node

The signal processing unit is directly connected with the sensor (including temperature and humidity sensor, toxic gas sensor), and the NH3 sensor selects the NH3 3E Sensoric 1000 SE sensor, and the H2S sensor selects the H2S 3E Sensoric 100 S sensor. NH3 and H2S sensor use the same principle, in the monitoring to the corresponding gas will produce a weak current signal, signal acquisition circuit to get the current signal into voltage signal and amplification, and then get the voltage analog signal transmission to the core processing unit. Temperature and humidity sensor uses the SHT11 type sensor, it is slightly different with the gas sensor, it outputs digital signal, without the need for analog conversion. SHT11 sensor with strong anti-interference ability, low power consumption, fast response, high accuracy. At the 14 resolution, the humidity was ±3 % RH and the temperature accuracy was ±0.4°C. The core processing unit is connected to the temperature and humidity sensor through the I2C bus to obtain the data.

The core processing unit selects STM32 microcontroller, which has the features of low power consumption, low cost and stable performance. In NH3/H2S acquisition node, the core processing unit obtains the voltage analog signal of the signal processing unit through A/D sampling, and after correction, compensation according to the sensor site temperature (different operating temperature, the corresponding current signal of 1 ppm gas) will be converted to the voltage signal to the gas collection value. The core processing unit of the air temperature and humidity acquisition node can obtain the numerical value of SHT11 transmission directly through the I2C bus.

The communication module is composed of Si4432, which is simple, communication distance, supports multi band and low power consumption. The acquisition node and gateway node are composed of Si4432, and the information is transmitted to the gateway node.

3.1.2 Design of Gateway Node

The gateway node is a bridge between the system application platform and the data acquisition node, which affects the stability of the system. The main function of gateway node: access to various information collection node data; establish the system application platform and long connection can maintain real-time communication; in wireless sensor network and communication network platform, the ability to parse and transform the protocol; configuration: intensity collection node address addressable configuration parameters the base configuration properties that can be configured. Gateway hierarchy is shown in Fig. 4.

The gateway node selects the S3C2440 processor of Samsung, which is equipped with the Linux 2.6 kernel and Yaffs2 file system. The gateway node is located in the center of the star network of Si4432, and the data obtained from each node is obtained. After data acquisition, the network joint points analysis data packets, the data is updated to real-time display in the interface of Qt, and the data packets are transmitted to the remote server according to the protocol. Communication unit Ethernet or GPRS two ways and the remote server through the design of the heartbeat packet to maintain a long connection state, uplink transmission data, downlink transmission. The physical map of the gateway node is shown in Fig. 5.

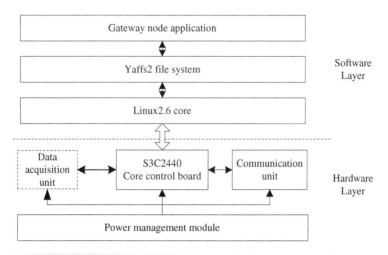

Fig. 4. Hierarchical structure of gateway node

Fig. 5. Physical picture of the gateway node

3.2 Production Management Information Perception

In livestock and poultry breeding enterprises in the production of information is still based on the traditional paper records, not only inefficient, error prone, and the data is difficult to timely collection and processing, resulting in a serious lag in production statistics. Livestock quality and safety traceability system requires the information of the whole industrial chain such as pig production, quality inspection, slaughter, transportation, freezing, sales and so on. In order to solve this problem, we designed a portable pig ear tag reading and writing equipment, and developed a set of APP software which named ZHINONGYUNDUAN based on the Android system, and it is necessary to use. Regardless of which link in the whole industry chain, when the need to enter the information, you can use a portable ear tag reader to read and write the device to read the information, the APP client software to enter the appropriate information, and then click on the preservation of information, you can save the collected data to the system. In this way, the individual and the production, quality inspection, slaughter, transportation, freezing, sales and other information links. The portable ear tag reading and writing equipment is made up of three parts: the core processor, the RFID read and write module and the Bluetooth module, and the structure of the device is shown in Fig. 6.

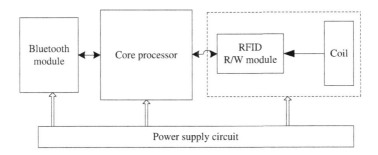

Fig. 6. Structure of portable ear tag reading and writing equipment

When using this system, the first portable ear tag information collection device (see Fig. 7(a)) Bluetooth module and handheld terminal device pairing, portable pig ear tag read-write device using RFID technology, through the non-contact way to obtain pig ear tag information, and ear tag information is transmitted to the APP via Bluetooth (Fig. 8(a)). The corresponding user input information in APP software (Fig. 8(b)), click save, relevant information can be transmitted to a remote server, the information flow is shown in Fig. 7.

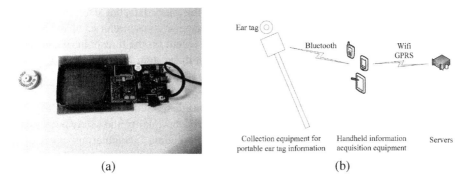

(a)   (b)

Fig. 7. Collection equipment for portable ear tag information

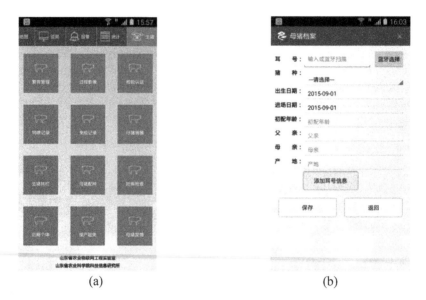

(a)   (b)

Fig. 8. UI of the ZHINONGYUNDUAN APP

4   Application of the System

The application of the system mainly includes: ① Data real-time monitoring: users can view the real-time data of the various pig breeding environment information acquisition nodes, and can view the production management information; ② Historical data query and export: environmental information collection node acquisition of data and the wisdom of the cloud APP access to the data stored in the Server SQL database, the user can choose according to their own needs to query the query data, but also can export the selected time period of data analysis; ③ Alarm management: users can set up the demand of air temperature and humidity, NH3/H2S gas content on the threshold, if the actual measurement value exceeds the threshold range of the system will automatically alarm to inform the user; ④ Equipment management: users can add/remove the environment information collection node, change the equipment (gateway node, environment information acquisition node) according to their own needs ID number.

For the realization of the traceability of pork products quality and safety process, the team based on information management of pig breeding, slaughtering, processing, storage and transportation and other production records, is developed for traceability terminal quality information of pork products (Fig. 9(a)) and electronic scale source (Fig. 9(b)). In the bar code generated by the special electronic scales, the information contained in the whole process of pig raising to pork production, quality inspection, slaughter, transportation, freezing and sales etc. The whole process can be obtained by scanning the two-dimensional code in the quality of pork.

(a)   (b)

Fig. 9. Pork product quality information traceability terminal

5   Brief Summary

This paper according to the actual needs of production, use the Internet of things technology, have developed a set of system for the management of pig farms. The system can collect the real-time data of the air temperature, humidity, H2S, NH3 in the environment of pig breeding according to the requirements of production, and obtain the production management information, and the information of the quality inspection, slaughter, transportation, freezing, sales and other information are linked with the pig individual, and realize the whole process of pork production. Although the network system of pig farms has been developed, it still has a large gap compared with the developed countries, which need to be improved in the future. The rapid development of the Internet of things technology will give the livestock and poultry breeding industry has brought a qualitative leap.

Acknowledgment. Funds for this research was provided by the National Science and Technology Plan Projects (2014BAD08B05-02), Shandong Province Major Projects of Independent Innovation (2014XGA13054), Shandong Province Key R&D Projects (2015GNC110024), Shandong Academy of Agricultural Sciences Technology Innovation Key Projects (2014CXZ09-1).

References

1. Shandong Provincial Bureau of Statistics. Shandong Statistical Yearbook 2014. China Statistics Press, Beijing (2014)
2. Li, X., Ai, H.: Design scheme of livestock and poultry farm. Agric. Netw. Inf. **8**, 28–30 (2012)
3. Lu, C.: Application and prospect of modern livestock and poultry breeding information technology. In: China Animal Husbandry and Veterinary Society Information Technology Branch of the 2014 Academic Seminar, pp. 194–205. China Agricultural University Press, Beijing (2014)
4. Zhu, M., Zhao, P., Liu, J., et al.: Technical support for the modernization of animal husbandry modernization. Chin. Livestock Ind. **2**, 20–23 (2015)

5. Xu, B., Shi, L., Liu, Y.: Study on the scale evolution model of pig breeding industry in China. Agric. Econ. Probl. **2**, 21–26 (2015)
6. Zhang, W., He, Y., Liu, F., et al.: Environmental monitoring system based on internet of things of large-scale livestock breeding. Agric. Mech. Res. **2**, 245–248 (2015)
7. Liu, Y., Yang, Z., Zhao, Y.: The application of the internet of things technology based on RFID in animal husbandry. Heilongjiang Anim. Husbandry Vet. **8**, 15–17 (2012)
8. Yao, X.: The application of the internet of things technology in the large-scale livestock and poultry breeding industry. Agric. Netw. Inf. **10**, 24–26 (2013)
9. Wu, L., Xu, G., Yang, L.: Study on the moderate scale of pig farms in the cost of environmental pollution control. Popul. Resour. Environ. Chin. **25**(7), 113–119 (2015)
10. Yang, B., Song, C., Cai, F., et al.: The design of the monitoring system of the digital aquaculture production process based on the internet of things. Agric. Equipment **2**, 183–184 (2015)
11. Xiong, B., Yang, L., Pan, X.: Research progress of the application of the internet of things technology in animal husbandry in China. In: China Animal Husbandry and Veterinary Society Information Technology Branch of the 2014 Academic Seminar, pp. 149–159. China Agricultural University Press, Beijing (2014)
12. Li, Y., Liu, Y., Ni, C., et al.: Temperature acquisition system based on SHT11 sensor for low power tree farm. Agric. Mech. Res. **1**, 204–208 (2013)

Research and Exploration of Rural and Agricultural Information Service – Taking Shandong Province as a Case

Jia Zhao, Jianfei Wang, and Wenjie Feng[✉]

Institute of Information Technology, Shandong Academy of Agricultural Sciences,
Jinan 250100, People's Republic of China
zhaojia9821@qq.com, 277059693@qq.com, fengwjcn@qq.com

Abstract. Rural and agricultural information service is the fundamental condition for agricultural information development. Moreover, it is a effective way to narrow the effectiveness of the "digital divide" and integrate the urban and rural development. In this study, taking Shandong Province as an example, the service platform, service tools, service support and service teams etc. are described. And, the rural agricultural information service model based on "Service platform as the core, service means as basic, service carrier as link, service team as support and public and market integration" is put forward in this study. This rural agricultural information service model is benefit for the agricultural information; accelerate the pace of rural agricultural information construction.

Keywords: Rural and agricultural · Information service · Research and explore

1 Introduction

With the accelerating rural agricultural informationization development, rural and agricultural information has led to support the direction of the transformation and upgrading of China's modern agriculture, and finally for the purpose of the application of information and services, and continuously infiltration. Rural agriculture information service is knowledge-intensive services of the integration of knowledge, intelligence, technology, information and funds in the foundation of the abundant information resources, convenient information network and modern information technology, which has the characteristics of wide demand, fast growth, high efficiency, energy saving, pollution-free, etc. Only by constantly sum up experience, explore new problems, new trends, new modes and new mechanisms of providing information services for rural and agricultural under the new situation, only continuous innovation, and to achieve better Agricultural information.

As a big agriculture province, Shandong province highly pay attention to the rural agricultural informatization work, and has been developing rapidly in the aspect of infrastructure, information resources, demonstration, application and service system construction etc. In 2010, Shandong province, in 2010, Shandong province positively research and explore rural agricultural information service, solve the problem of the last

D. Li and Z. Li (Eds.): CCTA 2015, Part II, IFIP AICT 479, pp. 557–562, 2016.
DOI: 10.1007/978-3-319-48354-2_57

kilometer of information service, effectively narrow the gap between urban and rural informatization development, thus was approved as the first batch rural agricultural informatization demonstration province pilot provinces jointly by national ministry of science and technology, the Organization Department of the Central Committee of the CPC, ministry of industry and information. Take Shandong province as an example to explore. After many years of the development in the leading mode of the government, rural agricultural information service gradually move towards to the parallel of public service and market service, rounded by different types user's needs, thus produced a large number of different forms and contents of information service organization, rural agricultural information service industry has been established. Which is mainly includes the service platform, service means, service carrier, service team, etc.

2   Rural and Agricultural Information Service Platform

2.1   Construction Ideas

To build a platform with Shandong features, industrial service focus on "the integration of resources" and "efficient service" "mechanism to explore" and other key nodes to break through and innovation. Accordance with the "platform shift" principle, all kinds of information resources integration and development of agriculture in rural areas are carried out, meanwhile, effective cohesion with advantage industry professional information service system and basic information service station is stressed. So that, the "low cost, convenient" service information villages and households are speeded up, personalized information needs of farmers are satisfied, and new rural construction and development of modern agriculture is serviced.

2.2   Platform Overview

Service platform relies on provincial party distance education network, and the development focus on ten content area and top ten industrial information service system. Scientific classification of column plate design is based on information services on websites. The columns and thematic of platform includes agricultural production, competitive industries, wisdom agriculture, commerce, village area, primary site, distance education, typical applications, convenience services. The platform named Qilu agricultural science and technology network (www.qlsn.cn) has been initially built and run, which provides service interfaces for various application systems taking advantages of real time and large capacity of the Internet. This platform becomes mainly rural agricultural information service window of Shandong Province.

3   Rural and Agricultural Information Service Means

3.1   Hotline Service

12396, which is jointly promoted by the Ministry of science and technology, industry and information technology, is the unified agricultural science and technology information service number. Shandong Province in December 2008 opened "12396" service hotline. Fixed and mobile telephone farmers live in any areas of Shandong could consult to the experts by calling 12396, and only need to pay the ordinary telephone costs. It's effective to solve the problem that farmers have difficulty to use agriculture technology for a long time.

However, with the development of science and technology, the traditional hotline service has been far from meeting the needs of farmers for information. We have merged the hotline and radio broadcast through coupler used multimedia speech fusion technology by making full use of 12396 hotline at the beginning of 2013. We are jointly with the Shandong radio and television broadcast channels to create the 12396 green voice in the inter broadcast. Farmers can participate in modern country live program by dialing 12396 after dialing 9. We have changed the traditional pattern one to one into one to N by strengthening the interaction between the host and the experts and the audience, and disseminated the knowledge of agriculture and technology to millions of households directly. At the same time, variety kinds of ways, such as microchannel, microblog, QQ group, were set to connect farmers and experts, to connect farmers and markets, to connect farmers and government.

3.2   Computer Remote Video Service

Video conference system, which was consider as the important application of interactive audio and video communication, is becoming more and more widely used with the development of computer communication network and audio and video coding and decoding technology. Combined efficient H.264/AVC video encoding algorithm and high level meeting of GIPS voice engine, the remote video conferencing system has many advantages, such as excellent video quality, high level of automatic echo suppression, automatic gain control, Mike volume automatic adjustment, noise suppression and so on. The B/S structure, advanced video encoding and transmission were adopted in this system. The remote video services with a low bandwidth clear and smooth image was achieved through the Internet. The farmer could land the system and connect the video anywhere after get the ID.

The software of system server-side is installed on the specialized video server of Service Platform which accessing to internet through 100 M fiber, installation and distribution process of the software is simple, and it is easy to use. It can achieve instant communications, people meeting, conference recording, file sharing and other functions after installation and it provide services through many ways, like experts agents, booking agents, emergency relief, teleconferencing, training, etc.

3.3 TV Set-Top Box Service

Customers use IPTV and remote control to synchronize browsing Qilu Agriculture network information, demand agricultural science and technology video, implement web surfing by "TV + set-top box" approach. Currently, IPTV provides expert auditorium, live television, information services, agricultural science and technology, industries, commerce, web surfing and other columns.

3.4 Cellphone Service

Mobile Client Service provide more timely targeted services, like policies and regulations, entrepreneurial wealth, rural alerts, agricultural science and technology, market, encyclopedias and other information services. Currently, Qilu rural integrated information services client, agricultural facilities things monitoring application client (Shed commander, pigs commander, chickens commander), Shandong Leisure Agricultural Services client, Prices for Agricultural courier client, mobile video client. Not only the farmers can directly scan two-dimensional code to download the client by log Qilu three rural network client download interface, but they can also directly download the client from the Android, Apple Store client.

Mobile Video Services, the customer can get online video support via phone video by download client. The famer can achieve two-way communication using mobile phone through the video system. Farmers can consult various types of agriculture-related information, ask experts for a prescription for the pest, keep abreast of the latest agricultural information and solve problems which they encounter in agricultural production faster, and all above this greatly improve the efficiency and quality of the solving of various types agriculture problem.

4 Rural and Agricultural Information Service Carrier

Basic information service station is an important carrier of rural and agricultural information service, and meanwhile, it is a window to understand the platform information and accept the information service for grassroots organization, agricultural enterprises and farmers. The basic information service stations achieve flat information service with a direct connection with the service platform to and the effective docking with the majority of farmers.

4.1 Construction Standards

Basic information service station constructions need to meet the following requirements and standards, namely: a fixed place, a set of information equipment, an information technology correspondent, a set of training system, a set of management system and a long-term mechanism, mainly divide into two major categories of professional information service stations and integrated information service stations.

As the important link of service platform and farmers, professional information service stations mainly relying on the existing levels of agricultural science and Technology Park, agricultural leading enterprises, professional cooperative organizations, professional associations, large breeding, agricultural capital construction management entity, are a necessary part of the whole rural agricultural information service chain. The integrated information service stations are the effective measures to promote the information to villages and households. Furthermore, it is an important part of rural and agricultural informatization construction. On the existing party members and cadres of modern distance education based on the grassroots service station, to select the service station with better hardware condition, staff members and facilities. To encourage the combination of public service and paid services in basic information service station, and the integrated services of technology, material and information, exploring a long-term mechanism to combine the public welfare and market.

4.2 Site Administration

According to the actual needs of the grassroots information service station, the service platform is designed to set up the "grassroots" structure and the contents of the plate, including member account management, site registration information, site information management and other functional modules. Through the member account management module, the system administrator can view and review the member's personal information. In the site registration information module, the system administrator can query, preview, approval of the site registration information submitted by the member, in the "information management" module, the system administrator can query, review the information submitted by the members of the site, the audit, withdrawal and other operations.

4.3 Service Models

According to local characteristics, the main models of information service stations was explored, such as agricultural technology station for technical guidance, agricultural materials sales department of information and material combination, the leading enterprises for leading-led, demonstration park for radiated and proceed, farmers' cooperatives organization for the market growth. In the promotion process, gradually formed a "integrated" service mode, and insisted on multi-pronged, multi-policy, so that the information spreaded to village and farmer to increase information efficiency.

5   Rural and Agricultural Information Service Team

The construction of rural agricultural information service team mainly includes operation team construction, promotion team construction and specialist team construction. Operation team construction mainly rely on the scientific research institutions, public welfare operation team construction consisted by government departments and marketization operation team consisted by professional company. Promotion team construction

includes basic science and technology information officer, distance education information officer and primary agricultural technology personnel and strengthens the technical
training and support for the messenger of information service station. Specialist team
construction consists of not only the authoritative experts come from the institutions of
higher learning and scientific research institutes, but also the technology experts who
have practice experiences and can solve practical problems. These experts who are
responsible for technical guidance covers the fields of planting, breeding technology,
storage processing, pest control, construction of modern agriculture and so on.

6   The Effect on Rural and Agricultural Information Service

As the foothold of agricultural informatization, the rural and agricultural information
service to grow out of nothing, by weakly to the process which is strong, is experiencing
a long development period. At present, the rural and agricultural information service is
accelerating the coverage of penetration, the changes from heavy construction to heavy
service, from single service to integrated services, from passive service to interactive
services are happening. Through continuous exploration and summary, the rural agricultural information service is more mature following the thinking "giving priority to
public welfare, the market as supplement, and The market in turn auxiliary public
welfare".

Fully integrated and use the existing information resources, and taking the service
platform as the core, the service carrier as a link and service team as the support, on the
basis of service means, improving the level and ability of public service, at the same
time, actively exploring the market service mechanism, accelerates the low cost and
high efficiency dissemination of information in vast rural areas, and to realize the zero
connection between farmers and information.

Acknowledgment.   This work was supported by the National Science and Technology Support
Program (2014BAD08B05-02), the Shandong Academy of Agricultural Sciences Youth Scientific
Research Fund Project (2014QNM25).

References

1. Report to the Eighteenth National Congress of the Communist Party of China. Firmly March
   on the Path of Socialism with Chinese Characteristics and Strive to Complete the Building of
   a Moderately Prosperous Society in all Respects. People's Daily, 8 November 2012
2. Li, D.: China's Rural Informatization Development Report. Publishing House of Electronics
   Industry (2014)
3. Ruan, H., Feng, W., Chen, Y.: Construction of Integrated Service Platform for Agricultural
   and Rural Informatization. China Agriculture Press (2015)

Research on Agricultural Development Based on "Internet +"

Wenjie Feng, Lei Wang, Jia Zhao, and Huaijun Ruan[✉]

Institute of Information Technology, Shandong Academy of Agricultural Sciences, Jinan 250100, People's Republic of China
fengwjcn@qq.com, 375901677@qq.com, zhaojia9821@qq.com, rhj64@163.com

Abstract. Agricultural modernization is an important way to construct modern agriculture. The development of intelligent terminal, mobile internet, communication technology, internet of things is being applied to agricultural production. It will greatly improve the level and degree of agricultural informatization, promote agricultural informatization development, improve agricultural production efficiency and improve agricultural production efficiency and promote the development of agricultural information technology and new technology.

Keywords: Internet + · Internet of things · Agricultural information

1 Introduction

"Internet +" is to use the Internet thinking, Internet technology to promote the development of various industries and the transformation and upgrading, regardless of which kind of traditional industries, there is the opportunity to "Internet +", in the process, the core is the internet. China's agricultural development is lagging behind the industry, the government has introduced many documents to promote agricultural reform, mobile Internet communication technology in rural applications is also becoming more and more extensive, agricultural information is also changing the whole process of agricultural industry, and promote the optimization and upgrading of the industrial chain.

2 The Internet + to Provide a New Platform

"Internet +" is to use the Internet thinking, Internet technology to promote the development of various industries and the transformation and upgrading, regardless of which kind of traditional industries, there is the opportunity to "Internet +", in the process, the core is the internet. China's agricultural development is lagging behind the industry, the government has introduced many documents to promote agricultural reform, mobile Internet communication technology in rural applications is also becoming more and more extensive, agricultural information is also changing the whole process of agricultural industry, and promote the optimization and upgrading of the industrial chain.

D. Li and Z. Li (Eds.): CCTA 2015, Part II, IFIP AICT 479, pp. 563–569, 2016.
DOI: 10.1007/978-3-319-48354-2_58

Internet industry is the symbol of the Internet technology, business model, organizational approach to become the standard configuration of various industries. Three key technologies: ubiquitous terminal applications, an unprecedented strong background cloud computing capabilities, and continuously upgrade the broadband network constitutes the basis of the industry internet. The core competitive advantage of the relevant enterprise is the lower cost of the sensor, data storage and faster data analysis capabilities, the intelligent machine, big data analysis and other fields have accumulated and in-depth study of the enterprise will get a good opportunity for development.

China's broadband construction has made great progress in 2013, the State Council issued the "broadband China" strategy and implementation plan, proposed broadband network to become the new era of China's economic and social development of strategic public infrastructure, accelerate the rapid and healthy development of broadband infrastructure, and increase the speed of access to rural households, rural broadband access. In the next few years, rural areas will become the main source of Internet users in China, the mobile phone has become the main platform for the popularization of the Internet and the realization of the information technology.

The Internet into China 20 years, profound changes bring a full range of economic development, agriculture, transportation, energy, integrated into the health and education industries, has accumulated to a touch of Fayin burst change time, the Internet has become a new force on behalf of the engine of economic development and innovation driven development. The future, things will connect everything, O2O accelerated to open a new business model, industry cross-border cooperation will continue to innovate, to consumer Internet industry migration evolution, the Internet will deconstruct each industry, stimulating industrial upgrading, the characteristics of almost all industries will present the pan internet.

3   The Main Problems of Agricultural Development

In recent years, with the rapid development of our economy and the improvement of people's life, the domestic demand for food is becoming more and more diversified, and other aspects of the food industry is also increasing. A lot of changes in the supply and demand, the price of food imports increased rapidly in 2014, China's domestic demand for soybean 87 million tons, of which 75 million tons, 11 million tons of soybean consumption of about 80 % tons, of which 14 million tons, 4 million tons of domestic output, 6 million 400 thousand tons of cotton consumption, the output is 1 million 300 thousand tons, imports accounted for 20 % of domestic consumption, the comprehensive grain self-sufficiency rate has exceeded 95 %. According to the "national food security needs of long-term planning outline" forecast that by 2020 China's grain gap may reach 32 million 500 thousand tons.

On imports of soybeans and other staple agricultural products, the high degree of dependence of imported food sources are too centralized, constitutes a serious threat to food security in china. In addition, grain production in China is still faced with many negative factors: the scarcity of water resources threaten agricultural irrigation; rapid growth in labor costs against the enthusiasm of farmers; yield slow growth; bijiexiaoying caused grain acreage to expand the limited development of agricultural production mode; lag; land pollution problems resulting in food production and crops has reached alarming proportions. According to research report released by the Chinese Academy of Sciences Institute of ecology in 2012, China is currently subject to cadmium, arsenic, chromium, lead and other heavy metal contaminated land area of nearly 20 million hectares. In January 6, 2015, the Ministry of Agriculture said China will start a staple potato strategy, not China is the food safety problem of choice under the grim situation.

With the continuous improvement of the degree of external dependence of agricultural products, the area of cultivated land is gradually decreasing, and the situation of the people and the land is much worse. According to the World Bank statistics, in 2014 China's agricultural population accounted for as high as 45.6 % of the total population, while the proportion of agricultural population in developed countries is generally low, such as the United States only 18.6 % of the population engaged in agricultural production. With that, the average per capita arable land area of China in 2014 is only 0.22 hectares, which is lower than the level of 1.1 hectares in the United States, even lower than the world average of 65.2 hectares.

The agricultural population continued to decrease and the phenomenon of aging is obvious in recent 10 years, with the rapid development of industrialization and urbanization and the difference between urban and rural infrastructure and social life, rural youth are flocking to the cities (towns), the younger generation is no longer willing to engage in agricultural production, the proportion of working age population is declining, the age structure of agricultural population is gradually changing with the aging population. According to the National Bureau of statistics released the "2013 national migrant workers monitoring survey report" shows that in 2013 the new generation of migrant workers (1980 and after the birth of) a total of 125 million 280 thousand people, accounting for 46.6 % of the total migrant workers, accounting for 1980 and the proportion of rural workers born after 65.5 %. In the statistical period, 87.3 % of the new generation of migrant workers not engaged in any agricultural production, most of the main working away from home. "Migrant workers" heat has not been back, plus the emergence of China's demographic dividend turning point, the agricultural labor force in particular, to continue to reduce the age of school-age.

The development of rural economy has stalled and the improvement of production efficiency has become more and more slowly. Since 1997, the average annual growth rate of agricultural added value was only 6.88 %, far below the 12.45 % growth rate of industrial growth and 14.21 % of service industry. And this is a relatively high proportion of rural labor and the lower agricultural production efficiency, the 2013 end of the total employment of agricultural population in China, and the proportion of the total employment population reached 34.8 %, while in 2011, only 1.2 % in the United Kingdom, Japan in 1990 has only 5.9 %. From the perspective of income, rural residents in 2013, the first time beyond the income of household management, accounting for 45.2 %, compared to the display of income, rural household management does not have the advantage. Agriculture needs to reform, and vigorously improve the level of agricultural intensification, in order to improve the efficiency of rural areas and farmers' income levels.

4   The Comprehensive Reform of Agricultural Development

With the development of land circulation and scale management, the new business entities such as cooperatives, family farms, and major industries are rapidly formed, which correspond to the emerging market players. The agricultural information industry is developing rapidly, and the modern agricultural products circulation, which is the representative of electronic commerce.

4.1   Land System Reform

In November 2014, the central office of the State Council issued "on the guide of rural land management rights orderly transfer of agricultural scale operation and development of opinions", documents based on register right, ownership, contract rights, management rights division of powers, to guide the orderly transfer of land management rights, the development of various forms of moderate scale management. Land transfer reform is the important part of the reform, the reform of state-owned enterprises, the important part of comprehensive deepening reform. Three right separation (ownership, contract rights, management rights) is a major practice of agricultural reform, China's rural areas are facing unprecedented a comprehensive reform.

4.2   Business Subject Reform

Cultivate new agricultural business entities is the basis of the development of modern agriculture. Under the background of industrialization and new urbanization, the rural labor force to urban and rural labor transfer to cities and towns and the two or three industries, the reduction of agricultural workers, and the development trend of land scale management, will promote the transformation of business entities, the future of cooperatives, family farms and modern agricultural enterprises will become the main force of agricultural management.

Country in the policy and financial support, will further accelerate the formation of new business entities. Policy support, the Chinese Communist Party in the third plenary proposed to encourage the development of professional cooperation, joint stock cooperation and other forms of farmers' cooperatives, guide the operation of the norms, focus on strengthening capacity building. Encourage local governments and private investors to set up a financing Guarantee Corporation to provide credit guarantee services for the new agricultural business entities. Increase the new occupation farmer and new agricultural business entities lead people's education and training". Financial support, in February 2014, the people's Bank of China promulgated the new agricultural management, such as family farms and other financial services guidance, requires financial institutions to increase the credit support for new agricultural business entities, such as family farms, and make an inventory of the amount of funds to support family farms and other new agricultural business development.

4.3   Agricultural Business Model Reform

The development of smart phones and wireless Internet technology to promote mobile phones become the main platform for rural information. 2014 rural Internet users use the proportion of mobile Internet access to 84.6 %, the development of mobile Internet technology to promote the rapid coverage of Internet technology in rural areas, mobile Internet platform as a rapid release of the advantages, but also to the Internet technology and all kinds of Internet access to rural areas can be.

Agricultural industry has a large market space, industrial backwardness, the information asymmetry is more serious, the large scale of the user, the transaction process is longer, the transaction costs are high, the transaction is highly sustainable and so on, so the potential of the Internet is huge. From the domestic and international agricultural development trend, the Internet and agriculture has begun to accelerate the integration of agriculture, the Internet era has arrived, the Internet is the agricultural industry chain comprehensive transformation, from agricultural sales, intermediary services, the transfer of land to agricultural production, sales of agricultural products, with the Internet thinking of agricultural enterprises from all sectors of the agricultural industry chain on the active layout try to use the Internet, explore various business models.

5   "Internet +" to the Reform of Agricultural Modernization

2015 central rural work conference to accelerate the modernization of agriculture as the theme, pointed out that we should adhere to the reform as the driving force to technology as the lead, to promote the development of agriculture, transfer mode, adjust structure. Further emphasis on increasing the intensity of reform and innovation. Reference to the development of modern agriculture, the scale, industrialization, modernization, industrialization and gradually formed, the future will further extend to the information, in order to build a modern agriculture, accelerate the transformation of agricultural development mode as the core of agricultural information into a bright spot. From the development of domestic and international, the Internet has begun to merge with the depth of agriculture, the Internet is a comprehensive transformation of agriculture in order to improve the efficiency of agricultural industry and agricultural informatization level.

5.1   Industrial Chain Business Model Change

Agricultural Internet era has come, the Internet is a subtle way to transform the whole process of agricultural industry chain, and promote the optimization of the industrial chain and improve the efficiency of the. Occur on the basis of effective promotion of large-scale operations, the Internet technology is expected to produce the whole process of agricultural products from production to effectively integrate, involving the whole process of agricultural product traceability system, cold chain fresh, brand agriculture, etc. In the Internet mode, the agricultural industry chain will form a new business model, to provide more space for the integration of agricultural and agricultural products market trillions.

5.2   Opportunities for Agricultural Service Platform

With the advance of modern agriculture, the acceleration of land transfer, the increase of the main body of the new intensive management, and the way of agricultural modernization. New agricultural enterprises put forward higher requirements on the agricultural product supply and service. Coupled with China's huge market capacity of agricultural fertilizer, seed, feed, only three kinds of size over one trillion yuan. In recent years, Internet companies are "going to the countryside", to seize the rural electricity market, in the objective to cultivate rural business, stimulate the vitality of rural electricity providers to help agricultural development of the electricity supplier. At present, agricultural electricity providers model is the main mode and the third party business platform of agricultural enterprise self mode etc.

5.3   Information Platform Based on Internet of Things

Things are recognized as the world's third wave of world information industry after the computer, Internet and mobile communication network. It is based on perception as the premise to realize the network of people and people, people and things, things and things. And the farm is through the sensing device to obtain crop information, on the basis of all kinds of network transmission, the central system for remote operation, information awareness - network transmission - decision support - remote control is the four basic chain of agricultural things. The essence of agricultural informatization is to transform the agricultural production from the traditional mode of labor to achieve efficient production by means of information, and to construct the Agricultural Internet of things that people and things are fully interconnected. From a broad sense, all agricultural information form of expression are in the category of agricultural things, the Internet of things is the soul of agricultural information. Throughout the country's latest application situation, the Agricultural Internet of things in the four areas of the main effect.

5.4   E-commerce Platform for Agricultural Products Circulation Changes

2015 rural work conference pointed out that the circulation of agricultural products to support innovation, business, logistics, trade, finance and other enterprises to participate in the construction of agricultural e-commerce platform. 2015, the Ministry of agriculture also rural e-commerce as an important means to improve production efficiency, change management mode, the main business of agricultural production and business skills training, organization of agricultural production and business entities and business enterprises docking, to carry out agricultural e-commerce pilot and other aspects of the work.

6   Conclusions

The arrival of the information economy has brought new infrastructure, new elements, the establishment of the new division of labor, the establishment of a new agricultural product circulation model to provide a possible. Under the catalysis of the Internet, the distribution mode of agricultural products is changing, the new mode of electronic commerce is the main form of the rapid rise, in the main form of circulation, the organization, the impact of upstream and downstream, etc.

Acknowledgment. Funds for this research was provided by the National Science and Technology Plan Projects (2014BAD08B05), Science and Technology Projects of Shandong Province (2013GNC21006), Independent Innovation Projects of Shandong Province (2014ZZCX07104).

References

1. The CPC Central Committee and State Council on promoting the steady development of agriculture in 2009 to develop the sustainable income of farmers if the. 2008 (12) (1) (2009)
2. Yan, X., Wang, W., Liang, J.: Beijing municipal facilities and agricultural application mode to build. J. Agric. Eng. (4), 149–154 (2012)
3. Li, D: The internet of things and the wisdom of agricultural engineering. Agric. Eng. (1), 1–7 (2012)
4. Zhang, M., Chen, P.: China's agricultural networking status quo, challenges and thinking about. China's Investment Sci. Technol. (9), 38–41 (2012)
5. Shi, L., Chen, Z.: Cover of the Chinese internet of things in the wisdom of the application of Agricultural Mechanization in agriculture research **6**, 250–252 (2013)
6. Yuan, C.: Based on the industrial economics perspective of China's Internet of things industry development analysis academic exchange (7), 115–118 (2011)

Research and Design of Shandong Province Animal Epidemic Prevention System Based on GIS

Jiabo Sun, Wenjie Feng[✉], Xiaoyan Zhang, Luyan Niu, and Yanzhong Liu

Institute of Scientific Information, Shandong Academy of Agricultural Sciences,
Jinan 250100, China
sjbsd@qq.com, fengwjcn@qq.com, 239491965@qq.com,
nly83412@126.com, 773716510@qq.com

Abstract. Animal husbandry has become one of the important pillars of agricultural economy. In the meantime, along with the rapid development of animal husbandry, all kinds of animal disease hazardous to health of livestock and poultry appear, such as High Pathogenic Avian Influenza (HPAI) and Foot and Mouth Disease (FMD). Therefore, establishment of animal epidemic prevention system is very necessary. Geographic Information System (GIS) has been widely used both in the field of disease control and prevention and livestock and poultry epidemic prevention for its strong ability of spatial analysis and visualization analysis. Despite all this, provincial livestock and poultry epidemic prevention system based on GIS is rare. To obtain animal epidemic information accurately and timely, and improve the mechanism of animal outbreaks, combing with the reality of Shandong province, an animal epidemic prevention and early warning forecast system is developed. The purpose of the system, the system structure, function module, and development platform are analyzed. Combined with GIS, handheld mobile GIS/GPS, and GPRS/CDMA, the system has proved its flexibility, stability, convenience, and easy extensibility, assuring effective implementation of animal epidemic monitoring.

Keywords: Animal epidemic prevention · GIS · Information system

1 Introduction

Animal by-products economic development is one of the important direction of our country's rural economic development. In recent years, animal husbandry in Shandong province keeps the sustainable and healthy development, for example, animal husbandry output value, the total output of meat, eggs and milk, livestock products export and other major economic indicators among the top. Animal husbandry has gradually transformed from subsistence to commercial, from traditional to modern, and has become the most dynamic and potential of the pillar industries of the rural economy, and is also an important way of farmers to get rich.

Meanwhile, animal disease has become an important factor which restricts animal husbandry economy rapid growth, prevents animal by-products to expand exports, and

© IFIP International Federation for Information Processing 2016
Published by Springer International Publishing AG 2016. All Rights Reserved
D. Li and Z. Li (Eds.): CCTA 2015, Part II, IFIP AICT 479, pp. 570–578, 2016.
DOI: 10.1007/978-3-319-48354-2_59

threatens quality and safety of animal origin. Major animal disease prevention and control not only directly affects the animal husbandry development and farmers' income, but also relates to people's physical health, public health security and social stability. Therefore, to strengthen animal epidemic prevention system and animal disease traceability system construction, improve animal epidemic prevention, monitoring, quarantine supervision system, establish animal disease risk assessment and early warning forecast mechanism is of great significance, which is also the necessary premise of to achieve animal husbandry quality security, disease security and ecological security.

The application of information technology in livestock and poultry epidemic prevention, supervision and management has become the trend of the development of animal husbandry industry. By introducing professional data management module, establishing basic database, executing digital management, livestock and poultry epidemic prevention information platform is developed. GIS is an auxiliary space data information management system, combining unique visual effect, the map geographical analysis function, and database operations together [1]. It is an important means of realizing digital decision-making management, and provides strong theoretical support and technical support for epidemic prevention and control from the perspective of spatial evolution process analysis and forecast. GIS is mainly used in the field of animal epidemic prevention, which can be summarized as the following respects: (1) production of livestock epidemic spatial distribution thematic map, which means visually displaying epidemic analysis results in the form of space distribution thematic map and statistical charts, on the basis of GIS spatial analysis and visual display; (2) forecast of animal epidemic situation and development trend, which means using the multi-factor comprehensive analysis and statistical prediction model to predict and evaluate popular trend, space accumulation and dispersion model, and outbreak impact in a certain time; (3) control and planning of Animal epidemic prevention resources, which means optimizing the resources distribution of epidemic prevention with the help of GIS; (4) animal husbandry and outbreak early warning, forecast, and information publishing, which means the use of GIS for risk factors analysis and positioning, and early warning in related areas.

Domestic and international numerous agriculture departments have been using GIS for animal epidemic monitoring and early warning. The United States established the National Animal Health Report System (NAHRS) in 1996. Australia's National Animal Health Information System (NAHIS) is fully functional and works well. Since 1991, New Zealand began to build information system for emergency animal disease control Epiman-IMS [2, 3]. In China, the national center for epidemiological studies was established and put into special funds to establish the national animal health information system in 2000, an integration of animal health information management and animal heath GIS. National SARS control and warning system was built in 2003, which supplied a modern means of prevention and control work. Chongqing city major animal epidemics GIS system was established in 2011. The application of GIS in the health filed at home and abroad has proved that animal husbandry GIS system is the inevitable trend for the scientific and automatic animal epidemic information management [4–6]. Although, there is a lack of domestic provincial animal husbandry outbreak early warning system based on GIS. The existing livestock and poultry epidemic monitoring mechanism in Shandong province is given priority to artificial investigation and reports submit, which is free from the technology of GIS, GPS, and big data, reducing the sensitivity, timeliness, and traceability of the epidemic monitoring.

2   System Framework

2.1   System Objectives

As the first step to develop the system, we identify the core requirements and objectives the system seeks to meet, which includes four aspects: (1) Integration of vector data, images, videos and statistical data, which is used to establish multi-source information database; (2) Development of livestock epidemic rapid acquisition subsystem based on portable mobile intelligent terminal (LERCS), which combines embedded GIS real-time positioning, wireless communication, video capture, barcode recognition technology together to achieve real-time collection, fast storage and instant report of epidemic information; (3) Development of provincial livestock epidemic monitoring and early-warning subsystem (LEMES), which demonstrates the popular trend and epidemic law, assess and predicts the harmfulness, and achieves rapid feedback and control in a timely manner of livestock epidemic; (4) Development of major livestock epidemic emergency decision support subsystem (LEEDSS), which supplies decision support for major animal epidemic response administration, and configures epidemic prevention resource optimally on the basis of LEMES.

2.2   GIS Platform and Development Mode

ArcEngine 9.3 second-development component, from ESRI' ArcGIS series products, is chosen as GIS development platform used on the Windows Server 2008 operation system. In consideration of multiple source data types, database is constructed on the basis of Oracle 9i. Visual C#.NET is selected as programming language.

There are three common development modes, which are Client/Server (C/S), Browse/Server (B/S), and Mobile client/Server (M/S). Here, we integrate C/S, B/S and M/S together, provide data sharing based on the unique data source, switch to appropriate mode when facing different users and application purposes, which guarantee the integrity and flexibility of the system.

2.3   Spatial Database and Non-spatial Database

Livestock epidemic multi-source information database contains two major data types:

(1) Spatial Database, which can be divided into basic geographic information and thematic geographic information about animal epidemic information, the former contains administrative boundary, roadway, and river system, while the latter contains the distribution of species group, livestock and poultry farms, disease communication media, and epidemic prevention resources.

(2) Non-Spatial Database, which covers a multitude of data types, for example, attribute data (the density of livestock and poultry, disease characteristics, traffic condition, environmental condition, et al.), image, and video.

2.4 System Architecture

As is shown in Fig. 1, the overall structure of provincial animal epidemic prevention system includes four levels: The first level is the foundation layer, including a series of hardware and software, such as host device, operating system, database, Internet, Intranet, VPN; The second level is the data layer, containing spatial database and non-spatial database which are established in the form of vector data, image data, video data and statistical data; The third level is the support layer, which supplies developing environment, GIS platform, workflow engine, security opponents, short message service to construct livestock epidemic early warning mechanism; The fourth level is the business

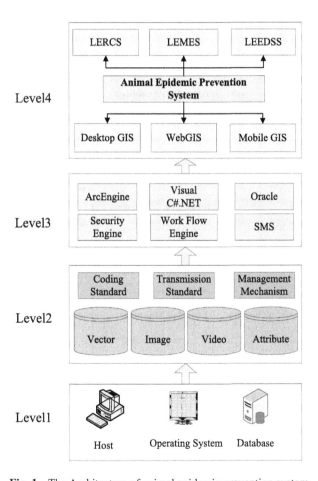

Fig. 1. The Architecture of animal epidemic prevention system

layer, respectively developed in the mode of C/S, B/S and M/S, and divided into three subsystems, LERCS, LEMES, and LEEDSS.

3   System Function

3.1   LERCS

This subsystem is based on the embedded terminal device and can be used to acquire and transmit typical epidemical information. The standards of epidemic information collection are in accordance with the existing epidemic reporting requirement. Information acquired by LERCS refers to disease information, epidemic information, and animal die information in the form of image, video, QR code, and character. Meanwhile, information acquired can be transmitted in two means, one of which is by connecting the terminal device to data server with USB, the other is uploading to data server remotely by GPRS network [7]. Location mapping can precisely localize epidemic sampling points. LERCS function module is shown in Fig. 2.

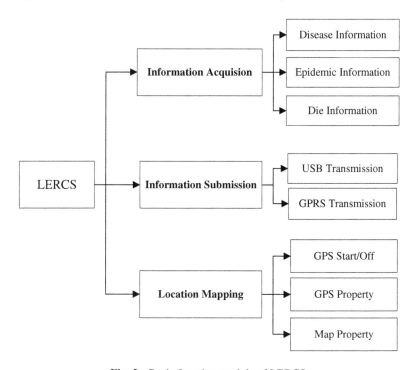

Fig. 2.   Basic function module of LERCS

3.2   LEMES

Based on the basic geographic information, the spatial distribution of modern animal husbandry demonstration county, standardization of livestock and poultry breeding demonstration county, and no prescribed animal epidemics demonstration area in Shandong province, can be visually displayed on the map. By receiving animal epidemic information from LERCS, this subsystem takes full advantage of GIS spatial analysis, statistical analysis, and thematic map display to predict epidemic information and improve the capability of disease prevention and control. LEMES function module is shown in Fig. 3.

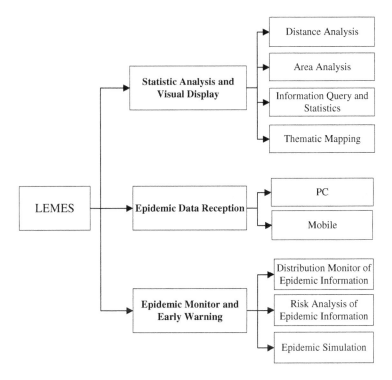

Fig. 3. Basic function module of LEMES

The most specific performance of statistic analysis and visual display is composed of distance analysis, area analysis, information query and statistics, and thematic mapping basically composite. Based on the above function, livestock and poultry relevant information can be queried, summarized, analyzed and fed back in the form of thematic map and statistical graph. After clear about data format, data type, data security safeguard, data transmission mechanism, epidemic information can be acquired accurately, timely, and diversely, which guarantees character, images, video and other related information about the outbreak field can be obtained by epidemic data reception function. Epidemic monitor and early warning function utilizes a series of GIS spatial analysis, such as density analysis, clustering analysis and dynamic analysis, to make an overall assessment of the distribution of epidemic prevention resources and breeding resources. Through the analysis and data-mining of real-time data and historical data, distribution, development and perniciousness of epidemic information can be monitored and predicted. At the same time, it is very feasible that by using the simulation method, changing the conditions of nature, geography, ecology, meteorology, transportation, virtually setting related factors such as population structure and breeding density, the key epidemic control and prevention area and the optimization control strategy can be determined [8].

3.3   LEEDSS

When there are major outbreaks, LEEDSS is able to acquire animal epidemical information accurately, timely and exhaustively, which enables a rapid definition of epidemical area of influence and grading of epidemic situation. After that, distance analysis can be utilized to summarize distribution of livestock and poultry farms and their attribute information in different level of distance range, dynamically display locations of disease management department, persons chiefly in charge and their contact information, and at last give an optimal path to the outbreak site. Meanwhile, base on mobile GIS and wireless transmission technology, outbreak sites can be located in the electric map.

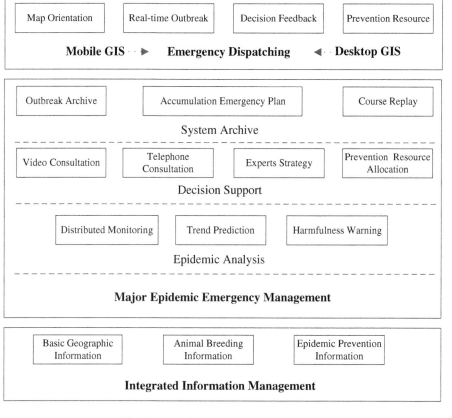

Fig. 4. Basic function module of LEEDSS

The scene of the real-time monitoring information will also be transmitted to emergency command staff. LEEDSS function module is shown in Fig. 4.

4   Conclusions

The characteristics of animal husbandry can be summarized as multi-subject, small scale modal, decentralized decision-making and scattered farmers. Due to subjective decision-making and traditional report filling method, in the actual epidemic prevention and control work, there will be a variety of problems, for example, relevant decisions is not conformed with actual fact, acquisition of epidemical information is not real-time. Thus, this paper builds up animal epidemic prevention system at the provincial level, respectively developing LERCS, LEMES, and LEEDSS in the three perspectives of epidemical information acquisition, epidemical early warning and monitoring, and epidemical prevention and control decision support. Relevant information about basic geographic information, animal breeding information, and epidemic prevention information is combined together to summarize animal husbandry and outbreak investigation and monitoring data of the whole province. This provides a scientific basis for government regulators, monitoring institutions at all levels, scientific research units, farmers about livestock and poultry epidemic prevention.

Provincial animal husbandry and outbreak early warning forecast system established in this study has presented limitations to some extent. In order to guarantee the accuracy of the early warning analysis, livestock epidemic prediction model needs to be further researched and improved with the combination of real-time and historical epidemical data, growth environment, breeding habits, and common epidemical disease for different kinds of livestock and poultry.

Acknowledgment. This work was supported by the National Science and Technology Support Program (2014BAD08B05-02).

References

1. Ma, J., Pan, Y., Shen, T., et al.: Spatial decision support system for controlling the outbreak and spread of animal epidemics. Comput. Appl. **27**(5), 1289–1292 (2007)
2. Li, L., Dong, J., Zhang, Z., et al.: Research and design of GIS based on animal disease information system. China Anim. Husbandry Vet. Med. **34**(4), 81–83 (2007)
3. Lu, C., Wang, C., Hu, Y., et al.: Digital monitor and control system for major animal epidemic disease in China. Jiangsu J. Agric. Sci. **21**(3), 225–229 (2005)
4. Preriffer, D.U., Hugh, M.J.: Geographical information system as a tool in epidemiological assessment and wildlife disease management. Revue Scientifique Et Technique De Loie **21**(1), 91–102 (2002)
5. Xie, L.: Design and implementation of animal husbandry information system based on GIS. J. Qinghai Normal Univ. (Nat. Sci.) **4**, 17–20 (2013)
6. Chen, J., Yuan, Y.: Design of a construction project for the emergency command system. J. WUT (Inf. Manage. Eng.) **27**(2), 122–127 (2005)

7. Zhang, J., Wang, X., Peng, Z., et al.: Design and implementation of intelligent terminal system for collecting animal epidemic disease data. Guizhou Agric. Sci. **39**(4), 230–233 (2011)
8. Jong, S.C., Jean, H.L., Jong, H.P., et al.: Design and implementation of a seamless and comprehensive integrated media device interface system for outpatient electronic medical records in a general hospital. Int. J. Med. Inform. **80**(4), 274–285 (2011)

Research and Design of Wireless Sensor Middleware Based on STM32

Jiye Zheng, Fengyun Wang, and Lei Wang[✉]

S&T Information Institute of Shandong
Academy of Agricultural Sciences (SAAS), Jinan 250100, China
jiyezheng@163.com, wfylily@163.com, 375901677@qq.com

Abstract. In order to make agricultural production become more convenient, also meet the requirement of the modern intelligent agriculture to make control equipment integrated and miniaturization, the paper designed a wireless sensor middleware based on STM32. This wireless gateway receives control command from the control software of agricultural IOT application which installed in the tablet computer through the WiFi/USART module, after the data processing and protocol conversion, then send the control signal to the wireless sensor network through the USART port. At the same time, if a particular model of agricultural sensor changes state, it also can timely feedback to the tablet computer to display through the gateway. This plan solved the problem of real-time monitoring of agricultural information and realized remote control. The experimental results show that, the wireless sensor middleware has the characteristics of light and handy, high control precision, data large quantity and high speed. It's very appropriate for intelligent agricultural system application.

Keywords: STM32 · Wireless sensor · Middleware · Zigbee

1 Introduction

With the rapid development of Internet of things, IOT middleware becomes one of the important research topics in related fields in recent years, also acquired a lot of research results, mainly divided into application service middleware, embedded middleware. Such as White [1] put forward a kind of typical middleware based on J2EE architecture, has a good scalability. paper [2] proposed wireless sensor network middleware based on mobile Agent, the middleware has data management, integration, application target adaptive control strategy, Wang fan design and implemented an IOT middleware, that can filter the heterogeneous network data, integrate ZIGBEE and RFID (Radio Frequency Identification, RFID) data into a unified format [3]. Paper [4–7] proposed database based middleware, it views the entire network as a distributed database, users use similar to SQL query command to get the data needed, query is distributed to each node through the network, the nodes determine whether the sensory data satisfy the query conditions, and decide to send data or not. Deng Yihua etc. Put forward embedded RFID middleware on smart RFID read-write device [8], it can realize the management of multiple read/write devices, strengthen the deployment flexibility. The domestic scholars make use of the gateway hardware to research IOT

© IFIP International Federation for Information Processing 2016
Published by Springer International Publishing AG 2016. All Rights Reserved
D. Li and Z. Li (Eds.): CCTA 2015, Part II, IFIP AICT 479, pp. 579–585, 2016.
DOI: 10.1007/978-3-319-48354-2_60

system [8–11], however, the research and design of STM32 based wireless sensor network middleware is still relatively rare.

The research combines the ZigBee wireless sensor network with the STM32 MCU, it has the advantages of simple structure, strong generalization ability and others. The objective of this study is to develop an easy method to implement the remote monitoring and control system for the agricultural production environment.

2   Wireless Gateway Hardware Platform Design

Wireless gateway is the data transfer station of the Agro-IOT control system and communication media, it is critical to the whole system design. The main frame of the hardware platform is shown as Fig. 1.

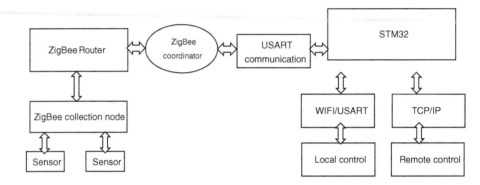

Fig. 1. Wireless gateway overall architecture diagram

The hardware involved in this paper include wireless sensor nodes and gateway hardware, hardware design is introduced as follows.

2.1   Wireless Sensor Nodes

CC2530 on-chip system is used for the wireless sensor nodes with very low cost and ultralow power consumption. It integrates the microprocessor module and a wireless transceiver module in one single chip. It has the excellent performance of RF transceiver adapting to IEEE 802.15.4 with extra high reception sensitivity and anti-interference. Its core is industry-standard enhanced 8051 microcontroller unit with code prefetching function. CC2530 has four different types of flash memory version i.e. CC2530F32/ 64/128/256 separately with 32/64/128/256 KB in-system programmable flash memory. The microcontroller has 8 KB RAM with data remaining capacity under various power supply. CC2530 has various operating modes such as active mode RX (CPU idle), active mode TX 1 dBm (CPU idle), power supply mode 1 (4us awake), power supply mode 2 (sleep timer), power supply mode 3 (external interruption), wide voltage range of power supply and so on which makes the energy consumption very low.

The wireless transceiver adopts zigbee technology. It has low power consumption, rapid response and large network capacity etc. features. Figure 2 shows the wireless sensor nodes used in the system.

According to the function, node types can be divided into three categories, namely the acquisition node (only responsible for data collection), routing nodes (responsible for data gathering and routing), and the coordinator node (network management). All kinds of nodes on the hardware adopts the above design, the coordinator needs a binding to the gateway via the serial port.

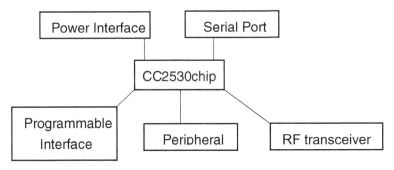

Fig. 2. Wireless sensor node hardware structure diagram

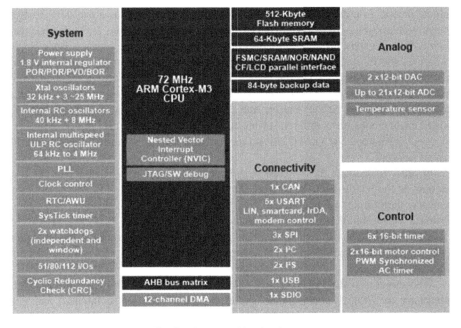

Fig. 3. Gateway Circuit Diagram

2.2 Gateway Hardware

Gateway is the key module of this system, it needs to bind with the coordinator, the gateway Circuit Diagram is shown in Fig. 3.

Gateway hardware uses STM32F103ZET which core is ARM Cortex-M3. F1-series is the first generation microcontrollers of STM32 that is considered as its mainstream of ARM MCU. With the development of technology, F1 series has many improvements such as CPU speed, size of internal memory and type of peripherals. The excellent peripherals and low-power, low-voltage operation make it with high performance. It is highly integrated with the simple structure and easily used tools which cost is very low. STM32 is control core of the wireless gateway, it mainly divides into five modules, include WiFi module, RF module, a central controller module, power module, JTAG emulation debugging module respectively.

3 Wireless Gateway Software Platform Design

Wireless sensor network middleware software platform are actually established on the embedded development board wildfire STM32 ISO with transplantation of Linux system, and then Boa Web server, database Sqlite3 was also set up in the Linux system.

The main task of the wireless sensor network gateway is as follows: access to information of each node in the wireless sensor network, processing a request from the Internet, the wireless sensor network configuration management, and data statistics and analysis, etc. The main task of wireless sensor network is to monitoring the state of the remote device. Literature [12] presents a so-called Cougar middleware system, Cougar views wireless sensor network as the distributed database and the network nodes are divided into clusters, cluster is a collection of multiple nodes, elects a node in the cluster as the cluster head. Paper [13] proposed a flexible wireless sensor network middleware system. The system using Web Service technology and middleware technology, the system views the wireless sensor network as middleware system, each node uses a different SOAP Engine, the coordinator node has a registration module, other sensor nodes have different event handlers, the user can use any SOAP proxy access coordinator node via TCP/IP protocol, and further access to the entire wireless sensor network.

In order to manage the wireless sensor network better and expand Internet connections to the wireless sensor network, through careful analysis of the wireless sensor network monitoring program, and combined with hardware gateway, this paper abstracts a new middleware model as shown in Fig. 4.

Middleware structure model has many modules include Cluster model controller, Command interpreter, Boa server, database controller and Sensor network coordination interface and so on.

Web server provides HTTP service to external applications, and internally access to interpreter by CGI; The interpreter responses to explain users command, and give orders to the cluster model control layer and accept its response; Model control layer abstracts the clusters for the wireless sensor network, and provides the control interface; Sensor network coordinator interface provides an interface for gateway to read

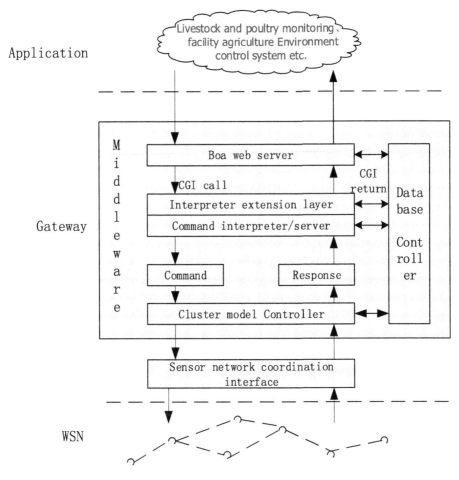

Fig. 4. Wireless sensor network middleware structure model

and write in wireless sensor network, the actual implementation is divided into the gateway and the coordinator two part, the coordinator and the sensor nodes' communication uses data collection protocol.

4 Result and Discussion

Wireless Sensor Middleware was used in livestock and poultry breeding, agricultural facilities accurate monitoring, Edible fungus production information collection and similar applications, the coordinator node on the packet is bound to the embedded gateway. Using TCP/IP protocol stack, the communication between monitoring instrument and upper computer was realized. The real time information on the upper computer is like Fig. 5.

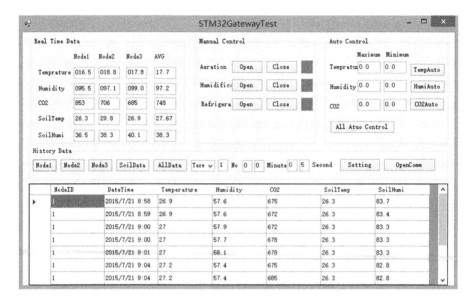

Fig. 5. Real time data from the remote sensors

5   Conclusions

The research analyzed the IOT middleware research status, and combining the wireless sensor network practice, proposed the Wireless Sensor Middleware Based on STM32. The software has two characteristics include embedded server and application services, it adopted the idea of components, reduced the coupling between modules, the configurable and customizable functionality improved reusability, lower development costs, solve the problems of mass data processing, and provide the visual configuration monitoring platform, the realization of Internet of things "content" of the administration and monitoring, eliminate information isolated island. This middleware application was used in Edible fungus production information collection system, and the results show that the feasibility and effectiveness of the above frameworks and methods.

Acknowledgment. Funds for this research was provided by the National Science and Technology Support Program (2014BAD08B05-02), Shandong Academy of Agricultural Sciences (SAAS) Youth Scientific Research Funds Project (2015YQN58), Independent Innovation Projects of Shandong Province (2014ZZCX07104).

References

1. White, S., Alves, A.,Rorke, D.: Web logic event server: a light-weight, modular application server for event processing. In: Proceedings of the Second International Conference on Distributed Event-based Systems, pp. 193–200. ACM Press, New York (2008)

2. Huang, H., et al.: Wireless sensor network based on mobile Agent middleware. J. Nanjing Univ. Nat. Sci. Edn. **44**(2), 157–163 (2008)
3. Wang, F.: Design and Implementation of Internet of Things Middleware Based on ZIGBEE and RFID. Beijing University of posts and telecommunications, Beijing (2011)
4. Deng, Y., Xie, S.: The design and implementation of embedded RFID middleware. Comput. Eng. Des. **7**, 1716–1718 (2008)
5. Bonnet, P., Gehrke, J., Seshadri, P.: Towards sensor database systems. In: Proceedings of the 2nd International Conference on Mobile Management, pp. 3–14. Springer, Berlin (2001)
6. Madden, S.R., Franklin, M.J., Hellerstein, J.M.: Tiny DB: an acquisitional query processing system for sensor networks. ACM Trans. Database Syst. **30**(1), 122–173 (2005)
7. Srisathapornphat, C., Jaikaeo, C., Shen, C.: Sensor information networking architecture. In: Proceedings of the International Workshop Parallel Processing, pp. 23–30. IEEE Computer Society Press, Washington (2000)
8. Ishiguro, M., Tei, K., Fukazawa, Y., Honiden, S.: A sensor middleware for lightweight relocatable sensing programs. In: International Conference on Computational Intelligence for Modelling, Control and Automation, Sydney Australia (2006)
9. Biswas, P.K., Qi, H., Xu, Y.: A mobile-agent-based collaborative framework for sensor network applications. In: IEEE International Conference on Mobile Adhoc & Sensor Systems, Vancouver, Canada, (2006)
10. Mohsen, S., Alkaee, T.M., Amirhosein, T.: A middleware layer mechanism for QoS support in wireless sensor networks. In: The International Conference on Networking, International Conference on Systems and International Conference on Mobile Communications and Learning Technologies, Morne, Mauritius (2006)
11. Heinzelman, W.B.: Middleware to support sensor network applications. IEEE Netw. **18**(1), 6–14 (2004)
12. Pedro, J.M., Daniel, M., Andreas, L.: TinyCubus: a flexible and adaptive framework for sensor networks. J. Inf. Technol. **47**(2), 87–97 (2005)
13. Endler, M., Schmidt, D. (eds.): IFIP International Federation for Information Processing. Federal University of Rio de Janeiro (2003)

Technical Efficiency and Traceability Information Transfer: Evidence from Grape Producers of Four Provinces in China

Lei Deng[1], Ruimei Wang[1(✉)], Weisong Mu[2], and Jingjie Zhao[3]

[1] College of Economics and Management, China Agriculture University,
Beijing 100083, China
dengl98919@126.com, h03109t@cau.edu.cn
[2] College of Information and Electrical Engineering,
China Agriculture University, Beijing 100083, China
wsmu@cau.edu.cn
[3] State Administration of Taxation,
Beijing Municipal Office, Beijing 100091, China
zhaojingjie006@163.com

Abstract. This paper attempts to estimate the technical efficiency of grape producers using stochastic frontier analysis (SFA) approach as well as survey data of 1388 farmers in Hebei, Liaoning, Shandong and Xinjiang Province in China, and clarify the relationship between technical efficiency and traceability information transfer. The results show that technical efficiency of grape producers ranges from 0.8 to 0.9 with a stable but low distribution; input of land and physical makes a great contribution to the increase of rural income which is followed by input of labor and agricultural machine; the level of technical efficiency has a significant impact on the traceability information transfer in the way that the higher the technical efficiency is, the higher the willingness for participants to transfer the traceability information will be. Therefore, the technical efficiency of grape producers should be increased to establish the grape traceable system, ensure the safety of grape products and improve the development of grape industry.

Keywords: Grape · Technical efficiency · Stochastic frontier analysis · Traceability information · Information transfer

1 Introduction

With the development of China economy and people living standard, the consumption and demand of vegetables and fruits are increasing rapidly, which, according to the data of population and consumption of urban and rural residents in China Statistical Yearbook, has been increased by 33.11 % in the past decade from 2003 to 2012. As a result, the consumption and demand of grape and its products are increased rapidly. This leads to a fast development of grape industry and, given its high benefit, grape industry become an important part of the rural economy [1]. Meanwhile, consumers pay more and more attention to the issue of food safety which makes the valid transfer of grape traceability information become an important guarantee of the healthy and stable development of grape industry. On the other hand, the valid transfer of traceability information can help participants achieve the improvement of supply chain performance, help consumers distinguish the high quality food from the low quality food and improve the quality and safety level of agricultural products to keep up with market demand and preferences [2]. The valid transfer, therefore, is important to the increase of rural income and the improvement of development of rural economy.

Nowadays, the grape traceability system in China is still imperfect, which means the willingness of all participants to transfer the traceability information is low and the cost of transfer this information is high. On the other hand, the productivity and efficiency of grape production are both low which leads to a depression of rural income with the increasing production cost. As a result, the cost of transfer traceability information becomes higher and significantly decrease the willingness of all participants to share the traceability information. A better understand of the relationship between technical efficiency and traceability information transfer is important and meaningful to the clarification of the impact productivity and efficiency on the willingness to share traceability information and to solve problems behind the establishment of traceability system.

A large number of studies have been conducted on agricultural traceability system, in which the authors pay attention to three areas including government, enterprises and consumers. The studies on government were devoted to the effect of government behaviors on the establishment of traceability system [3] and on the willingness of enterprises and consumers [4]; researches on the enterprises pay attention to the impact of traceability system on the clarification of responsibility and increasing of profit [5]; studies on consumers pay much attention to the recognition issues [6]. There are, however, few literatures on traceability information transfer of grape supply chain and the specific studies about the relationship between grape production efficiency and willingness to share traceability information are not found to be reported. Therefore, this paper pays attention to the technical efficiency of grape producers and analyses the willingness of all participants to share the traceability information with different level of technical efficiency to improve the grape safety level, increase the profit of all participants and achieve an advance in the improvement of the healthy and stable development of grape industry.

2 Materials and Methods

2.1 Data

The data used in this paper is mainly from a fieldwork conducted by our research team in Hebei, Liaoning, Shandong and Xinjiang Province in China, which are the major grape production areas [7]. 1388 farmers were interviewed through a face to face approach. The questionnaire mainly focuses on the physical input, input of agricultural mechanization and labors during the production of grape. Table 1 shows the distribution of samples we have interviewed.

Table 1. Distribution of sample

Province	Sample size	Samples account for total [%]
Hebei	376	27
Liaoning	338	24
Shandong	290	21
Xinjiang	384	28
Total	1388	100

According to the exist studies on the technical efficiency, this paper selects the physical input, input of agricultural mechanization and labors during the production of grape as the indicators for input. Given the fact that the variety, production and farm gate prices of grape are different in different areas, and farmers pay most attention to the economic outcomes [8], according to Liu Z., et al. (2000) [9], this paper select the sales revenue as the indicator of output. In order to distinguish the difference of areas, this paper regards the areas as dummy variables. Table 2 shows the statistical description of samples.

Table 2. Statistical description of input and output indicators

Variables	Unit	Mean	Standard deviation	Minimum	Maximum
Sales revenue (y)	CNY/667 m^2	9859.3	7821.55	1420	50000
Farm size (x_1)	667 m^2	15.76	34.09	0.7	350
Physical input (x_2)	CNY/667 m^2	2508.43	2137.76	794	14170
Mechanical input (x_3)	CNY/667 m^2	306.58	621.82	0	6050
Labor input (x_4)	CNY/667 m^2	2989	1283.37	1456	6364

Some interesting observations can be found in Table 2 are: Firstly, there is an obvious difference in the sales revenue in different regions, which may resulted from the different agricultural technology, level of management and production input. On the other hand, to a significant extent, the production, market and price risks taken by farmers has an impact on their outcomes; secondly, the labor input is high while the mechanical input is relatively low. This may result from the dependence of labor input in grape production and the absence of mechanical utilization which implies the low level of mechanization; thirdly, there is a diversity of the scale of grape production ranging from 233335 m^2 to 467 m^2 with an average farm size of 10507 m^2.

2.2   Estimation of Technical Efficiency

The Stochastic Frontier Analysis (SFA) and Data Envelopment Analysis (DEA) are the most popular ways used by many recent researches to evaluate the technical efficiency. When there are multi-input and multi-output with no specific formulation, the DEA approach is more useful than SFA approach in the estimation of technical efficiency. The advantages of DEA approach are its objectivity and diversity, while the conduction of a statistical test and exclusion of the impact of statistical errors seem to be impossible. On the contrary, the SFA approach can clarify and estimate the impact of stochastic and statistical errors on the production frontier as well as the effect of technical inefficiency on the real production. On the other hand, a SFA approach also can evaluate and find out the critical factors impacting on productivity and efficiency. Taking all these advantages of SFA approach into consideration, this paper choose SFA model to analyze the productivity and efficiency of grape production in China, the basic formula of our model is following:

$$y_i = \beta x_i + v_i - u_i$$

$i = 1, 2, \cdots, I$ is the number of farmers, y_i presents the output of grape production, x_i is the input during the production of grape, β is the estimated parameters, $v_i - u_i$ denotes the combined errors, v_i is the stochastic errors, u_i presents the technical inefficiency, both v_i and u_i are independent identically distributed, and $v_i \sim N(0, \sigma_v^2)$, $u_i \sim N(\alpha_i, \sigma_u^2)$.

According to Battese and Corra [10], we can let $\sigma^2 = \sigma_v^2 + \sigma_u^2, \gamma = \sigma_u^2/(\sigma_v^2 + \sigma_u^2)$ and estimate the production function and technical inefficiency function, then the parameters $\beta, \delta, \sigma^2, \gamma$ can be evaluated as well as technical efficiency of each producer, σ^2 presents combined variance, γ is the proportion of technical inefficiency account of combined errors. When $\gamma = 0$, stochastic errors make the greatest contribution to total errors and when $\gamma = 1$, technical inefficiency makes the greatest contribution to total errors.

The technical efficiency of each producer is calculated by the followed function:

$$TE_i = E(y_i|u_i, x_i)/E(y_i|u_i = 0, x_i)$$

$E(\cdot)$ is the mathematical expect. When $TE_i = 1$, there is no technical inefficiency, while when $TE_i < 1$, there is technical inefficiency.

The formulations of SFA are often set as the types of Cobb-Douglas and trans-log production function. Considering the production lead time of grape is very long, the production of grape is dependent on the current and past input, and the relationship among all inputs is unknown. Hence, the trans-log production function form is chose as a basic formula to evaluate and calculate the technical efficiency of each producer. The model is as follows:

$$Lny_i = \beta_0 + \sum_{n=1}^{4}(\beta_n Lnx_{ni}) + \frac{1}{2}\sum_{n=1}^{4}\sum_{m=1}^{4}(\beta_{nm}Lnx_{ni}Lnx_{mi}) + \beta_5 d_1 + \beta_6 d_2 + \beta_7 d_3 + v_i - u_i$$

y_i is the sales revenue of farmers, $n, m = 1, 2, 3, 4$ presents the number of inputs, x_1 denotes farm size, x_2 denotes the physical costs, x_3 denotes the agricultural machine costs, x_4 denotes the labor costs, d_1, d_2, d_3 are dummy varieties, $d_1 = 1$ denotes Hebei, $d_2 = 1$ denotes Liaoning, $d_3 = 1$ denotes Shandong, $d_1 = d_2 = d_3 = 0$ denotes Xinjiang.

3 Results and Discussion

This paper estimates the production function by using Frontier 4.1 software; the results are showed in Table 3.

Table 3. Results of parameter estimation

Variable	Coefficient	T value	Variable	Coefficient	T value
Lnx_1	-0.3262^{**}	-1.6823	Lnx_1Lnx_4	0.0099	0.4470
Lnx_2	0.2271^{*}	1.3636	Lnx_2Lnx_3	-0.0624***	-3.9893
Lnx_3	0.2109^{***}	2.3940	Lnx_2Lnx_4	-0.0002	-0.0093
Lnx_4	-0.0812	-0.4345	Lnx_3Lnx_4	0.0148^{*}	1.3598
$(Lnx_1)^2$	0.0253	0.8742	(d_1)	0.0268	0.4301
$(Lnx_2)^2$	-0.0020	-0.5984	(d_2)	-0.0134	-0.1845
$(Lnx_3)^2$	-0.0287^{***}	-4.2655	(d_3)	0.2233^{***}	3.3574
$(Lnx_4)^2$	0.0264^{*}	1.5947	Constant	8.2148^{***}	19.7581
Lnx_1Lnx_2	0.0206^{*}	1.4610	σ^2	0.7789^{***}	7.5275
Lnx_1Lnx_3	-0.0267^{*}	-1.3870	γ	0.6457^{***}	12.8243
Log likelihood function				-730.2154	
LR test				102.3675^{***}	
Total number of samples				1388	

According to the data showed in Table 3, the model we selected works well, and the inefficiency γ is significant at 1 % confidence level, which implies there is loss of technical efficiency in grape production. Most variables are successes to the test of significance and this implies the model we selected is appropriate and reasonable.

Figure 1 shows the distribution of technical efficiency of grape producers. According to this figure, the current technical efficiency of grape producers is low and there is room and potential for improvement. On the other hand, most grape producers located in the interval of 0.8-0.9 in technical efficiency, and the distribution is relatively stable.

Trans-log production function reflects the complex relationship among all inputs and coefficients are interpretable as elasticities of output evaluated at the sample mean. The formulation is as follow:

$$\frac{\partial Lny}{\partial Lnx_n} = \beta_n + 2\beta_{nn}Lnx_n + \sum\nolimits_{m=1}^{4} \beta_{nm}Lnx_m$$

$n, m = 1, 2, 3, 4$ denotes the number of input, $(\partial Lny)/(\partial Lnx_n)$ denotes the elasticity of n th input, β_{nm} denotes the coefficient of the n th and m th input. Table 4 shows the elasticity of each input in grape production.

According to Table 5, land input and physical input make a great contribution to the outcomes of grape producers while the contribution made by labor input is higher than that made by mechanical input, which implies that grape production is more dependent on the labor input.

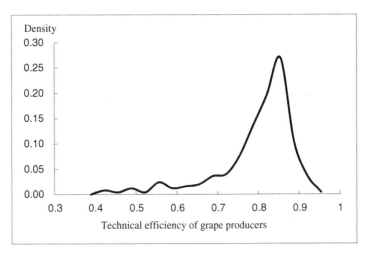

Fig. 1. Distribution of technical efficiency of grape producers

Table 4. Elasticity of each input

Variable	Land input	Physical input	Mechanical input	Labor input
Elasticity	0.2665	0.3439	0.1002	0.1622

Table 5. Name and definitions of variables

Variable	Definition	Variable	Definition
T	Technical efficiency	α	Constant of market demand curve
a	Volume of traceability information transfer	β	Price coefficient of market demand curve
c	Marginal cost of traceability information transfer	γ	Information transfer of market demand curve
q	market demand	P_H	Price of traceable grape
P_L	Price of non-traceable grape	C_P	Grape production cost
ε	Technical efficiency coefficient of traceability information transfer	C_I	Cost of traceability information sharing

4   Relationship Between Technical Efficiency and Traceability Information Transfer

A higher technical efficiency means that the outcomes of grape producers will be higher, that is to say, a higher sales revenue given the input series. There will be three dimensions that technical efficiency has impact on the transfer of traceability information: Firstly, a higher technical efficiency implies that grape production cost is lower and this will leads to a higher profit creation to conduct the share of traceability information; secondly, a higher technical efficiency means the output value of produced grape will be higher and famers will share the traceability information with much incentive to meet the demand and preference of consumers to expand the sales value of grape and increase their profit; thirdly, a higher technical efficiency means a lower cost of traceability information sharing, and encourage participants to share traceability information. All these conclusions can be analyzed through the model followed.

Table 5 provides the name and main definitions of each variable.

According to Table 5, the cost for grape producers to share the traceability information can be defined as $C_I = ac - \varepsilon T$, and market demand for grape can be defined as $q = \alpha - \beta P + \gamma C_I$. Therefore, the sales revenue of grape producers can be defined as two types, one is with traceability information sharing: $\pi_1 = P_H q - C_p - \gamma(ac - \varepsilon T)$, the other is without traceability information sharing: $\pi_0 = P_L q - C_p$. According to the analysis above, a higher technical efficiency will leads to a decrease of production cost and the willingness to share traceability information wi be relatively higher. When the technical efficiency is low, the fund that used to share the traceability information will be sequent low while the cost of traceability information transfer is also high. When consumers' willingness to pay for grapes with or without traceability information is a constant (that is to say, P_L and P_H are constants), this will result to relative lower outcomes when grape producers are sharing the traceability information than there is no traceability information sharing (that is to say, $\pi_1 < \pi_0$). On the contrary, when the technical efficiency of grape producers is high, the outcomes will be higher when they share traceability information. Therefore, the willingness of grape producers to share traceability information will be high.

5   Conclusions and Implications

This paper estimates the technical efficiency of 1388 grape producers based on the field survey in Hebei, Liaoning, Shandong and Xinjiang Province in China in 2013 using stochastic frontier analysis (SFA) approach. And a further study on the relationship between technical efficiency and traceability information sharing is conducted. The results show that the technical efficiency of most grape producers are located at the interval of 0.8 to 0.9 and the distribution is stable but the level of technical efficiency is low, which implies that the increase of technical efficiency still could make a great contribution to the rural income; land and physical input makes a great contribution to the increase of rural income which is followed by labor input and mechanical input. This is the result of the over dependency of grape production on labor and the fact that the level of mechanization is still low; the level of technical efficiency has a significant impact on the willingness of participants to share traceability information, which will significantly increase their outcomes. Therefore, the technical efficiency of grape producers should be increased to establish the grape traceable system, ensure the safety of grape products and improve the development of grape industry.

Acknowledgments. This study was supported by China Agricultural Research System (CARS-30), Humanities and Social Sciences Foundation of Ministry of Education of China (13YJCZH182) and National Key Technology R&D Program of the Ministry of Science and Technology (2014BAL07B05).

References

1. Ma, A.H., Guo, Z.J., Li, H.S., et al.: Development situation of grape industry in China. J. Hebei Agric. Sci. **12**, 6–9 (2009)
2. Schulz, L.L., Tonsor, G.T.: Cow-Calf producer preferences for voluntary traceability systems. J. Agric. Econ. **61**(1), 138–162 (2010)
3. He, Y., Sun, Y., Wu, Q.Y.: The role of government in the establishment of food traceability system. J. Chongqing Univ. Sci. Technol. (Soc. Sci. Ed.) **5**, 32–34 (2012)
4. Wu, L.H., Xu, L.L., Zhu, D.: Study of the main factors affecting enterprises' investiment on food traceability system: perspective from logistic model with a penalty function. Manag. Rev. **26**(1), 99–108, 119 (2014)
5. Golan, E.: Traceability for food safety and quality assurance: Mandatory systems miss the mark. Curr. Agric. Food Resour. Issues **4**, 27–35 (2003)
6. Meuwissen, M.: Consumer preferences for pork supply chain attributes. NJAS-Wagenigen J. Life Sci. **54**(3), 293–312 (2007)
7. Mu, W.S., Feng, J.Y.: The Research on Economic Issues of Grape Industry in China. China Agricultural University Press, Beijing (2010)
8. Liu, Y., Huang, J.K.: A multi-objective decision model of farmers' crop production. Econ. Res. (01), 148–157, 160 (2010)
9. Liu, Z., Zhuang, J.: Determinants of technical efficiency in Post-Collective chinese agriculture: evidence from farm-level data. J. Comparat. Econ. **28**(3), 545–564 (2000)

Erratum to: The Molecular Detection of *Corynespora Cassiicola* on Cucumber by PCR Assay Using DNAman Software and NCBI

Weiqing Wang[✉]

Beijing Vocational College of Agriculture, Beijing, China
weiqingfine@163.com

Erratum to:
Chapter 26 in: D. Li and Z. Li (Eds.)
Computer and Computing Technologies in Agriculture IX
DOI: 10.1007/978-3-319-48354-2_26

The original version of the paper starting on p. 248 was revised. The affiliation of the author was corrected.

The updated original online version for this chapter can be found at
DOI: 10.1007/978-3-319-48354-2_26

D. Li and Z. Li (Eds.): CCTA 2015, Part II, IFIP AICT 479, p. E1, 2016.
DOI: 10.1007/978-3-319-48354-2_62

Author Index

Ma, Zhongren   II-143
Meng, Lumin   II-163
Meng, Ying   II-76, II-100
Meng, Zhijun   I-348, I-528, II-117
Mu, Weisong   II-586
Mu, Yuanjie   II-548

Nengfu, Xie   I-437
Niu, Chong   I-425
Niu, Luyan   II-570
Niu, Yuguang   I-425

Ouyang, Jihong   II-417

Pan, Ligang   II-505
Pan, Yuchun   II-225
Pei, Zhiyuan   I-161, I-335, I-366
Peng, Bo   I-612
Peng, Hui   I-266
Peng, Yuli   I-208

Qi, Lijun   II-429, II-437
Qi, Yan   I-173, II-530
Qian, Man   II-194, II-202
Qiao, Xi   I-519
Qiao, Yuqiang   II-68
Qin, Jing   II-539
Qin, Xu   II-522
Qiu, Yun   I-275
Qu, Mei   I-19

Ren, Yanna   I-53
Ruan, Chengzhi   I-485
Ruan, Huaijun   II-548, II-563

Sa, Liangbing   II-269
Shang, Minghua   II-548
Shao, ChangYong   I-546, II-522
Shao, MingXi   I-546
Shen, Kejian   I-366, II-41
Shen, Tian   II-330
Shi, Binjie   II-269
Shi, Chunlei   I-612
Shi, Chunlin   II-1, II-133, II-217
Shi, Qinglan   I-19, I-197
Shi, Xiaohui   II-320
Shi, Yongle   II-1
Shu, Jing   II-522
Si, Xiuli   I-153, I-239

Song, Chaoyu   I-231
Song, Chuwei   I-459, I-502
Song, Yu'e   II-92
Su, Baofeng   I-246, I-399
Sui, Yuanyuan   II-155, II-178, II-455
Sun, Fangli   I-231
Sun, Guannan   I-366
Sun, Jiabo   II-570
Sun, Li   I-325
Sun, Lijuan   II-185
Sun, Longqing   I-82
Sun, Min   I-115
Sun, Ming   II-111
Sun, Xinxin   I-82
Sun, Yonghong   I-231
Sunwei   I-437

Tang, Shuping   I-485
Tang, Xin   II-522
Teng, Ling   I-300
Teng, Xiaowei   II-163
Thiravong, Sisavath   II-479, II-487
Tian, Hongwu   I-255
Tian, Ruya   I-375
Tian, Xiaojing   II-143
Tian, Xuedong   I-27
Tong, Ling   I-473
Tu, Xingyue   I-64

Wang, Bin   I-399
Wang, Bingbing   II-185
Wang, Chaopeng   II-202
Wang, Chengguo   II-92
Wang, Cong   II-287
Wang, Dan   II-346
Wang, Danqiong   I-325
Wang, Danyang   I-473
Wang, Dengwei   II-310
Wang, Dong   II-505
Wang, Fei   I-325, I-366
Wang, Fengyun   II-579
Wang, Fujun   II-548
Wang, Guowei   I-317, I-536, II-155, II-178, II-455
Wang, Haijun   I-325, II-41
Wang, Haiyang   I-292, I-587
Wang, Hui   I-231
Wang, Jian   II-127
Wang, Jianfei   II-557

Printed in the United States
By Bookmasters